Statistics in Psychology Using R and SPSS

Statistics in Psychology Using R and SPSS

Dieter Rasch
University of Life Sciences, Vienna, Austria

Klaus D. Kubinger • **Takuya Yanagida**
University of Vienna, Austria

A John Wiley & Sons, Ltd., Publication

This edition first published 2011

Registered office
John Wiley & Sons Ltd, The Atrium, Southern Gate, Chichester, West Sussex, PO19 8SQ, United Kingdom

For details of our global editorial offices, for customer services and for information about how to apply for permission to reuse the copyright material in this book please see our website at www.wiley.com.

Library of Congress Cataloging-in-Publication Data

Statistics in psychology using R and SPSS / Dieter Rasch, Klaus D. Kubinger and Takuya Yanagida.
p. ; cm.
Includes bibliographical references and index.
ISBN 978-0-470-97124-6 (cloth) – ISBN 978-1-119-97964-7 (E-PDF) – ISBN 978-1-119-97963-0 (O-book) – ISBN 978-1-119-95202-2 (E-Pub) – ISBN 978-1-119-95203-9 (Mobi)
1. Psychometrics. 2. SPSS (Computer file) I. Rasch, Dieter. II. Kubinger, Klaus D., 1949- III. Yanagida, Takuya.
[DNLM: 1. SPSS (Computer file) 2. Psychometrics–methods. 3. Statistics as Topic–methods. 4. Software. BF 39]
BF39.S7863 2011
150.72′7–dc23

2011020660

A catalogue record for this book is available from the British Library.

Print ISBN: 978-0-470-97124-6
ePDF ISBN: 978-1-119-97964-7
oBook ISBN: 978-1-119-97963-0
ePub ISBN: 978-1-119-95202-2
Mobi ISBN: 978-1-119-95203-9

Typeset in 10/12pt Times by Aptara Inc., New Delhi, India

Contents

Preface

This textbook contains, on the one hand, everything that is needed for a freshman statistician. On the other hand, it can also be used in advanced courses and in particular it can be used for empirical research work.

Within the Bachelor's curriculum it is only possible to demonstrate the correct use of the most important techniques. For the Master's curriculum, however, a certain understanding of these methods is necessary. For doctoral studies, understanding alone is not enough: a willingness to reflect critically on the statistical methods must be developed.

Since even for doctoral students a repetition of the basics of statistics on an elementary level is often useful, with this book they can be picked up individually where their powers of recollection end – if necessary at the beginning of the Bachelor education. And in contrast, Bachelor's students are often interested in the contents of a Master's curriculum or where the textbook leads. They can get a taste of that now.

Even lecturers will find something new in this textbook; according to our experience, 'statistics for psychologists' is not taught by professional statisticians but by psychologists, mostly by those at the beginning of their academic careers; anecdotes may at least help them didactically. These casual reflections can of course also be academically amusing for students.

Accordingly, the three to four mentioned target groups are guided through the book using distinctive design elements.

All examples given in this textbook refer to psychology as an empirical science. However, the topics covered in psychology are similar to those of (other) social sciences, above all sociology and educational science. So, of course, this textbook suits their framework as well.

The statistical methods that are recommended in this book and which can be used for answering the research questions posed by psychology as a science are often only practicable when using a computer. Therefore we refer to two software packages in this book. The program package **R** is both freely accessible and very efficient; that is why we continuously use **R** here. However, since in psychology the program package IBM SPSS Statistics is still preferred for statistical analyses most of the time, it is also illustrated using the examples; here we use version 19.

We try to present statistical knowledge as simply as possible using these program packages, and avoid formulas wherever reasonable. However, we did not completely avoid formulas because we also wish to help those readers interested in the theoretical background. As a matter of fact, more important than formulas is the procurement of appropriate applications and interpretations of statistical methods. And that is actually the main focus of this book.

We have refrained from citing the exact sources for the practical, everyday methods given, reserving that for methods that are new or uncommon.

With the hope that the reader may easily gather from this textbook all information relevant to his/her individual academic level.

This book includes an accompanying website. Please visit `www.wiley.com/go/statisticsinpsychology`

Dieter Rasch, Klaus D. Kubinger, and Takuya Yanagida
Rostock and Vienna

Acknowledgments

The authors wish to express their thanks to those who contributed in either the translation or in the programs for this book, especially:

Dr. Albrecht Gebhardt, Alpen Adria University of Klagenfurth, Austria,
who gave us access to the **R**-package OPDOE and assistance in programming the sequential triangular tests.

In translating the text from German into English we received help from (in alphabetical order):

Maximilian Alexander Hetzel, University of Vienna, Austria
Nina Heuberger, University of Vienna, Austria
Mag. Jürgen Grafeneder, University of Vienna, Austria
Mag. Bernhard Piskernik, University of Vienna, Austria
Sarah Treiber, University of Vienna, Austria
Mag. Alexander Uitz, University of Vienna, Austria

Because the authors and translators are not native English speakers, we are happy that we found help from (in alphabetical order):

Mag. Carrie Kovacs, University of Luxembourg
Mag. Renate Dosanj, University of Vienna, Austria
Sandra Almgren, Kremmling, Colorado, USA
Peter Loetscher, University of Vienna, Austria

We thank, for a lot of editorial work (in alphabetical order):

Bettina Hagenmüller, University of Vienna, Austria
Mag. Bernhard Piskernik, University of Vienna, Austria

We thank **Prof. Dr. Rob Verdooren**, **Wageningen, The Netherlands** for carefully reading many chapters and giving helpful remarks, **Dr. Maciej Rosolowski** from the **University of Magdeburg, Germany** for the **R**-programs for the principal component tests.

We further thank IBM SPSS STATISTICS for providing the most recent version 19 of the IBM SPSS STATISTICS program.

Last but not least, we thank **Mag. Joachim Fritz Punter**, **Medical University of Vienna, Austria** for producing all figures as a reproduction proof, but above all for editing the antecessor book.

Part I

INTRODUCTION

This textbook requires a multi-layered view of 'statistics in psychology'. Within the Bachelor's curriculum it is only possible to demonstrate the correct use of the most important techniques. For the Master's curriculum, however, a certain understanding of these methods is necessary: for the Master's thesis, where usually a scientific question has to be worked on single-handed but under supervision, the student has to refer to statistical analyses in literature concerning the topic, and if necessary to improve the choice of the method used for analysis. For doctoral studies, understanding alone is not enough; a willingness to reflect critically on the statistical methods must be developed. The statistical methods used in the doctoral thesis, which means the entrance to a scientific career, have to be oriented on state-of-the-art methodological developments; the ability to follow these developments requires profound knowledge as well as the aptitude to evaluate new statistical methods regarding their shortcomings.

Since even for doctoral students a repetition of the basics of statistics on an elementary level is often useful, with this book they can be picked up individually where their powers of recollection end – if necessary at the beginning of the Bachelor's education. And in contrast, Bachelor's students are often interested in the contents of a Master's curriculum or where the textbook leads. They can get a taste of that now.

Finally, even lecturers will find something new in this textbook; according to our experience 'statistics for psychologists' is not taught by professional statisticians but by psychologists, mostly by those at the beginning of their academic careers; anecdotes may at least help them didactically. These casual reflections can of course also be academically amusing for students.

Accordingly, the three to four mentioned target groups are guided through the book using distinctive design elements.

The running text, without special accentuation, is directed at all target groups. It is information essential for the further study of the textbook and its practical use – as is this introduction before Chapter 1. Also the terminology used in the book has to be conveyed in a standardized way. Finally, some contents, which should be familiar to doctoral students, are nevertheless aimed at all target groups because we think that repetition is useful.

Moreover, special symbols and labels on the outer edge of some pages signal the target group that the information is aimed at. Target groups other than the ones indicated with the symbol can skip these passages without being in danger of missing the respective educational aim.

The symbol Bachelor indicates that the material in these passages is aimed particularly at Bachelor's students since it deals only with the *Ability to Use*. The symbol Master on the outer edge indicates that here the reader finds an explanation of the underlying methods, without using a mathematical derivation that is too detailed; this is about *Understanding*. The symbol Doctor on the outer edge of the page announces that the shortcomings of the method will be discussed and that common misuses will be indicated; this is about *Critical Reflection*. Finally, the note **For Lecturers** signals didactically useful observations, entailing understanding of the respective topic in a very demonstrative way.

In order to bring all target groups together again, occasionally a *Summary* is given. At the beginning of every chapter a short description of its contents is given.

1

Concept of the book

In this chapter, the structure of the book and accordingly the didactic concept are presented to the reader. Moreover, we outline an example that will be used in several chapters in order to demonstrate the analytical methods described there.

In six sections this book conveys the methods of the scientific discipline of 'statistics' that are relevant for studies in psychology:

I. Introduction (Chapters 1 to 4)
II. Descriptive statistics (Chapter 5)
III. Inferential statistics for a single character (Chapters 6 to 10)
IV. Descriptive and inferential statistics for two characters (Chapter 11)
V. Inferential statistics for more than two characters (Chapters 12 and 13)
VI. Theory building statistical procedures (Chapters 14 and 15).

Chapter 1 explains the concept underlying our presentation of the methods. Furthermore an empirical example that will be used as an illustration in various parts of the book is provided.

Chapter 2 will demonstrate that quantifying and measuring in psychology is not only possible but also very useful. In addition we would like to give the reader an understanding of the strategy of gaining knowledge in psychology as a science; the approach however is similar to other scientific fields, which is why this book can be used in other fields too.

In Chapter 3 we will address the issue that empirical research is performed in several steps. For all scientific questions that are supposed to be answered by the study (as diverse as they might be regarding contents), exact planning, careful collecting of data, and adequate analysis are always needed.

Statistics in Psychology Using R and SPSS, First Edition. Dieter Rasch, Klaus D. Kubinger and Takuya Yanagida.

Within this context we wish the reader to realize that a study does not always have to include all the people that the research question is directed at. Out of practical reasons, most of the time only part of the group of interest can be examined; this part is usually called sample, whereas the group of interest is called the population. Chance plays an important role here. It will be shown that we have to make probability statements for the results of the statistical analysis; the probability calculus used for this is only valid for events for whose occurrence (or non-occurrence) chance is responsible. For example, a certain event might be that a specific person is part of the study in question. We will treat this topic in Chapter 4, as well as in Chapter 7. Since 'chance' often has a different meaning in everyday use as opposed to its general meaning in statistics and therefore in this book, we will point out at this early stage that a random event is not necessarily a rare or unanticipated event.

Finally, if data concerning one or more person(s) or character(s) that are of interest have been gathered within the framework of the study, they have to be processed statistically. The data in their totality are too unmanageable to be able to draw conclusions from them that are relevant for answering the scientific question. Therefore, special methods of data compression are necessary. We will deal with this issue in Chapter 5. The decision of which one of these methods is applicable or most appropriate is substantially based on the type of data: for example, whether they have been derived from physical measurements or whether they can only express *greater/less than* and *equal to* relations. In the latter case it is important to use methods that have been specially developed for this type of data.

Mathematical-statistical concepts are needed, especially for the generalization of study results; these will be introduced in Chapter 6. For readers who are unpracticed in the use of formulas, this chapter is surely difficult, although we try to formulate as simply as possible.

If the generalization of the study results is the aim, then a prerequisite for the use of appropriate methods is that the collected samples are random samples; information on this topic can be found in Chapter 7.

In Chapter 8 an introduction to statistical inference, in particular the principle of hypothesis testing, will be given. Because of the fact that random samples are used, it is necessary to take random deviations of the sample data from the population into account. Through hypotheses that have been formulated before data collection we try to find out as to what extent these deviations are systematic or can/must be traced back to chance. The aim is to either accept or reject a hypothesis based on the empirical data.

Chapter 9 pursues a similar objective, but this time the focus is on two populations that are compared with each other.

The implied separation between planning, data collection, and analysis is true for the classic procedure for empirical studies. In this book, however, we also want to promote a sequential approach. Thereby the gradual collection of data is constantly interrupted by an analysis. This leads to a process that looks like this: observe–analyze–observe–analyze . . .; this goes on until a predetermined level of precision is reached. This procedure is also described in Chapters 8 and 9.

Special methods are needed in studies that examine a certain character of the research unit (which in psychology often is a person or a group of persons) not only under constant conditions but also under varying conditions or when the study includes more than two populations. In Chapter 10 we cover situations where there are three or more different conditions or two or three treatment factors, with at least two values of each (treatment or factor levels).

In psychological research hardly ever is only one character used. If more than one character per person is observed, then a certain connection between them may exist; we refer here to statistical relationships. If these relationships are of interest, then the statistical methods described primarily in Chapter 11 are needed.

If there really are relationships between several characters – or if there is reason to think so – then one needs very special methods for comparing several populations. Chapters 12 and 13 describe these.

Finally Chapters 14 and 15 give an introduction into theory-building techniques that establish or test models regarding content.

The appendix of the book is split into three parts: Part A lists the data of Example 1.1 which will be illustrated below, and in part B one can find tables, helpful for some analyses; often it is faster and more convenient to look up a value than to calculate it with the help of some software. Appendix C contains a summary of the symbols and abbreviations. A complete list of references and a subject index are given at the end of the book.

Summary

We assume that empirical studies always yield data regarding at least one character. Optimally, planning takes place prior to any study. Data are used to answer a specific question. Statistics as a scientific discipline provides the methods needed for this.

The diverse statistical methods that are recommended in this book and which can be used for answering the research questions posed by psychology as a science are often only practicable when using a computer. Therefore we refer to two software packages in this book. The program package **R** is both freely accessible and very efficient; that is why we continuously use **R** here. However, since in psychology the program package IBM SPSS Statistics is still preferred for statistical analyses most of the time, it will also be illustrated using the examples. The appropriate use of such packages is not trivial; that is why the necessary procedures will be demonstrated by the use of numerical examples. The reader can recalculate everything and practice their use.

The program package **R** can be used for the planning of a study, for the statistical analysis of the data and for graphical presentation. It is an adaptation of the programming language **S** that has been developed since 1976 by *John Chambers* and colleagues in the *Bell Laboratories* (belonging to *Alcatel-Lucent*). The functionality of **R** can be enhanced through freely available packages by everybody and at will, and also special statistical methods and some procedures of C and Fortran can be implemented. Packages that already exist are being made available in standardized archives (*repositories*). The most well-known archive to be mentioned here is CRAN (Comprehensive **R** Archive Network), a server network that is serviced by the **R** Development Core Team. With the distribution of **R**, the number of **R** packages has increased exponentially: whereas there were 110 packages available on CRAN in June 2001, there were 2496 in September 2010. **R** is available, free, for Windows, Linux and Apple. With few exceptions, there are implementations for all statistical methods in **R**. With the means of the recently built package `OPDOE` (see Rasch, Pilz, Verdooren & Gebhardt, 2011), it is possible, for the first time, to statistically plan studies or to calculate the optimal number of examination objects and also to successively collect and analyze data in **R**.

The program package **R** is available for free at http://cran.r-project.org/ for the operating systems Linux, MacOS X and Windows. The installation under Microsoft Windows is initiated via the link 'Windows', from where the link 'base', which leads to the installation website, must be chosen. The setup file can be downloaded under 'Download R 2.X.X for Windows' (where X stands for the current version number). After executing this file, one is lead through the installation by a setup assistant. For the uses described in this book all the standard settings can be applied. SPSS as a commercial product must be acquired by purchase; normally universities offer inexpensive licenses for students. More on **R** can be found under www.r-project.org, and on SPSS under www.spss.com. In order not to unnecessarily prolong the explanation of the operational sequence in **R** or SPSS, we always assume that the respective program package, as well as the file that will be used, are already at hand and open.

In **R** the input window opens after starting the program; the prompt is in red: '>'. Here commands can be entered and run by pressing the *enter* button. The output is displayed in blue right below the command line. If the command is incomplete, a red '+' will appear in the next line in order to complete the command or to cancel the current command input by pressing the *Esc* button. An instruction sequence is displayed as in the following example:

```
> cbind(sub1_t1.tab, sub1_t1.per, sub1_t1.cum)
```

or also as

```
> cbind(sub1_t1.tab,
+        sub1_t1.per,
+        sub1_t1.cum)
```

or also as

```
> cbind(sub1_t1.tab,
+ sub1_t1.per,
+ sub1_t1.cum)
```

A special working environment in **R** is the `Workspace`. Several (calculation-) objects that have been created in the current session with **R** can be saved in there. These objects include results of calculations (single scores, tables, etc.) and also data sets. A workspace can be loaded with the sequence

```
File - Load Workspace...
```

For all the examples presented in this book the reader can download the `Workspace` '`RaKuYa.RData`' from the website www.wiley.com/go/statisticsinpsychology.

Since there are more data sets in our `Workspace`, the scores of single research units/persons have to be accessed by specifying the data set with a '`$`'; for example: `Example_1.1$native_language`. A useful alternative for the access is the command `attach()`, which makes the desired data set generally available; for example: `attach(Example_1.1)`. To minimize repetition, in the instruction sequences given throughout the book, we assume that the `attach()` command has already been run and therefore the relevant data set is active. For some examples we need special **R** packages; they must be installed once via the menu `Packages - Install Package(s)...` and then loaded for every session in **R** with the command `library()`. The installation of packages is done via the menu

```
Packages - Install Package(s)...
```

In SPSS the desired data frame can be opened via File – Open – Data... after starting the program. Then we write the instruction sequence as in SPSS handbooks; for example like this

```
Analyze
    Descriptive Statistics
        Frequencies...
```

For all examples in the book the reader can find the data in the SPSS folder 'RaKuYa' on the website www.wiley.com/go/statisticsinpsychology.

For figures that are shown as the results of the calculations for the examples, we use either the one from SPSS or the one from **R**. Only if the graphs differ between **R** and SPSS will we present both.

It is the concept of this textbook to present illustrative examples with content – that can be recalculated – from almost all subject areas concerning the planning and statistical analysis of psychological studies. A lot of the methods described in this book will be demonstrated using one single data set in order to not have to explain too many psychological problems. This will be introduced in Example 1.1.

Example 1.1 The goal is to test the fairness of a popular natural-language intelligence test battery with reference to children with Turkish native language[1,2] (see Kubinger, 2009a[3]).

The following characters were observed per child (see Table 1.1 and the data sheet in Appendix A; then see, for **R**, the respective data structure in Figure 1.1, and for SPSS the screen shot shown in Figure 1.2).

---

[1] Fairness is a specific quality criterion of psychological assessment methods (tests). A psychological test meets the requirement of fairness if the resulting test scores don't lead to a systematic discrimination of specific testees: for example because of sex, ethnic, or socio-cultural affiliation; see Kubinger, 2009b).

[2] The data originally applied to German-speaking countries; however, there was no socio-political difference when the data in the following analyses were interpreted as relating to English-speaking countries and some ethnic-minority groups.

[3] Due to copyright reasons the original data had to be slightly modified; therefore no deductions regarding content can be drawn from the data found in the data sheet in the appendix.

Table 1.1 The characters and their names in **R** and SPSS (including coded values).

Name of the character	Name in **R**	Name in SPSS	Coded values
testee number	no	no	
native language of the child	native_language	native_language	1 = 'German' 2 = 'Turkish'
age of the child	age	age	
sex of the child	sex	sex	1 = 'female' 2 = 'male'
gestational age at birth (in weeks)[4]	age_birth	age_birth	
number of siblings	no_siblings	no_siblings	
sibling position	pos_sibling	pos_sibling	1 = 'first-born' 2 = 'second-born' 3 = 'third-born' 4 = 'fourth-born' 5 = 'fifth-born' 6 = 'sixth-born'
social status (after Kleining & Moore [1968] according to occupation of father/alternatively of the single mother)	social_status	social_status	1 = 'upper classes' 2 = 'middle classes' 3 = 'lower middle class' 4 = 'upper lower class' 5 = 'lower classes' 6 = 'single mother in household'
urban/rural	urban_rural	urban_rural	1 = 'city (over 20 000 inhabitants)' 2 = 'town (5000 to 20000 inhabitants)' 3 = 'rural (up to 5000 inhabitants)'
marital status of the mother	marital_mother	marital_mother	1 = 'never married' 2 = 'married' 3 = 'divorced' 4 = 'widowed'

[4] The gestational age is the age of the (unborn) child counted from the day of supposed fertilization.

Table 1.1 (*Continued*)

Name of the character	Name in **R**	Name in SPSS	Coded values
test setting	test_set	test_set	1 = 'German speaking child' 2 = 'Turkish speaking child tested in German at first test date' 3 = 'Turkish speaking child tested in Turkish at first test date'
Everyday Knowledge, 1st test date (T-Scores)[5]	sub1_t1	sub1_t1	
Applied Computing, 1st test date (T-Scores)	sub3_t1	sub3_t1	
Social and Material Sequencing, 1st test date (T-Scores)	sub4_t1	sub4_t1	
Immediately Reproducing – numerical, 1st test date (T-Scores)	sub5_t1	sub5_t1	
Coding and Associating, 1st test date (T-Scores)	sub7_t1	sub7_t1	
Everyday Knowledge, 2nd test date (T-Scores)	sub1_t2	sub1_t2	
Applied Computing, 2nd test date (T-Scores)	sub3_t2	sub3_t2	
Social and Material Sequencing, 2nd test date (T-Scores)	sub4_t2	sub4_t2	
Immediately Reproducing – numerical, 2nd test date (T-Scores)	sub5_t2	sub5_t2	
Coding and Associating, 2nd test date (T-Scores)	sub7_t2	sub7_t2	

[5] Test scores are generally standardized to a certain scale; T-Scores are a very common method of standardization.

```
RGui - [R Console]
File Edit View Misc Packages Windows Help

> head(Example_1.1)
  no native_language age    sex age_birth no_siblings pos_sibling       social_status                       urban_rural
1  1          German   6 female        39           1  first-born      middle classes  City (over 20000 inhabitants)
2  2          German   7   male        40           2 second-born lower middle class  City (over 20000 inhabitants)
3  3         Turkish   8 female        38           3  third-born lower middle class Rural (up to 5000 inhabitants)
4  4         Turkish   9   male        36           2  first-born      middle classes  City (over 20000 inhabitants)
5  5          German   8 female        37           3 second-born  upper lower class  City (over 20000 inhabitants)
6  6          German   9   male        40           0  first-born      upper classes  City (over 20000 inhabitants)
  marital_mother                                             test_set sub1_t1 sub3_t1 sub4_t1 sub5_t1 sub7_t1
1        married                                German speaking child      52      56      43      55      50
2        married                                German speaking child      54      59      52      57      64
3        married  Turkish speaking child tested in German at first test date      46      55      47      40      50
4        married Turkish speaking child tested in Turkish at first test date      58      47      59      47      50
5       divorced                                German speaking child      37      50      38      36      50
6        married                                German speaking child      50      59      55      52      64
  sub1_t2 sub3_t2 sub4_t2 sub5_t2 sub7_t2
1      50      55      45      55      54
2      56      62      53      59      65
3      54      47      52      55      50
4      60      48      57      45      54
5      38      48      38      36      46
6      52      60      57      50      61
> tail(Example_1.1)
     no native_language age    sex age_birth no_siblings pos_sibling       social_status                      urban_rural
95   95         Turkish   9 female        39           1  first-born      middle classes City (over 20000 inhabitants)
96   96         Turkish   6   male        37           3  first-born      middle classes City (over 20000 inhabitants)
97   97         Turkish   8 female        39           0  first-born lower middle class City (over 20000 inhabitants)
98   98         Turkish   7   male        38           4 fourth-born       lower classes City (over 20000 inhabitants)
99   99         Turkish   9 female        37           3  third-born   upper lower class City (over 20000 inhabitants)
100 100         Turkish   6   male        41           4  first-born      middle classes City (over 20000 inhabitants)
    marital_mother                                             test_set sub1_t1 sub3_t1 sub4_t1 sub5_t1 sub7_t1
95         married Turkish speaking child tested in Turkish at first test date      60      56      52      28      54
96         married Turkish speaking child tested in Turkish at first test date      61      52      47      36      36
97        divorced Turkish speaking child tested in Turkish at first test date      58      50      67      70      57
98        divorced Turkish speaking child tested in Turkish at first test date      27      50      35      42      43
99         married Turkish speaking child tested in Turkish at first test date      60      53      62      35      33
100        married Turkish speaking child tested in Turkish at first test date      50      37      67      52      26
    sub1_t2 sub3_t2 sub4_t2 sub5_t2 sub7_t2
95       58      59      52      33      54
96       54      52      50      33      50
97       60      59      64      67      65
98       30      47      42      40      39
99       60      48      64      36      31
100      46      36      64      59      31
> |
```

Figure 1.1 Representation of the data structure of Example 1.1 in **R**.

In order to illustrate some statistical procedures we need other examples regarding content, but the data for these examples will not be found in Appendix A due to space limitations; however they are provided in the aforementioned `Workspace` and SPSS folders respectively. For the recalculation of the examples as well as for later calculations with the reader's own data, we will also provide the **R** instruction sequences, so that they don't have to be typed out. They can be found on the website www.Wiley.com. For beginners in **R** these are simply listed in order in a PDF file; for those readers already experienced in the use of **R** they are in a syntax editor for **R**; that is, `Tinn-R` (www.sciviews.org/Tinn-R/).

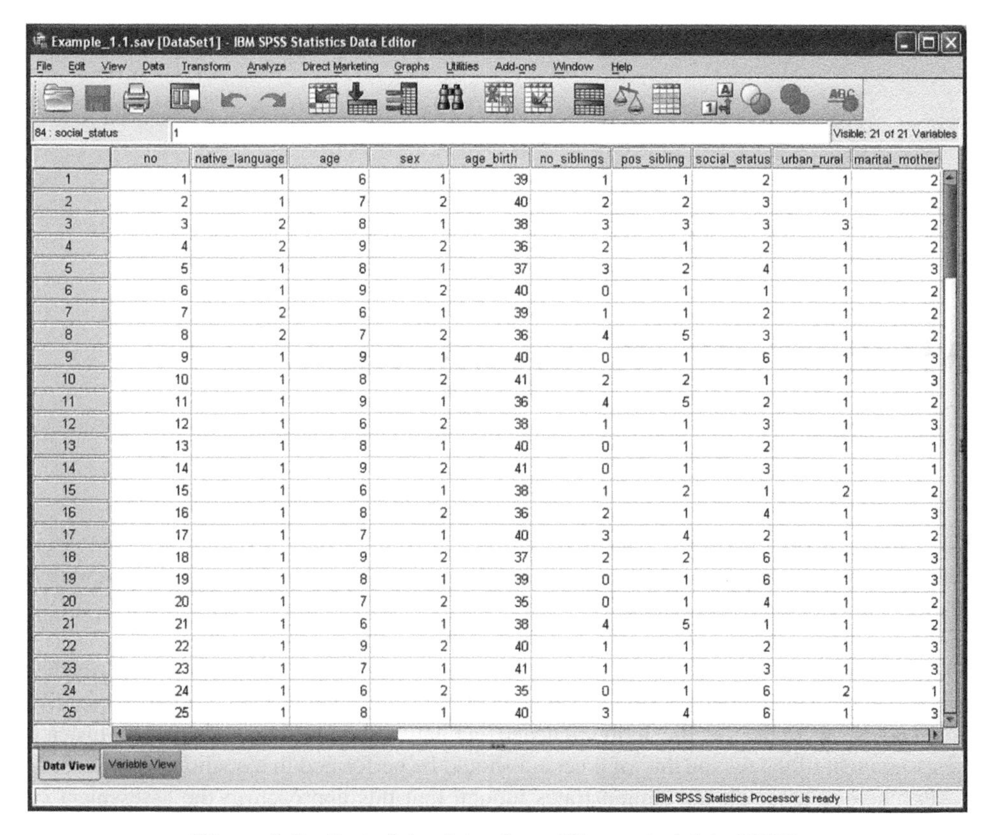

	no	native_language	age	sex	age_birth	no_siblings	pos_sibling	social_status	urban_rural	marital_mother
1	1	1	6	1	39	1	1	2	1	2
2	2	1	7	2	40	2	2	3	1	2
3	3	2	8	1	38	3	3	3	3	2
4	4	2	9	2	36	2	1	2	1	2
5	5	1	8	1	37	3	2	4	1	3
6	6	1	9	2	40	0	1	1	1	2
7	7	2	6	1	39	1	1	2	1	2
8	8	2	7	2	36	4	5	3	1	2
9	9	1	9	1	40	0	1	6	1	3
10	10	1	8	2	41	2	2	1	1	3
11	11	1	9	1	36	4	5	2	1	2
12	12	1	6	2	38	1	1	3	1	3
13	13	1	8	1	40	0	1	2	1	1
14	14	1	9	2	41	0	1	3	1	1
15	15	1	6	1	38	1	2	1	2	2
16	16	1	8	2	36	2	1	4	1	3
17	17	1	7	1	40	3	4	2	1	2
18	18	1	9	2	37	2	2	6	1	3
19	19	1	8	1	39	0	1	6	1	3
20	20	1	7	2	35	0	1	4	1	2
21	21	1	6	1	38	4	5	1	1	2
22	22	1	9	2	40	1	1	2	1	3
23	23	1	7	1	41	1	1	3	1	3
24	24	1	6	2	35	0	1	6	2	1
25	25	1	8	1	40	3	4	6	1	3

Figure 1.2 Part of the data view of Example 1.1 in SPSS.

References

Kleining, G. & Moore, H. (1968). Soziale Selbsteinstufung (SSE): Ein Instrument zur Messung sozialer Schichten [Social Self-esteem (SEE): An Instrumnet for Measuring the Social Status]. *Kölner Zeitschrift für Soziologie und Sozialpsychologie*, *20*, 502–552.

Kubinger, K. D. (2009a). *Adaptives Intelligenz Diagnostikum - Version 2.2 (AID 2) samt AID 2-Türkisch* [Adaptive Intelligence Diagnosticum, AID 2-Turkey Included]. Göttingen: Beltz.

Kubinger, K. D. (2009b). *Psychologische Diagnostik – Theorie und Praxis psychologischen Diagnostizierens* (2nd edn) [Psychological Assessment – Theory and Practice of Psychological Consulting]. Göttingen: Hogrefe.

Rasch, D., Pilz, J., Verdooren, R. L., & Gebhardt, A. (2011). *Optimal Experimental Design with R*. Boca Raton: Chapman & Hall/CRC.

2

Measuring in psychology

This chapter deals with several methods of data acquisition that are used in psychology. The methods for psychological assessment and the methods primarily for answering research questions have to be distinguished.

Within the field of psychology, the claim of conducting *measurements*, e.g. to measure 'psyche' or psychological phenomena, is often adamantly refuted. The attempt to measure or to quantify would not allow for the specific, individual, and qualitative characteristics of a person. Instead, the assessment of the personality of a person should be performed in a qualitative way.

Psychology as a science demonstrates though that this approach to the assessment of a person, regarding a specific *character* (within psychology: *trait/aptitude*), is limited to a pre-scientific level. While it can lead to important assumptions on causal relations, it never allows for binding generalizations. On the contrary, measurements that are conducted under defined abstractions can relate a person's personality to an objective framework.

Bachelor

Statistical data calls for a useful bundling of what is to be measured. Not everything that is measurable regarding a certain character can be compared in depth, i.e. individually, but the whole essential part of the information has to be compressed. A factually acceptable abstraction of the available information has to be made. For example, this abstraction could be that all 35-year-old women are viewed equally regarding their age, irrespective of whether one of them has a biologically 'young' body caused by practicing competitive sports or another one has a biologically 'old' body because she lived in war zones for some years.

We can be sure that *measuring* in psychology is valuable for psychological case consulting as well as for research on the evaluation of psychological treatments, and especially for basic psychological research.

Statistics in Psychology Using R and SPSS, First Edition. Dieter Rasch, Klaus D. Kubinger and Takuya Yanagida.

Master Doctor

Although there are measurement techniques in psychology that follow the methods of natural science, measurements of psychic or mental phenomena are additionally based on specific scientific methods. One thing, however, is common to all natural sciences, psychology included: measuring means the ascertainment of the interesting character's value for the research unit (in psychology this is mostly a person). This happens as an assignment of numbers or signs in such a way that these assignments (measuring values), represent empirical factual relations. That is, the assignment relations must coincide with the empirical (obviously) given relationships of the research units (discussed in detail in Chapter 4).

Master

Although important to note here but not explicitly a distinct measurement technique are the measurements of physiological psychology: its first sub-specialty, neuropsychology, studies the relationship between behavior and the activity of the central nervous system by the means of electrophysiological methods (e.g. EEG, electroencephalography). Its second sub-specialty, psycho-physiology, investigates the relationships between behavior and the activity of the vegetative nervous system by the means of physical methods (e.g. measurement of electrodermal activity, EDA). Its third sub-specialty, chemical psychology, explores the relations between behavior and chemical substances, which are either brought into the organism from outside (pharmaco-psychology) or are built inside the organism (endocrine psychology, neuro-chemopsychology, psycho-genetics) by the means of chemical methods.

2.1 Types of psychological measurements

Some measurement techniques used in psychology are standard methods of *psychological assessment*; they are used in case consulting of clients but also in research. For these techniques specific psychometric quality criteria apply. Other measurement techniques are used specifically in research. Some of them are also used in fields other than psychology, like sociology or market- and opinion research.

2.2 Measurement techniques in psychological assessment

An extensive introduction to psychological assessment can be found in Anastasi & Urbina (1997) and Kubinger (2009b).

2.2.1 Psychological tests

The term 'psychological test' subsumes all achievement tests, including intelligence tests, as well as so-called objective personality tests.[1]

[1] If the word 'test' is used without an adjective, the reader should be able to tell from the context if a psychological or statistical test is meant; however, we prefer the use of the more comprehensive term 'psychological assessment tool' when the current explanations aren't limited to (psychological) tests.

Master

Example 2.1 Intelligence test

An example for an intelligence test is the subtest *Verbal Abstraction* from the intelligence test battery AID 2 (*Adaptive Intelligence Diagnosticum, Version 2.2*; Kubinger 2009a). 'How are a candle and a torch alike?' or 'How are an airplane and a bird alike?' are 2 (very) simple items out of 15 that are presented to the children. The measurement problem is whether the number of solved items alone is a representative and fair measure allowing all the testees to be put into a fair relationship regarding their ability. For example, is child *A*, who solves the first of the aforementioned items but not the very difficult one 'hunting and fishing', really as 'intelligent' as child *B*, who doesn't solve the first of the aforementioned items but the last, very difficult one? Specific mathematical statistical methods, developed by *psychometrics*, can answer this question (see, as an introduction, Kubinger 2009b, and more precisely Kubinger, 1989). They can also solve even greater measurement problems in testing. In the AID 2, for example, not all children are given the same items, but items meeting their ability, as demonstrated in preceeding items, are selected (so-called adaptive testing).

Master

Example 2.2 Objective personality tests

Objective personality tests register individual stylistic features ('cognitive styles') while performing a (achievement) task. The *Gestalt test* (Hergovich & Hörndler, 1994), for example, differentiates between 'field dependent' and 'field independent' persons. A field-dependent perceptive style occurs when the perceptive environment (field) has a strong influence on the perception target; whereas a field-independent perceptive style occurs when the perception is centered on the perception target. The measurement problems in this test are similar to the aforementioned example: can the number of solved items, in which a figure disguised in confusing line drawings has to be found, put all the testees in a fair relation regarding their field in/dependency?

2.2.2 Personality questionnaires

Personality questionnaires include not only the well-known questionnaires, often called tests, where one rates oneself and one's typical behavior patterns, but also tools of assessment by others where a third party rates a person as well as 'tests' of interest.

Master

Example 2.3 Personality questionnaires

The internationally most renowned personality questionnaire is the MMPI-2 (*Minnesota Multiphasic Personality Inventory-2*; Butcher, Dahlstrom, Graham, Tellegen, & Kaemmer, 1989). 'I find it hard to keep my mind on a task or job' and 'I have had periods of days, weeks, or months when I couldn't take care of things because I couldn't "get going" ', are two out of more than 500 statements that have to be rated 'true' or 'false'. The sum of the 'true' answers is supposed to indicate the degree of psychasthenia (i.e. a psychological disorder associated with fear and obsessive ideas). We will see below that, in contrast to tests, the measurement problem here is not only aggravated but ultimately completely unsolvable.

Master

Example 2.4 Tools of assessment by others

An often-used tool of assessment by others is the HAMD (*Hamilton-Depression-Rating-Scale*; Hamilton, 1980). Again the problem remains: whether a collection of questions or the scores resulting from them can really picture the empirically given relations. For example, is patient *Z*, who has been rated as a person with suicidal thoughts ('Wishes he were dead or any thoughts of possible death to self') but without any sleeping problems by the psychiatrist or clinical psychologist in charge, as depressive as patient *W* without suicidal thoughts but with occasional '... difficulty falling asleep – i.e. more than $^1/_2$ hour' and '...being restless and disturbed during the night'? The calculation procedure prescribed by the method, however, presumes this. Of course this presumption would be testable with the above-mentioned methods of psychometrics, but it has not been done yet.

Master

Example 2.5 'Tests' of interest

The above-stated measurement problems of tools of assessment by others also hold for 'tests' of interest, or, even worse, unsolvability becomes a problem, as described above for the questionnaires.

2.2.3 Projective techniques

Projective techniques are psychological assessment tools that try to uncover the personality structure and the motives for action of a person by ambiguous material or stimuli.

Master

Example 2.6 Projective techniques

Projective techniques imply that a person, when confronted with ambiguous material or stimuli, unconsciously reacts with his or her own feelings, thoughts, and attitudes and transfers them into the material. However, the psychometric groundings of projective techniques often do not justify these conclusions. The *Rorschach inkblot test* (see Exner, 2002) is the most well known. As long as the observations of how and as what the testee interprets the (symmetric) inkblots are only of a qualitative nature, the procedure doesn't have a measurement function: users claim that with some disorders it is only through this technique that there is the possibility to be able to talk to the patient, because they open up more easily than if they were asked direct questions. Of course the technique is useful (for the experienced user) in order to hypothesize about behavior-determining conditions or typical facets of a human. However, the technique becomes problematic when the very controversial quantification rules are used. Here the argument mentioned earlier that the numerical relations don't depict the empirical ones doesn't even have to be used, because the technique doesn't even claim to reliably measure any specific character. In order to draw consulting conclusions we have to ask ourselves for example the following question. In which way is person *U*, with 5 holistic answers (that means that the answer can be traced to the registration/description of the whole figure), 3 answers concerning details, and 5 answers concerning small details, better/inferior, stronger/weaker etc. than person *V* with 8 holistic answers, 6 answers concerning details, and 12 answers concerning small details?

2.2.4 Systematical behavior observation

Systematical behavior observation focuses on category-system-bound observations of a person's behavior; it is not the arbitrary or casual observation that is mainly based on subjective impressions.

Master **Example 2.7** Systematical behavior observation

Similar to projective techniques, the rule is: as long as only a (qualitative) impression is drawn from a systematical behavior observation that leads to hypotheses on behavior-determining conditions, then the technique has no measurement function. For example in the AID 2 there is a supplemental sheet for the observation of work habits, which is used for the qualitative evaluation of work and contact behavior. Although there are no generally binding quantitative category systems, their use is not connected with measurement problems (for example for the characterization of verbal behavior in social contacts or as a method for the analysis of self-management in everyday life): the counting of observed character categories is similar to physical measurements – for example how often somebody uses the word 'please'.

2.3 Quality criteria in psychometrics

As mentioned earlier, specific quality criteria are relevant for measurement methods in psychological assessment; they define the quality of the collected data: *objectivity*, *reliability*, *validity*, *standardization*, and *unfakeability*.

Objectivity means that the result of the measurement is independent from the diagnosing psychologist. *Reliability* alludes to the degree of formal accuracy of the measurement, i.e. precision. *Validity* refers to the correctness of the measurement with regards to the content, which means that the character that is desired to be measured is actually the one that is measured. *Standardization* permits the placement of the individual measurement result of a person within the distribution of all results of a population. *Unfakeability* means that a measurement instrument doesn't allow individual control of type and content of the desired information. More on psychometric quality criteria can be found in Kubinger (2009b), for example.

Tests usually meet the requirement of objectivity, especially when they are administered in a group setting or via a computer. The exactness of the measurement differs from test to test. In general, tests measure less exactly than physical measurement instruments. Sometimes their validity has not been analyzed, sometimes it is unsatisfactory, but sometimes it is also provided. A (up-to-date) standardization is generally available, and tests are essentially unfakeable.

Personality questionnaires, however, are extremely fakeable: they are very transparent concerning their measurement intentions, and most people will be inclined to answer to their advantage (e.g. socially desired). Consequently personality questionnaires are rarely valid and hardly ever accurate. In a group or computer setting they can be considered objective. This, just as the usually given standardization, doesn't make these kinds of measurement instruments any more useful.

Projective techniques are considerably less fakeable than personality questionnaires because their measurement intention is less transparent. However there are hardly any studies concerning the validity and reliability of projective techniques. A standardization is rarely available.

In systematical behavior observation, objectivity is the primary problem. Apart from unconscious, mostly nonverbal experimenter effects, there are typically observation and categorization mistakes (i.e. something relevant is not recorded, or a behavior is misinterpreted and then coded in an incorrect way). However, systematical behavior observation has a fundamental advantage as concerns validity: contrary to personality questionnaires, here real-life behavior instead of verbal behavior is recorded. The measurement accuracy depends on the representativity of the chosen observation situation; the fakeability depends on how disturbing or impressing the observer is. Normally there is no standardization.

2.4 Additional psychological measurement techniques

Although the measurement methods described below can nowadays be found in practical case consulting as well as being special techniques in psychological assessment, they primarily come from research.

2.4.1 Sociogram

Starting from a graphic visualization of all positive and negative relations between persons in a small group, the *sociogram* tries to measure person-specific as well as group-specific characters.

Master **Example 2.8** Sociogram

With a sociogram it is possible to make topic-related quantifications, for example performance- or sympathy-related quantifications of a person's status of selection or rejection; this being based on the observed individual preferences or objections of all members of the group. For the group as a whole, a group cohesion score can be calculated. The measurement itself is unproblematic as for systematical behavior observation, because it is based solely on counts. If the quality criteria discussed above are applied to this method, then the sociogram performs better than for example the systematical behavior observation concerning objectivity, but worse as regards validity. Here again, the verbal behavior is measured instead of the actual behavior. Moreover the generalizability of the result of the measurement is questionable because of the heavily limited representativity of the behavior, which is caused by a limitation to a specific composition of the group.

2.4.2 Survey questionnaires

Survey questionnaires are mainly used for the observation of opinions and attitudes that are present in the population.

Master **Example 2.9** Survey questionnaires

There are numerous variants for the formal arrangement of survey questionnaires. Open questions, which are questions with *free response format,* offer the respondent the possibility of freely choosing the words for the answers themselves;

however the evaluator will sooner or later group the individual answers into categories. So-called closed questions only offer the respondent the possibility of choosing between a smaller or greater number of given response options ('fixed response format'). If only two response options exist – *dichotomous response format* (*forced choice response format*) – many respondents complain that they are overstrained with the decision they are being asked to make. If there are more than two response options – *multiple choice response format* – then arbitrary or random answers are encouraged. It is also conceivable that some response options overlap, so that multiple entries are possible. Finally one must distinguish between different types of response options, particularly whether they are numerical, gradual, or qualitative (i.e. yes, no, maybe) differentiations. Depending on this there are different measurement problems. They can range from those found in psychological tests to those in conventional personality questionnaires; in extreme cases every single question is evaluated and interpreted on its own.

2.4.3 Ratings

Ratings are a subjective judgment concerning a character that is perceived as being continuous.

Master **Example 2.10** Ratings

For some specific contents, ratings are another formal arrangement possibility for questionnaires. In most cases any gradation is possible and can be made by marking a line between two poles with a cross (so called *analogue scale-response format*). If the rating is demanded on a computer, the gradation can be accomplished in a similar way by clicking with a mouse. Despite the ostensible metric scaling, ratings only provide less/greater-or-equal-relations: although the indications themselves are metric, that is to say physically measurable, extensive literature concerning psycho-physics has shown that humans are not able to estimate equal distances as in metric scales. If the character in question is rated globally then only this problem occurs. If, however, several ratings per person are to be combined, then measurement problems arise that are even more complicated than the ones that apply to tests. Nevertheless they are solvable with the help of the methods of psychometrics.

2.4.4 Q-sort

The method of *Q-sort* is also a tool for subjective judgment; several objects that are to be compared (persons, activities, situations) are represented by cards, which must be divided into given categories.

Master **Example 2.11** Q-sort

The method of Q-sort most of the time demands that the allocation into categories follows a predetermined frequency distribution in order to make use of the whole spectrum of categories. The name comes from early studies about personality theory where there was a typification of people by means of the so-called Q-technique in factor analysis (for factor analysis see Section 15.1). The quantification of the sorting regarding the character of interest also brings in the repeatedly discussed measurement problems.

2.4.5 Semantic differential

The *semantic differential* is a special case of ratings.

Master **Example 2.12** Semantic differential

In a semantic differential, the response options are reduced to a seven-level scale between two poles on the one hand, and on the other hand 24 predetermined polarities (i.e. small–big, weak–strong) define something like a standard, no matter which content is rated. This is because the denotative meaning of a term, which is the rationally derived meaning, is not of as much interest as the connotative one, which is the associative-emotional meaning. In various studies it became clear that these 24 pairs of opposites always measure three things: appraisal, potency, and activation of the content in question. Corresponding to the views in Example 2.10, ratings only provide less/greater-or-equal relations; this must be taken into consideration during such common statistical analyses as the ascertainment of differences or dependencies. Apart from that, the semantic differential is free of measurement problems.

2.4.6 Method of pair-wise comparison

In an early effort in psychology to determine the functional connection between a physical stimulus and its perception, psycho-physicists developed several methods of measurement that were used in other contexts later on. In this way ratings can be subsumed here. Here the *method of pair-wise comparison* has become especially important.

Master **Example 2.13** Method of pair-wise comparison

Instead of having an object rated directly by a person concerning the character of interest, the measurement using the method of pair-wise comparison is done indirectly, by rating two objects concerning only the relation more/less than (perhaps: same as). If the pair-wise comparisons are analyzed with special mathematical-statistical methods of psychometrics, then metric values result.

2.4.7 Content analysis

Originally developed as a method for the systematical and quantitative description of written texts, *content analysis* is related to verbal systematic behavior analysis because it also categorizes verbal communicational content.

Master **Example 2.14** Content analysis

In content analysis counts are used with the aim of making something comparable to other texts or communicators. This can refer to syntax or semantics. Again this count does not lead to measurement problems.

2.5 Statistical models of measurement with psychological roots

It's not so much that measurements in psychology are limited to the methods described up to this point as that some methods within statistics have taken root, having been developed from problems in psychology and mostly by psychologists. Much more is measurable with these higher methods of analysis.

Typifications have already been mentioned: for example, with the help of few or many characters, an allocation to several typical groups of persons is possible (see Chapter 15). Again with the help of a few or many characters, any object (i.e. of the existing psychotherapeutic schools) can be positioned in a multidimensional space of characters by means of the aforementioned factor analysis, in which a system consisting of as few as possible, not directly observable (orthogonal) dimensions is taken or sought as a basis.

The mathematical *information theory* measures the extent of uncertainty that is present during the search for information about unknown contents. It is determined according to how much 'either/or' information has to be successively received in order to fully know the content. Within the framework of psychological communication theory, corresponding methods for the registration of the information transference of humans (i.e. information content of the intervention of a session, of a psychologist, or a psychotherapeutic school) are treated. And finally there are special methods for the fundamentally problematic measurement of change, the models by Fischer (1977). With them the effects of treatments that happen simultaneously but with different intensity can be separated and quantified independently from each other.

Summary

Quantifying or *measuring* serves the aim of making research units comparable with respect to some character – and therefore serves for the increase of scientific insight. There are different techniques in psychology to measure psychic phenomena. The various measurement methods of psychological assessment differ regarding their *psychometric quality criteria.*

References

Anastasi, A. & Urbina, S. (1997). *Psychological Testing* (7th edn). Upper Saddle River, NJ: Prentice Hall.

Butcher, J. N., Dahlstrom, W. G., Graham, J. R., Tellegen, A., & Kaemmer, B. (1989). *The Minnesota Multiphasic Personality Inventory-2 (MMPI-2): Manual for Administration and Scoring*. Minneapolis, MN: University of Minnesota Press.

Exner, J. E. (2002). *The Rorschach: Basic Foundations and Principles of Interpretation: Volume 1.* Hoboken, NJ: John Wiley & Sons, Inc.

Fischer, G. H. (1977). Linear logistic models for the description of attitudinal and behavioral changes under the influence of mass communication. In W. H. Kempf & B. H. Repp (Eds.), *Some Mathematical Models for Social Psychology* (pp. 102–151). Bern: Huber.

Hamilton, M. (1980). Rating depressive patients. *Journal of Clinical Psychiatry, 41*, 21–24.

Hergovich, A. & Hörndler, H. (1994). *Gestaltwahrnehmungstest* [Gestalt Test] (Software and Manual). Frankfurt am Main: Swets Test Services.

Kubinger, K. D. (1989). Aktueller Stand und kritische Würdigung der Probabilistischen Testtheorie [Critical evaluation of latent trait theory]. In K. D. Kubinger (Ed.), *Moderne Testtheorie – Ein Abriß samt neuesten Beiträgen* [Modern Psychometrics – A Brief Survey with Recent Contributions] (pp. 19–83). Munich: PVU.

Kubinger, K. D. (2009a). *Adaptives Intelligenz Diagnostikum - Version 2.2 (AID 2) samt AID 2-Türkisch* [Adaptive Intelligence Diagnosticum, AID 2-Turkey Included]. Göttingen: Beltz.

Kubinger, K. D. (2009b). *Psychologische Diagnostik – Theorie und Praxis psychologischen Diagnostizierens* (2nd edn) [Psychological Assessment – Theory and Practice of Psychological Consulting]. Göttingen: Hogrefe.

3

Psychology – an empirical science

This chapter is about the importance of statistics and its methods for psychology as a science. It will be demonstrated that for the gain of scientific insight in psychology, empirical studies are needed. An example describes the statistical approach answering the scientific question that a study is based on. Important statistical terms, which will be clear in context, will be introduced.

Empirical research starts with a *scientific question*. Its concluding answer leads to a gain of insight. The way from the question to the gain of scientific insight is often intricate, not trivial from the outset, and different according to question. That is why the following section is about a general strategy for gaining insight in an empirical science.

A first example will be based on a question that can act as representative for many other scientific questions in applied psychological research. It is about the psychological consequences of a hysterectomy; that is, the short- and middle-term condition of women whose uteruses had to be removed due to medical reasons. It will be demonstrated that this question can only be answered by means of an empirical *research study*. In this context it will be shown that statistically well-founded planning, as well as an operationalization of the psychological phenomenology (what will be investigated) are both crucial.

During the careful (research) *planning of a study*, which is necessary to answer a question, the focus is on the balance of concurrent demands of optimality. What the adequate solution strategy generally looks like will be demonstrated with a second example. It does not allude at all to a psychological question, but to an everyday trivial one. Even here it becomes clear that statistics as a scientific discipline is able to adequately work on and answer complex questions which come from – in the first instant – very easy-sounding questions but have finally been stated more precisely.

Statistics in Psychology Using R and SPSS, First Edition. Dieter Rasch, Klaus D. Kubinger and Takuya Yanagida.

3.1 Gain of insight in psychology

Psychology as a science deals with the (long-life development of the) behavior and experience (consciousness) of humans as well as with the respective causative conditions. 'The goals ... are to describe, explain, predict, and control behavior' and 'seek[s] to improve the quality of each individual's and the collective's well-being' (Gerrig & Zimbardo, 2004, p. 4).

From that perspective, there is a need for empirical studies in psychology in order to gain scientific insight. Using rules and methods from natural sciences, systematic observations of a character must be made and they have to be related to *treatment factors* that are controlled as far as possible. The actually realized values of our observations we call *observed* (*measurement*) *values*/*outcomes* of a character.

Master
Doctor

Example 3.1 The psychological consequences of a hysterectomy will be assessed

According to clinical psychology, physical illnesses are connected with mental aspects most of the time. Some aspects are coping strategies or the prevention of psychic crises or psychic disorders. After undergoing a hysterectomy there is reason to fear that patients suffer from lasting psychic crises, for example in the sense of a massive loss of self-esteem, especially concerning self-esteem as a woman.

Let's make the assumption that the cause for the given question is an unsystematic, subjective, or selective perception of some of these patients' advisors. At least in terms of health policy or maybe even in economic terms it appears appropriate to research this question.

For the sake of simplicity let's assume that earlier research has provided a psychological assessment tool that can measure the self-esteem or the 'psychic stability' of a person in a valid way. Let's simply call that tool Diagnosticum Y. Then we can begin to design a study. Usually one thinks about the first group of patients that comes along. The most easily reachable, reasonably sized group (i.e. 30), as concerns the relation of workload and gain of insight, could be psychologically tested with Diagnosticum Y after surgery.

People critical of this empirical research design would at once argue:

1. It is an arbitrary selection of patients. The institution may have systematically chosen patients who were too old: in other institutions the mean age of such patients might be substantially lower. There could be patients with comparatively low educational level, divergent ethnic origin, long-term single status and much more.

2. Every result would be meaningless because one would not know which test scores women without surgery have in Diagnosticum Y (that is to say the ('normal'/total) population) as well as which test scores the women of the study would have had before surgery.

According to this, one should design to examine women from the 'healthy' (that is the (not-yet) positively diagnosed) part of the population at the same time, too. They form the *comparison group* compared to the *target group*. Initially, here again the first group of women that is available will be chosen – (presumably similar in number to the first group).

People critical of this empirical research design would argue:

3. More than ever, this is an arbitrary selection of persons. In the group of 'healthy' women, that is to say the comparison group, there could be people with systematically different psychosocial characteristics (age, education level, social status, ethnic origin, relationship) as compared to the patient group; that is, the target group.

4. The two group sizes seem to be too small; hardly anybody will dare to draw generally binding conclusions because of the possibly observable differences between the two arbitrarily examined groups. Therefore no general conclusion about the psychic consequences of a hysterectomy can be made. A compulsory psycho-hygienic need for action for coping or prevention cannot be deduced from this – but nobody is interested in the differences between the two concrete groups (except the patients themselves and their family members).

Therefore, one must basically reflect on the choice of women that should be examined. Perhaps one comes to the conclusion that a so-called representative group – subsequently termed sample – of the population is hard to survey (for more details see Chapter 4). Not only the comparability of the psychosocial characteristics is questionable, but also especially the circumstances under which women from the 'healthy' population are willing to undergo the examination (i.e. only if they are paid and presumably not in a hospital) or which women are willing at all (i.e. those with especially high self-esteem or a special degree of 'psychic stability'). As a consequence one will design to examine the patient group with Diagnosticum *Y* not only after surgery (note that here it is important to think about the exact time of examination after surgery; preferably not right after surgery has been performed but shortly before discharge) but also before surgery (note that in this case one must think about the exact time of examination before surgery; preferably not right before surgery is performed but a short time before hospital admission). The respective results could be individually compared and from this the psychic consequences of a hysterectomy could be estimated.

People critical of this empirical research design would argue:

5. Before surgery, presumably no patient will have a test score in Diagnosticum *Y* that can serve as a comparable value typical of the time before an illness with indication of hysterectomy.

6. And even if this were the case, a change towards loss of self-esteem or 'psychic stability' as a result of a surgery would hardly be surprising, because every surgery means a massive intrusion into a human's 'bio-(psycho-socio-) tope'.

Thus one has to specify the question: it is less about the examination of the psychic consequences of surgeries (in a selective way that is a specific surgery indication), but rather about the examination of the consequences of a specific surgery that is of interest (namely hysterectomy), preferably compared with other surgeries (that are less related to the role/functioning as a woman). Accordingly, an empirical research design is indicated that also includes, apart from a group of patients after hysterectomy, a group of patients with surgery that is comparable regarding severity (from a medical point of view; i.e. gallbladder surgery). Both groups would be examined with Diagnosticum *Y* after surgery.

Critics would again object to the choice of the sample:

7. Neither sample has been chosen in a representative way as concerns all the patients, for whom conclusions should be made. We actually want insight that refers to all hysterectomy patients (compared to patients with gallbladder surgery) in the Western civilization or at least the English-speaking countries. Our findings should be applicable for the typical age of such patients, for their typical psychosocial characteristics but also especially for patients in the conceivable future.
8. The choice of gallbladder surgery, out of all surgeries that are comparable regarding severity, is arbitrary and therefore may not be suitable.
9. The sample size is still not plausible.

Consequently, preliminary studies have to show that the first patient group that comes along, namely the one from a specific institution, really is typical regarding specific criteria – especially regarding the aforementioned psychosocial characteristics. Otherwise the research design has to be designed as a *multi-center study*. If necessary one has to take care to pick the patients representatively regarding the calendar month of their surgery, in order to take into consideration seasonal variations of what is examined with Diagnosticum *Y* (self-esteem, 'psychic stability'). Also, at least through literature, the choice of gallbladder surgery as being typical for all other surgeries that are comparable regarding severity must be proven. Finally the number of investigated women should be considered in detail (see Chapter 8 and the subsequent ones).

The starting point is the just-confirmed question: 'Are the psychic consequences of a hysterectomy graver than those of surgery with comparable severity?'

Critics of the current empirical research design would now have one final grave argument:

10. Women, who fall ill such that a hysterectomy is indicated, are different from the start (maybe from the time of birth) from women who undergo gallbladder surgery during their lifespan; for example the former could have a systematically different personality structure and as a consequence – under corresponding environmental conditions – a vulnerability to illnesses of the uterus must be suspected.

Regarding this point of criticism, we ultimately have nothing to offer: this empirical research design is a classical *retrospective study* (in experimental psychology: an *ex-post-facto* design); that means that the allocation of patients to the two samples did not happen, as in an *experiment*, by chance (see Chapter 4) before the exposure to different conditions; but the grouping of the patients was done afterwards (after falling ill), and therefore by definition unable to be influenced by the examiner. Differences between patients after indications for hysterectomy or gallbladder surgery cannot, if once established, necessarily be traced back to the group criterion, instead it can never be ruled out that the differences have been there all along.

The gain of scientific insight in psychology starts, as in all other empirical sciences, with a deductive phase. Besides a general description of the problem, this phase also comprises: the specification of the aim of the study; the exact definition of the *population* of the units of research for which insights (from a subset, the sample) concerning the scientific question have to be gained; the exact definition of the required accuracy of the final conclusion; and the selection or construction of (optimal) *designs of the study*. Then the investigation and the collection of data connected with it are carried out. Afterwards, an inductive phase follows, beginning with the statistical evaluation of the data and the subsequent interpretation of the results. The latter can lead to new questions that initiate further empirical research.

3.2 Steps of empirical research

Empirical research can be divided into seven steps:

1. Exact formulation (specification) of the scientific question.
2. Definition of certain precision requirements for the final conclusions, required for answering the scientific question.
3. Selection of the statistical model for the planning and analysis of the study.
4. (Optimal) planning of the study.
5. Realization of the study.
6. Statistical analysis of the collected data.
7. Interpretation of the results and conclusions.

The three first steps, however, cannot just be completed one after the other. The specification of the *precision requirements,* for example, can only be accomplished if one knows how the data will be analyzed later on.

The exact formulation of the scientific question is important, because in contrast to imprecise questions in common speech – which will be understood even if they are posed in the wrong manner – a lack of precision in research will not lead to the desired gain of insight.

For Lecturers:

The answer of the former publisher of the *ZEIT,* Marion Gräfin Dönhoff, to the question 'Do you mind if I smoke in your company?' is quite subtle – 'I don't know, nobody ever dared to' – here, she actually answered two questions: the posed ('Do you stand people smoking?') as well as the intended ('May I please smoke?') one.

Doctor

Example 3.2 A manufacturer wants to state the mean fuel consumption for a specific car model.[1,2]

First we have to point out that the posed question cannot be asked for every single car, but only for the car model as a whole; in the given case all produced cars of a specific model form the population (see also Chapter 4). Next, we think about factors on which fuel consumption depends. At the same time we have to refrain from looking at the just-bought individual car, which for example as a Friday Car might have certain defects. We find that fuel consumption depends on the driving style of the driver (i.e. high- or low-revving typical driving style); on the route (i.e. Sacramento–Reno over the Sierra Nevada vs. San Diego–Los Angeles along the Pacific coastline); on whether one drives in the city, on a highway or on an expressway; on whether one gets into a traffic jam or slowly moving traffic; and perhaps on many other factors. Therefore we ask ourselves: for which situation should the statement in the prospectus be valid? For example, one could state the consumption for the most important situations in a chart. This is unusual, and because of the amount of information it may be daunting. Instead of eliminating the mentioned influences (context factors) with special experimental adjustments (see Section 7.2.2), one could conduct several test drives under some arbitrarily chosen conditions (i.e. 150 miles each) and determine the fuel consumption; we will show later on why this would be an inappropriate approach.

Now we consider that instead of a single number it may be advantageous to state a range (a confidence interval; for more details see Chapter 8) for the average/mean fuel consumption, averaged particularly concerning all influences – for most cars produced this range should be true. During the construction of such a confidence interval one has to fix the relative frequency (more exactly, the probability; see Chapter 6), $1 - \alpha$, with which the mean fuel consumption really is in that range. One will be anxious to keep α (very) small and therefore keep $1 - \alpha$ (very) large. At the same time, however, we don't want to have too large an interval: the statement of, for example, 'between 15 and 40 miles per gallon' would only lead to disapproval in future customers; it hardly contains surprising information. From this we learn to keep the range within admissible boundaries; we could for example determine that it should not be larger than 3 miles per gallon.

With this we have accomplished a great deal from the seven steps of empirical research. The exactly formulated scientific question now is: 'Within what boundaries is the mean fuel consumption of the specific car model expected to be?' Also part of the analysis has been determined; namely: from the collected data (here, measurement values in the natural sense; that is fuel consumption in miles per gallon), a confidence interval will be calculated with the accuracy of α at a determined width.

---

[1] In a modified way this question was posed as a consulting problem to the first author of this book; the consulting led not only to a hardly ever published design of the study but also to a statistical analysis that even nowadays is rarely found in any statistics book.

[2] The chosen non-psychological example can easily be transformed by the means of an analogy into a psychological one: a sport psychologist, who gives a certain treatment ('mental training') to long jumpers in a competitive sports center, wants to publish the mean-achieved training performance.

Let us now come to the choice of the statistical model and the empirical research design: a statistical 'model', which is an assumption, is needed as a mathematical explanation for the collection of the data. Due to reasons that will be explained later on (see Chapter 7) we will aim to have a random sample of $a > 1$ cars taken for test drive purposes from the total of all cars produced. With every single one of these cars, n test drives with predetermined distance will be made, and the amount of fuel consumed will be measured. These measurement values will be termed y_{iv}, with i being any (fixed) number of the car that can take any value between 1 and a. And v is the number of the test drive with car i. For simplicity (and as we will see later also because it is, in a well-defined way, optimal) we make the same number of test drives with every car, so that v runs from 1 to n. Symbolically, our future data will have the form: y_{iv}; $i = 1, 2, \ldots, a$; $v = 1, 2, \ldots, n$, which are $a \cdot n$ measurement values. The future data structure therefore has already been determined at the time of planning the study. Now we want to specify a statistical model for y_{iv}. Therefore we assume that the measurement values y_{iv} fluctuate around a mean μ. Out of all the reasons why we don't get the same measurement value every time (these are the so-called causes of variation), we only can/want to look at possible differences between the cars (inter-individual causes of variation). Accordingly, deviations between the single measurement values can be traced back to the effect a_i of the respective car i as well as to a measurement error e_{iv}. We model the observed measurement values y_{iv} through a random variable $\boldsymbol{y}_{iv}$ (note the difference between y_{iv} and $\boldsymbol{y}_{iv}$: random variables are here made distinguishable from non-random quantities through bold letters). Chance has an effect because we want to assume that the cars in the study have been randomly taken from the population of cars produced.[3] That is why also the effects $\boldsymbol{a}_i$ of the cars must be modeled or described through random variables. Hence the model equation is:

$$\boldsymbol{y}_{iv} = \mu + \boldsymbol{a}_i + \boldsymbol{e}_{iv} \qquad (i = 1, 2, \ldots, a;\ v = 1, 2, \ldots, n) \tag{3.1}$$

More about this model and its side condition can be found in Section 10.4.1.3.

With that, essentially all of the first three steps of empirical research have been completed. For planning the study it is now necessary to optimally determine the two parameters a and n.[4] We can either minimize the amount $a \cdot n$ of test drives (that is the size of the study) or the cost of the study. At this point we don't want to explain in detail how we get to the solution, we only state the result here: with $a = 14$ cars, according to the price and precision requirements (not given here), the test track has to be driven $n = 12$ times.

As soon as the study has been carried out according to this design, we then only have to analyze the data and interpret the results, as described in Chapter 10.

[3] Basically all cars have the same opportunity of becoming part of the study, for example by having a lottery to decide which cars will actually be picked.

[4] Neither the calculation for a confidence interval, nor the optimal design of the study could be found in literature at the time of the aforementioned consulting. That is why at that time two colleagues were asked to develop something appropriate (see now Herrendörfer & Schmidt, 1978).

For Lecturers:

Non-statistical reasoning often leads to unjustified generalizations in the interpretation of observations. This can be well demonstrated with the following humorous example. Three Continental Europeans travel through Scotland in a train. One is very uncritical, one is critical, and the third one is statistically educated. They see three black sheep standing on a hill. The uncritical one says to the others: 'See, in Scotland the sheep are black.' Next the critical one says: 'One cannot say that in such a general way. What one can say is that there are at least three black sheep in Scotland.' Then the third one says: 'Even that doesn't have to be true, we can only say that there are at least three sheep in Scotland that are black on one side.'

Most of the time in psychological studies one will have to deal with the collection of more than one single character. One then has to decide in favor of one character that is the most interesting in order to complete the first four steps of empirical research.

Summary

For the gain of insight in psychology, a statistically (meaning: derived from statistics as a science) founded *design of the study* is needed. For this purpose, a definition should be given for the population (of persons), for whom findings will be recorded (using a subset/sample of it). Regarding content, the needed observations must be carried out in a way that relevant context factors are controlled. The way of sampling, as well as the size of the sample, depend on fundamental rules of statistics. The collection of data regards the ascertainment (often measurement) of observable phenomena, which leads to actually realized values of our observations; we call them *observed* (*measurement*) *values/outcomes*.

References

Gerrig, R. J. & Zimbardo, P. G. (2004). *Psychology and life* (17th edn). Boston: Allyn & Bacon.

Herrendörfer, G. & Schmidt, J. (1978). Estimation and test for the mean in a model II of analysis of variance. *Biometrical Journal, 20*, 355–361.

4

Definition – character, chance, experiment, and survey

In this chapter, quantitative and qualitative, and continuous and discrete characters, as well as factors, will be distinguished, and different scale types of observed values/outcomes will be described. The term 'chance' will be defined more precisely. Studies will be divided into experiments and surveys.

Carrying out a study for the gain of scientific insight within psychology entails conducting observations with respect to at least a single specific character. For this, there are basically two strategies. Both of which are based on the principle of *chance*. Fundamentally chance is the phenomenon in gambling that is responsible for the outcome of the game (e.g. the outcome when throwing a die, namely the number 1, 2, ..., 6, or the outcome of a lottery). For empirical scientific purposes, one could also say that chance means all influences on *events* or observations of interest, which are either not ascertainable or which we don't want to ascertain.

Bachelor The term 'chance' should not be confused with its meaning in everyday use; there chance often means 'seldom' or 'unexpected'. In this book an event that happens by chance is an event that can, but doesn't have to happen.

Bachelor **Example 4.1** Arbitrariness and randomness

The reader might ask within his/her wider range of friends (preferably exactly 18 persons) for the first number between 3 and 20 that comes to mind. If chance were the only cause of their respective responses, we would basically expect that every number between 3 and 20 would occur the same number of times, which is about once. Experience has shown, however, that up to half of the persons

Statistics in Psychology Using R and SPSS, First Edition. Dieter Rasch, Klaus D. Kubinger and Takuya Yanagida.

will say 17; also other prime numbers and odd numbers tend to occur relatively more frequently than other numbers. Subsequent arguments explain that people look for a random number and that they try not to pick any 'typical' number like '4' or '8'. This example documents that, in this case, arbitrariness is at work, not chance.

In contrast to events that are dependent on chance, thus being *random* events, other events we call (strictly) deterministic.

Master

Example 4.2 Drawing from a deck of playing cards

When drawing from a set of well-mixed playing cards, the result is a random event, for example the 10 of diamonds. If somebody searches for the queen of hearts and takes it, then this result is an arbitrary event because it has been determined by the will of that person. Finally if somebody draws the first card from a newly unwrapped deck, it is a systematic event because, when packed, decks usually have the ace of spades on top.

Master
Doctor

When several isolated observations (occurrences, events; for example the throwing of two dice and the observation of the resulting number of spots on top) occur, we talk about chance if there is either no connection at all or an irrelevant (inter-) connection between them. If there is a great mass of events (for example the opening of buds on a blossoming apple tree) then chance is the product of an accumulation of non-ascertainable influences, resulting without any rule in a chain of events where one cannot predict in which order the buds will open.

Master
Doctor

Students and lay people (professionals in neighboring disciplines), especially psychotherapists with a particular school of thought, often express the following opinion: 'Chance-based coincidences don't happen.' What they mean is that all events happening around one person have their reason (cause) in that person's history; and if not in that person's own history then in that of some other 'players'. In the end this belief in causality implies that the past and the future of the universe are principally predictable. The past could be completely reconstructed and the future could be predicted up to the smallest detail. Despite the fact that this attitude towards life may motivate a client to strengthen his/her internal locus of control of reinforcement belief,[1] such belief in causality is rather crude.

Werner Heisenberg's Uncertainty Principle shows that the reference to chance does not have its origin in a lack of theoretically available information, but that chance is a principle (Heisenberg, 1927); this principle includes the notion that it is generally impossible to determine simultaneously both the position and momentum of an elementary particle in some accurate manner, and this is not a question of methodological failings, but a principle.

[1] Greatly simplified, locus of control of reinforcement belief means: 'by whom or what a person thinks that his/her life course is determined'.

The two strategies for obtaining observations of a character are the following:

1. The behavior of certain chosen persons (in general: *research units*) is (psychologically) examined with respect to some traits/aptitudes without them being influenced by the examiner. This is what we call a *survey*. If sampling by chance, that is *random sampling*, occurs, we speak of a *random sample*. Since we don't look at samples other than random samples in this book, we often only talk about *samples*. Rarely is there a complete inventory ascertainment of the population or, in other words, a *census* of the *population/universe*, as with statistical almanacs.
2. The persons (research units) receive systematic *treatments* after they have been selected for a sample; for example they are first *randomly* allocated to several groups, then every group receives a different (psychological) intervention, and finally they are psychologically examined. This procedure is referred to as an *experiment*.

For psychological experiments we can give the following simplified definition: in an experiment the researcher observes the behavior (verbal and nonverbal actions, reactions) of his/her participants (persons; very often then called *subjects*) under strictly established and controlled conditions which were varied by him/her intentionally and specifically in order to test the conditions' influence on the assessed behavior. Thereby randomization of the subject's condition assignment is indispensable, and replication of the procedure must be essentially possible; controlling conditions aims for a reduction in variability of confounding sources.

In both cases, survey and experiment, the collected data for the character of interest (for example (*test*) *scores* on a psychological assessment tool; in general – observed values/outcomes) are recorded and then statistically analyzed. If an experiment or a survey consists of several consecutive steps, where every step is based on the result of the preceding one, then we call this procedure sequential.

Generally the observed data depend on chance in two ways: on the one hand the research units have been chosen at random and assigned to an experimental condition by chance, respectively. On the other hand the same research units observed under nearly identical sampling circumstances don't lead to the same outcomes, but by chance to (slightly) different ones. The latter is due to measurement errors (see in Section 2.3 the quality criterion of reliability).

Once the data have been collected by means of random sampling and randomization within an experiment, respectively, we can analyze them with statistical methods. Then we reflect upon all the results concerning our scientific question that we could possibly obtain. First we look at the underlying population: a population is the set of objects (persons,[2] schools, hospitals, etc.), for which an empirical study using a subset (sample) is supposed to make a conclusion regarding certain characters. There must be an operational definition for the population that allows for an assertion of whether a certain object belongs to or not.

---

[2] In the case of experiments we talk about subjects, within psychological assessment about *testees/examinees*.

Example 3.1 – continued

The psychological consequences of a hysterectomy are to be assessed.

It is important, while formulating the scientific question, to clearly identify the group of people or, in other words, the population that one wants to observe or whose aptitudes/traits one wants to quantify. In this example all hysterectomy patients could be the population, or those aged below 40 years, or those who come from Central Europe, or those aged below 40 years that live in urban areas, etc. Therefore populations must be defined exactly with regards to space, time, and content.

Often not only the objects, that are the (potential) research units, are called population or sample, respectively, but also the set of (potential) outcomes themselves.

Real populations are finite; the *population size* is termed N – the *sample size* is termed n. In principle, it is possible to carry out a census, meaning that one examines all elements of the population. In statistics, populations are often assumed to be infinite (in practice we can use conclusions that are valid under the assumption of infinite populations for $N > 1000$ and $\frac{n}{N} < 0.1$).[3]

Master

Research units can be distinguished more accurately for both experiments and surveys. In an experiment, the object from a given population that is randomly assigned to a certain condition (such conditions might be particular treatments) is what we call an *experimental unit*. In a survey, the element of the population that becomes part of the sample is termed *survey unit*. Of course also the experimental units have to be (randomly) selected as a sample from a well-defined population.

Up to now we have talked about characters without properly defining what is meant by the term: a character is that specific feature, which is the research objective of a study; it is (in-)directly deduced from the scientific question. Apart from socio-graphic factors, in psychology these are mostly *traits*, *aptitudes* and all sorts of behavior patterns (see Example 1.1).

Characters have thus various *measurement values*. The result of ascertainment of a character, that is the *outcome*, is in psychology often called an investigational result, also observational value or observed (measurement) value. Since the result of a statistical analysis of a study can also be called an 'investigational result' (when indicated result of the experiment) it is preferable to speak of outcomes as observed values, or for short, simply as observations. By the way, in psychology there is the special case of a test score, which is the (action or) reaction shown in a psychological test (in general: a psychological assessment tool) that has been scored in a certain way.

Values of characters are ascertained by means of a 'measuring instrument' (in a broader sense) with the help of a certain *scale*. More precisely: an unambiguous assignment rule

---

[3] N stands for the extent of finite populations on the one hand, but on the other hand often also stands for the total number of research units in a study that consists of several groups. However n is always the number of research units sampled from a population as a single sample.

of numbers or symbols to the different shapes of a character is termed *scaling* in general and a *scoring rule* in psychology. The set of numbers or symbols that are available, and between which a more, or less, differentiated relational system is defined, is called a scale. Therefore a scale mirrors (mostly in numerical terms) the (empirically detectable) relations between research units. It makes the relations accessible to mathematical operations. The term scaling includes measurement of a character. The multitude of characters that are potentially interesting to us now differ regarding their respective *scale type*.

Master If we empirically compare (put in relation to one another) two objects, let's say two sticks, then we expect from the scaling that is available to us, in this case ideally (metric) measurement, that the assigned scale values (measurement values) represent the same relation. If we observe that stick Z is double the length of stick U – by holding stick U two times against stick Z – then an adequate scale (cm scale) will represent this; for example $U : Z = 20 : 40\,cm$. If we empirically compare two objects, for example two pupils, and observe that pupil V can regularly correctly answer the questions concerning the subject material that the teacher addresses to him (and attends to his schoolwork according to instructions and excels at written tests), whereas pupil W answers the questions concerning subject material either incorrectly or not at all (and sometimes only attends to his schoolwork after being asked to do so several times and sometimes fails written tests), then we expect from the chosen scale, the assignment of grades, a correspondingly identical relation of measurement values; for example grade A for pupil V and grade D for pupil W.

Example 4.3 Typical empirical relations

In psychology we can transform empirical relations into numerical ones as in the following examples:

1. Spouse B earns *twice as much* as spouse C.
2. In a perception experiment (signal detection with presentation time appropriate to the subject's age), senior citizen D makes, in comparison to senior citizen E, *three times more* mistakes than senior citizen F, in comparison to senior citizen E.
3. The level of education (assessed by the highest completed schooling) of person G is *higher than* person H's.
4. Patient I is never married, patient J married, patient K divorced, and patient L widowed.

Every assignment of numbers to persons can logically only mirror those relations that are empirically given and observable; and those numbers are transformable as long as they adequately depict the empirical relations.

Below we will have a look at four types of scales, which differ with respect to the number and the kind of represented relations between the research units. A superior (higher-order) scale type also includes the relational qualities of the lower-order scale types.

4.1 Nominal scale

The *nominal scale* depicts the fewest empirical relations. The scale values consist of signs (symbols, letters, names) that – even if they are sometimes numbers – can only express discrepancies. In particular, there is no indication of order between the signs.

Master One recognizes a nominal-scaled character by the fact that the feasible scale values are arbitrary as long as the discrepancies between them remain. The sex of persons, for example, is assessed by a nominal scale. It does not matter whether female persons are characterized by a Venus symbol, *f* for female or 0, and male persons by the Mars symbol, *m* for male or 1.[4]

Example 4.3 – continued
In Case 4 (patient *I* is never married, patient *J* married, patient *K* divorced, and patient *L* widowed), we only need arbitrary names (nominal) for *I*, *J*, *K*, and *L*, because the only thing that matters is that the discrepancies are visible; e.g. the names *n*, *m*, *d*, and *w* are just as possible as 1, 2, 3, and 4 or *a*, *b*, *c*, and *d*.

Bachelor **Example 4.4** Marital status

In Example 1.1 we find the nominal-scaled character *marital status of the mother*. This character's different shapes cannot at all be empirically relationally ordered; there is no general chronological order in the course of a person's life.

In Example 5.2 we will identify the absolute frequency of the different scale values found in the $n = 100$ children; in Table 5.3 one can see that 9 mothers are never married, 65 are married, 22 are divorced, and 4 are widowed. Of course we can report these results in any other sequence. In any case, the value 'widowed' is not logically relegated to 'divorced'.

When a (nominal-scaled) character has only two values, we also talk about *alternative characters* or *dichotomous/binary* characters.

4.2 Ordinal scale

The *ordinal scale* creates symbols or numbers that express a rank order of the character's different shapes. In ordinal-scaled characters the number of possible scale values is generally independent from the set of research units (cf. e.g. the grading scale). Also, full rankings (where every research unit out of *n* gets a rank from 1 to *n* and thus the number of signs is dependent on the set of research units) that are derived from metric data (which is how physics mainly scales; see below for more details) represent ordinal scales.

---

[4] By the way, experience has shown that when female researchers use numbers they tend to code female as 1 and male as 0, and male researchers the other way around, although a relation of order does not exist for psychological questions.

Master

One recognizes an ordinal-scaled character by the fact that the scale values may be arbitrarily chosen as long as the empirical rank order of them is captured.

For Lecturers:

Most statistics text books within psychology claim that a rank-scaled character is recognizable in that its scale values may be transformed monotonically (e.g. $y \rightarrow y^2$). For students this is indeed a very easily understood interpretation, but it is by no means exact. It is only correct if one postulates that negative values do not occur – see for instance, however, the scale values '−1', '0', and '1' for the response options in a questionnaire: 'no', 'I don't know', and 'yes'; they do not allow the monotonic transformation $y \rightarrow y^2$.

Example 4.3 – continued
In Case 3 (the level of education of person G is *higher than* that of person H), the possible scale values of G and H are located on an ordinal scale because only the relation '>' is relevant; e.g. the scale values 5 and 4 would be as appropriate as the (admittedly strange-looking) values 10.3 and 8.6.

Bachelor
Master

School grades form an ordinal scale; in ascending order the character's different shapes are, for instance, 'needs improvement', 'satisfactory', 'great', and 'excellent' or 4.0 to 1.0 (in descending order of achievement).

Bachelor

Example 4.5 The scale type of the character *social status* in Example 1.1

In our example, the character *social status* is operationally defined via the occupation of the father, or in the case of single mothers via their occupation. The various occupations have been ordered (in groups) as follows: 'upper classes', 'middle classes', 'lower middle class', 'upper lower class' and 'lower classes'. However the way the character *social status* has been observed here leads to an additional scale value, which is 'single mother in household'. It does not fit into this order either at the beginning, or at the end, or anywhere in between. When considered like that, this is a nominal-scaled character. If we want to convey the basically given (rank-) order to the recipient of the results, we have to omit all the cases with the scale value 'single mother in household'.

Nominal scale and ordinal scale are often summarized by the term 'non-metric scales'. And one terms both types as *qualitative data*. In contrast, data that stem from a scaling of one of the two scale types discussed below, *interval scale* and *ratio scale,* are called quantitative data.

Doctor

Within psychology one usually counts ordinal-scaled characters as quantitative characters too. This, after all, is because scale values in rank scales mirror gradations or gradual discrepancies, which means that they express an order regarding a 'more or less' quantity. This differs from nominal-scaled qualitative characters, where the scale values have a different qualitative meaning; they are not one-dimensional but represent several dimensions that cannot be related to each other. Below we will use the term *quantitative character*, divergently from its usage in psychology, but in concordance with the terminology in other natural sciences and mathematical statistics, only for interval- and ratio-scaled characters.

4.3 Interval scale

The interval scale consists of numbers. The positive differences between them are interpreted as the distance between the character's different shapes. Besides the equality of the values, the equality of the distances is also given. However, there is no absolute zero point – if the interval scale had an absolute (that is to say 'natural') zero point then it would be a ratio scale!

Master An interval-scaled character is recognized by the fact that the measurement values allow for a linear transformation of the form $y \rightarrow a + by$ ($b > 0$) without violating the empirically given relations. One example for such a transformation is the conversion of temperatures from Fahrenheit into Celsius:

$$\text{Celsius} = (\text{Fahrenheit} - 32) \cdot \frac{5}{9}$$

or with $b = 5/9$, $a = -160/9$, Celsius $= y$, and Fahrenheit $= x$:

$$y = \frac{5}{9}x - \frac{160}{9}$$

Doctor The measurement of temperatures in degrees Celsius or Fahrenheit happens on an interval scale. Both scales are convertible into each other by a linear transformation. Particularly with the existence of two scales (Celsius and Fahrenheit) instead of a single one, it is obvious that temperatures are usually described without an absolute zero point; the numerically existing zero of the Celsius scale is not a 'natural' zero because it has been arbitrarily located at the temperature where water passes from a fluid into a solid state. There is, however, a natural zero that the temperature cannot fall below; it is defined by the kelvin scale, which is a ratio scale. Length and mass also have an absolute zero.

Example 4.3 – continued
In Case 2 (in a perception experiment – signal detection with presentation time appropriate to the subject's age – senior citizen D makes, in comparison to senior citizen E, *three times more* mistakes than senior citizen F, in comparison to senior citizen E), the possible measurement values for D, E, and F are scaled at an interval scale level. This is because the measurement value difference (concerning E) must always express the ratio 3 : 1, without the measurement values being able to take into account the chosen difficulty of the signal detection (i.e. depending on the presentation time); for example the test scores 7, 1, and 3 would be as adequate as 9, 3, and 5, or 20, 8, and 12. The reason why we cannot interpret the ratio of the measurement values regarding content is that those ratios do not represent empirical relations.

In the given signal detection experiment, only some of all possible signals are used, but we could instead have used, for example, two signals that are very difficult to discern; then we might really have observed the values 9, 3, and 5 instead of 7, 1, and 3. If we had used a few easy, medium, and difficult ones in addition, then we could have got the results 20, 8, and 12, or would have expected a corresponding result if really fair measurements apply.

Doctor

The scientific attempt to scale 'intelligence' is a good example for psychologists, showing how to adequately and not artificially interpret an interval-scaled character. Described very simply, the *intelligence quotient* (IQ) is defined as follows. The test scores (mostly the number of solved problems) obtained in a pertinent intelligence test from a very large *standardization sample* are first averaged and second put in relation to the extent of all the differences between them: the mean estimated for the population is termed μ; the corresponding extent of all differences is termed σ (see more precisely in Chapter 6). Then the test score of every single testee, for example V with the test score y_V, is linearly transformed:

$$IQ_V = \frac{15\,(y_V - \mu)}{\sigma} + 100$$

In this, the specification of the two constants 15 and 100 is arbitrary, especially the specification of the additional 100; this could have been specified differently in an equally plausible way, for example 0 instead of 100. Therefore it makes sense to form the quotient between differences – for example in such a way that the difference concerning the IQ between person V with $IQ_V = 110$ and person W with $IQ_W = 120$ is actually twice as much as the difference between person P with $IQ_P = 90$ and person S with $IQ_S = 95$. It is, however, not empirically founded to judge that person W with $IQ_W = 120$ is one-third more intelligent than person P with $IQ_P = 90$, although arithmetically $120/90 = 4/3$. If the additive constant were set to 0, then the test scores of the persons V, W, P, and S would be 10, 20, -10, and -5; the quotient of the differences between V and W or P and S would still be 2, but the quotient between W and P would have changed to $(-)2$! Therefore the intelligence quotient is interval-scaled.

4.4 Ratio scale

A ratio scale has all the relational qualities of an interval scale and in addition an absolute zero point. Therefore the equality of relations (proportions, quotients) is presupposed.

Master

A ratio-scaled character is recognizable by the fact that the character values can only be transformed in the way $y \rightarrow ay$, $(a \neq 0)$; otherwise the empirically ascertained relations do not match.

Example 4.3 – continued
In Case 1 (spouse B earns *twice as much* as spouse C), the feasible measurement values taken by B and C lie on a ratio scale, because there must always be the ratio 2 : 1; e.g. 10 family allowances and 5 family allowances, \$1750 and \$875, £1120 and £560.

Doctor

As previously mentioned, the measurement of temperature in kelvin uses a ratio scale. In psychology, ratio-scaled characters can be found principally where physical measurements such as age of a person are given. However, from a psychological point of view, not all physical measurements represent empirical relations that correspond to a ratio scale. For example, measurements of reaction times that are mandatorily obtained by the means of a specific psychological assessment tool within the personel recruitment for a job in surveillance, are indeed fundamentally based on a metric scale (time scale of the clock); the psychological quality of the numerical relations of the measurement values is ambiguous, yet: extreme values (for example missed stimuli or very delayed reactions) can – as concerns the achievement potential of a testee – hardly establish the same ratio as the concrete numerical ratio of the observed reaction times. Even the quotients of differences in reaction times must be viewed critically regarding their content. The choice of whether the researcher insinuates that the data are ratio-, interval-, or only ordinal-scaled appears to be based solely on his/her subjective opinion; oriented on methodic fundamentals, it is, however, possible to empirically determine the scale type of data by implementing the scaling techniques of psychometrics (see Kubinger, 2009b).

For Lecturers:

An example taken from sports illustrates how arbitrarily and improperly non- or pre-scientific scaling procedures in everyday life are applied. In the (alpine) ski world cup, best-ranked athletes are rewarded with world cup points and the sum of gained points determines the winner at the end of the season. It would be easy to contrast racer *R* with racer *S*; that is to say to determine the empirical relation of their speed over the whole season by matching corresponding video recordings (provided that both have contested the same races). The rules, however, provide another scaling which represents numerical relations that can contradict the empirical ones; from that point of view the scaling is unfair: The winner of a race gets 100 points, the second 80, the third 60, the fourth 40, etc. Supposing *R* is 0.01 seconds slower in each of five races than the winner *S*, and always in second place; and wins the sixth race, being 2 seconds faster than *S*, who places fourth. *S* would then win the world cup with 550 points, 50 points ahead of *R* with 500 points, although the matching of the video recordings shows that, calculated over the whole season, *R* was 1.95 seconds faster than *S*. Such an unfair scaling is irrelevant as long as athletes and viewers freely accede to such authoritarian rules for the sake of amusement – in psychological assessment in case work, however, there exists a demand for fairness and objectivity. In those instances such an unfair scaling would not be justifiable, for ethical reasons!

Master

A superior scale can always be downgraded into a lower-ranked one. Thus, as measures of achievement in a speed-skating competition, instead of stating the times of the three best racers, one can award a gold medal for the shortest time, and silver and bronze for the second and third best times. In long jumping this is applicable in an analogous way, except that the gold medal here is awarded for the highest numerical value reached. From this one can see how senseless it would be to position medal ranks on a number line. In one example gold is on the left side of the others and in the other it is on the right side. If one doesn't know what is behind the ranks in a quantitative way, then not even the direction is clear!

Doctor

For example, concerning the measurement of temperature, temperatures below –40 degrees Fahrenheit and above 120 degrees Fahrenheit might happen, although with normal outdoor thermometers we can't measure them. Statements of temperature are partly made using the measurement values, that is to say quantitatively, but partly in the form of 'below –40 degrees' and 'above 120 degrees'. This corresponds to a mixture of ordinal and interval scale. Such cases occur during the measurement of reaction times (at least in the upper level) as well as during the measurement of lifetime, especially when the researcher cannot wait for the death of the research units of a sample (e.g. for scientific questions of survival after clinical treatments). Statistically such data are analyzed either at a lower-ranked scale – that is to say that the scale values are represented by an ordinal scale – or the given sample is treated as a *censored sample* (for a precise description of this procedure see Rasch, Herrendörfer, Bock, Victor, & Guiard, 2008).

For their authorized use, statistical analysis methods hardly ever need the high requirements of the ratio scale for the character of interest. Instead they require just an interval scale. That is why we don't have to distinguish between the two scale types in the following. We will only talk about whether a character meets the requirements of an interval scale or not, and mean by this that the character at least meets these requirements. A ratio scale perhaps beyond that is nonessential for our observations. Therefore we will simply talk about quantitative characters.

4.5 Characters and factors

Irrespective of the scale type, quantitative characters have to be divided into *discrete* and *continuous characters*. A quantitative character is discrete if it only has very specific measurement values; for example only natural numbers. If however, within a specific interval, all real numbers are possible measurement values, then the character is continuous.

For Lecturers:

Lem (1971) made clear in one of his utopian novels just how extensive even the amount of natural numbers is. Here we give a modified account from a passage of his work:

> Ijon Tichy is in a cosmic hotel with an endless amount of rooms – which have just been taken by an endless number of dentists that have come from all over the universe in order to take part in a conference. He witnesses the arrival of an endless number of psychologists that want to take part in a cosmic psychology conference. Of course they also need rooms but the hotel manager explains that all rooms have been taken. Ijon Tichy suggests now that every dentist should just

move into the room with double the room number of the one they originally had. Since, after the multiplication by two, these are all even room numbers, all the rooms with uneven room numbers are free: there is also an endless amount of them and the psychologists can then move into those.

Master Depending on the scale type that underlies the sampled data, different statistical analysis methods are applicable; that is to say, appropriate. For example, the mean (Section 5.3.1) can only be used to suitably describe quantitative data in a short and concise way. The mean is therefore absolutely proper to represent the children from Example 1.1 regarding their test scores in the subtest *Everyday Knowledge*. The 'middle' *social status* or moreover the 'middle' *marital status of the mother* are senseless quantities, as there are no empirical differences that can be observed; at best a rank order is given; the mean would unwarrantedly insinuate very specific quantities.

If a character is designated to several (treatment) conditions within a study, then that character is commonly termed a *factor*. Completely analogous to characters, one can talk about quantitative and qualitative factors. The values of a character that is a factor aren't called scale or measurement values, but instead *factor levels* or, for short, *levels*. A particular difference between character and factor is the following: Whereas a character is generally postulated as being determined by chance, a factor can also be modeled as being not random.

Sometimes there are factors that are not (primarily) of interest for a specific question and therefore are not part of the design of the study: they are confounding factors and are called *noise factors*. If they were not held constant or adequately taken into account, the study's results would be *biased* – one says that the results have a *bias*.

The choice of which statistical method will be used for the analysis of a study depends closely on which conditions are investigated and which noise factors have to be taken into account, as well as if, when, and how the influence of unknown noise factors can be minimized.

Summary

For the *research units* in an empirical study (either a *survey* or an *experiment*) we get observed (measurement) values (within psychological assessment mostly *test scores*) for the character in question. These stem from different *scale types* according to which empirical relations they represent. These scale types distinguish between *quantitative* and *qualitative characters*; the latter have to be divided into *nominal*- and *ordinal-scaled* ones.

References

Heisenberg, W. (1927). Über den anschaulichen Inhalt der quantentheoretischen Kinematik und Mechanik. *Zeitschrift für Physik*, *43*, 172–198 (for an English translation see J. A. Wheeler & H. Zurek (1983). *Quantum Theory and Measurement* (pp. 62–84)).

Kubinger, K. D. (2009b). *Psychologische Diagnostik – Theorie und Praxis psychologischen Diagnostizierens* (2nd edn) [Psychological Assessment – Theory and Practice of Psychological Consulting]. Göttingen: Hogrefe.

Lem, S. (1971). *Dzienniki Gwiadowe* [Die Stern-Tagebücher des Ijon Tichy]. Frankfurt am Main: Suhrkamp (for an English translation see The Star Diaries: Further Reminiscences of Ijon Tichy (1985). Philadelphia: Harvest Books).

Rasch, D., Herrendörfer, G., Bock, J., Victor, N., & Guiard, V. (2008). *Verfahrensbibliothek Versuchsplanung und -auswertung. Elektronisches Buch* [Collection of Procedures in Design and Analysis of Experiments. Electronic Book]. Munich: Oldenbourg.

Part II

DESCRIPTIVE STATISTICS

Descriptive statistics describe data independently of whether they stem from a population or a sample. The data can be described numerically or graphically. When it comes to obtaining scientific findings, however, descriptive statistics serve at the most as a first step. But *inferential statistics* are of more importance: with the help of these, conclusions can be made about the population, on the basis of a sample.

In mathematical statistics, populations are often considered or 'modeled', respectively, as infinitely large; this proves especially worthwhile if the size of the elements, N, is 'sufficiently' large, e.g. $N > 1000$; we will demonstrate later that sizes of 120 are often sufficient from a practical point of view.

In many cases, it is not possible to observe all elements of the population. Then, one has to be content with a part of it, i.e. a sample. In doing so, a randomly sampled part from the population in question is required so that the methods of inferential statistics introduced in Chapter 7 can be applied. Various approaches to random sampling will be described in Chapter 6. Although this requirement may be irrelevant for descriptive statistics, the results of descriptive statistics alone are not sufficient for today's research projects. The objective is to extrapolate the results from the sample to the population as is done in extrapolations during political elections, where the overall result is predicted after the first vote counts – in inferential statistics we use the term 'to estimate' instead of the term 'to extrapolate'.

We assume in the following that the observed values $y_1, y_2, \ldots, y_n$ of a single character, or the observations $y_{11}, y_{12}, \ldots, y_{1n_1}; y_{21}, y_{22}, \ldots, y_{2n_2}; \ldots; y_{m1}, y_{m2}, \ldots, y_{mn_m}$ of m characters, respectively, are given. The objective is then to clearly illustrate or to concisely describe the essential information that is covered with these data.

Bachelor

Hence, the individual observations of a sample or of a population should be compressed appropriately. They cannot be considered in detail, i.e. individually,

> and cannot be completely disclosed to those interested in the research study. Instead they should be compressed to the essentials of the information in their entirety.[1]

Thus, *statistics* are ascertained and *tables* and *figures* are generated.

[1] In the first years of the twentieth century, articles in the scientific journal *Biometrika* can be found in which all outcomes are listed.

5

Numerical and graphical data analysis

In this chapter, descriptive statistics will be introduced as a contrast to inferential statistics. Data which have been collected in order to answer a certain scientific research question are mostly very complex in their structure; their essential information is therefore not immediately ascertainable. There are several methods for the 'compression' of data which will be described in the following. One of them is to list the frequencies of the observed character's measurement values in tables or figures. Another method is to calculate statistics that characterize the data briefly and concisely.

Data analysis can be carried out for a single character alone; however is often performed simultaneously for several characters. If we consider Example 1.1, several characters have been acquired, for example *age of the child*, *sex of the child* and *gestational age at birth*. We will deal first with the case of analyzing each character separately. Then, we will briefly discuss the special features which occur in the simultaneous analysis of two characters. The simultaneous analysis of two or more characters will, however, be discussed in more detail in Chapter 11.

5.1 Introduction to data analysis

Even statistical lay people are used to tables and figures as a method of data analysis. They would probably also apply the best known of all statistics, the *(arithmetic) mean*.

Statistics in Psychology Using R and SPSS, First Edition. Dieter Rasch, Klaus D. Kubinger and Takuya Yanagida.

Bachelor **Example 5.1** Number of weeks of pregnancy for children with Turkish as a native language, in Example 1.1

We want to describe the character *gestational age at birth* (in weeks) briefly and concisely. This should be accomplished by applying the mean. However, we first have to identify those children encoded with the value 2 = 'Turkish' in the character *native language of the child*.

In **R**, we first enable access to the database *Example_1.1* (see Chapter 1) by using the function `attach()`. Then we type

```
> summary(age_birth[native_language == "Turkish"], na.rm = TRUE)
```

The first argument (arguments serve to specify functions; they are set in parentheses and separated from each other by commas) in the function `summary()` refers to the character *gestational age at birth*, namely the vector `age_birth`. Since the analysis is to be limited to children with 'Turkish' as a native language, we add to this argument: `[native_language == "Turkish"]`. Because we would like to exclude any missing values, we use the additional argument `na.rm = TRUE`. Generally, the function `summary()` displays the main results of the analysis.

After the whole command is typed in, we confirm it by pressing the '↵' key and get the result (shortened output):

```
Min.    Mean    Max.
35.00   38.52   41.00
```

In SPSS, we select

Analyze
  Compare Means
    Means...

As a consequence, the window in Figure 5.1 appears. There we select the character gestational age at birth and move it (by clicking the arrow-button) to the field Dependent List:. Afterwards, we select the character native language of the child to drop and drag it (analogously) to the field Independent List:. Next we open the window shown in Figure 5.2 by clicking Options.... In this window, we select Minimum in the panel Statistics: and move it to the panel Cell Statistics:. We proceed in a similar fashion with Maximum. Finally, we click Continue and return to the window shown in Figure 5.1, where we press OK to get the results (see Table 5.1). For the time being, we won't consider the row German or the column Std. Deviation.

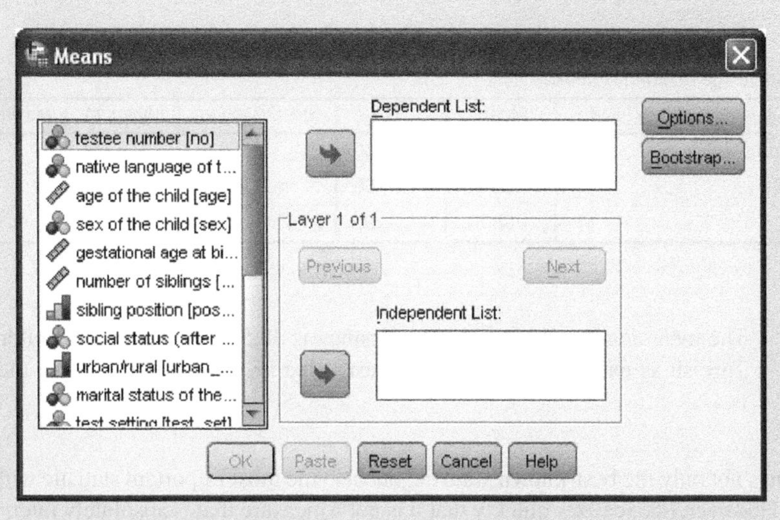

Figure 5.1 SPSS-window for calculating some basic statistics.

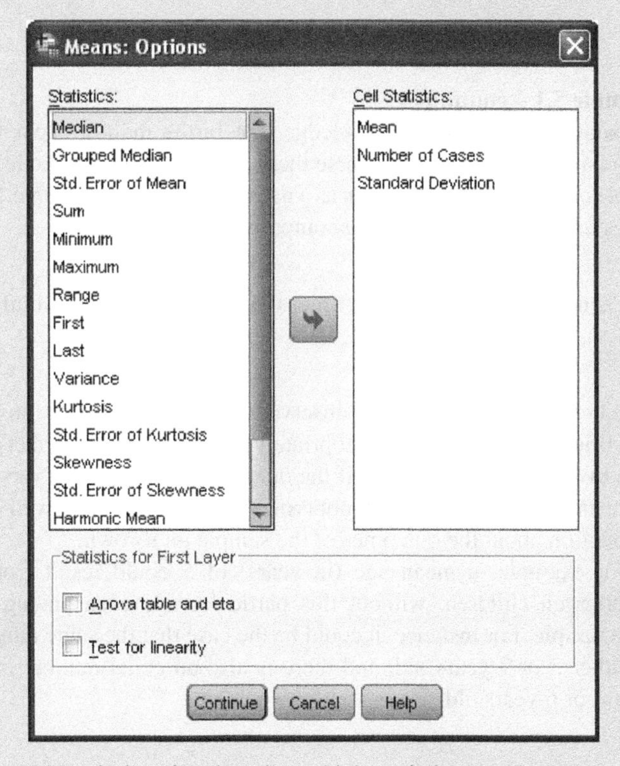

Figure 5.2 SPSS-window for choosing which statistics to calculate (e.g. the mean).

Table 5.1 SPSS-output for the statistics in Example 5.1.

Report

gestational age at birth (in weeks)

native language of the child	Mean	N	Std. Deviation	Minimum	Maximum
German	38.50	50	2.013	32	41
Turkish	38.52	50	1.568	35	41
Total	38.51	100	1.795	32	41

The mean number of weeks of pregnancy is 38.5 in the sample of children with Turkish as native language; the lowest value is 35 weeks (rounded); the largest is 41.

The mean is not only the best-known statistic, but also the most important statistic within psychology. However, one realizes quickly that it is not a measure that is absolutely interpretable. One cannot draw a conclusion from the calculated result based solely on the concrete value; instead it should be compared relatively to additional information gained: either compared to another mean from a different sample (population) or compared to a (generally well-known) 'standard'.

Bachelor **Example 5.1 – continued**

Nowadays it is common knowledge that births mainly occur between weeks 35 and 40 of pregnancy; of course there are exceptions, as Table 5.1 shows. The sample of children with German as a native language in Example 1.1 immediately comes to mind as an interesting comparison.

A single statistic is not always sufficient in order to describe the essential information of the data concisely.

Master A mean is only typical for the observations of a sample – it thus only describes them briefly, concisely and appropriately – if all values are in fact distributed very close to it. On the other hand, if the outcomes are dispersed very widely around the mean, the mean is actually not typical; it, then, does not well enough provide information about the outcomes of the sample on its own.

For example, a mean age (in years) of 5 could result from a sample of kindergarten children, without this particular age even having occurred once in the sample. For instance, it could be the case that the same number of children are either 3 or 7 years old, and there is also an equal number of children aged either 4 or 6 years old.

In order to be able to evaluate the informative value, or the degree of typicalness, of a sample's mean, we also need, in addition to a *measure of location* (i.e. the 'position' in the measured range with the outcomes dispersed around it), a second statistic: namely a *measure of scale*.

For Lecturers:

The meaning of dispersity is illustrated by a poem from *List* (see Krafft, 1977):[2]

A man who of statistics had heard
Thought of the mean and found it absurd.
He doesn't like it; he is opposed,
Though an example has that thinking deposed.

A hunter wants to shoot a duck
And with his first shot he tries his luck.
The shot flies fast out of the barrel,
Lands well in front; for the duck it's no peril.

A second shot goes with a loud crack
But lands a good distance toward the back.
The hunter speaks without fear or dread,
With belief in the 'mean',
Statistically speaking, the duck is dead.

But had he been smart and taken buckshot
(We say this only to give him thought)
He could have multiplied his chances.
The shot would miss, yet the duck falls in strife,
For standard deviation shortens his life.

Especially in the case that two samples are to be compared, the calculation of a measure of scale in addition to a measure of location is absolutely necessary.

Master In the example of kindergarten children with a mean age of 5 years, it could be the case, for example, that a second sample with exactly the same mean includes mostly observations of 5 years old but does not include either 3 or 7 years old at all – and only relatively few of either 4 or 6 years old. Obviously, in this case, the two samples could not be interpreted as equal with regard to the age.

Bachelor **Example 5.1 – continued**
Table 5.1 shows that the mean of children with German as a native language is equal to the mean of children with Turkish as a native language. However, the observations in the first sample fluctuate between 32 and 41, and in the second sample between 35 and 41.

5.2 Frequencies and empirical distributions

A simple method which leads directly to a compression of the given data's information without substantial loss of information is to count how often the various measurement values of a

[2] Freely elaborated by Sandra Almgren using the German original.

nominal- or ordinal-scaled character occur – we often call the values of qualitative characters *categories*, too. As a result, each category is characterized by an *(absolute) frequency* of its occurrence. The information that gets lost with this method is only the position of each element within the original series of outcomes.

Bachelor Especially in the case of extensive data, the classification of the character's measurement values into *classes* is required in order to visually represent quantitative characters. In doing so, it is often useful to aggregate the information to a few classes. However, this entails a certain loss of information: without any knowledge of the original data, there is likewise no knowledge of the exact outcomes. Given the clarity of the illustration, this, however, can be accepted.

Master Strictly speaking, classes of a character's measurement values are always given, namely

- in discrete, interval-scaled characters in the form of the different measurement values;
- in continuous, quantitative characters because of the limitations of the measuring precision within a certain measuring range.

Bachelor The observed frequencies per measurement value or class are called absolute frequencies. *Relative frequencies* result from dividing the absolute frequencies by the number of all observations (research units). They are needed if different data sets of different sizes are supposed to be compared with regard to all measurement values. The set of all relative frequencies per measurement value or class is called a *frequency distribution* or, simply, *distribution*. It is a case of an *empirical distribution* as opposed to a *theoretical distribution*, which means a certain assumption (see Section 6.2).

For quantitative and also ordinal-scaled characters, it is possible to calculate *cumulative frequencies*. These show the absolute or the relative frequencies with which this or any preceding (smaller) measurement value (class) occurs. The set of all these relative cumulative frequencies is called the *empirical distribution function* of the character.

All of the above-mentioned frequencies can be displayed in tables or figures. Often, both are produced together.

5.2.1 Nominal-scaled characters

In the case of nominal-scaled characters, the calculation of cumulated frequencies does not make sense. Instead, we only count the numbers of observations in the different measurement values (categories); this is done both in the form of absolute and relative frequencies. A *pie chart* is a suitable means to illustrate this. In it, the areas of the segments of a circle represent the relative frequencies of the measurement values (categories) of the investigated character.

Bachelor **Example 5.2** *Marital status of the mother* as well as *sex of the child* in Example 1.1

Both are nominal-scaled characters. We are interested in the frequencies of the corresponding categories.

Using **R**, we enter the following four commands:

```
> sex.tab <- table(sex)
> marital.tab <- table(marital_mother)
> print(sex.tab)
> print(marital.tab)
```

i.e. we apply the function `table()` to the characters *sex of the child* (`sex`) and *marital status of the mother* (`marital_mother`). The respective results are assigned to the objects `sex.tab` and `marital.tab`, respectively, using the command '`<-`' consisting of the symbols '<' and '-'. Next, we set these objects as arguments into the function `print()`.

This results in:

```
sex
female   male
    50     50

marital_mother
never married        married        divorced         widowed
            9             65              22               4
```

Now we ascertain the relative frequencies by typing

```
> prop.table(sex.tab)
> prop.table(marital.tab)
```

i.e. we apply the function `prop.table()` using the relative frequencies stored in `sex.tab` and `marital.tab` as arguments.

This yields:

```
sex
female   male
   0.5    0.5

marital_mother
never married        married        divorced         widowed
         0.09           0.65            0.22            0.04
```

In order to create a pie chart of the character *marital status of the mother*, we type

```
> pie(marital.tab, main = "Pie chart")
```

i.e. we apply function `pie()` to the object of interest, `marital.tab`, and add a title to the resulting pie chart using the command `main = "Pie chart"`.

As output, we get the pie chart shown in Figure 5.3.

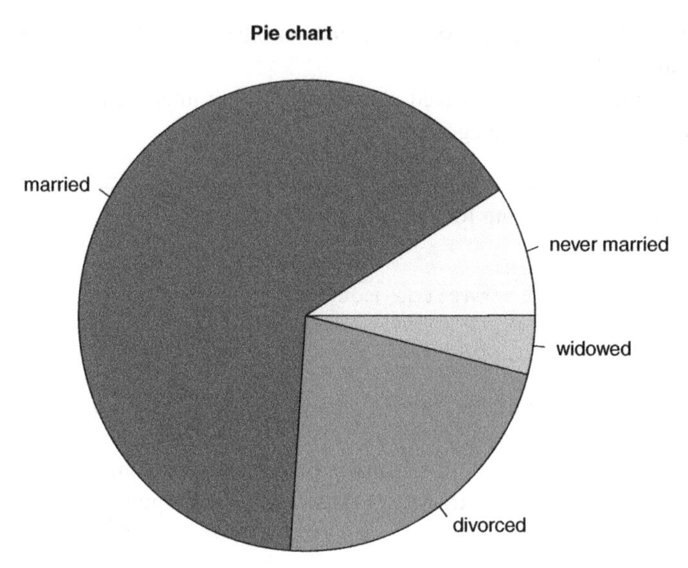

Figure 5.3 **R**-output showing a pie chart of the character *marital status of the mother* from Example 1.1.

In SPSS, we select

Analyze
    Descriptive Statistics
        Frequencies...

from the menu to see the window shown in Figure 5.4. Now we select the character sex of the child and move it to the panel Variable(s):

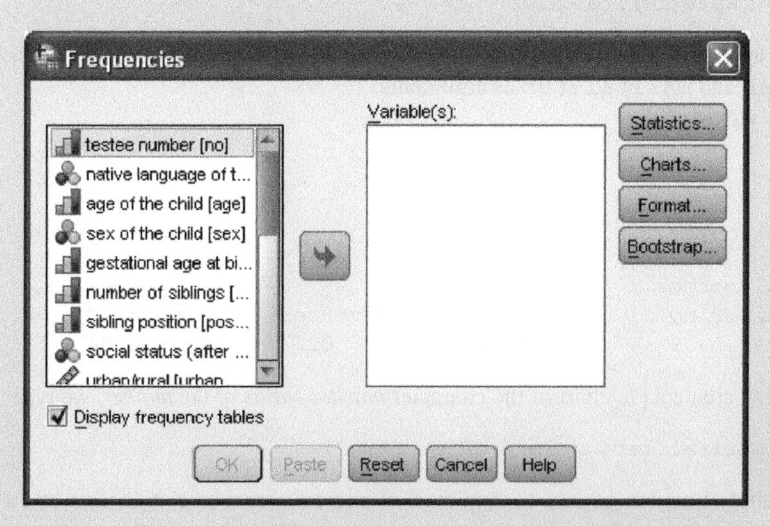

Figure 5.4 SPSS-window for calculating empirical frequency distributions.

At the moment, we do not want to calculate any statistics. We therefore ignore the Statistics... button (among others) and just click OK. In the resulting output, we disregard the column Cumulative Percent due to the fact that there is no ranking of the sexes and, for this reason, cumulative frequencies (in SPSS called Cumulative Percent) are completely meaningless; for the currenty relevant part of the output, see Table 5.2.

Table 5.2 SPSS-output showing the empirical frequency distribution of the character *sex of the child* in Example 1.1 (shortened output).

sex of the child

		Frequency	Percent	Valid Percent
Valid	female	50	50.0	50.0
	male	50	50.0	50.0
	Total	100	100.0	100.0

We do the same for the character marital status of the mother and see Table 5.3 as a result.

Table 5.3 SPSS-output showing the empirical frequency distribution of the character *marital status of the mother* in Example 1.1 (shortened output).

marital status of the mother

		Frequency	Percent	Valid Percent
Valid	never married	9	9.0	9.0
	married	65	65.0	65.0
	divorced	22	22.0	22.0
	widowed	4	4.0	4.0
	Total	100	100.0	100.0

Via

Graphs
    Chart Builder...

we come to the Chart Builder (see Figure 5.5). Next we select Pie/Polar in the panel Choose from: of the (already activated) Gallery tab. A pie chart symbol appears in the window on the right, and we drag and drop it into the Chart preview above. Now a pie chart with the axes Slice by? and Angle Variable? appears (see Figure 5.6), as does the additional window Element Properties (not shown here), which we ignore (e.g. close) for the time being. We continue by selecting the character marital status of the mother and moving it to the panel Slice by?. Now we click OK to obtain the pie chart shown in Figure 5.3.

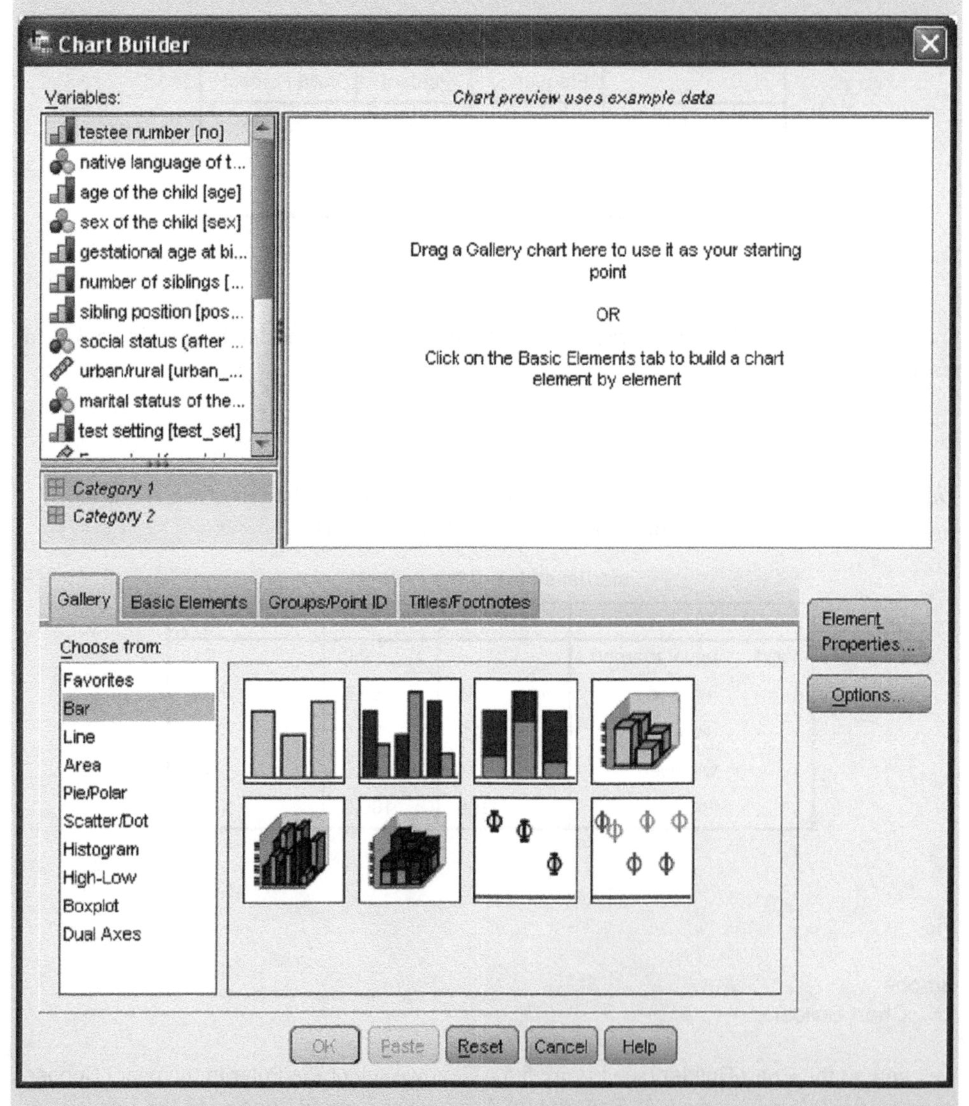

Figure 5.5 SPSS-window Chart Builder.

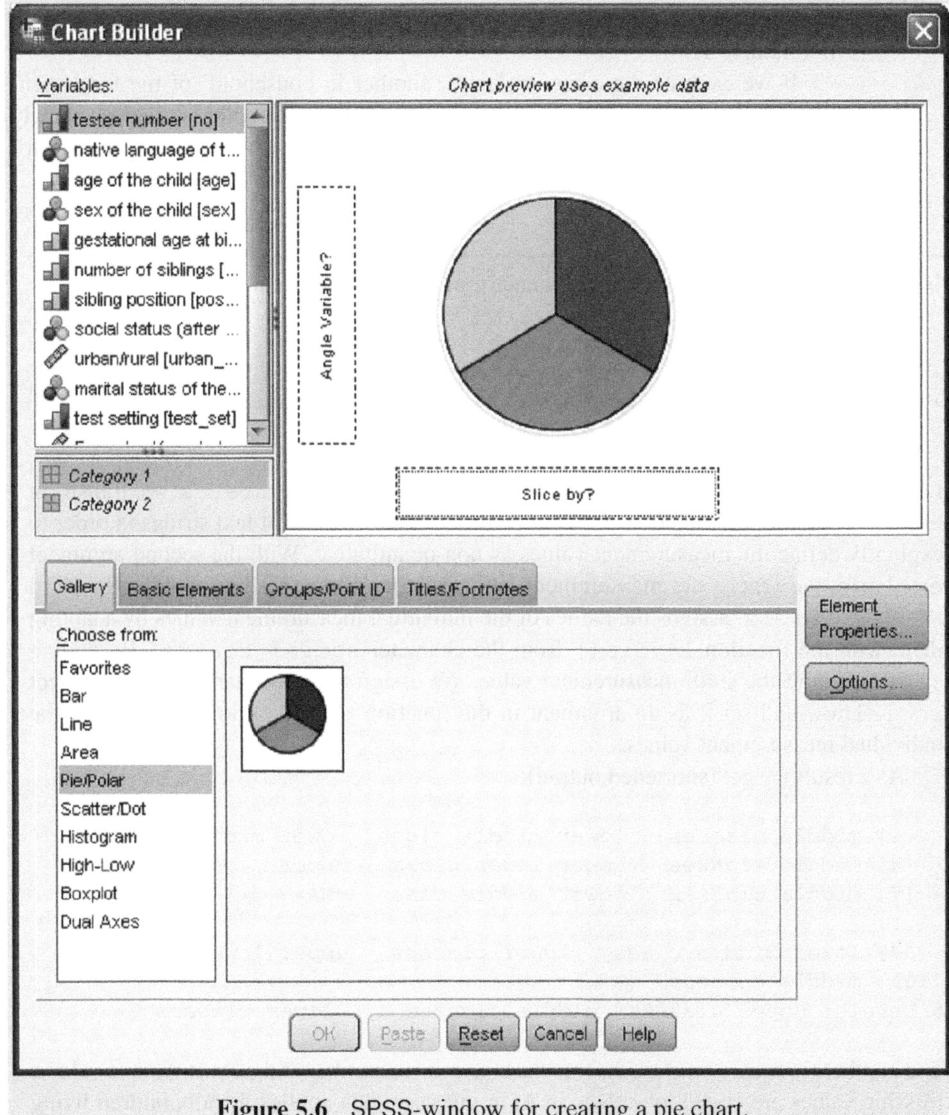

Figure 5.6 SPSS-window for creating a pie chart.

5.2.2 Ordinal-scaled characters

In the case of ordinal-scaled characters, we prefer to use the *bar chart* and the *dot diagram* instead of the pie chart, as it is not possible to represent the rank order of the observations with it.

Bachelor **Example 5.3** *Social status* without the category 'single mother in household' in Example 1.1

If we exclude the category 'single mother in household' of the fundamentally nominal-scaled character *social status*, a new character is created, namely an ordinal-scaled character. We now want to work with this new character for statistical analysis.

In **R**, we first create a new character by entering the following commands:

```
> socialnew <- factor(as.character(social_status),
+                     exclude = "single mother in household",
+                     levels = levels(social_status)[-6])
> print(socialnew)
```

i.e. we apply the function `factor()` to the character `social_status`; we transform this character with the function `as.character()` to a vector of text strings in order to explicitly define the measurement values as non-quantitative. With the second argument `exclude`, we ignore the measurement value 'single mother in household'. The third argument, `levels`, assigns the names of the individual measurement values by adapting them with the function `levels()` from the character `social_status`. By adding `[-6]`, we omit the sixth measurement value. We assign the new character to the object `socialnew` and set it as an argument in the function `print()` in order to print the individual measurement values.

As a result we get (shortened output):

```
  [1] middle classes   lower middle class  lower middle class
  [4] middle classes   upper lower class   upper classes
  [7] middle classes   lower middle class  <NA>
  ...
 [97] lower middle class   lower classes   upper lower class
[100] middle classes
5 Levels: upper classes  middle classes ... lower classes
```

The numbers in square brackets refer to the consecutive number of the measurement values. Missing values are coded as <NA> in **R**; in our case, this applies for all children living with their single mother.

To ascertain the empirical frequency distribution of the character `socialnew`, we use a command similar to that in Example 5.2

```
> soc.tab <- table(socialnew)
> soc.pro <- prop.table(soc.tab) * 100
> soc.cum <- cumsum(soc.pro)
> cbind(soc.tab, soc.pro, soc.cum)
```

i.e. again we use the functions `table()` and `prop.table()` to calculate the absolute and the relative frequencies, respectively. We multiply the relative frequencies by `100` to gain percentages and assign both results to new objects. Furthermore, we sum up the percentages by using the function `cumsum()`; we use the percentages as an argument, and then we assign the result to the object `soc.cum`. Finally, we combine the resulting vectors with the help of the function `cbind()` to a matrix.

As a result we get:

```
                     soc.tab     soc.pro     soc.cum
upper classes             12   13.636364    13.63636
middle classes            30   34.090909    47.72727
lower middle class        25   28.409091    76.13636
upper lower class         16   18.181818    94.31818
lower classes              5    5.681818   100.00000
```

In order to illustrate these results graphically, we type

```
> windows(width = 10, height = 6)
> barplot(soc.tab, xlab = "social status",
+         ylab = "Absolute frequency",
+         main = "Bar chart")
```

i.e. we demand a new output-window for our chart using the function `windows()` and define its size as `10` by `6` inches. Next, we use the object `soc.tab` as the first argument in the function `barplot()` and add as other arguments `xlab`, `ylab`, and `main = "Bar chart"` in order to set the labels of the abscissa and the ordinate, and in order to give a title.

As a result, we get the bar chart shown in Figure 5.7.

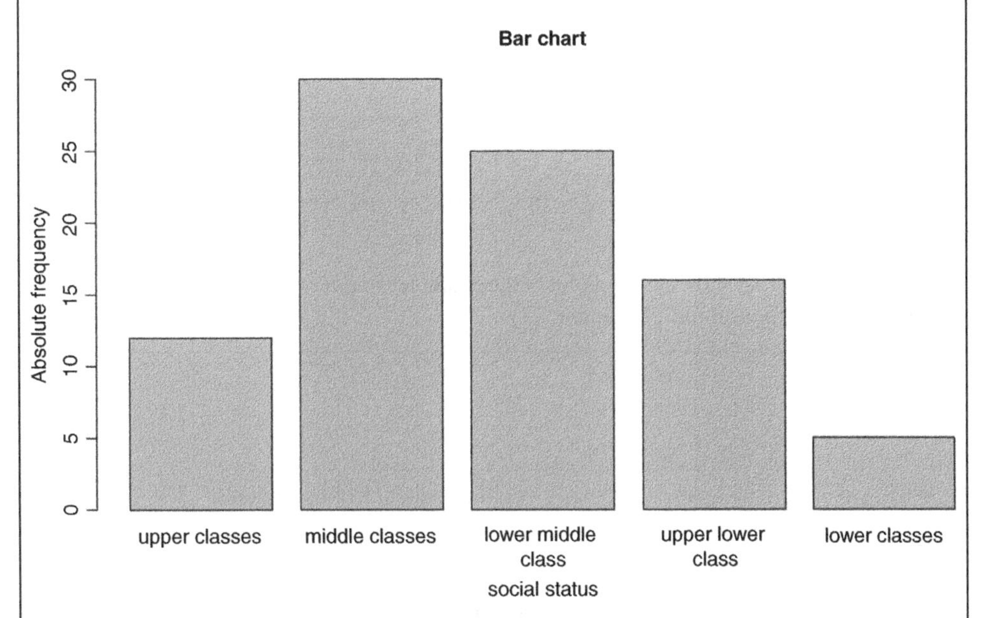

Figure 5.7 R-output showing a bar chart of the character *social status* (without the category 'single mother in household'; Example 1.1).

Now we produce a dot plot by typing

```
> dotchart(c(soc.tab), pch = 16, main = "Dot plot",
+          xlab = "Frequency")
```

i.e. we apply function `dotchart()` and set `soc.tab` as the first argument, after transforming the respective table to a vector using the function `c()`. With the second argument, `pch = 16`, we configure our dots as small, filled circles. Furthermore, `main = "Dot plot"` adds a title to the plot and `xlab = "Frequency"` labels the abscissa.

This command results in the dot plot shown in Figure 5.8.

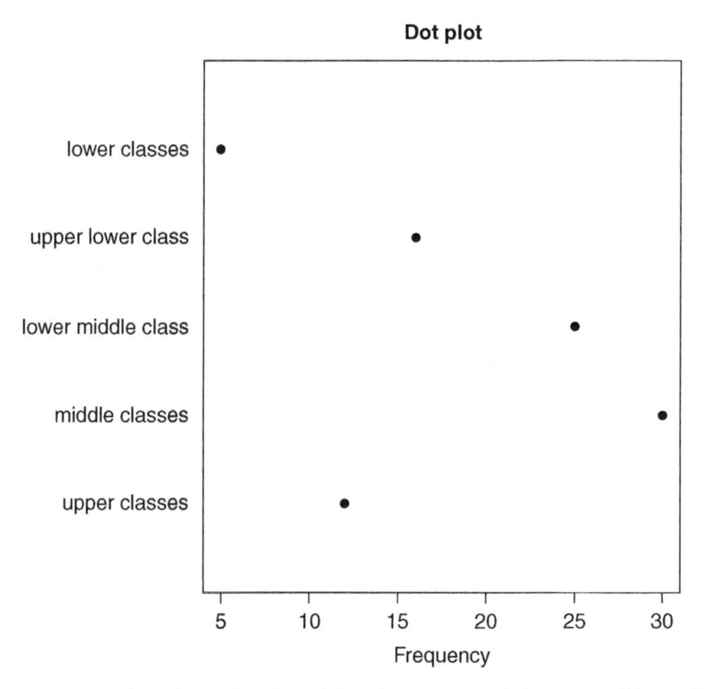

Figure 5.8 **R**-output showing a dot plot of the character *social status* (without the category 'single mother in household'; Example 1.1)

In SPSS, we can create a new character using the sequence of commands:

Transform
    Compute Variable...

In the following pop-up window, we type socialnew as Target Variable, select the character social_status, and move it to the panel Numeric Expression: (see Figure 5.9). Now socialnew equals social_status, but we need to add a few conditions. For this we click If... and choose

Include if case satisfies condition: in the subsequent window (Figure 5.10). We want to include all research units with a social_status of less than or equal to 5 (6 denotes the measurement value single mother in household). We apply the same procedure as in the window before, but now we move social_status to the right panel and type '<=' and '5' either on the keyboard or click the respective buttons (in Figure 5.10 we have already done this). A click on Continue brings us back to the previous window, and clicking OK leads us back to the SPSS Data View, where we find an additional column containing the new character socialnew (a section of this can be seen in Figure 5.11). Missing values correspond to children living with their single mothers. Finally, we define the type of scale of the new character by switching to the SPSS Variable View and clicking on the column Measure in order to choose the option Ordinal from the pull-down list (see Figure 5.12).

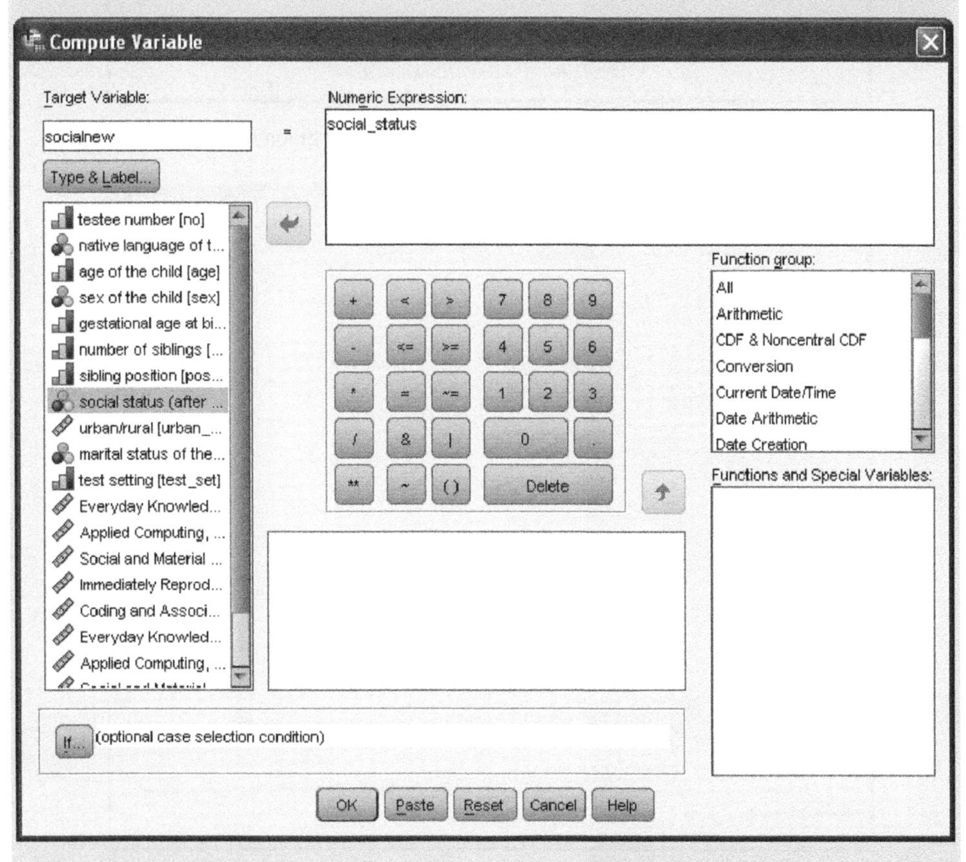

Figure 5.9 SPSS-window for computing a new character.

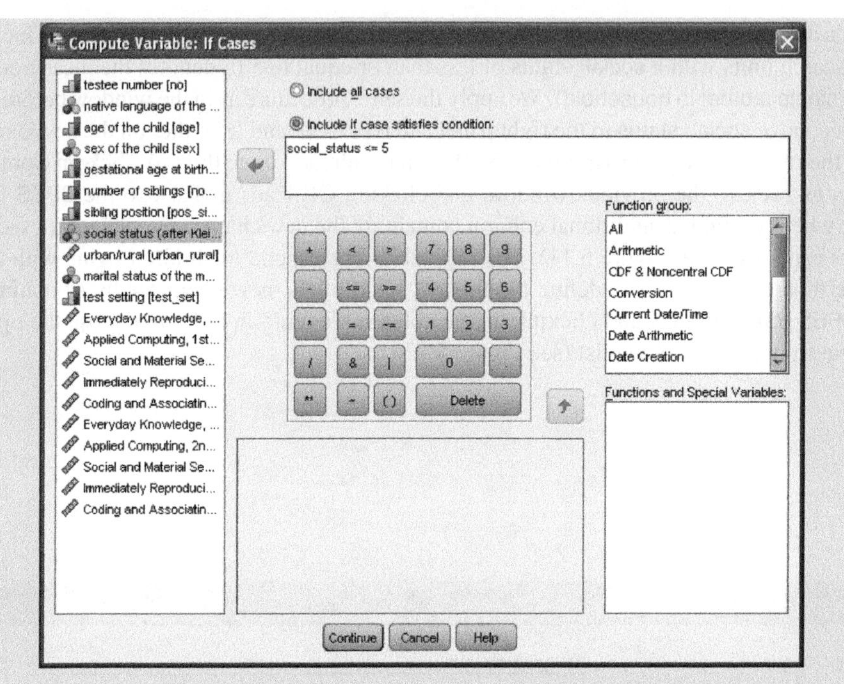

Figure 5.10 SPSS-window for computing a new character given qualifying conditions.

*Example_1.1.sav [DataSet1] - IBM SPSS Statistics Data Editor

	sub4_t2	sub5_t2	sub7_t2	socialnew
1	45	55	54	2.00
2	53	59	65	3.00
3	52	55	50	3.00
4	57	45	54	2.00
5	38	36	46	4.00
6	57	50	61	1.00
7	43	45	46	2.00
8	64	67	69	3.00
9	57	57	61	.
10	57	48	65	1.00
11	36	35	46	2.00
12	59	53	61	3.00
13	59	53	73	2.00
14	67	36	42	3.00
15	59	64	73	1.00
16	48	38	31	4.00
17	62	53	61	2.00
18	52	69	50	.
19	36	50	58	.
20	43	42	54	4.00
21	52	57	69	1.00
22	42	62	54	2.00
23	36	43	46	3.00
24	52	50	61	.
25	45	42	27	.

Data View | Variable View

IBM SPSS Statistics Processor is ready

Figure 5.11 SPSS Data View showing the newly computed character socialnew in Example 5.3 (section).

	Name	Type	Width	Decimals	Label	Values	Missing	Columns	Align	Measure
1	no	Numeric	8	0	testee number	None	None	8	Right	Ordinal
2	native_langu...	Numeric	8	0	native language of the child	{1, German}...	None	13	Right	Nominal
3	age	Numeric	8	0	age of the child	None	None	8	Right	Ordinal
4	sex	Numeric	8	0	sex of the child	{1, female}...	None	10	Right	Nominal
5	age_birth	Numeric	8	0	gestational age at birth (i...	None	None	10	Right	Ordinal
6	no_siblings	Numeric	8	0	number of siblings	None	None	10	Right	Ordinal
7	pos_sibling	Numeric	8	0	sibling position	{1, first-born...	None	10	Right	Ordinal
8	social_status	Numeric	8	0	social status (after Kleini...	{1, upper cl...	None	13	Right	Nominal
9	urban_rural	Numeric	8	0	urban/rural	{1, City (ove...	None	10	Right	Scale
10	marital_mot	Numeric	8	0	marital status of the mother	{1, never ma...	None	10	Right	Nominal
11	test_set	Numeric	8	0	test setting	{1, German ...	None	8	Right	Ordinal
12	sub1_t1	Numeric	8	0	Everyday Knowledge, 1st...	None	None	8	Right	Scale
13	sub3_t1	Numeric	8	0	Applied Computing, 1st t...	None	None	8	Right	Scale
14	sub4_t1	Numeric	8	0	Social and Material Sequ...	None	None	8	Right	Scale
15	sub5_t1	Numeric	8	0	Immediately Reproducing...	None	None	8	Right	Scale
16	sub7_t1	Numeric	8	0	Coding and Associating, ...	None	None	8	Right	Scale
17	sub1_t2	Numeric	8	0	Everyday Knowledge, 2n...	None	None	8	Right	Scale
18	sub3_t2	Numeric	8	0	Applied Computing, 2nd t...	None	None	8	Right	Scale
19	sub4_t2	Numeric	8	0	Social and Material Sequ...	None	None	8	Right	Scale
20	sub5_t2	Numeric	8	0	Immediately Reproducing...	None	None	8	Right	Scale
21	sub7_t2	Numeric	8	0	Coding and Associating, ...	None	None	8	Right	Scale
22	socialnew	Numeric	8	2		None	None	11	Right	Nominal

Figure 5.12 SPSS Variable View after creation of the character socialnew in Example 5.3.

Next, we proceed analogously to Example 5.2 (Analyze – Descriptive Statistics – Frequencies...) in order to ascertain the empirical frequency distribution of socialnew and obtain the output shown in Table 5.4. In SPSS, the values of the empirical frequency distribution are multiplied by 100 and hence express percentages; they are labeled Cumulative Percent.

Table 5.4 SPSS-output showing the empirical frequency distribution of the character socialnew from Example 5.3.

socialnew

		Frequency	Percent	Valid Percent	Cumulative Percent
Valid	1.00	12	12.0	13.6	13.6
	2.00	30	30.0	34.1	47.7
	3.00	25	25.0	28.4	76.1
	4.00	16	16.0	18.2	94.3
	5.00	5	5.0	5.7	100.0
	Total	88	88.0	100.0	
Missing	System	12	12.0		
Total		100	100.0		

For graphical illustration of the data, we use the sequence of commands from Example 5.2 (Graphs - Chart Builder...) to reach the SPSS Chart Builder shown in Figure 5.5. This time, we select Bar in the panel Choose from: located under the tab Gallery. Next, we drag and drop the symbol Simple Bar (this label appears when the cursor hovers over the symbol) into the Chart preview and move the character socialnew to the panel X-Axis?. By confirming these settings with OK, we obtain a graph analogous to Figure 5.7. To create a dot plot, we select Scatter/Dot from the Gallery (see Figure 5.5) and drag and drop the symbol Summary Point Plot into the Chart preview. We move the character socialnew to the panel X-Axis? and select Count in the pull-down menu Statistic:, found in the window in Figure 5.13 (if you initially closed this window, as we recommended, you can reopen it by clicking Element Properties...). By clicking Apply, we confirm this selection; a further click on OK yields the chart similar to Figure 5.8.

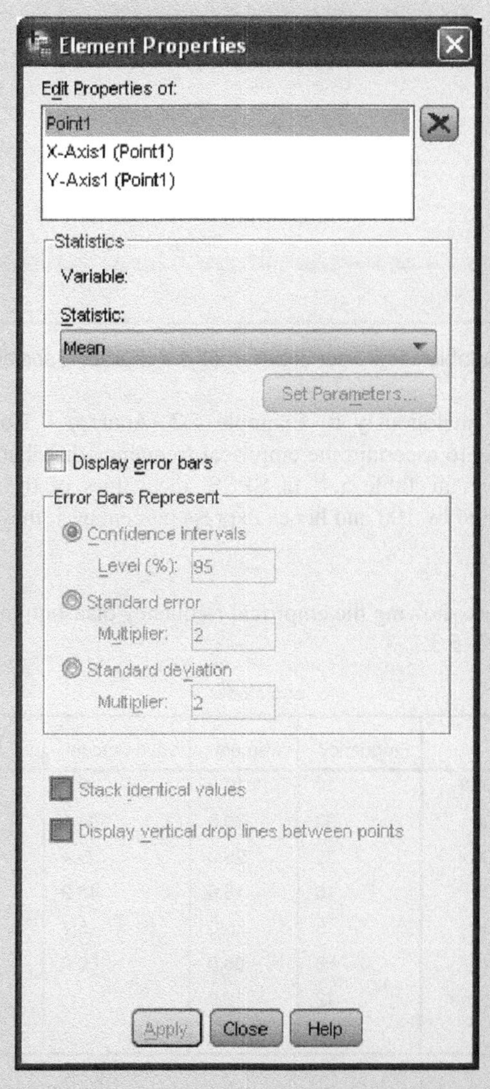

Figure 5.13 SPSS-window to ascertain absolute frequencies in charts.

If an ordinal-scaled character is given, then both the bar chart and the dot diagram are to be viewed critically. Firstly, the bars suggest that the frequencies refer to an interval on the abscissa – which is not at all the case; secondly, both figures, Figure 5.7 as well as Figure 5.8, imply that the distances between the ordered categories are of equal size – which is not necessarily true.

5.2.3 Quantitative characters

In quantitative characters, there are rarely equal outcomes, if there is a sufficiently high measuring precision; thus, several measurement values are often pooled together into classes. In this case, the *histogram* is an appropriate means to illustrate the empirical frequency distribution.

Bachelor In principal, the researcher does not have to deal with pooling the outcomes into the individual classes when illustrating the data with a histogram; this is done automatically by computer programs.

Bachelor **Example 5.4** The empirical distribution function of the character *Everyday Knowledge, 1st test date* in Example 1.1 – without children with Turkish as a native language who were tested in the German language at the first test date

Since an important question of the study is also to what extent children with Turkish as a native language are disadvantaged by being tested in the German language, about half of these children were tested in German at the first test date while the other half were tested in Turkish. We now disregard the data from those children with Turkish as a native language who were tested in the German language at the first test date. We are interested in the character *Everyday Knowledge, 1st test date*.

Therefore, we first have to exclude the children with the coded value 'Turkish speaking child tested in German at first test date' in the character *test setting*.

In **R**, we first need to select the children in question. To do this, we enter the commands

```
> sub1_t1.set <- sub1_t1[test_set == "German speaking child" |
+ test_set ==
+ "Turkish speaking child tested in Turkish at first test date"]
```

i.e. include only those research units of the character *Everyday Knowledge, 1st test date* (`sub1_t1`) that meet the condition specified in the square brackets. This condition is that the value of the character *test setting* (test_set) has to be equal to either one or (the symbol '|' indicates a logical 'or') the other specified value for the case to be included. We assign the resulting data to the object sub1_t1.set. Now we type:

```
> sub1_t1.tab <- table(sub1_t1.set)
> sub1_t1.pro <- prop.table(sub1_t1.tab) * 100
> sub1_t1.cum <- cumsum(sub1_t1.pro)
> round(cbind(sub1_t1.tab, sub1_t1.pro, sub1_t1.cum), digits = 1)
```

i.e. we apply the functions table(), prop.table(), and cumsum() as in Example 5.3 and assign each result to a new object. With the function cbind(), we combine these objects to form a table and round() the values to a precision of one decimal place (digits = 1). This yields:

```
   sub1_t1.tab sub1_t1.pro sub1_t1.cum
25           1         1.4         1.4
27           1         1.4         2.7
31           2         2.7         5.4
33           2         2.7         8.1
35           2         2.7        10.8
37           2         2.7        13.5
41           3         4.1        17.6
46           4         5.4        23.0
48           1         1.4        24.3
50          14        18.9        43.2
52           2         2.7        45.9
54           6         8.1        54.1
56           2         2.7        56.8
58           4         5.4        62.2
60          13        17.6        79.7
61           4         5.4        85.1
63           1         1.4        86.5
65           6         8.1        94.6
69           3         4.1        98.6
71           1         1.4       100.0
```

In SPSS, we use

Data

  Select Cases ...

to open a window (Figure 5.14) where we can choose If condition is satisfied from the field Select. Next we click If... and, in the following window, we type the text seen in Figure 5.15 (the symbol '|' indicates a logical 'or'). By clicking Continue, we return to the previous window and confirm our selection with OK. Figure 5.16 shows a section of the SPSS Data View including the just-created 'filter variable', filter_$. To compute the empirical frequency distribution and the empirical distribution function, we proceed analogously to Example 5.2 (Analyze – Descriptive Statistics – Frequencies...) and select as Variable(s): the character Everyday Knowledge, 1st test date. By the way, the test scores of *Everyday Knowledge* (as for other subtests of the intelligence test battery used here), are scaled in T-Scores (see Footnote 5 in Table 1.1; for more details, see Example 5.5). The result is presented in Table 5.5, with the values of the empirical distribution function given in the column Cumulative Percent.

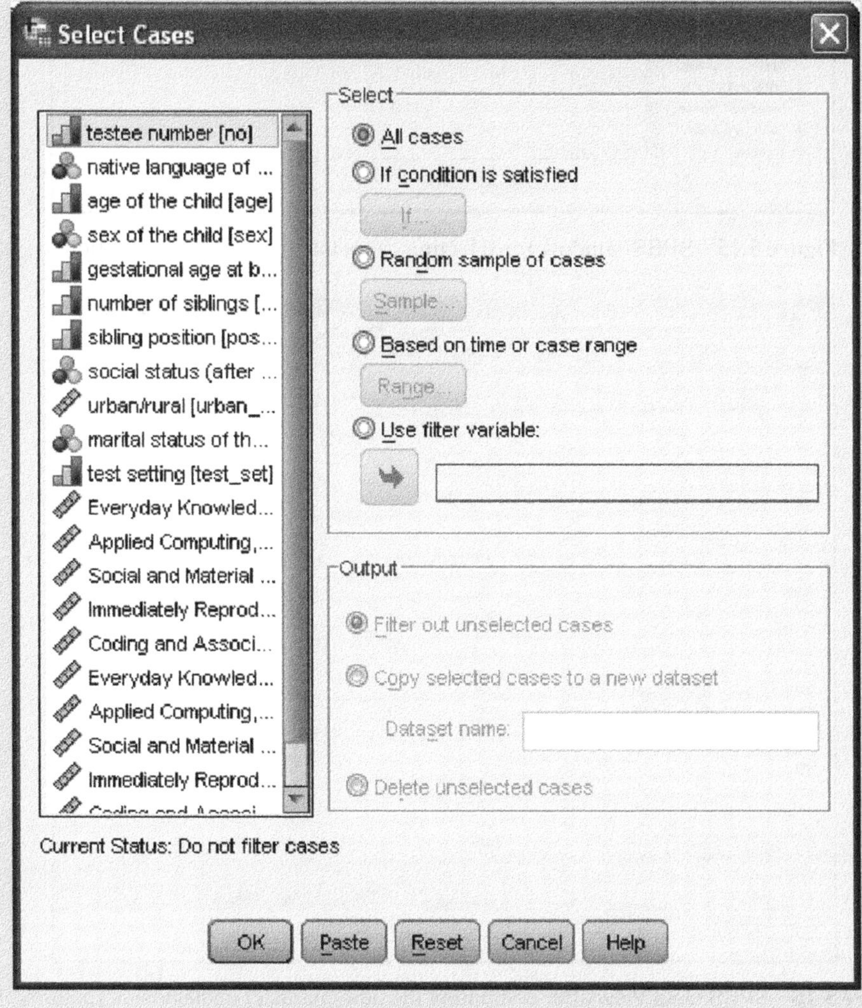

Figure 5.14 SPSS-window for selecting cases.

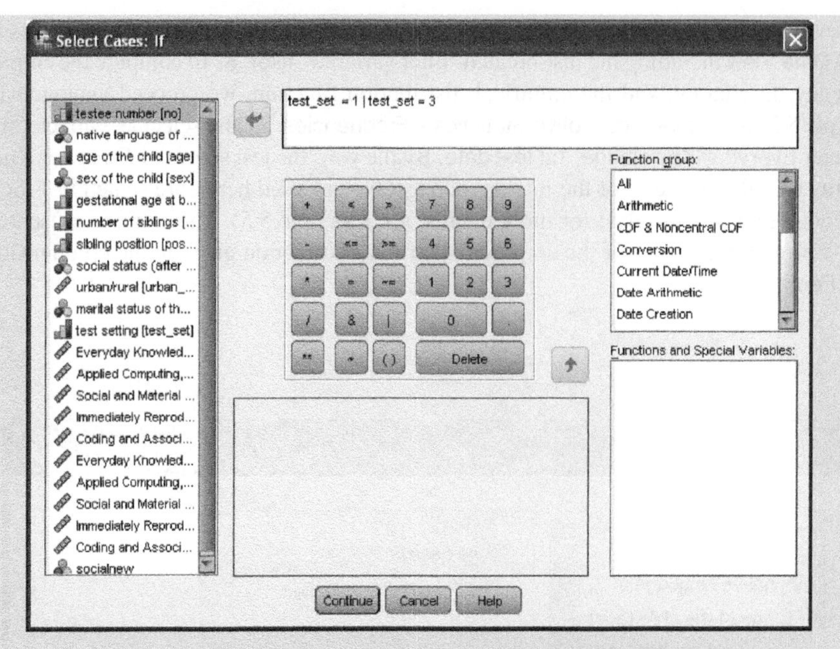

Figure 5.15 SPSS-window for selecting cases that meet certain conditions.

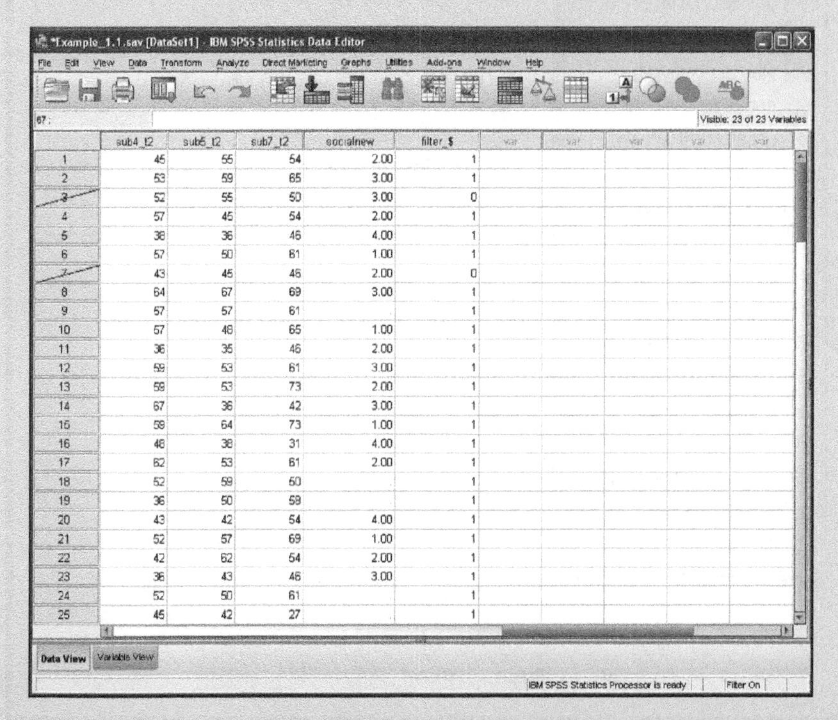

	sub4_t2	sub5_t2	sub7_t2	socialnew	filter_$	var	var	var	var	var
1	45	55	54	2.00	1					
2	53	59	65	3.00	1					
3	52	55	50	3.00	0					
4	57	45	54	2.00	1					
5	38	36	46	4.00	1					
6	57	50	61	1.00	1					
7	43	45	46	2.00	0					
8	64	67	69	3.00	1					
9	57	57	61	.	1					
10	57	48	65	1.00	1					
11	36	35	46	2.00	1					
12	59	53	61	3.00	1					
13	59	53	73	2.00	1					
14	67	36	42	3.00	1					
15	59	64	73	1.00	1					
16	48	38	31	4.00	1					
17	62	53	61	2.00	1					
18	52	59	50		1					
19	36	50	59		1					
20	43	42	54	4.00	1					
21	52	57	69	1.00	1					
22	42	62	54	2.00	1					
23	36	43	46	3.00	1					
24	52	50	61		1					
25	45	42	27		1					

Figure 5.16 SPSS Data View after computing the new character socialnew in Example 1.1 (section).

Table 5.5 SPSS-output showing the empirical frequency distribution of the character *Everyday Knowledge, 1st test date* from Example 1.1.

Everyday Knowledge, 1st test date (T-Scores)

		Frequency	Percent	Valid Percent	Cumulative Percent
Valid	25	1	1.4	1.4	1.4
	27	1	1.4	1.4	2.7
	31	2	2.7	2.7	5.4
	33	2	2.7	2.7	8.1
	35	2	2.7	2.7	10.8
	37	2	2.7	2.7	13.5
	41	3	4.1	4.1	17.6
	46	4	5.4	5.4	23.0
	48	1	1.4	1.4	24.3
	50	14	18.9	18.9	43.2
	52	2	2.7	2.7	45.9
	54	6	8.1	8.1	54.1
	56	2	2.7	2.7	56.8
	58	4	5.4	5.4	62.2
	60	13	17.6	17.6	79.7
	61	4	5.4	5.4	85.1
	63	1	1.4	1.4	86.5
	65	6	8.1	8.1	94.6
	69	3	4.1	4.1	98.6
	71	1	1.4	1.4	100.0
	Total	74	100.0	100.0	

Master

A histogram consists of rectangles on the abscissa, which are constructed from all *class intervals*; their areas (and, given equal class intervals, also their heights) correspond to the respective (relative or absolute) frequencies. The selected standard measure of these frequencies is shown in the ordinate.

In the case of a *cumulative staircase*, rectangles are constructed from all class intervals; however, now their areas (and, given equal class intervals, also their heights) correspond to the respective (relative or absolute) cumulative frequencies. The selected standard measure of these frequencies is shown in the ordinate.

Master

Even if there is no pooling of the outcomes to classes in the case of quantitative characters, it should be noted that for practical reasons only discrete measurement values are observed; each in itself already represents a class. For example, the measurement of Mr. *K*'s height yields the value of 185 cm; however, as there is little requirement for precision with regards to the subject to be measured, and likewise as the measurement instrument is inaccurate, this measurement value is representative for a whole interval, namely for the interval 184.95 to 185.049999...

Bachelor **Example 5.4 – continued**

The resulting empirical frequency distribution of the character *Everyday Knowledge, 1st test date* (still) has a relatively large number of classes; namely 20. However, in the following we want to use fewer classes for the graphical illustration by using a histogram and a step function.

In **R**, we create a histogram by typing

```
> hist(sub1_t1.set, breaks = 10, main = "Histogram",
+       xlim = c(20, 80), ylim = c(0, 20),
+       xlab = "Everyday Knowledge, 1st test date (T-Scores)",
+       ylab = "Frequency")
```

i.e. we use the already created object `sub1_t1.set` as the first argument in the function `hist()` and `breaks = 10` as the second argument determining that there should be 10 classes. The argument `main = "Histogram"` defines the title of our histogram. Then `xlim` and `ylim` define the represented value range of the abscissa and the ordinate, respectively; for this, we use the function `c()` to declare the minimum and maximum. Finally, `xlab` and `ylab` label the axes.

This results in the histogram shown in Figure 5.17a.

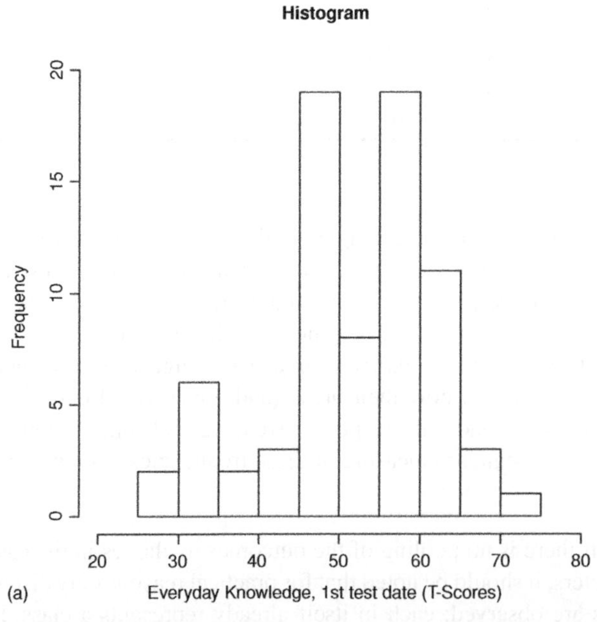

Figure 5.17 a **R**-output of the histogram in Example 5.4 (the character *Everyday Knowledge, 1st test date* presented in 10 classes).

To ascertain the empirical cumulative distribution function, we type

```
> sub1_t1.ecdf <- ecdf(sub1_t1.set)
> plot(sub1_t1.ecdf, verticals = TRUE, do.points = FALSE,
+       main = "Empirical cumulative distribution function",
+       xlab = "Everyday Knowledge, 1st test date (T-Scores)",
+       ylab = "Cumulative relative frequency")
```

i.e. we use the object `sub1_t1.set` as the first argument in the function `ecdf()` and assign the result to object `sub1_t1.ecdf`. Then we submit this object as the first argument to the function `plot()`. With `verticals = TRUE`, we choose to plot a cumulative staircase, and we prevent the output of points by setting `do.points = FALSE`. Finally, `main`, `xlab`, and `ylab` again define the text of the chart title and label the axes.

As a result, we get the chart shown in Figure 5.18.

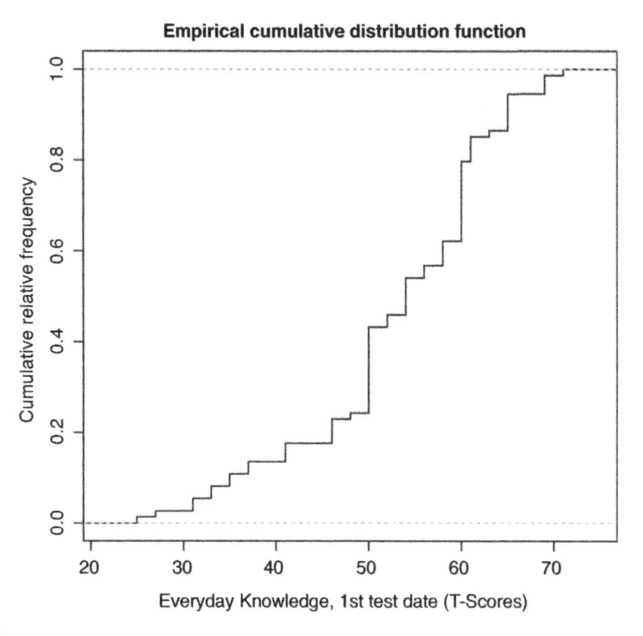

Figure 5.18 **R**-output of the cumulative staircase in Example 5.4.

In order to produce a histogram in SPSS, we start the SPSS Chart Builder (see Figure 5.5) using the same sequence of commands (Graphs - Chart Builder...) as in Example 5.2. Next, we select Histogram in the Gallery tab; we drag and drop the symbol Simple Histogram (on the left side) into the Chart preview and move the character Everyday Knowledge, 1st test date to the field X-Axis?. Instead of letting SPSS decide on the number of classes, we want to define 10 classes. To do this, we click Element Properties... and a window very similar to that in Figure 5.13 appears. There we choose Set Parameters... and the window shown in Figure 5.19 appears. Next, we select Custom in the field Bin Sizes and set Number of intervals: to 11 (in Figure 5.19, we have already done so). Finally, we click Continue, followed by Apply and OK. The result is the histogram shown in Figure 5.17b.

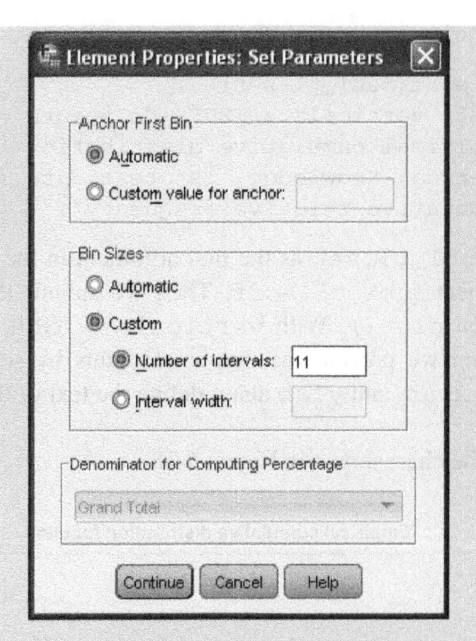

Figure 5.19 SPSS-window to define the number of histogram classes.

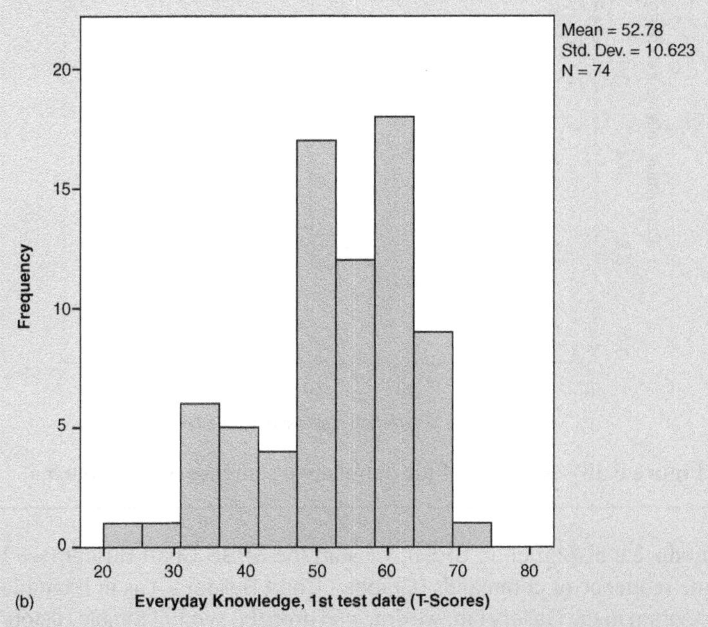

Figure 5.17 b SPSS-output of the histogram in Example 5.4 (the character *Everyday Knowledge, 1st test date* presented in 10 classes).

Compared with our solution, the automatic SPSS routine would set more classes.

Now we look at the empirical cumulative distribution function. In the SPSS Chart Builder in Figure 5.5, we select Line in the Gallery tab and drag and drop the symbol Simple Line into the Chart preview. We then move the character Everyday Knowledge, 1st test date to the field X-Axis?. A click on Element Properties... gets us to the window shown in Figure 5.13, where we select Cumulative Count from the pull-down menu Statistic: in the panel Statistics. In the field Interpolation, we change the Type: to Step and confirm all settings with Apply, followed by OK. This yields a chart similar to Figure 5.18.

There are obviously certain differences in the histogram in **R** and SPSS. In general, the histogram is more illustrative than the step function.

Often it is a matter of clearly illustrating and concisely describing, respectively, two or more characters simultaneously. Ultimately, it is only possible to illustrate in tabular form with two characters. Even graphical illustrations are only really successful in the case of exactly two characters. For more than two characters, a diagram only works well if one of the characters is compared to all the others and these other characters are 'related' with regards to content.

Bachelor **Example 5.5** The difference between the first and second test date in the subtest *Everyday Knowledge* for children with German as a native language per *age of the child* (Example 1.1)

The two characters *Everyday Knowledge, 1st test date and Everyday Knowledge, 2nd test date* will be compared graphically.

In **R**, we first select the children with German as a native language by typing

```
> age.ger <- age[native_language == "German"]
> sub1_t1.ger <- sub1_t1[native_language == "German"]
> sub1_t2.ger <- sub1_t2[native_language == "German"]
```

i.e. we filter the characters *Everyday Knowledge, 1st test date* (sub1_t1), *Everyday Knowledge, 2nd test date* (sub1_t2), and *age of the child* (age) for those children who have German as a native language [native_language == "German"]; then we assign the respective data to new objects. Now we type

```
> table(age.ger)
```

i.e. we submit the argument *age of the child* (age.ger) to the function table().

As a result, we get:

```
age.ger
 6  7  8  9
13 11 13 13
```

It can be seen that children aged 6, 7, 8, and 9 years are in the sample. Next, we want to explore the distribution of the scores among the four age groups. To do this, we type

```
> sub1_t1.mean <- tapply(sub1_t1.ger, age.ger, mean)
> sub1_t2.mean <- tapply(sub1_t2.ger, age.ger, mean)
> sub1.mean <- cbind(sub1_t1.mean, sub1_t2.mean)
```

i.e. we successively submit the characters sub1_t1.ger and sub1_t2.ger to the function tapply() calculating the mean of the children's age (age.ger). Each result is assigned to a new object, which we combine to a matrix using the function `cbind()`. To illustrate the data graphically, we type

```
> matplot(sub1.mean, type = "b", lty = 1:2, pch = 18, col = 1:2,
+         xlab = "age", ylab = "T-Score", ylim = c(35, 65),
+         axes = FALSE)
> axis(1, at = 1:4, labels = 6:9)
> axis(2, at = seq(from = 35, to = 65, by = 5))
> legend("topright", col = 1:2, lty = 1:2,
+        c("Everyday Knowledge, 1st test date",
+          "Everyday Knowledge, 2nd test date"))
```

i.e. we use the object `sub1.mean` as the first argument in the function `matplot()`. With `type = "b"`, we define that the values be presented as points and linked by lines (due to `lty = 1:2`, the first line is continuous and the second dashed). `pch = 18` defines the shape of the points, while `col = 1:2` sets their colors to black and red. Furthermore, we label the axes using `xlab` and `ylab`, defining the presented range of values (`35` to `65` T-Scores) of the ordinate using `ylim`, and preventing the (default) drawing of the axes with `axes = FALSE`. We adjust the axes manually with the function `axis()`, identifying the affected axis with the first argument (`1` for the abscissa and `2` for the ordinate). Using the argument `at`, we establish the location of the tick-marks, namely `1` to `4` on the abscissa and `35` to `65` incremented by `5` on the ordinate – for the latter, we apply the function `seq()`. Finally, we add a legend using the function `legend()`, and define its position through the first argument `"topright"`. The further arguments `col` and `lty` correspond to the arguments with the same names in the function `matplot()`. Using `c()`, we add the two labels `"Everyday Knowledge, 1st test date"` and `"Everyday Knowledge, 2nd test date"`.

As a result, we obtain the graph shown in Figure 5.20.

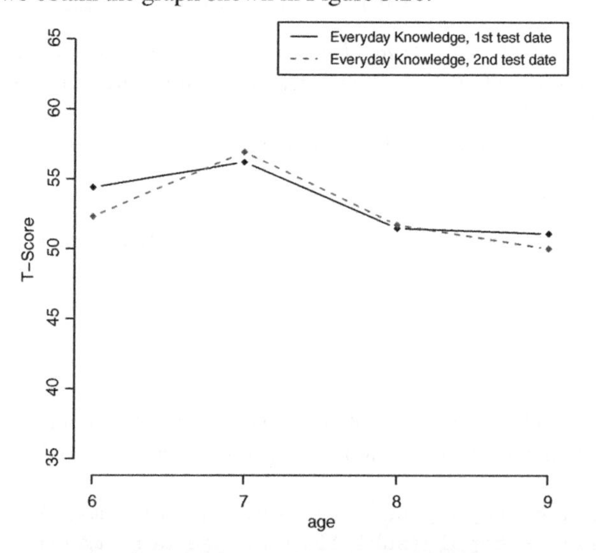

Figure 5.20 **R**-output showing the improving mean of *Everyday Knowledge, 1st test date* and *Everyday Knowledge, 2nd test date* across the *age of the child* for children with German as a native language in Example 1.1.

Analogously to Example 5.4, in SPSS we first make sure that only children with German as a native language are considered in the analysis: we insert the condition native_language = 1 into the field shown in Figure 5.15 and confirm this entry by clicking Continue, followed by OK. Next, we start the SPSS Chart Builder in Figure 5.5 with the same sequence of commands (Graphs – Chart Builder...) as in Example 5.2. There we select Dual Axes in the Gallery tab and drag and drop the symbol Dual Y Axes with Scale X Axis (on the right) into the Chart preview. To define the ordinate, we move the characters Everyday Knowledge, 1st test date and Everyday Knowledge, 2nd test date to the field Y-Axis? on the left and right side of the chart, respectively. Finally, we move the character age of the child to the field X-Axis?. By clicking on Element Properties..., we open a similar window to that shown in Figure 5.13. In this window, we select Point1 in the panel Edit Properties of: and Mean in the pull-down menu Statistic:. A pop-up window appears, which we confirm with OK. We then Apply our settings. To avoid a distorted measurement scale, we select (still in the window Element Properties) Y-Axis1 (Point1) in the panel Edit Properties of: and remove the check marks of Minimum and Maximum in the field Scale Range. Now we can set the correct scale range manually by entering 35 and 65 in the textbox Custom. Again, we Apply these changes and select the same course of action for Y-Axis2 (Point2). Finally, we press OK and obtain a graph analogous to that in Figure 5.20, except that the points are not linked.

There are no constant changes in the test scores between the first and second test date. And obviously, the level of the test performances, in other words the means of test scores, is not equal across all ages in the given sample.

Summary

The graphical illustration of data is different for discrete, ordinal-scaled, and qualitative characters from that for (continuous) quantitative characters. There is the *pie chart* for nominal-scaled characters, which is not appropriate for ordinal-scaled and quantitative characters. There is the *dot diagram* and the *bar chart* for ordinal-scaled characters. And for quantitative characters, there is the *histogram*. All the described graphical methods can be applied for the observations of a population as well as of a sample.

5.2.4 Principles of charts

Charts serve as the visual expression of information from extensive data sets, especially for those persons who are not capable of grasping the quantitative meaning of numbers from tables. Therefore, in principal, charts must be designed simply and must be immediately comprehensible. Charts which strive only for impressive effects, but detract from or even hide the essential information, are therefore counterproductive.

Master

Three-dimensional block diagrams (bar charts), derived from the histogram, are inappropriate: first, it is difficult, even for people with a high spatial ability, to put the volumes represented in a perspective view into fair relation with one another. Second, often actually only the areas of the front views represent the relations in question, and not at all the three-dimensional blocks with respect to their volumes – this is the case if the surface areas of the blocks are not quadratic. Such artificial effects increase if, instead of a three-dimensional presentation, the rectangular areas cast a shadow like a slab of wood.

Master If histograms are confused with bar charts, either the impression of a continuous character is given while in actual fact it concerns a qualitative or a discrete quantitative character, or, vice versa, the impression of a discrete character is given although it is in actual fact a continuous one.

Master Sometimes the dots are connected by straight lines in the dot diagram. The result is a *frequency polygon*. This should be avoided because it again implies a continuity of the character; but it is also misleading in the case of a continuous character: the resulting sub- (non-rectangular) areas do not match the actual frequencies.

It is important that the standard measures of the abscissa and the ordinate are shown and, ideally, start at zero. If the latter is not possible on account of the content, the compression of the scale should be clearly displayed at the point of intersection of ordinate and abscissa; unfortunately, this is not possible in SPSS.

Bachelor For example, if the development of the average intelligence quotient (IQ; see the so-called Flynn effect in, for example, Kubinger, 2009b) over time is to be illustrated, each of the smallest measurement values (perhaps 55 for the IQ, perhaps 1920 for the calendar year) is hardly of relevance for the illustration.

5.2.5 Typical examples of the use of tables and charts

Tables are of key importance for the standardization of a test within psychological assessment. If there is also an interest in the course of the test scores across the age groups, this can be illustrated best with a chart.

Bachelor **Example 5.6** Standardization of a test

The test scores of psychological tests are generally put in relation to the population and then interpreted, or, more precisely, put in relation to a relevant reference group. For example, in aptitude assessment, the reference group is defined as that group of persons to which the testee currently belongs, should belong, or wants to belong in the future and with which he/she should be compared. Standardization therefore means the determination of the distribution of the test scores in the relevant test within the reference group.

Therefore it is a matter of the (cumulative) frequencies of all test scores in a particular test in a 'large' standardization sample. An empirical distribution function is hence determined: for each measurement value the summed (relative) frequency must be ascertained; this represents outcomes that are smaller than or equal to the respective value. By defining this empirical distribution function as valid for the entire reference group in question, it becomes a theoretical distribution function. Each future testee can therefore be put in relation to this reference group.

This can be demonstrated using the test statistic '*Range* of Intelligence' in the AID 2. This is determined by the difference between a testee's largest and smallest T-Score in the 14 subtests of this intelligence test battery. Table 5.6 shows the observations of the $n = N = 977$ persons from the standardization sample. However, the class intervals have been manually chosen in such a way that specific cumulative (relative) frequencies result. This is in order to be able to directly assign future test scores to the *percentile ranks* that correspond to the values of the most frequent T-Scores by using this table: percentile ranks in psychological assessment correspond to cumulative frequencies in statistics.

Table 5.6 The empirical/theoretical distribution function of the test score '*Range* of Intelligence' in the AID 2 for transformation into percentile ranks (and T-Scores) (taken from Kubinger, 2009a, p. 208).

T-Score	Percentile rank	*Range*
19	0.1	7
27	1.1	13
34	5.5	17
38	11.5	20
41	18.4	22
43	24.2	23
45	30.9	25
46	34.5	25
47	38.2	26
48	42.1	27
49	46.0	28
50	50.0	29
51	54.0	29
52	57.9	30
53	61.8	31
54	65.5	32
55	69.2	34
57	75.8	36
59	81.6	37
62	88.5	40
66	94.5	44
73	98.9	51
81	99.9	59

Bachelor **Example 5.7** Intellectual development between 6 and 15 years

As an example for the subtest *Everyday Knowledge* of the intelligence test battery AID 2, we want to investigate the course of the (average) test scores regarding

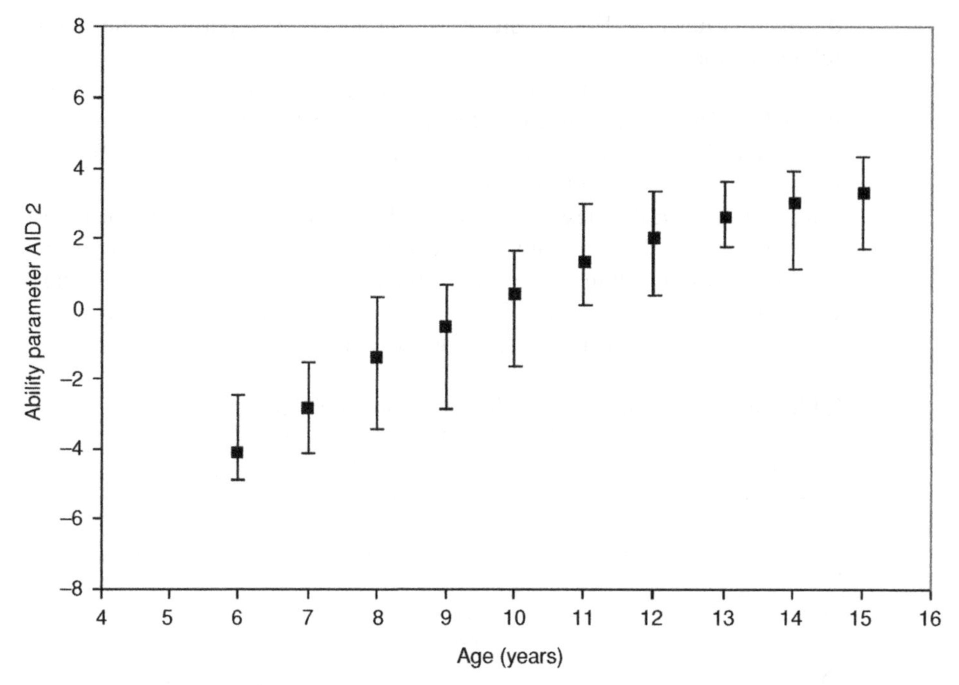

Figure 5.21 Average ability parameter (according to the Rasch model) in the subtest *Everyday Knowledge* of AID 2 across the ages (male standardization sample 1995–1997; taken from Kubinger & Wurst, 2000, p. 68); the average ability parameters per age correspond to a T-Score of 50, the given interval limits correspond to the T-Scores of 41 and 59.

the age. In doing so, the test scores are the so-called 'ability parameters', as they result for each testee when a psychological test is calibrated in accordance with the Rasch model (see in detail in Section 15.2.3.1). Section 2.2.1 has already indicated that certain statistical methods of psychometrics allow the generation of interval-scaled test scores that indeed adequately represent the underlying empirical relationships between different testees (see for details e.g. Kubinger, 2009b). Figure 5.21 shows the course of intellectual development on the basis of one of the two parts of the data from the standardization sample; namely the male standardization sample. Besides the average ability parameters, those ability parameters that belong to the cumulative frequency of 18.4% and 81.6% in the sample, which correspond to T-Scores of 41 and 59, have also been entered.

A clear trend can be recognized, whereby the slope of the curve flattens across the age with time. Additionally, the standardization sample seems to have, in fact, certain irregularities with respect to the population – this population certainly cannot be taken by census. When establishing a standardization table according to Table 5.6, it is therefore advisable to 'smooth' the observed curve. One way of smoothing is the application of a regression analysis, which is described in Chapter 11; the technique of smoothing, however, cannot be discussed here in further detail.

5.3 Statistics

Statistics, or more precisely 'descriptive' statistics are quantities that characterize or represent a set of numbers. We usually use the term 'statistic' for the case of statistically describing samples (of size n). In contrast, we use the term 'parameter' when it is a matter of a statistical description of (empirical and theoretical) populations (of size N; see further in Section 6.4).

For quantitative characters, the calculation of a measure of location and a measure of scale is mostly sufficient in order to characterize the data. The mean, as the most popular measure of location, is often calculated without any special statistical knowledge. Its corresponding measure of scale, which here describes the variability of the outcomes, is the square root of the *variance*, that is the *standard deviation*.

5.3.1 Mean and variance

In a finite population (as well as for a theoretical distribution), we use the symbol μ for the mean; in a sample we use the symbol $\bar{y}$. In the population, we call the observations or, put more precisely, the (theoretically) observable outcomes (i.e. the character's measurement values realized by the individual population units) Y_v, $v = 1, 2, \ldots, N$. In the sample, the observations are denoted by y_v, $v = 1, 2, \ldots, n$. By using the symbol $\sum$ (sigma sign) for summing up several values, we obtain the Formulas (5.1) and (5.2): below the sigma sign, the current index is specified with its smallest value; above the sigma sign, the largest value of the index is specified. The sigma sign is to be read as follows: 'create the term after the sigma sign for all values of the index (here: v) and sum up all elements of the resulting sequence'.

$$\mu = \frac{1}{N}\sum_{v=1}^{N} Y_v = \frac{1}{N}(Y_1 + Y_2 + \cdots + Y_N) \tag{5.1}$$

$$\bar{y} = \frac{1}{n}\sum_{v=1}^{n} y_v = \frac{1}{n}(y_1 + y_2 + \cdots + y_n) \tag{5.2}$$

Although the two formulas look different, the calculation is entirely equal in terms of concrete numerical values.

This is not the case as concerns the variance. In contrast to the usual definition for a sample (see below), in the population the variance is defined as the arithmetic mean of the squared difference of the individual outcomes from the mean (of the population) – we use the symbol σ^2:

$$\sigma^2 = \frac{1}{N}\sum_{v=1}^{N}(Y_v - \mu)^2 \tag{5.3}$$

Master In the first instance it may see completely incomprehensible for laypersons, as to why the arithmetic mean of the squared difference of the individual outcomes from the mean is chosen in statistics as the function of a measure of scale. It would be a lot more straightforward to average the absolute values of the differences of the outcomes from the mean: $\sum |Y_v - \mu|/N$. However, squaring the differences

$(Y_v - \mu)$ has, in any case, the same function: namely, to ignore the direction of the deviation from the mean. Furthermore, it is precisely the variance defined above that describes the normal distribution which is so essential for statistics (see in detail in Section 6.2.2). Finally, many derivations in mathematical statistics are a lot easier to obtain by using squared differences.

The variance has the squared measurement unit of the outcomes themselves as the unit of measure or the dimension. This is not very easy to interpret; for example, in the case of the character intelligence quotient, the measurement unit of the variance would be: the squared intelligence quotient. Therefore, it is better to use the (positive) square root of the variance for the description of data, i.e. the standard deviation. The standard deviation of a population is denoted by σ.

Master Figure 5.22 shows the histogram for the population's test scores in a fictional psychological test. In this example it is possible to visually extract the mean, without having to calculate anything. And hence intuitively it can clearly be seen that the mean μ does in actual fact describe the observations typically: the position of the recorded area of outcomes is clearly localized with it.

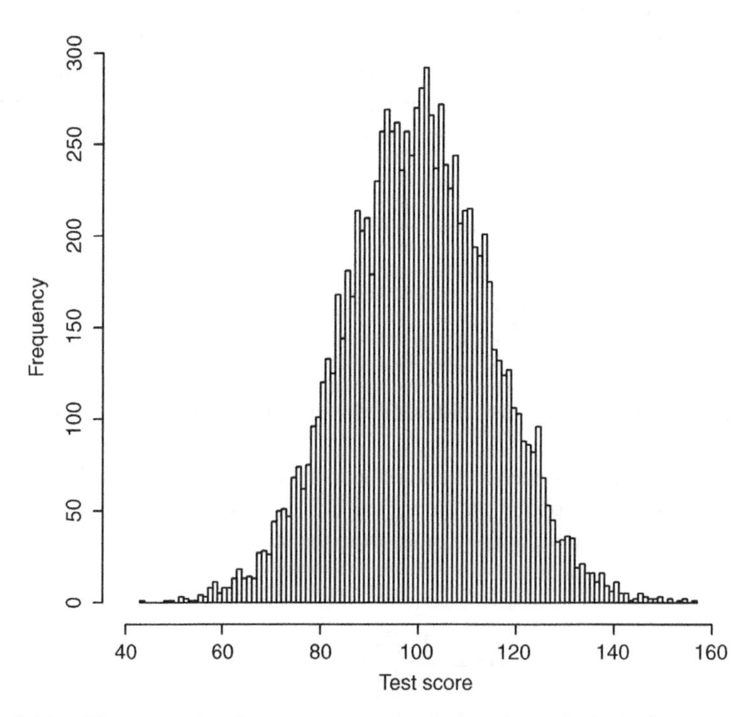

Figure 5.22 Histogram for the test scores of a fictional psychological test ($\mu = 100$).

Master The physical analogue to the mean is the centroid of a geometrical object or of a mass distribution: the object can be balanced (kept in balance) at this point. Correspondingly, the set of all outcomes in Figure 5.22 is 'in balance' at the point

$\mu = 100$. Additionally it can also be seen from this figure how much the standard deviation contributes to the characteristic of the distribution of the outcomes. The standard deviation ascertains the extent of dispersion of the outcomes around the mean. Thus, a population can be described very accurately with two statistics: a measure of location and a measure of scale; given a normal distribution (see Section 6.2.2) it can in fact be described completely with these two statistics. Vice versa, it is also quite clear from these considerations that the description of a population using the mean, alone, is completely inadequate. A classic empirical example at the turn of the nineteenth to the twentieth century shows that although the approximate average age of death was more or less the same for men and women, the dispersion of the age of death was much larger for women; i.e. a relatively large number of women died very young (mainly as a result of childbirth fever), even though quite a lot of women also became very old – much older than most men became.

The physical analogue to the variance in statistics is the moment of inertia of a rotating object. That is the extent of the resistance, provided by a rotating object against the change of the rotation speed. It is easy to comprehend that the further away the 'mass' lies from the centroid, the greater the force that is needed to change the rotation speed when an object rotates. For example, the 'runout' deceleration at the end of a 'game of flying' with young children (where the adult spins around on their own axis as quickly as possible several times, holding the child by its hands) lasts much longer if the father instead of the mother plays the game; this is because usually the father has longer arms.

In the case of a sample, the symbol s^2 is used for the variance; however, it is often not defined completely analogously to the variance of the population. If the variance is needed merely for descriptive purposes regarding a sample, then an analogy, similar to $\bar{y}$ instead of μ and n instead of N, i.e. hence s^2 instead of σ^2, would be logical and totally comprehensible for mathematically and statistically lay people. However, if one refers to *inferential statistics* (see Chapter 7), then a slightly different formula can be derived for the sample variance:

$$s^2 = \frac{1}{n-1} \sum_{v=1}^{n} (y_v - \bar{y})^2 \tag{5.4}$$

Master

The divisor $n - 1$ instead of n results from the fact – as we shall see later – that otherwise the mean (expressed statistically: the *expected value*) of all possible sample variances, calculated by drawing samples from the population, would not correspond to the actual variance in the population.

Again, we use the (positive) square root of the variance, $s = \sqrt{s^2}$, as the (sample) standard deviation.

5.3.2 Other measures of location and scale

For several reasons, the mean and the variance are not always appropriate in order to describe the data adequately.

One of the main reasons for this is that both nominal- and ordinal-scaled characters exclude their calculation by definition.

Bachelor This can be understood immediately in the case of nominal-scaled characters, as the different measurement values merely represent different names or qualities. For example, there is nothing to be calculated for the character *color* with regard to the names 'red', 'green', and 'blue'. For ordinal-scaled characters, the different measurement values can only express a rank order, even if they are represented in numbers. The extent of the distances between these numbers is not based on content; however, equality justified by content for numerically equal distances between different measurement values must be guaranteed in order to use mean and standard deviation.

Master In the case of ordinal-scaled characters, even if the measurement values are represented by numbers, the calculation of, for example, the mean is inappropriate; this can easily be seen in the following. Any transformation of measurement values that keeps the empirically given rank order is permissible in the case of ordinal-scaled characters (see Section 4.2). For example, the informational content remains the same, whether we use '1', '2', '3', '4', '5', or we use '1', '4', '9', '16', '25' as school grades: the greater-smaller-equal relations remain the same. But obviously, the relation of the means of two students would change if for the one 9, 9, 16, 4 (mean 9.5) instead of 3, 3, 4, 2 (mean 3.0) is now observed and for the other 1, 1, 25, 16 (mean 10.75) instead of 1, 1, 5, 4 (mean 2.75) is observed.

Another reason is then given if the data of a qualitative character are not distributed more or less symmetrically. In this case, the mean is not typical for all observations. For example, *outliers* in the data – i.e. rather rarely occurring but extremely deviating outcomes – affect both the mean and the variance to an unreasonably severe degree.

Doctor In addition to the statistics described here, further statistics that are robust against outliers have been developed; their introduction at this point would be beyond the scope of discussion (see, however, Reimann, Filzmoser, Garrett, & Dutter, 2008; Hoaglin, Mosteller, & Tukey, 2000; Maronna, Martin, & Yohai, 2006).

Bachelor **Example 5.8** Mean and standard deviation in the case of outliers

Given $n = 10$ employees of a work team, all earn \$2500 except for one who earns \$2000 and another who earns \$3000. The mean, $\bar{y}$, of the salary is therefore \$2500, with a standard deviation of $s = \$235.70$. If one employee with \$2500 is replaced by a new one, who, however, in turn earns \$20 000, then the average income of the team increases to $\bar{y} = \$4250$, of course without the income of the others changing; for the others, the new mean is obviously not appropriate. Also the variability of the income is, with $s = \$5539$, seriously greater now, although the differences seem to be very minimal for nine people.

In the case of ordinal-scaled characters, as well as in the case of a qualitative character which is not distributed very symmetrically, the *median* is an appropriate measure of location. It is based on the outcomes being ranked. The practical approach to ordering the outcomes according to their actual values is quite simple. A *rank* is assigned to each outcome; namely the ranks 1 to n for the outcomes arranged according to their values. This becomes rather difficult if the number of observations is very large. Furthermore, a fundamental problem arises if several research units obtain the same value; that is if there are tied ranks (*ties*). In this case, the same rank is assigned to each observation which has an identical measurement value; namely the mean of all ranks which would have been assigned for these observations if they had to have different values. Statistical methods based on ranking the outcomes carry out this ranking in SPSS and **R** without the user having to do anything. Hence, in practice, it is hardly ever necessary to assign ranks on one's own.

Bachelor We illustrate the assignment of ranks for the grades of the first student mentioned above: 3, 3, 4, 2. The best grade achieved (measurement value) is 2; rank 1 is therefore assigned to this observation. Then, there are two grade 3s; both these observations receive the ranks 2 and 3 in sum. Since we have to assign the same rank to both observations, we calculate the mean of both ranks, i.e. $\frac{2+3}{2} = 2.5$. The outcome 4 therefore corresponds to rank 4.

Master The necessary procedure of ranking and assignment of ranks seems to be complicated when put formally.

Example 5.9 Illustration of ranking without any reference to content

Given that we have character y and $n = 8$ research units with outcomes y_v, $v = 1, 2, \ldots, 8$ as follows: $y_1 = 13$, $y_2 = 3$, $y_3 = 87$, $y_4 = 50$, $y_5 = 0$, $y_6 = 16$, $y_7 = 21$, $y_8 = 55$. We now arrange these outcomes according to their size: 0, 3, 13, 16, 21, 50, 55, 87. In order to know which of the so-arranged outcomes belongs to which observation or research unit, respectively, we apply so-called order statistics: we denote them as $y_{(l)}$, $l = 1, 2, \ldots, 8$; the index is now placed in brackets, and y_v and $y_{(l)}$ must not be confused. Thus, we obtain: $y_{(1)} = y_5 = 0$, $y_{(2)} = y_2 = 3$, $y_{(3)} = y_1 = 13$, $y_{(4)} = y_6 = 16$, $y_{(5)} = y_7 = 21$, $y_{(6)} = y_4 = 50$, $y_{(7)} = y_8 = 55$, $y_{(8)} = y_3 = 87$. The index l in the order statistic $y_{(l)}$ indicates the rank of observation y_v. For example, $y_6 = 16$ has rank 4; thus, 16 is the fourth largest outcome – it is therefore $y_{(4)} = y_6$. Finally, we can also specify the rank (order) statistics $r(y_v)$, $v = 1, 2, \ldots, 8$; they determine which rank corresponds to observation y_v. We obtain: $r(y_1) = 3$, $r(y_2) = 2$, $r(y_3) = 8$, $r(y_4) = 6$, $r(y_5) = 1$, $r(y_6) = 4$, $r(y_7) = 5$, $r(y_8) = 7$; as said before, for example, $y_6 = 16$ receives rank $r(y_6) = r(16) = 4$. If several outcomes are equal, the assignment of ranks gets a little more complicated. This can even occur with continuous characters on account of the rounding. Let us take, as a variation to our numerical example, $y_4 = y_8 = 55$. Let us suppose further that the first five order statistics have already been assigned, so that for $y_4 = y_8$ the order statistics $y_{(6)}$ and $y_{(7)}$ come into question, i.e. ranks 6 and 7. Of course, we do not want to assign two different ranks to the same measurement value. One of several possibilities to avoid such an assignment is the use of the so-called 'mid-rank method' in statistics. It is applicable for any number of observations with equal measurement values: we calculate the mean of all consecutive ranks that would be assigned to the same values, and assign this mean as a rank to the observations in question. In doing this, equal values obtain the same rank. In the given case, we therefore have to assign rank 6.5 for both y_4 and y_8. Moreover, if we had $y_1 = 0 = y_2 = 0 = y_5 = 0$, we would have to assign the rank $\frac{1+2+3}{3} = 2$ to them.

For Lecturers:

One way of avoiding the assignment of different ranks to the same measurement value can be found in certain sports competitions, where athletes with the same performance result receive the same rank; however, this time it is the best available rank. Subsequent ranks are not assigned (according to this approach, there may be two gold medals, silver is not awarded and the third-best receives the bronze medal).

Master

Example 5.10 The ranks of the observations regarding the character *social status* for employed single mothers in Example 1.1

In the following, we are only interested in children (and mothers respectively) with the measurement value 'single' as concerns the character *marital status of the mother*. We want to determine the ranks with regard to the *social status* for these children. This is obviously only possible if the category 'single mother in household' is disregarded; then *social status* is in fact, as already mentioned, an ordinal-scaled character.

In **R**, we again use the character `socialnew` from Example 5.3, where the measurement value `'single mother in household'` is already filtered out. Next, we type

```
> rank(socialnew[marital_mother == "never married"], na.last = NA)
```

i.e. we use the character `socialnew` as the first argument in the function `rank()`, but restrict the research units to those cases with mothers who have never married, with `[marital_mother == "never married"]`. Additionally, using the argument `na.last = NA`, we exclude all missing values (due to `'single mother in household'`) from the analysis.

As a result, we get:

```
[1] 3.5 6.0 1.0 3.5 3.5 3.5
```

Now we ascertain the respective positions of the research units in question within the data set. We type

```
> which(socialnew != "NA" & marital_mother == "never married")
```

i.e. we use the function `which()` to select those research units which do not have a missing value in the character `socialnew` (defined using the operator '`!=`') and also (because of the operator '`&`') do have an unmarried mother.

This yields:

```
[1] 13 14 31 34 48 77
```

For better readability, we have summarized the results in Table 5.7.

To ascertain ranks in SPSS, we first need to select those children with unmarried mothers. Therefore, we follow the procedure from Example 5.4 (Data – Select Cases...) to open the window shown in Figure 5.15. Now we define the condition marital_mother = 1 and click Continue followed by OK. Next, we apply

Transform
  Rank Cases...

and in the resulting window (not shown here), we move the character socialnew from Example 5.3 to the field Variable(s):. A click on OK creates the new character Rsocialn (see the last column in SPSS Data View – not shown here).

For better readability, we have summarized the results in Table 5.7.

Table 5.7 The ranks of observations in the character *social status* for employed single mothers in Example 1.1.

Observation y_i (with i from the file)	Order statistic	Rank statistic
$y_{31} = 1$	$y_{(1)} = 1$	1
$y_{13} = 2$	$y_{(2)} = 2$	3.5
$y_{34} = 2$	$y_{(3)} = 2$	3.5
$y_{48} = 2$	$y_{(4)} = 2$	3.5
$y_{77} = 2$	$y_{(5)} = 2$	3.5
$y_{14} = 3$	$y_{(6)} = 3$	6

The median is therefore determined as follows: it is just that measurement value which corresponds to the rank located exactly in the middle of all ranks; thus, for n observations, this is rank $\frac{n+1}{2}$. If n is odd, the median takes on a specific outcome. If n is even, the median is located between two outcomes. In the latter case, any value between the two given measurement values is proper; in general, however, the mean of the two values is chosen. Hence, the measurement value above and below which each half of the observations is located, respectively, is considered as typical for a distribution: there are just as many smaller outcomes than the median as there are larger ones.

Master Formally, the median is defined as follows: it corresponds to the *order statistic* which is located exactly in the middle, i.e.

$$Md = y_{\left(\frac{n+1}{2}\right)} \text{ if } n \text{ is odd, and } Md = 0.5 \cdot \left(y_{\left(\frac{n}{2}\right)} + y_{\left(\frac{n}{2}+1\right)}\right) \text{ if } n \text{ is even} \tag{5.5}$$

Usually, the *interquartile range* or the *semi-interquartile range* is assigned to the median as a measure of scale. However, this is only appropriate in the case of a quantitative character. No adequate differences as regards content can be generated for ordinal-scaled characters; thus, a measure of scale is pointless for them by definition. On the other hand, the interquartile range proves to be quite useful if the distribution of the data is not at all symmetrical.

The interquartile range *IQR* is also based on the ranking of the observations. It is defined illustratively as the difference between the third and first *quartile*; in analogy to the median, the values of the quartiles refer to one-quarter or three-quarters of the outcomes' values located below and three-quarters or one-quarter above. Nowadays, it is hardly ever calculated by hand. The semi-interquartile range is $\frac{IQR}{2}$.

Master Formally, the interquartile range has to be determined by the so-called *P*-quantiles (see Section 6.3). These are limits of intervals of the order *P* on the number line, according to the theoretically possible measurement values of a character. With these *P*-quantiles the number line is divided into exactly $Q = \frac{1}{P}$ non-overlapping intervals, with each of them containing the same number of observations (namely $\frac{n}{Q}$). If n is not divisible as a whole by Q, an almost uniform partition of the observations is aimed for. That is, the number of elements in each of the Q intervals is $\frac{n}{Q} = nP$ or $\frac{n}{Q}$ is the nearest, smallest integer to nP. The jth *P*-quantile, $q(j; P), j = 1, 2, \ldots, Q-1$, is for integer values of $\frac{n}{Q}$ defined as

$$q(j; P) = \frac{y_{(jQ)} + y_{(jQ+1)}}{2} \tag{5.6}$$

For a non-integer value of $\frac{n}{Q}$, the *P*-quantile is determined by $q(j; P) = y_{(z)}$, with z being the largest integer number which is smaller than jQ. Thus, the interquartile range can be described as $IQR = q(3; 0.25) - q(1; 0.25)$.

Moreover, the median corresponds to the value $q(2; 0.25)$.

The interquartile range can be interpreted as that maximum difference of outcomes which is shown by 50% of the sample; namely for those research units with ranks located more in the middle. It has the same unit of measurement as the character.

Master **Example 5.9 – continued**

If we had had an odd number of observations, e.g. $n = 11$, then the median would have resulted as the order statistic $\frac{n}{2} = 5.5$; that is the value $y_{(6)}$. In our example with $n = 8$, however, any value between $y_{(4)}$ and $y_{(5)}$ is the median; hence, it is not that obvious, and generally it is defined analogously to Formula (5.6); thus in the example,

$$Md = q\,(1;\ 0.5) = q(2;\ 0.25) = \frac{y_{(4)} + y_{(5)}}{2}$$

Since $y_{(4)} = 16$ and $y_{(5)} = 21$, we obtain: $Md = \frac{16+21}{2} = 18.5$.

For the quartiles we need the quantiles of $P = 0.25$; Q is thus 4. As n is divisible as a whole by 4, we proceed according to Formula (5.6). It gives $q\,(1;\ 0.25) = \frac{y_{(2)} + y_{(3)}}{2} = \frac{3+13}{2} = 8$ and $q\,(3;\ 0.25) = \frac{y_{(6)} + y_{(7)}}{2} = 52.5$. Thus, the interquartile range is $52.5 - 8 = 44.5$.

Example 5.11 Do children with German as a native language differ from children with Turkish as a native language regarding *social status* (Example 1.1)?

As already mentioned several times, the character *social status* becomes ordinal-scaled by excluding the category 'single mother in household'. We now want to characterize both groups of mothers and children, respectively, shortly and concisely regarding the above-mentioned character, by using the median. Finally, we compare the results of the samples of children with German as a native language and children with Turkish as a native language.

In **R**, we type

```
> tapply(as.numeric(socialnew), native_language, median,
+         na.rm = TRUE)
> tapply(as.numeric(socialnew), native_language, quantile,
+         probs = c(0.25, 0.5, 0.75), type = 6, na.rm = TRUE)
```

i.e. we again apply the function `tapply()`, using the character `socialnew` as the first argument, which we first convert to a numeric value using the function `as.numeric()`. As a second argument, we use the character `native_language` for a comparison of both German and Turkish with respect to the functions `median()` or `quantile()` as a third argument. Finally, we exclude missing values from the analysis by setting `na.rm = TRUE`. Furthermore, we specify the function `quantile()` by defining the respective P-quantiles with `probs = c(0.25, 0.5, 0.75)`, and setting the proper algorithm `type = 6`.

As a result, we get:

```
German Turkish
     2       3

$German
   25%    50%    75%
  1.75   2.00   3.00

$Turkish
25%  50%  75%
  2    3    4
```

In SPSS, we proceed analogously to Example 5.4 (Data - Select Cases...) until opening the window shown in Figure 5.14. There we select All cases to again analyze the total sample. In order to ascertain the median and the interquartile range for the children with a native language of German on the one hand and Turkish on the other, we first need to split these groups in the analysis. To do this, we use

Data
    Split File...

and select Compare groups from the window shown in Figure 5.23. Then we mark the character native language of the child and move it to the panel Groups Based on:. After clicking OK, all subsequent analyses will be applied to two separate groups defined by the children's native language.

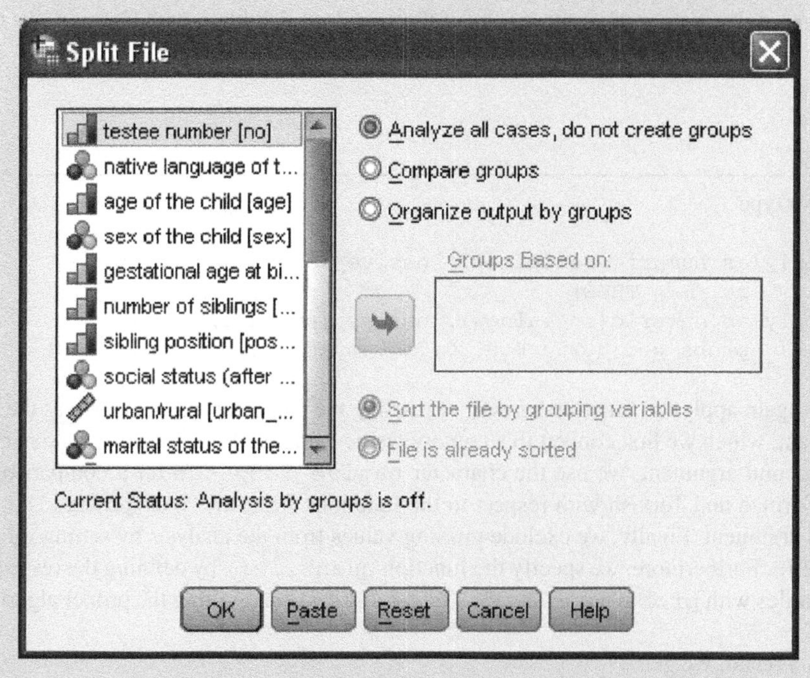

Figure 5.23 SPSS-window for splitting data into groups.

Now we apply the same sequence of commands (Analyze – Descriptive Statistics – Frequencies...) as in Example 5.2. In the resulting window (Figure 5.4), we move the character socialnew to the panel Variable(s):. A click on Statistics... opens the window shown in Figure 5.24, where we tick Median and Quartiles in the fields Central Tendency and Percentile Values, respectively. A click on Continue and OK yields the results shown in Table 5.8.

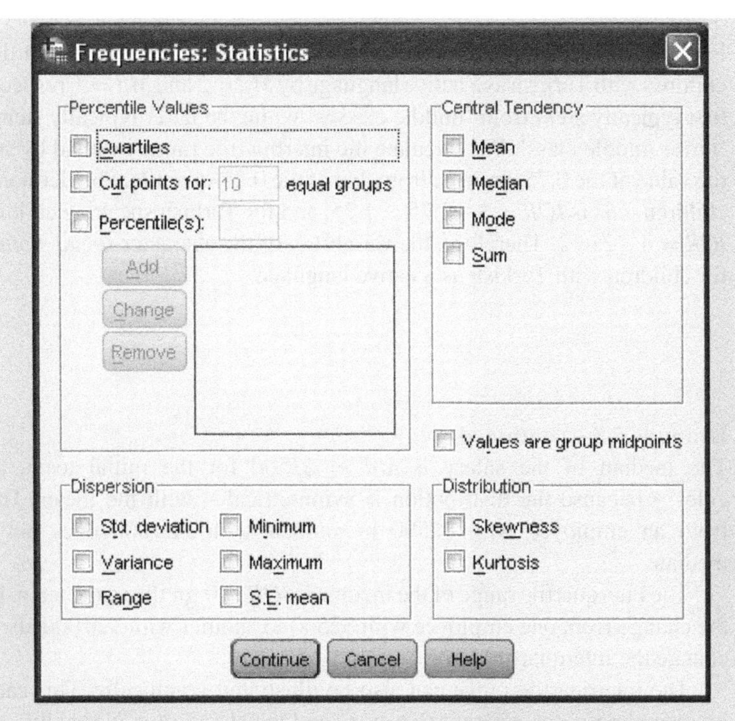

Figure 5.24 SPSS-window Frequencies: Statistics.

Table 5.8 SPSS-output showing median and the 0.25 as well as the 0.75 quantile of the character socialnew for children with German as a native language in contrast to those with Turkish in Example 1.1.

Statistics

socialnew

German	N	Valid	42
		Missing	8
	Median		2.0000
	Percentiles	25	1.7500
		50	2.0000
		75	3.0000
Turkish	N	Valid	46
		Missing	4
	Median		3.0000
	Percentiles	25	2.0000
		50	3.0000
		75	4.0000

Regarding the median, children with German as a native language differ from children with Turkish as a native language by $Md = 2$ and $Md = 3$, respectively; the first typically stem from 'middle classes' while the latter typically stem from the 'lower middle class'. We calculate the interquartile range by hand by subtracting the value of the 0.75-quantile from that of the 0.25-quantile. For German speaking children, this is $IQR = 3 - 1.75 = 1.25$, and for Turkish speaking children, this is $IQR = 4 - 2 = 2$. Therefore, the variability of the character *social status* is larger for children with Turkish as a native language.

Bachelor **Example 5.8 – continued**

The median of the salary is $Md = \$2500$ for the initial team, and coincides – because the distribution is symmetrical – with the mean. The change from an employee with \$2500 to another with \$20 000 does not alter the median.

The interquartile range of the income is $IQR = 0$ for the initial team. Logically, the change from one employee with \$2500 to another with \$20 000 also does not change the interquartile range.

The interquartile range can also be illustrated graphically. This can be done by extending a dot diagram to a *box-(and-whiskers)-plot*. Normally, this is particularly illustrative if a character has to be compared with regard to the median and the interquartile range as concerns different levels of a factor, i.e. another character.

Bachelor **Example 5.5 – continued**

We now want to construct a boxplot for the character *Everyday Knowledge, 1st test date*, still classified according to the four ages and limited to the sample of children with German as a native language.

In **R**, we type and enter

```
> boxplot(sub1_t1.ger ~ age.ger, main = "Boxplot", xlab = "age",
+         ylab = "T-Score")
```

i.e. we apply the function `boxplot()` and use the '~' command within the first argument to show that the character *Everyday Knowledge, 1st test date* (`sub1_t1`) should be analyzed in reference to the character *age of the child* (`age`). The further arguments `main`, `xlab`, and `ylab` set the title of the plot and label the axes.

As a result, we get the plot shown in Figure 5.25.

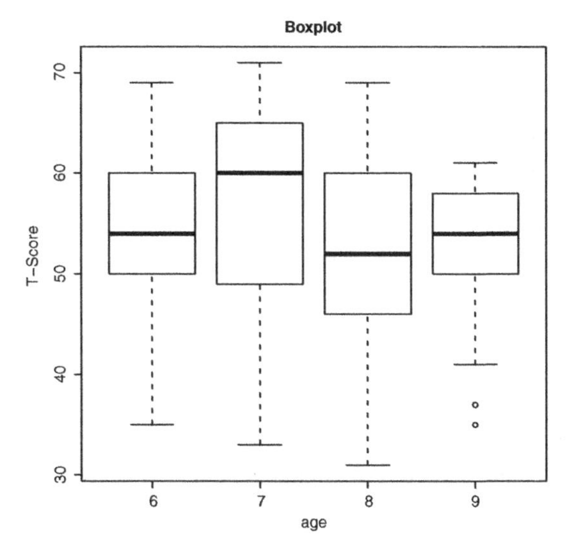

Figure 5.25 R-output showing a boxplot of the character *Everyday Knowledge, 1st test date* for the children with German as a native language (Example 1.1).

In SPSS, we apply the same sequence of commands (Data - Split File...) as in Example 5.11 to reach the window shown in Figure 5.23, where we select Analyze all cases, do not create groups. Now we proceed analogously to Example 5.2 (Graphs - Chart Builder...) until we reach the SPSS Chart Builder (Figure 5.5). There we choose Boxplot in the Gallery tab and drag and drop the symbol Simple Boxplot (on the left side) into the Chart preview. Then we move the character Everyday Knowledge, 1st test date to the field Y-Axis? and age of the child to the field X-Axis?. Finally, we click OK to obtain a plot similar to Figure 5.25.

In the boxplot, the minimum and maximum observed measurement values of *Everyday Knowledge, 1st test date* restrict the vertical line per level of the factor *age of the child*. Between these two boundaries, a rectangle is shown, basically of any width, whose length is derived from the interquartile range. A line parallel to the abscissa represents the median within these rectangles. Particularly extreme outcomes are displayed separately as outliers.

Figure 5.25 can be interpreted as follows: The medians obviously differ only slightly; this is not surprising as the subtest *Everyday Knowledge* is, like all other subtests, standardized in T-Scores per age class, i.e. with a mean of 50 and a standard deviation of 10 (see Figure 5.20, with no substantial differences in the respective means, either). As concerns the interquartile range, there are, in comparison, larger differences – from about 5 T-Scores to about 13 T-Scores per age; according to standardization, the interquartile range should be twice-times two-thirds of the standard deviation, which would thus be about 13 T-Scores. This demonstrates that, in the given sample of six- and eight-years-olds, the performances are less dispersed than expected; this suggests a certain *floor effect*, i.e. only a few children attain the worst possible test scores.

Analogously, boxplots can be created in which the mean is used instead of the median. Additionally the values of the mean plus one-times the standard deviation and mean minus one-times the standard deviation are entered instead of the quartiles.

Bachelor **Example 5.4 – continued**

To obtain a more accurate impression of the distribution, in analogy to quartiles, so-called deciles are to be calculated for the character *Everyday Knowledge, 1st test date*; these deciles divide the number of observations into 10, instead of 4, equally large groups, as would be the case in terms of quartiles.

In **R**, we ascertain the deciles by typing

```
> quantile(sub1_t1.set, probs = seq(from = 0.1, to = 0.9, by = 0.1))
```

i.e. we use the character *Everyday Knowledge, 1st test date* (`sub1_t1`) as the first argument in the function `quantile()` and then define *P*-quantiles (`probs`) by applying the function `seq()`, which is set to generate a sequence of numbers from `0.1` to `0.9`, incremented by `0.1`.

As a result, we get:

```
10%  20%  30%  40%  50%  60%  70%  80%  90%
35.6 46.0 50.0 50.0 54.0 58.0 60.0 60.4 65.0
```

In SPSS, we proceed analogously to Example 5.2 (Analyze – Descriptive Statistics – Frequencies...) to open the window shown in Figure 5.4, where we click on Statistics... This brings us to the window shown in Figure 5.24, where we select Cut points for: (dividing into 10 equally large groups is set by default) and click on Continue, followed by OK. We obtain the result shown in Table 5.9.

Table 5.9 SPSS-output showing the deciles of the character *Everyday Knowledge, 1st test date* for all children except those with Turkish as a native language being tested in Turkish on the 1st test date (Example 1.1).

Statistics

Everyday Knowledge, 1st test date (T-Scores)

N	Valid	74
	Missing	0
Percentiles	10	35.00
	20	46.00
	30	50.00
	40	50.00
	50	54.00
	60	58.00
	70	60.00
	80	61.00
	90	65.00

There are slight differences between **R** and SPSS.

Other measures of location and scale found in psychology are: The *mode* and the *range*. The mode is defined as the measurement value whose observed frequency is greater than those of the two neighboring values – thus distributions may have several modes. Such distributions are called *multimodal distributions*, in contrast to *unimodal distributions*; *bimodal distributions* occur relatively frequently. The range is the difference between the largest and the smallest realized measurement value. It hence describes the maximum difference that is shown by all outcomes. Thus, it is extremely dependent on outliers; a single extreme value changes the range tremendously. Unlike the standard deviation and the interquartile range, it cannot become any smaller by adding other research units. The mode may also be used for nominal-scaled characters. The range does not make any sense in the case of ordinal- (and also nominal-) scaled characters.

Bachelor

Example 5.8 – continued
The mode of the salary is \$2500 for the initial team; the replacement of one employee earning \$2500 with another earning \$20 000 does not change the mode. The range for the initial team is \$1000; the replacement of one employee earning \$2500 with another earning \$20 000 changes the range to \$18 000.

5.3.3 Statistics based on higher moments

So far, we have seen that the description of quantitative data by the mean and standard deviation (variance) possibly works well. However, problems arise if the distribution of the outcomes varies considerably from a normal distribution. In order to quickly identify such a case without any graphical illustration, the use of two further statistics is recommended, namely a *measure of skewness* and a *measure of kurtosis*. The two best-known statistics of such measures are simply called *skewness* and *kurtosis*, respectively.

Master
Doctor

The skewness quantifies the asymmetry of a distribution with respect to the mean; the kurtosis quantifies whether the distribution has a higher or a flatter peak in relation to the normal distribution. Both statistics emerge from the so-called k^{th} *order central moments* – that is, essentially, the sum of the differences between each outcome's value and the mean raised to the kth power ($k = 1, 2, \ldots$) – more precisely, the arithmetic mean of it. The skewness γ_1 in the population and the skewness g_1 in the sample, respectively, are based on the third-order central moment and refer to the cube of the standard deviation. In the ideal case of a symmetrical distribution, γ_1 and g_1, respectively, equal 0. For instance, as concerns the normal distribution, $\gamma_1 = 0$. However, a skewness of zero does not necessarily mean that the distribution is symmetrical. In the case of $\gamma_1 > 0$ and $g_1 > 0$, respectively, we talk about a *positive skewness*, i.e. a *right-skewed distribution*, which we can also call a *left-steep distribution*. In the case of $\gamma_1 < 0$ and $g_1 < 0$, respectively, we talk about a negative skewness, which is also called a left-skewed or a right-steep distribution.

Example 5.12 The skewness of the two characters *gestational age at birth* and *number of siblings* in Example 1.1

We can design a histogram and a bar chart, respectively, for both characters of interest, analogously to Example 5.4 (see Figure 5.26).

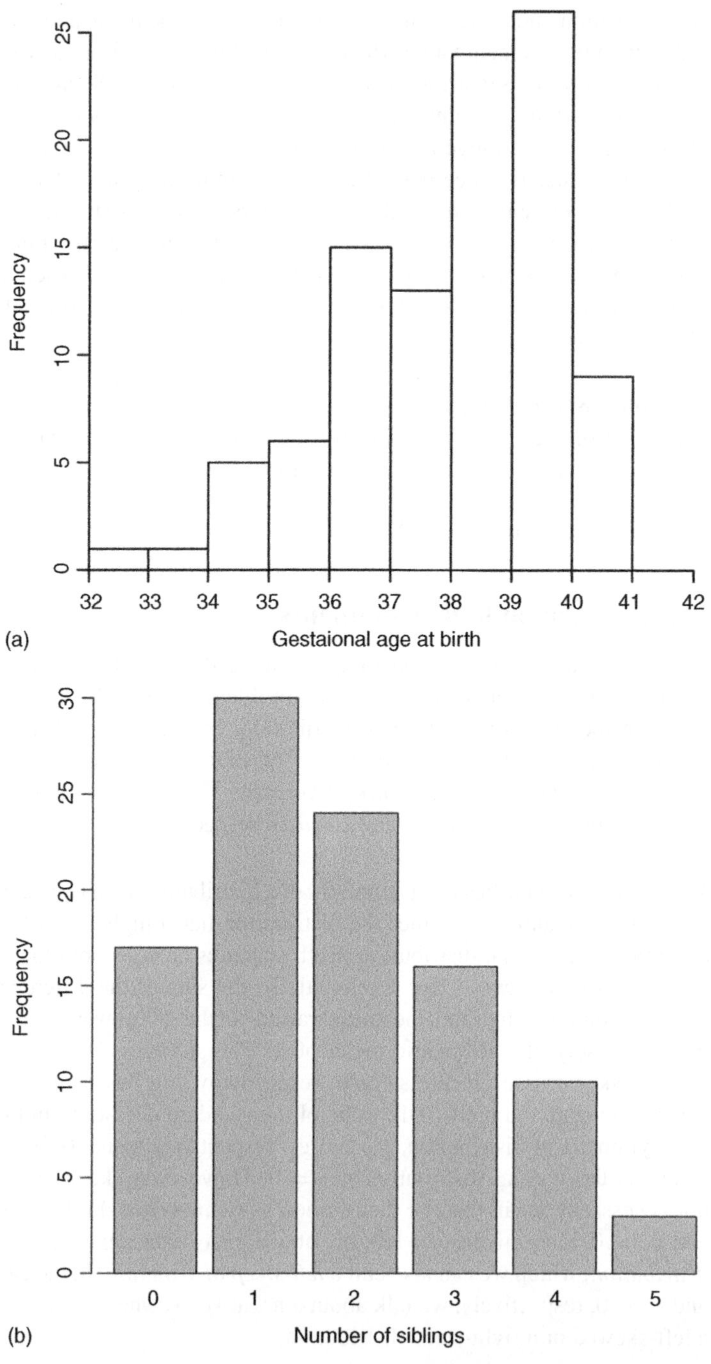

Figure 5.26 (a) The histogram for the character *gestational age at birth* in Example 1.1; (b) the bar chart for the character *number of siblings* in Example 1.1.

The character *gestational age at birth* has a left-skewed distribution; the character *number of siblings* a right-skewed distribution.

The kurtosis γ_2 in the population, and also the kurtosis g_2 in the sample, are based on the fourth-order central moment and refer to the standard deviation to the power of 4 – and additionally, are reduced by the value 3; this last adjustment gives the value of 0 in the case of a normal distribution. Above all, the kurtosis reveals if there are more or fewer outcomes' values at the ends of a distribution in comparison to the case of a normal distribution. It describes the 'concentration' of the outcomes' values around the points $\mu + \sigma$ and $\mu - \sigma$ or $\bar{y}+ s$ and $\bar{y}- s$, respectively; the turning points of the curve of the normal distribution are located at these positions (see Section 6.2.2).

Bachelor

Example 5.4 – continued
For an even better characterization of the distribution, both the skewness and the kurtosis will be calculated for the character *Everyday Knowledge, 1st test date*.

In **R**, we apply the package `QuantPsyc`, which we load after its installation (see Chapter 1) using the function `library()`. Next, we type

```
> norm(sub1_t1.set)
```

i.e. we apply the function `norm()` to the character *Everyday Knowledge, 1st test date*.
As a result, we get (shortened output):

```
            Statistic
Skewness -0.72621310
Kurtosis  0.03147714
```

In SPSS, we proceed analogously to Example 5.2 (Analyze – Descriptive Statistics – Frequencies...) to open the window shown in Figure 5.4, where we click the button Statistics... In the resulting window (Figure 5.24), we select Skewness as well as Kurtosis, click on Continue and OK, and obtain the results $g_1 = -0.726$ and $g_2 = 0.031$.

As the skewness is, in absolute value, smaller than 1 and also besides this it is negative, a slightly left-skewed distribution is given.

In a quantitative character with a multimodal distribution, all measures of location, i.e. mean, median, and mode(s), are not at all typical for the given outcomes – fortunately, such distributions are rare in psychology. For unimodal distributions which are, however, asymmetrical, the mean is located to the right of the median if there is a right-skewed distribution, and if there is a left-skewed distribution the mean is located to the left of the median. Therefore, the median, rather than the mean, is then more typical for the outcomes. The same applies in the case of outliers.

Summary
Data can be analyzed with *statistics*. Above all, one should distinguish between *measures of location* and *measures of scale*. For nominal-scaled characters, only a single measure of location is appropriate, i.e. the *mode*. For ordinal-scaled characters, the *median* is also an appropriate measure. Measures of scale are solely designed for quantitative characters. *Standard deviation* (and *variance*) are associated with the *mean*; the *interquartile range* is associated with the median. All measures of location mentioned are more suited for unimodal distributions. If the (unimodal) distribution is asymmetrical, the median should be favored instead of the mean; the same applies if there are outliers.

5.4 Frequency distribution for several characters

If two qualitative characters are to be statistically analyzed together, the first with r, the second with c measurement values (categories), then there are $r \cdot c$ combinations of categories that can be illustratively contrasted with each other in a table – for example with regard to the observed frequency in the sample (or the population). The (two-dimensional) table then consists of $r \cdot c$ cells. In the case of more than two characters, there are multivariate frequency tables.

Such tables with $r \cdot c$ cells might even be designed if one of both characters is quantitative.

Bachelor **Example 5.13** Two-dimensional frequency table for the characters *native language of the child* and *number of siblings* in Example 1.1

The character *native language of the child* is a nominal-scaled, namely a dichotomous character; the character *number of siblings* is a quantitative character with a very limited number of measurement values.

In **R**, we type

```
> addmargins(table(no_siblings, native_language))
```

i.e. we use the characters *number of siblings* (`no_siblings`) and `native_language` as arguments for the function `table()`. The function `addmargins()` demands the inclusion of row and column totals in the resulting table.

This yields a result identical to Table 5.10.

A grouped bar chart works well for graphical illustration of the frequency distribution of two qualitative characters; hence we enter the commands

```
> barplot(table(no_siblings, native_language), beside = TRUE,
+         main = "Grouped bar chart",
+         xlab = "native language of the child",
+         ylab = "number of siblings",
+         ylim = c(0, 20), axisnames = FALSE, legend = TRUE)
> mtext(0:5, side = 1, line = 0.15, at = 1.5:6.5)
> mtext(0:5, side = 1, line = 0.15, at = 8.5:13.5)
> mtext("German", side = 1, line = 1.5, at = 4)
> mtext("Turkish", side = 1, line = 1.5, at = 11)
```

i.e. we apply the function `barplot()` to the function `table()`, defining that the latter should generate a two-dimensional frequency using the characters *number of siblings* (`no_siblings`) and `native_language`. The other arguments of `barplot()` ensure that that the frequencies will be shown next to one another (`beside = TRUE`), define the chart title (`main`), and label the axes (`xlab` and `ylab`). Finally, `ylim` fixes the presented range of measurement on the ordinate, `axisnames = FALSE` repress the labeling of the axes according to the character *native language of the child*; the last argument, `legend = TRUE`, produces a legend with default settings. We use the function `mtext()` to label the abscissa (`side = 1`), namely with the values 0 to 5 (`0:5`) for the number of siblings as well as `"German"` and `"Turkish"` for the native language of the child; with `at =` we set the labels to the pertinent position of the abscissa.

As a result, we get the chart shown in Figure 5.27.

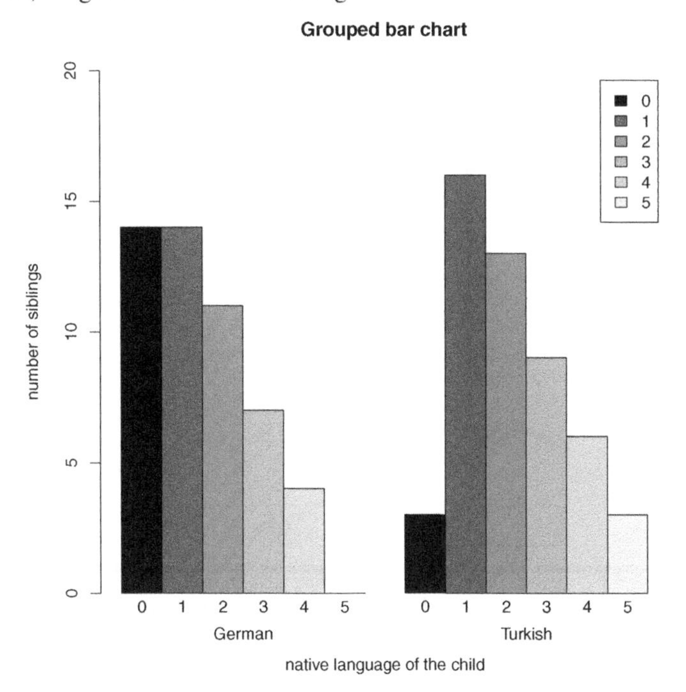

Figure 5.27 **R**-output of the graphical illustration of a two-dimensional frequency distribution in Example 5.13.

In SPSS, we use the sequence of commands

Analyze
    Descriptive Statistics
        Crosstabs...

to obtain the window shown in Figure 5.28. At this point, the decision needs to be made as to which of the two characters in question should appear in the rows and which in the columns. It is advisable to place the character with the larger number of measurement values in rows. Thus, we move the character number of siblings to the panel Row(s): and native language of the child to Column(s):. By clicking OK, we obtain the results shown in Table 5.10.

To graphically illustrate the frequency distribution of two qualitative characters, we use the sequence of commands already known from Example 5.2 (Graphs - Chart Builder...) in order to open the SPSS Chart Builder (Figure 5.5). There we select Bar in the Gallery tab and drag and drop the symbol Clustered Bar into the Chart Preview. Then we move the character native language of the child to the field X-Axis? and the character number of siblings to Cluster on X: set color. Finally, we click OK and obtain a chart similar to that shown in Figure 5.27.

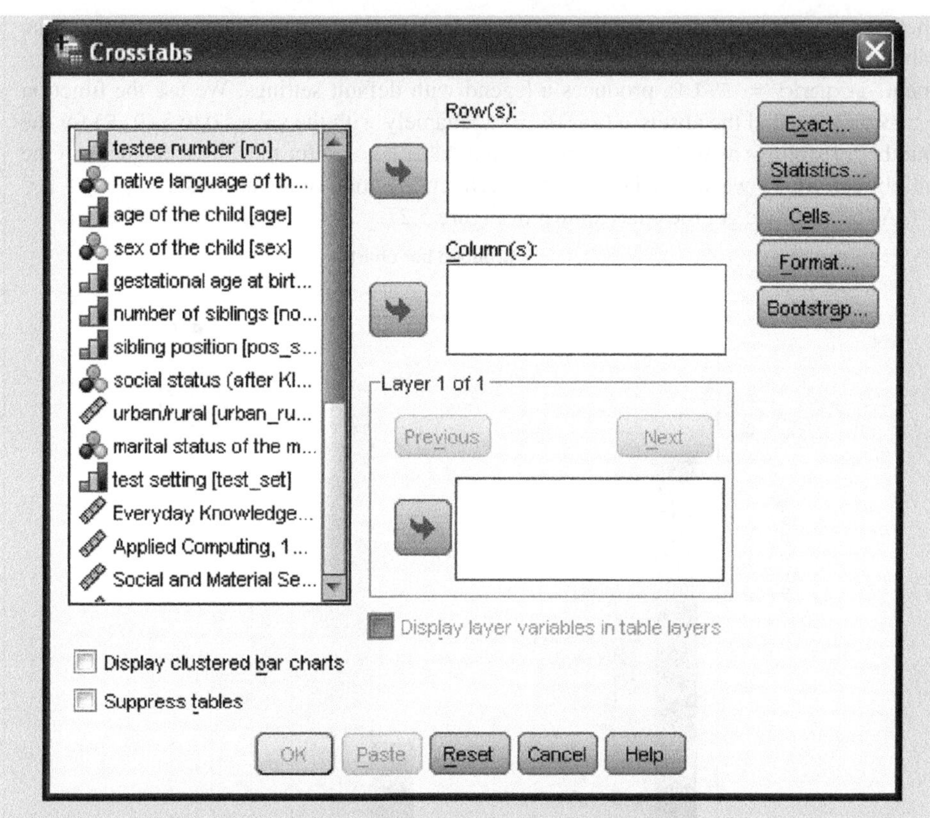

Figure 5.28 SPSS-window for the creation of a two-dimensional frequency table.

Table 5.10 SPSS-output showing a two-dimensional frequency table of the characters *native language of the child* and *number of siblings* (Example 1.1).

number of siblings * native language of the child Crosstabulation

Count

		native language of the child		Total
		German	Turkish	
number of siblings	0	14	3	17
	1	14	16	30
	2	11	13	24
	3	7	9	16
	4	4	6	10
	5	0	3	3
Total		50	50	100

A trend can be noticed that children with Turkish as a native language tend to have more siblings than children with German as a native language. How such an impression can be quantified and evaluated with respect to its statistical importance will be discussed in Chapter 11.

References

Hoaglin, D. C., Mosteller, F., & Tukey, J. W. (2000). *Understanding Robust and Exploratory Data Design*. New York: John Wiley & Sons, Inc.

Krafft, O. (1977). Statistische Experimente: Ihre Planung und Analyse [Statistical experiments: planning and analysis]. *Zeitschrift für Angewandte Mathematik und Mechanik*, *57*, 17–23.

Kubinger, K. D. (2009a). *Adaptives Intelligenz Diagnostikum - Version 2.2 (AID 2) samt AID 2-Türkisch* [Adaptive Intelligence Diagnosticum, AID 2-Turkey Included]. Göttingen: Beltz.

Kubinger, K. D. (2009b). *Psychologische Diagnostik – Theorie und Praxis psychologischen Diagnostizierens* (2nd edn) [Psychological Assessment – Theory and Practice of Psychological Consulting]. Göttingen: Hogrefe.

Kubinger, K. D. & Wurst, E. (2000). *Adaptives Intelligenz Diagnostikum - Version 2.1 (AID 2)* [Adaptive Intelligence Diagnosticum – Version 2.1 (AID 2)]. Göttingen: Beltz.

Maronna, R. A., Martin, D. R., & Yohai, V. J. (2006). *Robust Statistics: Theory and Methods*. New York: John Wiley & Sons, Inc.

Reimann, C., Filzmoser, P., Garrett, R. G., & Dutter, R. (2008). *Statistical Data Analysis Explained*. New York: John Wiley & Sons, Inc.

Part III

INFERENTIAL STATISTICS FOR ONE CHARACTER

Descriptive statistics provides the basis for inferential statistics. The latter tries to infer from the sample data to the respective population. For this, further background knowledge is needed: that is, *probability theory* as well as sampling procedures based on chance.

6

Probability and distribution

In this chapter the terms probability, random variable, parameter, and theoretical distribution are introduced, and the binomial and the normal distribution are described. Further, we discuss the method of estimation of a parameter.

6.1 Relative frequencies and probabilities

Random sampling means to take elements from a population in such a way that each element has an equal *probability* of being chosen; in other words: each element has an equal chance of being sampled.

The term 'probability' has to be defined precisely in the following, due to the fact that in everyday language 'probable' is often used quite differently, mostly in the sense of a relatively low uncertainty; for instance, 'It will probably rain today.'

As already introduced in Chapter 4, when throwing a die it is easy to understand that in the long run the outcome 6 occurs in one-sixth of the cases; more exactly we have a relative frequency for the result 6 of near to $\frac{1}{6}$. The difference between the relative frequency and $\frac{1}{6}$ decreases the more often we throw the die. Relative frequencies have, in many practical cases, the property that their values differ strongly for different samples and small values of n. However, for larger values of n they differ only slightly. Formally we say, with growing $n \to \infty$ or $n \to N$, that the relative frequencies of an event E_k fluctuate decreasingly; they are nearly constant.

Statistics in Psychology Using R and SPSS, First Edition. Dieter Rasch, Klaus D. Kubinger and Takuya Yanagida.

Bachelor **Example 6.1** The relative frequency of 'number of heads' for tossing a coin 10 and 100 times

If we toss a coin $n = 10$ times in d runs, it may happen that even the event 'number of heads: (only) $k = 2$ times' sometimes occurs or, vice versa, the event 'number of heads: $k = 8$ times'. The results of 'number of heads: $k = 3$ times' and 'number of heads: $k = 7$ times' will occur more frequently, and very often we will find the event 'number of heads: $k = 4$ times', 'number of heads: $k = 5$ times', or 'number of heads: $k = 6$ times'. If we do the same with $n = 100$ tosses, then the events 'number of heads: $k = 20$ times' and 'number of heads: $k = 80$ times' each occur extremely seldom, but relatively often will the events 'number of heads: $k = 40$ times' through 'number of heads: $k = 44$ times' occur, or the events 'number of heads: $k = 56$ times' through 'number of heads: $k = 60$ times'. And very clearly the most frequently realized events will be 'number of heads: $k = 45$ times' through 'number of heads: $k = 55$ times'. The reader may try this, whereby it may be sufficient to use $d = 20$ runs and instead of $n = 10$ tosses vs. $n = 100$ tosses only $n = 10$ vs. $n = 30$ coin tosses, to find out that the relative frequency of 'number of heads' for both these second cases is fairly constant.

In **R**, we define a new function by typing

```
> sim.bin1 <- function(n, r, prob = 0.5) {
+   h <- table(rbinom(r, size = n, prob = prob))
+   f <- h/r
+   cat("\n", "count obverse", "\n")
+   cat("\n", paste("n =", n, ", r =", r, "\n", "\n"))
+   return(cbind(h, f = round(f, 2)))
+ }
```

i.e. we use the function `function()` and declare the number of tossed coins `n`, the number of runs `r` and the predefined probability of heads `prob = 0.5` as its arguments. The sequence of commands between the braces specifies the algorithm of the new function and won't be further explained at this point. We assign this function to the object `sim.bin1`. Next we type

```
> set.seed(123)
> sim.bin1(n = 10, r = 20)
```

i.e. we set the arbitrary starting number `123` using the function `set.seed()` in order to enable repeatability of the results, and apply the previously created function `sim.bin1()` to toss `n = 10` coins in `r = 20` runs. That is, the line `> set.seed(123)` is only for didactic reasons, so that the reader obtains the same results.

The results, plus those for a simulation with `r = 100` are presented in Table 6.1. In addition to this, Table 6.2 shows the results of a rerun with `n = 30` coin tosses. If the reader retries the program and drops the line `> set.seed(123)`, the results will differ, as when tossing real coins.

Table 6.1 Absolute (h_k) and relative (f_k) frequencies of the occurrence of k times obverse, by $r = 20$ and $r = 100$ runs, each with $n = 10$ coin tosses.

	20 runs		100 runs	
k	h_k	f_k	h_k	f_k
0	0	0	1	0.01
1	0	0	0	0
2	2	0.10	4	0.04
3	1	0.05	12	0.12
4	3	0.15	20	0.20
5	6	0.30	24	0.24
6	2	0.10	24	0.24
7	4	0.20	10	0.10
8	2	0.10	4	0.04
9	0	0	1	0.01
10	0	0	0	0

Table 6.2 Absolute (h_k) and relative (f_k) frequencies of the occurrence of k times obverse, by $r = 20$ and $r = 100$ runs, each with $n = 30$ coin tosses.

	20 runs		100 runs			20 runs		100 runs	
k	h_k	f_k	h_k	f_k	k	h_k	f_k	h_k	f_k
0	0	0	0	0					
1	0	0	0	0	16	2	0.10	13	0.13
2	0	0	0	0	17	1	0.05	14	0.14
3	0	0	0	0	18	2	0.10	7	0.07
4	0	0	0	0	19	2	0.10	3	0.03
5	0	0	0	0	20	2	0.10	3	0.03
6	0	0	1	0.01	21	0	0	1	0.01
7	0	0	0	0	22	0	0	1	0.01
8	0	0	0	0	23	0	0	0	0
9	0	0	0	0	24	0	0	0	0
10	2	0.10	4	0.04	25	0	0	0	0
11	0	0	2	0.02	26	0	0	0	0
12	1	0.05	11	0.11	27	0	0	0	0
13	2	0.10	12	0.12	28	0	0	0	0
14	2	0.10	13	0.13	29	0	0	0	0
15	4	0.20	15	0.15	30	0	0	0	0

The phenomenon, that the relative frequency of an event E_k with $n \to \infty$ or $n \to N$ is nearly constant, led to the fact that the corresponding constant was (once) called probability (of the event E_k). Within the scope of this book this given (imprecise) definition is sufficient.

Master Nowadays probability theory is introduced abstractly and mathematically (to be more exact, with measure theory); i.e. the probability $P(E)$ for any event E is defined as a real number between 0 and 1 with certain (mathematical) properties; most of them corresponding with those for relative frequencies. The higher the probability of an event, the more confident we can be that this event will occur.

Master Unfortunately in everyday language we often hear: '... with a probability bordering on certainty'! As a matter of fact, a number can never approach certainty. Correctly, it must be expressed as follows: 'There is an event which is most likely to happen.' So, this event has a probability near 1. In contrast, events which are very unlikely to happen have a probability near 0.

Some probability theory terms will now be introduced by means of relative frequencies because of their above-mentioned properties and their level of understanding. Generally, we denote the relative frequency of an event E with $f(E)$. In the case where the event E does not occur, we denote this as the event $\bar{E}$ and say that the events E and $\bar{E}$ are *complementary events*. It is apparent that the relative frequency of a complementary event $\bar{E}$ of E is $f(\bar{E}) = 1 - f(E)$. A *certain event* C always occurs; its relative frequency is therefore $f(C) = 1$, and with this we get, for the *impossible event* I $(I = \bar{C}), f(I) = 0 = 1 - f(C)$.

Master Like the events heads or tails in coin tossing, the events 'male' and 'female' are complementary (the possibility that the coin lands vertically on its edge will be theoretically excluded). In contrast, the events 'pupil X scores an $IQ < 100$' and 'pupil X scores an $IQ > 100$' are not complementary, because the test result 'pupil X scores an $IQ = 100$' is also possible.

Bachelor **Example 6.2** Attitude of women and men towards psychotherapy

In a survey, 211 women and 198 men are asked about their attitude towards psychotherapy: 'When you have personal problems, do you then visit a psychotherapist?' The result of the survey is as follows:

	'yes, visit' (A)	'no, not visit' ($\bar{A}$)	Sum
women (B)	106	105	211
men ($\bar{B}$)	82	116	198
Sum	188	221	409

We call the event 'yes, visit' A (positive attitude) and the event 'no, not visit' $\bar{A}$ (negative attitude). The relative frequency (not taking the sex into account) of positive attitude is $f(A) = \frac{188}{409} = 0.4597$. We label the event 'women' with B, and the event 'men' with $\bar{B}$. We then obtain the relative frequency (independent of attitude) for women with $f(B) = \frac{211}{409} = 0.5159$. Analogously, the relative frequency of a negative attitude is $f(\bar{A}) = 0.5403$ and the relative frequency of men is $f(\bar{B}) = 0.4841$.

Besides these events A and B, sometimes the relative frequency (in the following, the probability) of A given B is of interest, i.e. the relative frequency of A, if the event B has already occurred: $f(A \mid B)$. This relative frequency is called *conditional relative frequency*. The events $(A \mid B)$ and also $(B \mid A)$ are called *conditional events*. If $f(B \mid A) = f(B)$ and $f(A \mid B) = f(A)$, then the events A and B are called mutually independent; they are *independent events*. Otherwise we have *dependent events*.

Bachelor **Example 6.2 – continued**

We now consider the question: 'What is the relative frequency of a positive attitude for women?' That is, what is the conditional relative frequency $f(A \mid B)$ of a positive attitude (A), given that the person considered is a woman (Event B). Amongst the 211 women interviewed, we found 106 with a positive attitude; thus $f(A|B) = \frac{106}{211} = 0.5024$.

We see that A and B are dependent because $f(A) = \frac{188}{409} = 0.4597 \neq f(A|B) = \frac{106}{211} = 0.5024$. Sex and attitude towards psychotherapy are dependent on one another. However, the small difference between $f(A) = 0.4597$ and $f(A|B) = 0.5024$ leads to the conclusion that the degree of dependency is not very high.

If we now turn from relative frequencies to corresponding probabilities – imagining that n becomes larger and larger until the entire (infinite) population is included – then the interpretation is actually the same as with the relative frequencies (a probability is always between 0 and 1).

Master Nowadays probabilities are often written as percentages; this is not wrong, if we understand 10% as $\frac{10}{100} = 0.1$. This practice stems from the fact that in everyday life, percentages are used more often than relative frequencies; the former are actually well-known among laypersons.

In many cases a probability statement of a particular event refers to several options of realization (measurement values), all having the same probability of occurring and all coinciding with this particular event. For instance the event 'an odd outcome in throwing a die' occurs for three equally likely outcomes – if the die is a fair one. In these cases the probability of the event E, namely $P(E)$, can easily be calculated. We denote the number of all realization options by e, and the number of those which coincide with the event E by g; then we have

$$P(E) = \frac{g}{e} \tag{6.1}$$

Master Probability theory has its roots in the mathematical treatment of gambling in the seventeenth and eighteenth centuries. The classical definition of probability in Formula (6.1) goes back to *Pierre-Simon Laplace*.

Bachelor **Example 6.1 – continued**

Similarly, the same is true for tossing a die; the relative frequency of tossing '6', $f(6)$, tends with growing n toward $\frac{1}{6}$, and using Formula (6.1) leads, due to $e = 6$ and $g = 1$, directly to $P(6) = \frac{1}{6}$.

For Lecturers:

In statistics the probability of an event is interpreted as the relative frequency of the occurrence of this event in a long run of observations under equal conditions. Contrary to this is the interpretation as 'degree of belief': the higher the probability for an event, the more we believe in its occurrence. Such degrees of belief can be expressed as *odds ratios*. If for instance someone says 'I receive €9 if the event E occurs, and pay €1 if not', then his/her 'subjective' probability for this event is 0.9 (namely $\frac{9}{9+1} = 0.9$).

Bachelor **Example 6.1 – continued**

Given that a coin collector arranges several coins with their heads facing upwards, then the event 'head' would not occur by chance. Therefore, using the theory of probability would obviously be useless. The theory of probability cannot make statements about events which are caused by influencing factors other than chance – in our example the event occurs because of the collector's arbitrary arrangement.

For probabilities, the same rules for calculation are valid as for relative frequencies. With the symbols used above for the latter, we get: $P(C) = 1$, $P(I) = 0$, $P(\bar{E}) = 1 - P(E)$. The *conditional probability* is denoted by $P(A \mid B)$ and $P(B \mid A)$, respectively, and defined analogously to the conditional relative frequency.

In the following we often need further rules for considering two events, A and B, at the same time. On the one hand we ask for the probability that either event A or event B occurs. On the other hand, we ask for the probability that event A as well as event B occurs. For the first case, the symbol $P(A \cup B)$ and for the second case the symbol $P(A \cap B)$ is used. Without proof (it can be obtained with relative frequencies; there are the same relationships), we have as a rule for the first case the so-called addition theorem:

$$P(A \cup B) = P(A) + P(B) - P(A \cap B) \tag{6.2}$$

and for the second case the so so-called multiplication theorem:

$$P(A \cap B) = P(A) \cdot P(B|A) = P(B) \cdot P(A|B) \tag{6.3}$$

If A and B cannot occur simultaneously (we say they are *mutually exclusive events*), then $P(A \cap B) = 0$ and we get $P(A \cup B) = P(A) + P(B)$. If A and B are independent, then that means we have $P(B \mid A) = P(B)$ and $P(A \mid B) = P(A)$, respectively; so then $P(A \cap B) = P(A) \cdot P(B)$ follows.

Master

Example 6.3 Evidence for the multiplication and the addition theorem from trials with a deck of cards

Take a deck of cards with four suits (Hearts, Diamonds, Clubs, Spades); each suit includes the same 13 ranks (Ace = 1, 2, 3, 4, 5, 6, 7, 8, 9, 10, Jack, Queen, King). We are interested now in event A, Queen, and event B, Hearts. The reader might shuffle the deck carefully – to guarantee randomness – and then draw a card from anywhere in the deck. After replacing the drawn card and shuffling again, we repeat this procedure, say a 100 times. Each time the result (event) obtained is registered as to whether it is or is not Queen of Hearts. We can expect that, in 100 such runs, the Queen of Hearts is seldom drawn. Should the reader draw any card a 1000 times in the aforementioned way, then the relative frequency of Queen of Hearts, i.e. $f(A \cap B)$, is near to $\frac{1}{52}$. We also get this result from Formula (6.1) with $g = 1$ and $e = 52$ or from the special case of Formula (6.3): $P(A \cap B) = P(A) \cdot P(B \mid A) = P(A \cap B) = P(A) \cdot P(B) = \frac{1}{13} \cdot \frac{1}{4}$.

Similarly, we can determine the probability of event A, Queen, in the first drawing, and in a second drawing – without replacement of the first card – the event X, (also) Queen. We compute: $P(A \cap X) = P(A) \cdot P(X \mid A) = \frac{1}{13} \cdot \frac{3}{51} = \frac{3}{663}$.

Analogously, we can proceed to determine the probability of observing either the event A, Queen, or the event B, Hearts. Computationally, we obtain, from Formula (6.2), $P(A \cup B) = P(A) + P(B) - P(A \cap B) = \frac{1}{13} + \frac{1}{4} - \frac{1}{52} = \frac{16}{52} = \frac{4}{13}$.

Similarly, we can proceed to determine the probability for either the event A, Queen, or the event W, King. Computationally, from a special case of Formula (6.2), we obtain $P(A \cup W) = P(A) + P(W) = \frac{1}{13} + \frac{1}{13} = \frac{2}{13}$.

Summary

The phenomenon 'chance' can be explained best by the results of gambling. The occurrence or non-occurrence of any *event* follows the rules of *probability theory*. Important is the *probability*, that a particular event as well as another particular event occurs, but also the probability that one of the two events occurs (for mutually exclusive as well as for non-mutually exclusive events), and, finally, the so-called *conditional probability;* i.e. the probability that a particular event occurs, given that a particular other event has already occurred.

6.2 Random variable and theoretical distributions

Now we interpret a character in such a way that the occurrence of each of its measurement values is considered to be a random event.

That is, we model each character y with a so-called *random variable* $\boldsymbol{y}$. The term 'variable' indicates that we think of a mathematical function. A variable takes on certain values; in our case a certain outcome of a character does not depend deterministically (and exclusively) on any mathematical function, but mainly on chance: that is our 'model'. Generally we denote such a chance-based (i.e. random) variable by a bold print letter; the realized value which a random variable $\boldsymbol{y}$ takes in a particular case is denoted by the same, but not bold, printed letter, *y*. This letter gets a suffix in order to show the number of an observation/realization (that is to say, outcome) in the sample.

For Lecturers:

While we use bold printed letters in this book to symbolize any random variable, probability theory and theoretical statistics often use capitals instead. However in application of statistics, capitals can not be used consistently at all: for instance the calculated value *F* of an *F*-test (see Section 10.4.1) is just a fixed value and not a random variable.

Master

'Modeling' means that we make certain assumptions about a character in question. There is, above all, an assumption about the way in which the (theoretically observable) outcomes are distributed throughout the different measurement values – within the population. The corresponding distribution is called the *theoretical distribution*. Such an assumption is necessary in order to make inferences from a sample to the population. Some assumptions of an underlying distribution can even be tested. And some statistical procedures are in some way robust so that a particular assumption may be violated.

Bachelor

Example 6.2 – continued

Take the random variable $\mathbf{y}$ used to model the character *attitude*; there are two measurement values (functional values) of $\mathbf{y}$: y_1 means positive attitude, and y_2 means negative attitude.

Analogous to characters, we distinguish between *discrete random variables* and *continuous random variables*. Hence there is also a theoretical *distribution function* $F(y)$. The distribution function $F(y)$ is, for each real y on the respective axis (here the abscissa), the probability that the random variable $\mathbf{y}$ takes a value smaller than or equal to that value y.[1]

Master

The distribution function $F(y)$ is the generalization of the sum of the frequencies in Section 5.2. In the case of a discrete random variable, the distribution can be characterized by the distribution function as well as by the *probability function.* The latter gives the probability of the occurrence of each measurement value (as a function of these values). It corresponds to the relative frequencies. Summarizing the values of the probability function for all measurement values smaller than or equal to a certain number y, the value of the distribution function for this value y is the result. This corresponds to the cumulative relative frequencies and is defined for each real value y; that is, it is not only defined for the (observable) measurement values of the character in question. The graph of the distribution function of discrete random variables is a *step curve*. In the case of a continuous

[1] In German literature, 'smaller' is often used instead of 'smaller than or equal'.

character, the first derivative of $F(y)$, $\frac{dF(y)}{dy} = f(y)$, is called a *density function* or the *density* of the random variable $\mathbf{y}$.

6.2.1 Binomial distribution

The following type of random variable often arises. Consider n identical, independent runs of an experiment (in other words: consider n independent research units); let $P(E) = p$ be the probability of the occurrence of event E. The event E in the following is called 'success'. The number of successes is then a random variable $\mathbf{y}$ with possible values $k = 0, 1, \ldots, n$. The probabilities are

$$P(k \text{ successes of } n) = p(\mathbf{y} = k) = \mathrm{P}(k) = \binom{n}{k} p^k (1-p)^{n-k}, \quad \text{for } k = 0, 1, 2, \ldots, n \tag{6.4}$$

Formula (6.4), considered for all values of k, delivers the probability function. The symbol $\binom{n}{k}$ is called the *binomial coefficient*, which is defined as

$$\binom{n}{k} := \frac{n!}{k!\,(n-k)!}$$

with $n! := n \cdot (n-1) \cdot (n-2) \cdot \ldots \cdot 3 \cdot 2 \cdot 1; 0! := 1$. $\binom{n}{k}$ is to be read as 'n choose k'. Generally, the binomial coefficient gives the number of different arrangements and combinations, respectively, of k equal objects on the one side and $n - k$ other, but again equal objects, on the other side in n positions.

For Lecturers:

Let us consider a family with four children, Anne (A), Beth (B), Carl (C), and David (D), to derive the binomial coefficient.

As concerns the order of birth, there are then 24 different possible combinations:

ABCD	ABDC	ACBD	ACDB	ADBC	ADCB
BACD	BADC	BCAD	BCDA	BDAC	BDCA
CABD	CADB	CBAD	CBDA	CDAB	CDBA
DABC	DACB	DBAC	DBCA	DCAB	DCBA

Generally, given n children we have $n \cdot (n-1) \cdot (n-2) \cdot \ldots \cdot 3 \cdot 2 \cdot 1 := n!$ combinations; therefore in our case $4 \cdot 3 \cdot 2 \cdot 1 = 24$. Now we have n options to site the first born; each of these n options can be combined with $n - 1$ options for the second born, and so on. In mathematics (combinatorics) we call this the number of *permutations*.[2]

[2]Coming from the Latin, *permutare*, meaning to interchange something.

Let us now consider how many combinations exist for four siblings comprised of two girls and two boys. We now inquire about the number of *permutations with replication*, namely to cover $n = 4$ positions by two groups, with replications in each group, namely two girls (Anne, Beth), $k = 2$, and two boys (Carl, David), i.e. $n - k = 2$. If we knew the looked-for number of permutations with replication $P_{k;\,n-k}$, then it follows: from each number $P_{k;\,n-k}$ we obtain $k! \cdot P_{k;\,n-k}$ permutations, if we additionally distinguish between the girls (which means it is actually important who the older one is – Anne or Beth). If we try the same for the boys (it is also important whether Carl or David is older), then instead of the $k! \cdot P_{k;\,n-k}$ permutations obtained so far, we have $(n - k)! \cdot k! \cdot P_{k;\,n-k}$ permutations; however, this number amounts to n!, which is just the number of different combinations from the very beginning. That is, we get $(n-k)! \cdot k! \cdot P_{k;\,n-k} = n!$; in other words $P_{k;n-k} = \frac{n!}{k!(n-k)!} := \binom{n}{k}$.

Master

Example 6.4 All options of filling out a lottery ticket

The question is: How many different ways exist in a lottery to mark 6 of 49 available numbers? For this we have exactly

$$\binom{n}{k} = \binom{49}{6} = \frac{49!}{6! \cdot (49-6)!} = \frac{49 \cdot 48 \cdot 47 \cdot 46 \cdot 45 \cdot 44 \cdot 43!}{6 \cdot 5 \cdot 4 \cdot 3 \cdot 2 \cdot 1 \cdot 43!}$$
$$= \frac{49 \cdot 48 \cdot 47 \cdot 46 \cdot 45 \cdot 44}{6 \cdot 5 \cdot 4 \cdot 3 \cdot 2 \cdot 1} = 13\,983\,816 \text{ options.}$$

Only one of them contains all 6 of the winning lottery numbers.

A random variable $\mathbf{y}$ with the probability of Formula (6.4) is called *binomially distributed* and follows a *binomial distribution* with the determinants n and p.

For Lecturers:

Our example with the family with four children, Anne, Beth, Carl, and David, may serve for the derivation of the probability of $\mathbf{y} = k$; $\mathbf{y}$ being a binomially distributed variable. We look for the probability of having, amongst $n = 4$ children, $k = 2$ girls (and by this $n - k = 2$ boys). The probability for the birth of a girl can be stated as follows: $p = \frac{1}{2}$.

Let us now consider a fixed sequence of girls, *G*, and boys, *B*, for instance *GGBB*; then the probability of exactly this sequence, due to the multiplication theorem, is $p \cdot p \cdot (1 - p) \cdot (1 - p) = p^2 \cdot (1 - p)^{4-2} = \frac{1}{16}$. Generally, we obtain that a certain sequence of n children with k girls and $n - k$ boys has the probability $p^k \cdot (1 - p)^{n-k}$. As already mentioned above, in total, $\binom{n}{k}$different sequences exist. All these $\binom{n}{k}$sequences fulfill the event of interest: k times G and $n - k$ times B. Therefore the probability P(k girls out of n siblings) is, according to the addition theorem, $\binom{n}{k} p^k(1 - p)^{n-k}$.

Master

Example 6.5 Probability of drawing Hearts from a complete deck of cards

Again we consider a deck of cards with four suits (Hearts, Diamonds, Clubs, Spades), each suit having the same 13 ranks (Ace = 1, 2, 3, 4, 5, 6, 7, 8, 9, 10, Jack, Queen, King). Given that we draw six cards randomly with replacement, the probability of drawing the event Heart in an arbitrary draw is $\frac{1}{4}$. The number of Heart cards in the six drawings, $\boldsymbol{k}$, is binomially distributed with $n = 6$ and $p = \frac{1}{4}$. Therefore the probability for all possibilities $k = 0, 1, 2, \ldots, 6$ is:

$$P(\boldsymbol{k}=0) = \binom{6}{0}\left(\frac{1}{4}\right)^0\left(\frac{3}{4}\right)^6 = \left(\frac{3}{4}\right)^6 = \frac{729}{4096} = 0.1779785$$

$$P(\boldsymbol{k}=1) = \binom{6}{1}\left(\frac{1}{4}\right)^1\left(\frac{3}{4}\right)^5 = 6\left(\frac{1}{4}\right)\left(\frac{3}{4}\right)^5 = \frac{1458}{4096} = 0.355957$$

$$P(\boldsymbol{k}=2) = \binom{6}{2}\left(\frac{1}{4}\right)^2\left(\frac{3}{4}\right)^4 = 15\left(\frac{1}{4}\right)^2\left(\frac{3}{4}\right)^4 = \frac{1215}{4096} = 0.2966309$$

$$P(\boldsymbol{k}=3) = \binom{6}{3}\left(\frac{1}{4}\right)^3\left(\frac{3}{4}\right)^3 = 20\left(\frac{1}{4}\right)^3\left(\frac{3}{4}\right)^3 = \frac{540}{4096} = 0.1318359$$

$$P(\boldsymbol{k}=4) = \binom{6}{4}\left(\frac{1}{4}\right)^4\left(\frac{3}{4}\right)^2 = 15\left(\frac{1}{4}\right)^4\left(\frac{3}{4}\right)^2 = \frac{135}{4096} = 0.0329590$$

$$P(\boldsymbol{k}=5) = \binom{6}{5}\left(\frac{1}{4}\right)^5\left(\frac{3}{4}\right)^1 = 6\left(\frac{1}{4}\right)^5\left(\frac{3}{4}\right)^1 = \frac{18}{4096} = 0.0043945$$

$$P(\boldsymbol{k}=6) = \binom{6}{6}\left(\frac{1}{4}\right)^6\left(\frac{3}{4}\right)^0 = \left(\frac{1}{4}\right)^6\left(\frac{3}{4}\right)^0 = \frac{1}{4096} = 0.0002441$$

We see that $k = 1$ is the most probable case and therefore just one is the highest expected number of Hearts drawn. The mode of the theoretical distribution of $\boldsymbol{k}$ is 1. Since there are no further possible combinations, the sum over all probabilities equals 1.

In **R**, we obtain these probabilities more conveniently by typing

```
> probfct <- dbinom(0:6, size = 6, prob = 0.25)
```

i.e. we apply the function `dbinom()` to ascertain the density and set `0:6` to include all possibilities from k is 0 to 6 as the first argument; with `size = 6`, we set the number of trials to n is 6 and set p with `prob = 0.25`, and assign the result to the object `probfct`. Next we compute the distribution function by typing

```
> distfct <- pbinom(0:6, size = 6, prob = 0.25)
```

i.e. we set the same arguments in the function `pbinom()` and assign the result to the object `distfct`. For graphical illustration of the probability as well as distribution function, we type

```
> plot(0:6, probfct, xlab = "Value",
+        ylab = "Probability function")
> lines(0:6, probfct, type = "h")
```

i.e. we set the values on the abscissa (`0` to `6`) in the function `plot()` as the first argument, and the corresponding values of the probability function `probfct` as the second. The further arguments `xlab` and `ylab` label the axes. To improve the readability of the resulting chart we amend it with the function `lines()` using the same coordinates, and with `type = "h"` stipulate that the lines be vertical histogram-like.

As a result we get the chart shown in Figure 6.1.

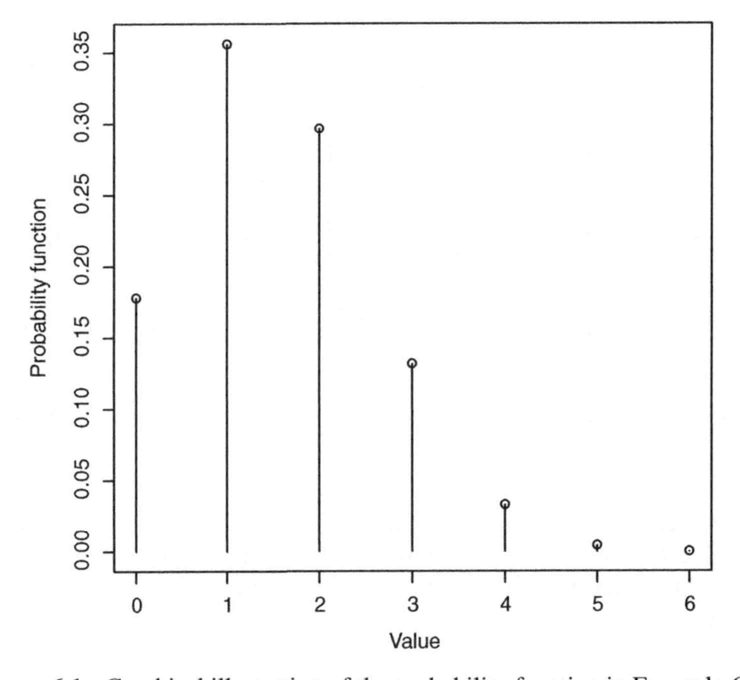

Figure 6.1 Graphical illustration of the probability function in Example 6.5.

To visualize the distribution function, we type

```
> plot(0:6, distfct, ylim = c(0, 1), ylim = c(0, 1), xlab = "Value",
+        ylab = "Distribution function")
> lines(0:6, distfct, type = "s")
```

i.e. again we apply the functions `plot()` and `lines()` and set their arguments analogously. Only this time we use the values of the distribution function `distfct` and set `type = "s"` to get stair steps.

As a result we get the chart shown in Figure 6.2.

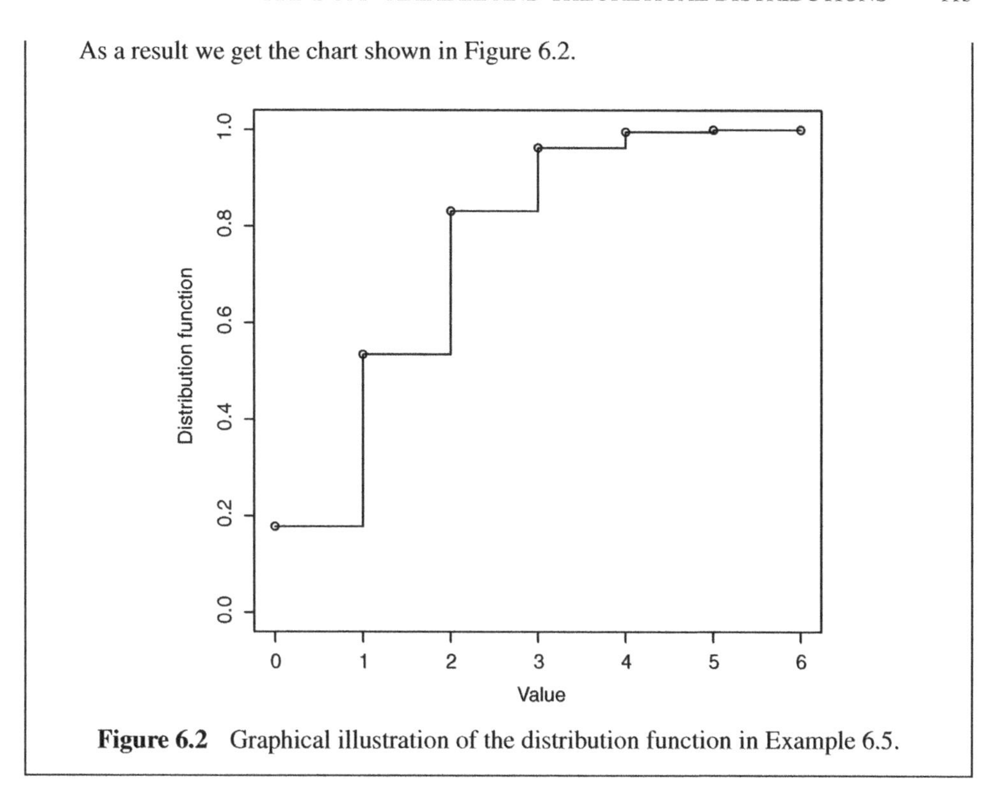

Figure 6.2 Graphical illustration of the distribution function in Example 6.5.

In SPSS, there is a more convenient way to obtain the probabilities than computing them manually. After starting SPSS we type the values of column value shown in Figure 6.3 in a new data sheet. Next we select the same command sequence (Transform – Compute Variable...) as in Example 5.3 to open the window in Figure 5.9. We input probfct as Target Variable: and select PDF & Noncentral PDF from the panel Function group:. Subsequently a list in the panel Functions and Special Variables: appears and we select PDF.Binom from it. The name stands for *probability density function* of a binomial distribution. Next we move this function to the field Numeric Expression: and set as its argument (value, 6, 0.25). We confirm these settings by clicking OK and get the probabilities in question (see second column in Figure 6.3). To ascertain the distribution function we proceed analogously, but name the Target Variable: distfct and select CDF.Binom from the CDF & Noncentral CDF group list. We set the same arguments as before and once again click OK to get the result (see third column in Figure 6.3).

Up next we'd like to visualize the probability function and the distribution function. First, we need to set the measurement scale of value, probfct, and distfct by changing to the SPSS Variable View and selecting (click to open a pull-down list) Scale in the column Measure (see Figure 5.12). Now we start the SPSS Chart Builder shown in Figure 5.5. There we select Scatter/Dot in the Gallery tab and drag and drop the symbol Simple Scatter into the Chart Preview. Next we move value to the field X-Axis? and probfct to Y-Axis?. Finally we click OK and a graphical illustration of the probability function (not shown here) appears, which we double-click. Subsequently the Chart Editor (Figure 6.4) pops up. In it, we double-click

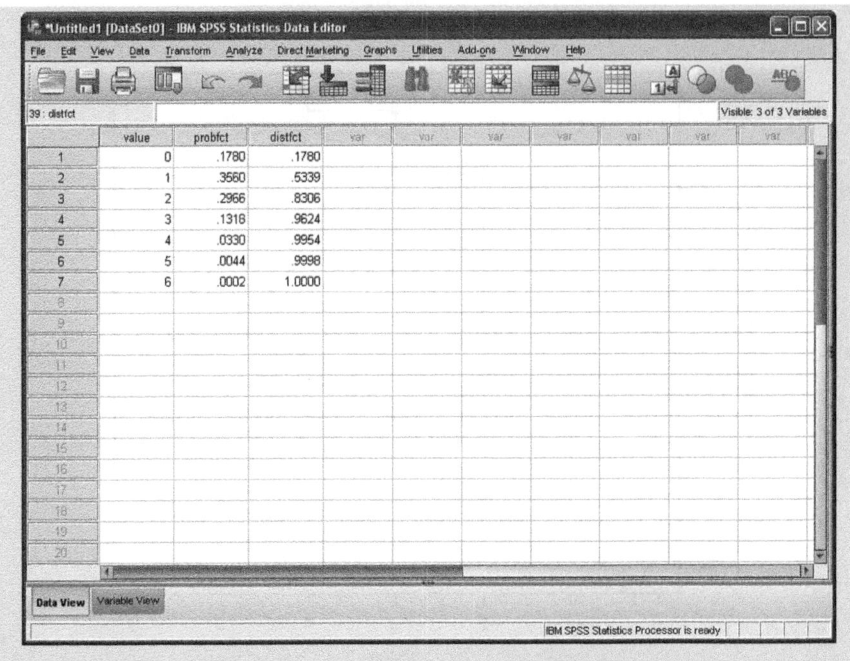

Figure 6.3 SPSS Data View in Example 6.5.

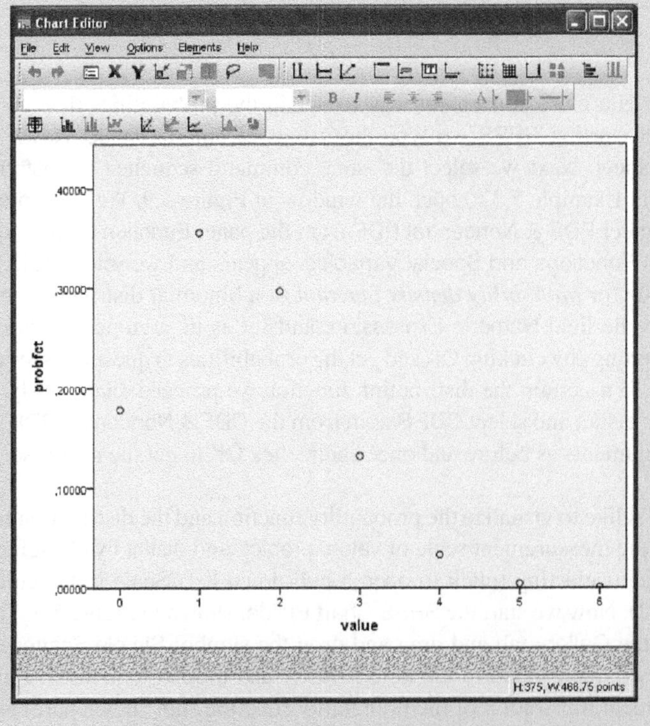

Figure 6.4 SPSS Chart Editor.

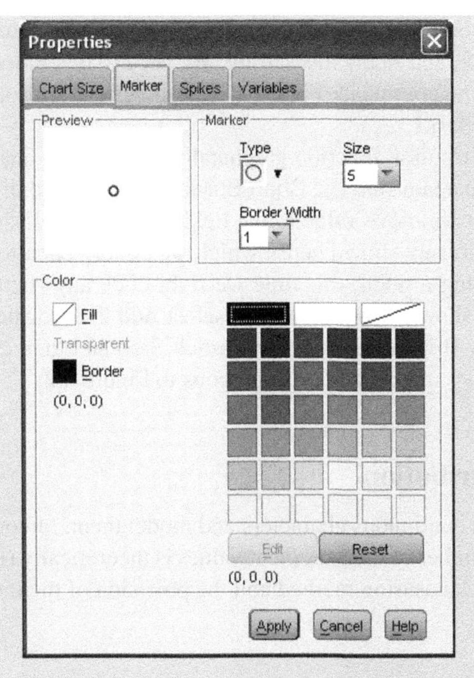

Figure 6.5 SPSS-window for defining properties of points.

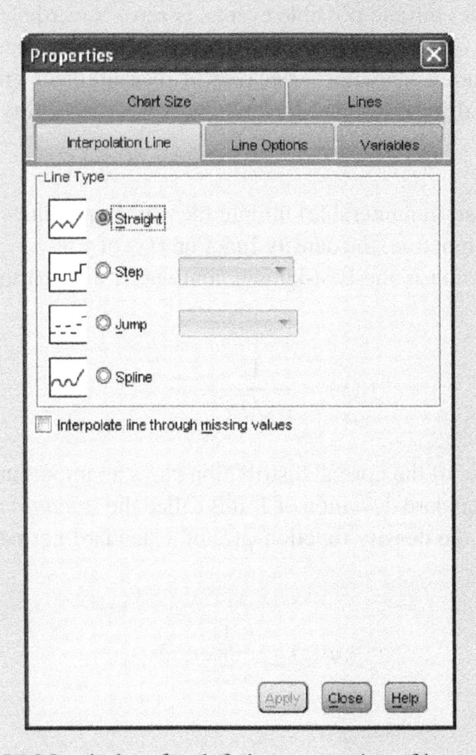

Figure 6.6 SPSS-window for defining properties of interpolation lines.

any of the seven points in the chart and get to the window in Figure 6.5. Next we activate the Spikes tab, tick Floor, and Apply this setting. We need to tick Floor and Apply once again so that SPSS executes the command. Finally we close the two still-open windows and get a chart analogous to Figure 6.1.

To illustrate the distribution function graphically, we proceed analogously to the probability function. We once again start the Chart Builder and drag and drop Simple Scatter into the Chart Preview. Next we move value to the field X-Axis? and distfct to Y-Axis?. Clicking OK gets us to a new chart (not shown here), which we double-click to start the Chart Editor. Instead of double-clicking a point, this time we right-click anywhere on the chart to open a pull-down menu (not shown) from which we select Add Interpolation Line. This results in another pop-up window (Figure 6.6), where we tick Step and then click Apply. Finally we close all pop-up windows and get a chart analogous to Figure 6.2.

6.2.2 Normal distribution

Frequently we deal with continuous characters and model them, of course, with a continuous random variable. The number of measurement values is theoretically (innumerably) [3] infinite, even if, due to the given measuring method and the precision of the instruments, this will not be apparent in practice.

Master The probability of each event for a continuous random variable, that is, with (innumerable) infinite possible events, is zero. Nevertheless, at each observation one of them occurs. This becomes obvious taking into account that each outcome of a continuous character – because of measurement imprecision – represents a (very) small interval; and for this interval a probability different from zero is given.

Therefore, for all of these (innumerable) infinite measurement values of a character we use, instead of a probability function, the density function $f(y)$ of $\mathbf{y}$ at y.

The *normal distribution* is the best-known continuous distribution. Its density function is:

$$f(y) = \frac{1}{\sigma\sqrt{2\pi}} e^{-\frac{1}{2\sigma^2}(y-\mu)^2} \tag{6.5}$$

In statistics a special case of the normal distribution plays an important role; namely the case with a mean of 0 and a standard deviation of 1. It is called the *standard normal distribution.* Its density function, that is the density function $\varphi(z)$ of a standard normally distributed random variable $\mathbf{z}$, is:

$$\varphi(z) = \frac{1}{\sqrt{2\pi}} e^{-\frac{z^2}{2}} \tag{6.6}$$

[3] 'innumerable' means: there are more values than natural numbers.

'Standardized' means that y is transformed to z: $z = \frac{y-\mu}{\sigma}$. We also say that z is $N(0, 1)$ distributed or, generally, y is $N(\mu, \sigma^2)$ distributed. Therefore the determinants of the normal distribution are μ (mean) and σ^2 (variance).

Bachelor **Example 5.4 – continued**

We considered the empirical distribution function of the character *Everyday Knowledge, 1st test date* in Example 1.1, restricted to the sample of children with German as a native language, as well as to children with Turkish as a native language but tested in Turkish at the first date.

In **R**, we have already illustrated the frequencies shown in Table 5.5 graphically (Figure 5.17a). Next we amend this chart with a normal curve based on the mean and standard deviation of the sample. The children with native language German, as well as those with native language Turkish who have been tested in Turkish on the first test date, have already been selected and assigned to the object `sub1_t1.set`.

```
> hist(sub1_t1.set, freq = FALSE, main = "Histogram",
+       xlim = c(20, 80),
+       xlab = "Everyday Knowledge, 1st test date (T-Scores)",
+       ylab = "Relative frequency")
```

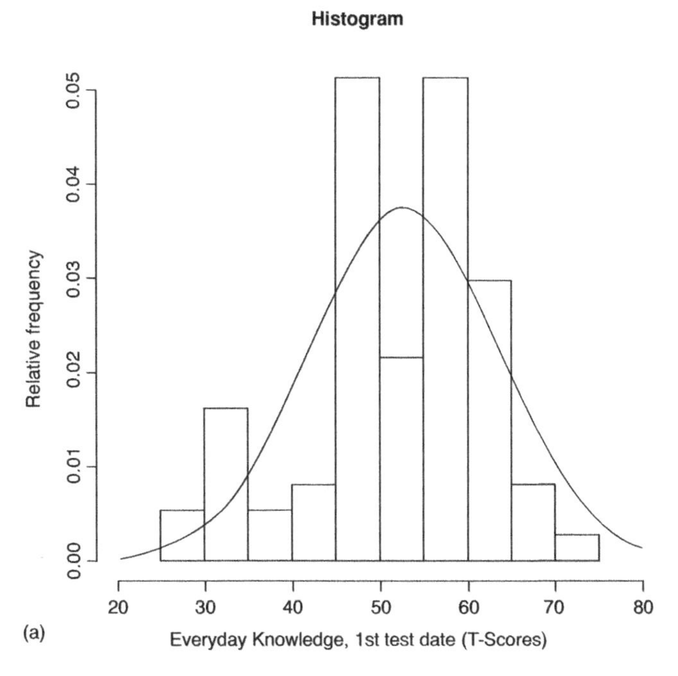

Figure 6.7a **R**-output showing a histogram along with the density function of the best-fitting normal distribution in Example 5.4.

i.e. we apply the function `hist()` with analogous arguments as before in this example (see Chapter 5), with the exception that with `freq = FALSE` we graphically illustrate the relative frequency. Next we type

```
> x <- seq(from = 0, to = max(sub1_t1.set),
+           length.out = length(sub1_t1.set))
> curve(dnorm(x, mean(sub1_t1.set), sd(sub1_t1.set)), add = TRUE)
```

i.e. we apply the function `seq()` to create a numerical sequence from `0` to the largest observed value of the character *Everyday Knowledge, 1st test date* (`sub1_t1.set`). With `length.out` we set the length of the sequence to be equal to the number of observation values of the character `length(sub1_t1.set)`. The resulting values are assigned to the object `x`. Finally we set the density function, ascertained with the function `dnorm()` based on object `x` as well as on the mean and standard deviation of character *Everyday Knowledge, 1st test date*, as the first argument in function `curve()`. The additional argument `add = TRUE` inserts the resulting normal curve into the existing chart.

As a result we get the chart shown in Figure 6.7a.

In SPSS, we have already illustrated the frequencies shown in Table 5.5 graphically (cf. Figure 5.17b). Next we would like to amend this chart with a normal curve based on the mean and standard deviation of the sample. We start the SPSS Chart Builder (Figure 5.5)

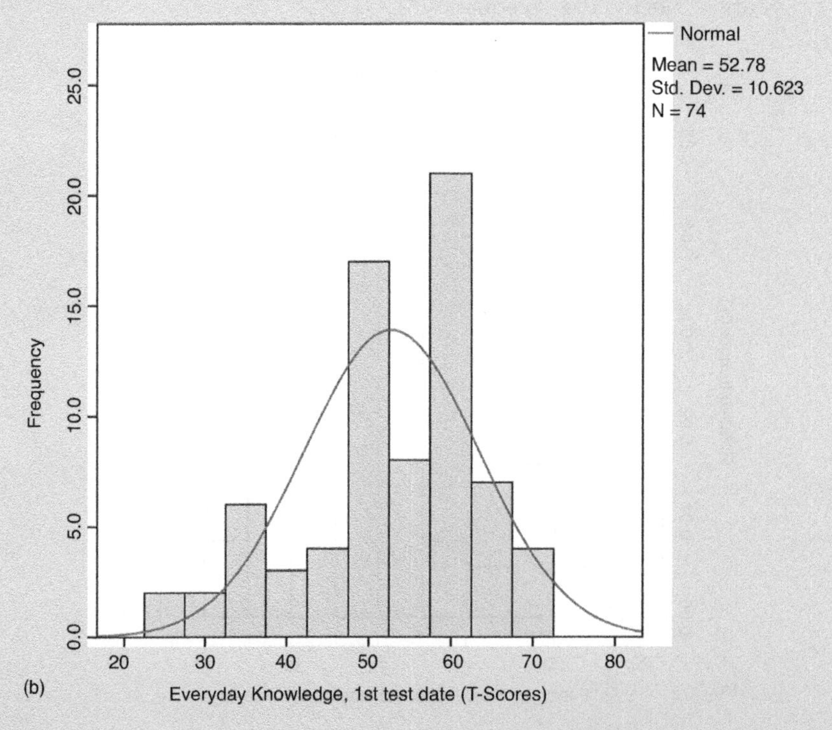

Figure 6.7b SPSS-output showing a histogram along with the density function of the best-fitting normal distribution in Example 5.4.

with the same command sequence (Graphs – Chart Builder...) as in Example 5.2. In it we select Histogram in the Gallery tab, and drag and drop the symbol Simple Histogram into the Chart Preview. Then we move the character Everyday Knowledge, 1st test date to the field X-Axis? and click Element Properties... to open the corresponding window. Here we tick Display normal curve, Apply this change and Close the window. A final click on OK and the chart shown in Figure 6.7b appears.

Besides the histogram given in Section 5.2.3, which shows the empirical distribution of the character *Everyday Knowledge, 1st test date*, the curve of the density function of a theoretical distribution is given; namely the one resulting if the character were modeled by a normally distributed random variable. The mean and variance of both distributions are the same.

Master **Example 6.6** Density and distribution function of the normal distribution

In **R**, we can ascertain the density function $f(y)$ and the distribution function $F(y)$ for any normal distribution, given that we provide the y-values. To do this, we enter the commands

```
> q <- seq(from = -3, to = 3, by = 0.1)
> densfct <- dnorm(q, mean = 0, sd = 1)
> distfct <- pnorm(q, mean = 0, sd = 1)
```

i.e. we use the function `seq()` to generate a numerical sequence from -3 to 3 incremented by `0.1`, and assign the resulting values to the object `q`. Next we apply the functions `dnorm()` and `pnorm()` to ascertain the density function and the distribution function of the normal distribution. For both functions we set the y-values in the object `q` as the first argument; with `mean = 0` and `sd = 1` we set the mean as well as the standard deviation. The results are assigned to the objects `densfct` and `distfct`. In order to illustrate these functions graphically, we type

```
> plot(q, densfct, type = "l")
```

i.e. we apply the function `plot()` and set the y-values in the object `q` as the first argument and the density function $f(y)$ in the object `densfct` as the second. With `type = "l"` we stipulate that the coordinates given by the prior arguments will be sequentially linked by lines. As a result we get a chart analogous to Figure 6.9. Next we type

```
> plot(q, distfct, type = "l")
```

i.e. we simply replace the density function by the distribution function `distfct`; as a result we get a chart analogous to Figure 6.10.

In SPSS, given that we create a variable (new column in the SPSS Data View) containing the y-values, we can, analogously to Example 6.2, ascertain the density function $f(y)$ and the distribution function $F(y)$. Therefore we apply the functions PDF.Normal and CDF.Normal, in each case with the arguments (quant, mean, stddev); this means that we input the y-values

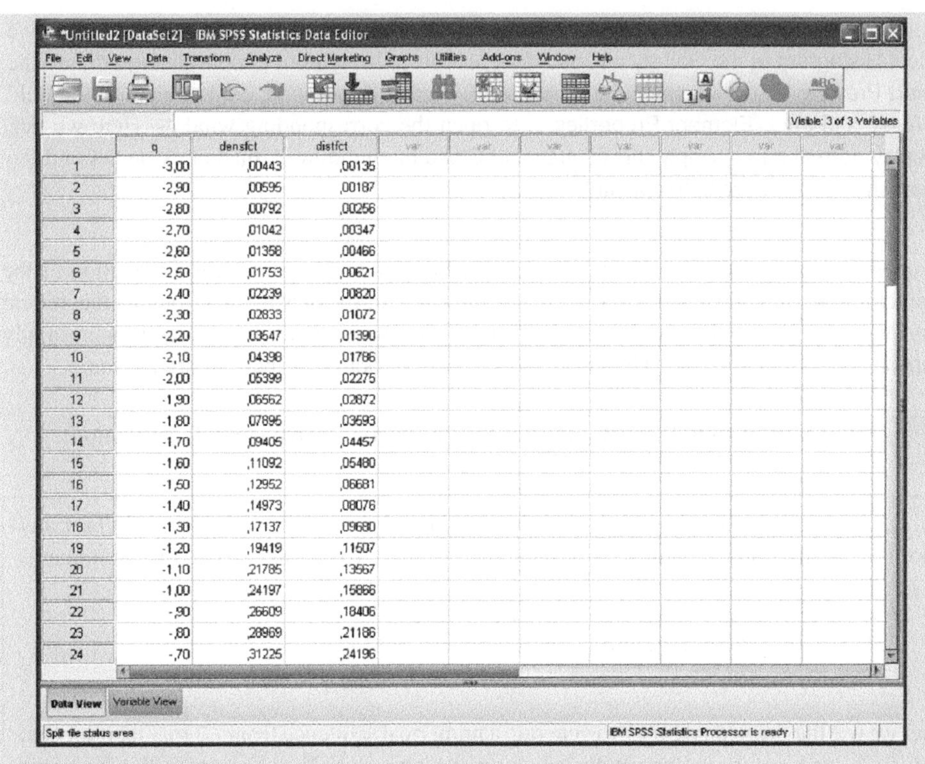

	q	densfct	distfct
1	-3,00	,00443	,00135
2	-2,90	,00595	,00187
3	-2,80	,00792	,00256
4	-2,70	,01042	,00347
5	-2,60	,01358	,00466
6	-2,50	,01753	,00621
7	-2,40	,02239	,00820
8	-2,30	,02833	,01072
9	-2,20	,03547	,01390
10	-2,10	,04398	,01786
11	-2,00	,05399	,02275
12	-1,90	,06562	,02872
13	-1,80	,07895	,03593
14	-1,70	,09405	,04457
15	-1,60	,11092	,05480
16	-1,50	,12952	,06681
17	-1,40	,14973	,08076
18	-1,30	,17137	,09680
19	-1,20	,19419	,11507
20	-1,10	,21785	,13567
21	-1,00	,24197	,15866
22	-,90	,26609	,18406
23	-,80	,28969	,21186
24	-,70	,31225	,24196

Figure 6.8 SPSS Data View showing the density function and the distribution function of the standard normal distribution.

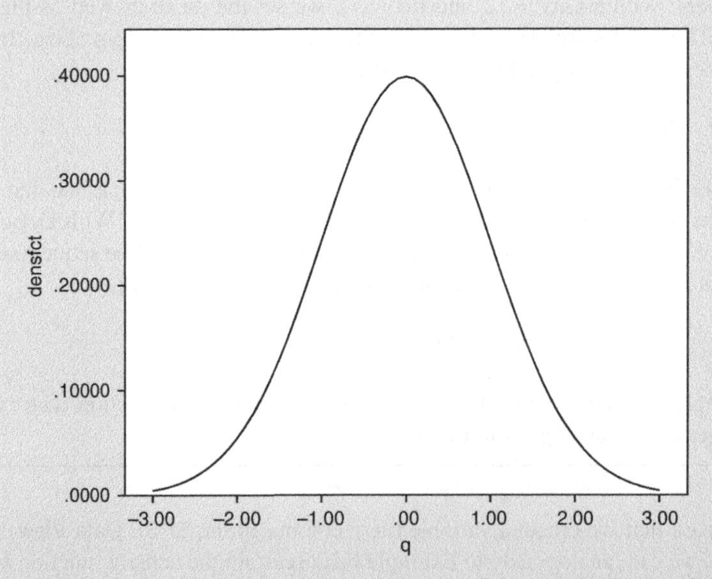

Figure 6.9 SPSS-output showing the density function of the standard normal distribution.

(provided in a variable) as quant and set the mean and standard deviation. In Figure 6.8 we have already done that for q containing a numerical sequence from −3 to 3, mean 0 and standard deviation 1, thus a standard normal distribution. The values of the density function and the distribution function are shown in the second and third columns, respectively.

In order to illustrate this data graphically we start the SPSS Chart Builder (Figure 5.5) with the same commands (Graphs – Chart Builder...) as in Example 5.2. Then we select Line in the Gallery tab and drag and drop the symbol Simple Line into the Chart Preview. Next we move the character q to the field X-Axis? and densfct to Y-Axis?. Finally, we click OK to obtain the chart shown in Figure 6.9. If we drag distfct instead of densfct to the field Y-Axis? we would obtain the distribution function of the standard normal distribution (Figure 6.10) after clicking OK.

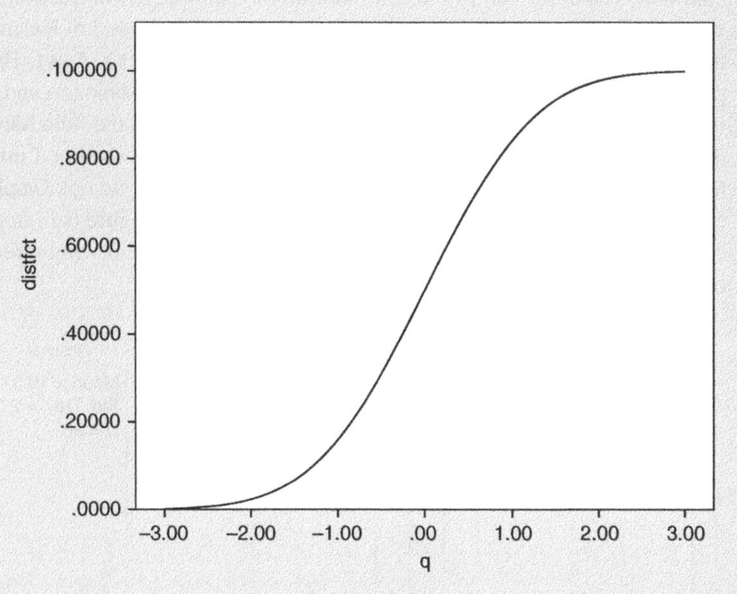

Figure 6.10 SPSS-output showing the distribution function of the standard normal distribution.

The normal distribution is important for several reasons.

1. Many populations in the real world have characters that are actually distributed in a way that can be approximated well with the normal distribution.

Master

There is a simple explanation for why so many characters have nearly normally distributed outcomes. This is because many characters are the result of the co-action of several (independent) random components; take intelligence for instance – we test a lot of partial performances in several intelligence tasks. Now, put simply, probability theory discloses, according to the so-called *central limit theorem,* that when a random variable is the sum of many independently but arbitrarily distributed random (sub-) variables, then the density function approaches the normal distribution as the number of (sub-) variables becomes larger.

Master **Example 6.7** Distribution of the sum of integers when throwing three dice

That the density function of a random variable as a sum of several independent, random (sub-) variables with any distribution tends to be normally distributed can be demonstrated by throwing three dice at once. The reader might do this $n = 30$ times. The random variable $\boldsymbol{y}$ might then be the sum of the integers from all three dice ($3 \leq y \leq 18$), for each throw. According to experience, the resulting histogram offers an image which comes quite close to the density of a normal distribution – despite there being just three summands. Figure 6.11 shows a numerical example, that we obtained after entering our outcomes (9, 8, 16, 8, 10, 11, 9, 7, 12, 10, 10, 13, 6, 8, 15, 12, 10, 11, 9, 11, 11, 6, 9, 13, 9, 12, 11, 12, 10, 13) as variable dice drawing in SPSS. We construct the histogram with the same commands as described in the continuation of Example 5.4 in Section 5.2.3. The x-axis of the histogram has been treated as described at the end of Example 5.5. In the window similar to that shown in Figure 5.13, we select X-Axis1 (Bar1) in the panel Edit Properties of: and remove the check marks for Minimum and Maximum in the field Scale Range and then insert the number 3 into the field Minimum and the number 18 into the field Maximum. Then we insert the number 1 into the field Major Increment (after removing the check mark). Finally, we tick Display normal curve. After Apply and a click on OK the chart shown in Figure 6.11 appears. The resulting reader's distribution of the sum of integers from throwing three dice will of course differ from the one given here.

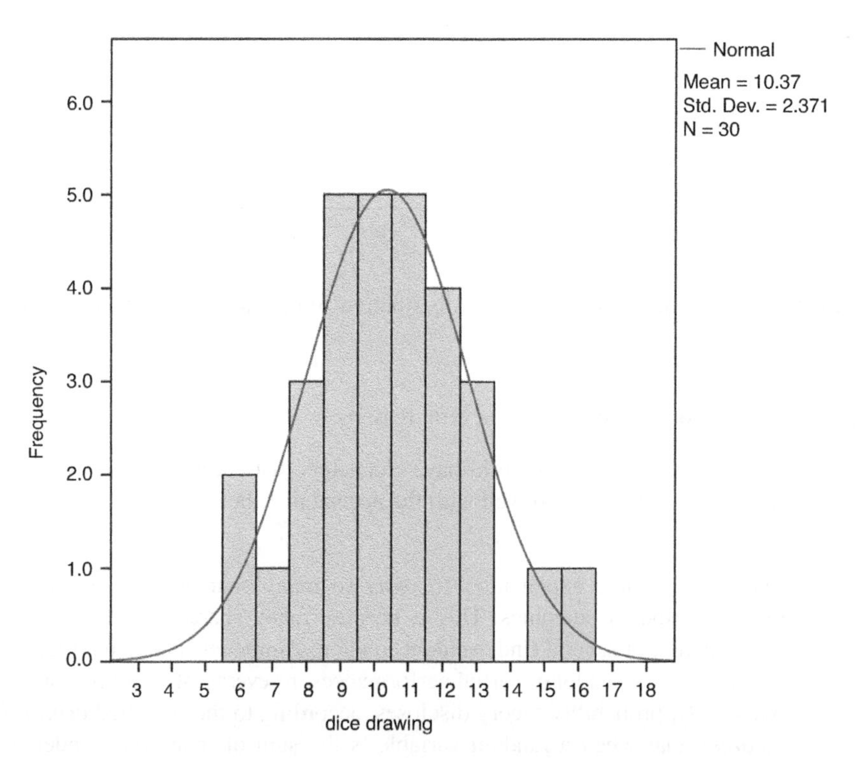

Figure 6.11 Histogram and density function of the best fitting normal distribution in Example 6.7.

2. Many statistical tests are derived under the assumption that the random variable of interest follows a normal distribution.

Master

However, for most statistical tests the assumption of a normally distributed random variable does not actually matter. Firstly, due to the central limit theorem, even if the distribution of $\mathbf{y}$ in the population is not normal, the distribution of the mean $\bar{\mathbf{y}}$ (that is if we consider all possible $\bar{y}$ of all possible samples of size n) tends for $n \to \infty$ toward a normal distribution – as a common rule of thumb $n > 30$ is enough to gain a good approximation of the normal distribution. However, in the end what is of relevance for the derivation of the respective statistical tests is: their assumption of normally distributed random variables is only stated to be sure, using simple means, that a normal distribution of the mean $\bar{\mathbf{y}}$ is given. Secondly, most statistical tests are robust; this means that the assumption of a normally distributed random variable can be violated for most of them with relatively minor effects (Rasch & Guiard, 2004). Hence, these tests are generally applicable.

3. Many distributions tend with increasing sample size, that is with $n \to \infty$, toward a normal distribution.

Master

The *t*-distribution (see Section 8.2.2) tends toward a normal distribution when the number of degrees of freedom (in principle the sample size n) becomes large. An analogous statement is valid for the χ^2-distribution (see Chapters 8 and 9). But also some discrete distributions tend with growing sample size toward a normal distribution. If, for instance concerning a binomial distribution, n becomes large, then the probability that a binomially distributed random variable $\mathbf{k}$, with the determinants n and p, is smaller than or equal to a number k tends to the probability that a standard normally distributed random variable $\mathbf{z}$ is smaller than or equal to a number $z = \frac{(k-np)}{\sqrt{np(1-p)}}$; that is, $P(\mathbf{k} < k) \cong P(\mathbf{z} \leq \frac{k-np}{\sqrt{np(1-p)}})$.

Summary

In statistics, a character of interest is modeled by a *random variable*, making a plausible assumption about the distribution of its measurement values. The most important *theoretical distribution* in this context is the *normal distribution*, which is uniquely determined by the mean and the standard deviation. Many distributions can, at least given certain conditions, be approximated by the normal distribution. The *standard normal distribution* has mean 0 and standard deviation 1.

6.3 Quantiles of theoretical distribution functions

For the application of statistical methods it is regularly of importance to find probabilities by using some given distribution function. For instance, for a binomial distribution we could be interested in the probability of at least $k = 8$ successes, i.e. following Formula (6.4), $P(\mathbf{y} \geq 8) = \sum_{k=8}^{n} P(k)$. Or, for instance, for a (standard) normal distribution we would like

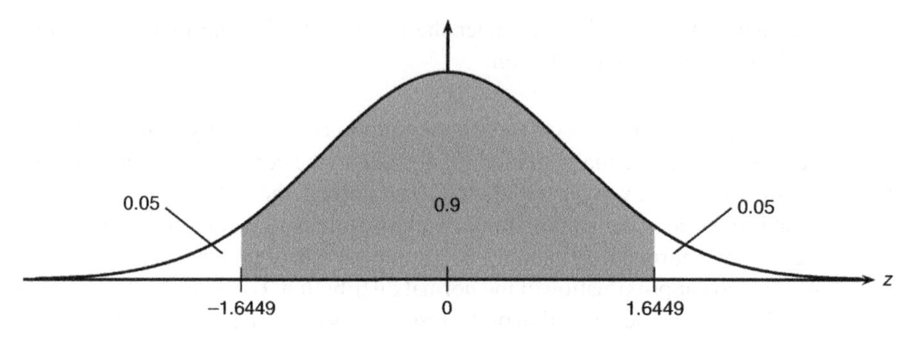

Figure 6.12 Density function of the standard normal distribution with common quantiles.

to know the probability of $z \geq z = 2.58$. Then tables (such as Table B1 in Appendix B) offer such probabilities, or one calculates them with SPSS or **R**.

The problem often is considered the other way round: we look for the value y_P, for which the distribution is split into two parts, so that $F(y_P) = P$ and $1 - F(y_P) = 1 - P$, respectively, where P is a certain probability. In the following we generally call y_P the *P-quantile*. Figure 6.12 gives an example of the density function of the standard normal distribution including common quantiles.

Master

We can illustrate the situation graphically due to the fact that the P-quantile is defined by $F(y_P) = P$. The area under the curve of the density function $f(y)$, to the right of y_P, is equal to $1 - P$; to the left of y_P it is equal to P. That is, y_P is that value on the abscissa, for which the distribution function $F(y)$ takes the value P. Figure 6.13 coincides with Figure 6.10 of the distribution function's curve of

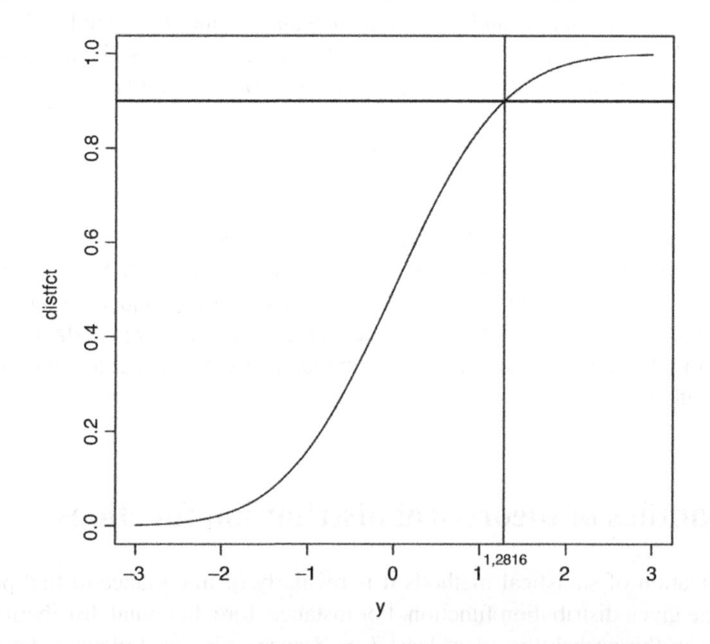

Figure 6.13 The 0.9-quantile of the standard normal distribution.

the standard normal distribution, but now additionally gives the 0.9-quantile on the abscissa (here the y-axis). For this, the ordinate's value of 0.9 has to be found, from where a parallel line to the abscissa must be drawn long enough to cross the curve; another line is plotted perpendicular to this point of intersection; the value $y_{0.9} = 1.2816$ is the result, when the abscissa is met.

Master

Example 6.8 Distribution of the character intelligence quotient

Nowadays, the term intelligence quotient (IQ) belongs to everyday knowledge, even if a layperson cannot interpret the exact value of an IQ correctly. The IQ is defined such that the (theoretical) outcomes, i.e. IQ-values, are normally distributed (see Figure 6.14). For each IQ-value on the abscissa, there is a corresponding certain value on the ordinate. For an arbitrarily small interval around such a point, the population's probability (relative frequency) can be ascertained for an IQ-value within that interval. *David Wechsler* (see for instance Kubinger, 2009b) intentionally, though arbitrarily, set the mean to $\mu = 100$ and the standard deviation to $\sigma = 15$. Consequently, in the interval $90 \leq IQ \leq 110$ lie about 50% (to be exact, 49.5%) of the persons in the population. Further, 68.3%, i.e. about two-thirds of the persons of the population, realize values within $\mu \pm \sigma$.

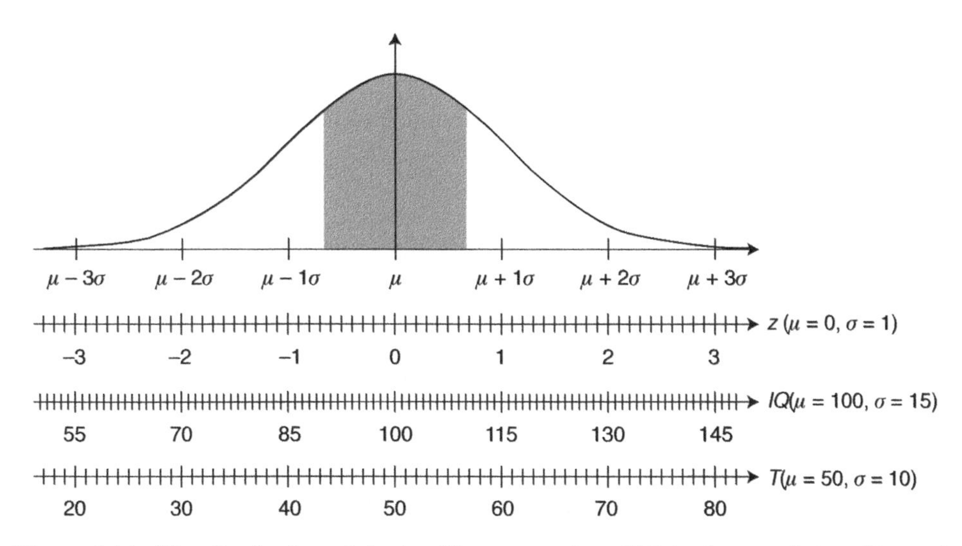

Figure 6.14 The distribution of the intelligence quotient (IQ) in the population. The scale of so-called T-Scores will be used quite often throughout the text (Example 1.1); the scale of z-values, used in psychological assessment corresponds to the P-quantiles, $z(P)$, of the standard normal distribution.

6.4 Mean and variance of theoretical distributions

Besides the graphical illustration of a theoretical distribution, analogous to the statistics of empirical distributions, so-called *parameters* (of a theoretical distribution) can be defined. These parameters are the target of estimations based on proper statistics from samples.

Master

In mathematical statistics, as regards theoretical distributions, most of the time the term *expectation* is used instead of the term mean: that is, the value which we most probably 'expect' for a single, randomly chosen unit (from the population). Even if that value means a contradiction from the content point of view (for instance in coin tossing we cannot expect the value 3.5, because it is an impossible event), in the long run, that is, in sum, our 'expectation' differs least from the actual realized outcome.

For a discrete random variable $\boldsymbol{y}$ with the measurement values $y_1, y_2, \ldots, y_r$ and the corresponding probabilities $p_1, p_2, \ldots, p_r$, the expected value, the expectation of $\boldsymbol{y}$ is defined as:

$$\mu = E(\boldsymbol{y}) = y_1 p_1 + y_2 p_2 + \cdots + y_r p_r = \sum_{l=1}^{r} y_l p_l \tag{6.7}$$

For a continuous random variable with the density function $f(y)$, the expected value of $\boldsymbol{y}$ is defined as:

$$\mu = \int_{-\infty}^{\infty} y f(y) dy \tag{6.8}$$

For the sake of simplicity, in this book we often use 'mean', though the 'expectation' is meant.

Master

Example 6.5 – continued

Given the probabilities $P(k = 0) = 0.1779785$. $P(k = 1) = 0.355957. \ldots, P(k = 6) = 0.0002441$, the following expectation results: $E(\boldsymbol{y}) = 0 \cdot 0.1779785 + 1 \cdot 0.355957 + 2 \cdot 0.2966309 + 3 \cdot 0.1318359 + 4 \cdot 0.0329590 + 5 \cdot 0.0043945 + 6 \cdot 0.0002441 = 1.5$; i.e. we expect that if we draw a card from a deck six times (with replacement) that (on average) a Heart results 1.5 times. This is intuitively obvious, because four different suits are randomly distributed throughout the six cards.

Master

The variance of a theoretical distribution is defined as

$$\sigma^2 = E(\boldsymbol{y} - \mu)^2 = E[\boldsymbol{y} - E(\boldsymbol{y})]^2 \tag{6.9}$$

6.5 Estimation of unknown parameters

In Section 6.2 we indicated that for given research questions the character of interest has to be properly modeled. This includes the determination of the modeled theoretical distribution's parameters. As a consequence, we *estimate* the parameters of the modeled distribution. For instance, $\bar{y} = \frac{1}{n}\sum_{v=1}^{n} y_v$ (Formula (5.2)) for μ, and $s^2 = \frac{1}{n-1}\sum_{i=1}^{n}(y_i - \bar{y})^2$ (Formula (5.4)) for σ^2 are appropriate *estimates*.

Master

An estimate is the result of an *estimation*; it is a realization of an *estimation function*. Estimation functions are sometimes called estimations but would be better called *estimators*. Bear in mind that an estimation function is not deterministic; that is, its functional value, the estimate, does not always result in the same value. The estimate depends instead on chance, as we are obviously dealing with a random variable. The estimator describes all theoretically possible estimates: imagine drawing a (random) sample infinitely often. Therefore all estimates have a variance and of course a mean.

For Lecturers:

The Dutch language makes a clear differentiation between estimator and estimate; the former is called *schatter*, which is a synonym for 'rifleman', and the latter is referred to as *schatting*, which is a synonym for 'shot'. Of course shots from a rifleman vary around the target (that is, the parameter to be estimated).

Master

A common principle to obtain an estimate $\hat{\theta}$ of an unknown parameters θ is the *least squares method*, which goes back to *Legendre* (and *Carl Friedrich Gauss*). That value $\hat{\theta}$ is chosen as estimate for the parameter θ, which minimizes the sum of the squared deviations between all the outcomes and the parameter θ.

In the case of the mean of a theoretical distribution of random variable $\boldsymbol{y}$, that is μ, we have to minimize $\sum_{v=1}^{n}(y_v - \mu)^2$. This leads to

$$\hat{\mu} = \frac{\sum_{v=1}^{n} y_v}{n} = \bar{y}$$

as the estimate. In terms of random variables we get the estimator

$$\hat{\boldsymbol{\mu}} = \frac{\sum_{v=1}^{n} \boldsymbol{y}_v}{n} = \bar{\boldsymbol{y}}$$

We see that the sample mean is a proper estimator for the mean of a theoretical distribution (or a population) – an intuitively evident result.

The empirical mean as an estimator for the mean of a theoretical distribution has special properties. It is *consistent* and *unbiased*. By consistency and lack of bias we mean the following: generally, an estimator $\boldsymbol{S}$ for the parameter θ is consistent with regard to θ, if the estimate $S = \hat{\theta}$ tends, for $n \to \infty$, toward the parameter θ. An estimator $\boldsymbol{S}$ of the parameter θ is called unbiased with regard to θ, if for all θ the following holds: $E(\boldsymbol{S}) = \theta$. The difference $E(\boldsymbol{S}) - \theta = w_n(\theta) \neq 0$ is called the bias of the estimator – the suffix n indicates that the bias may depend on n.

Master

Example 6.9 Consistency and unbiasedness of the average age of pupils in primary schools

Given a sample of size n of pupils in the third grade of primary school, we calculate the average age in months. Then with additional sampled pupils, as $n \to N$ the newly calculated mean comes nearer to the mean age of all pupils in the third grade, i.e. the mean μ of the population. Therefore, this is a consistent estimator: the estimate tends with $n \to N$ ($n \to \infty$) toward the parameter μ. If we obtain not only a single sample (of size n), but many samples, namely m, theoretically infinitely many, that is $m \to \infty$ (all with size n), and we calculate for each sample the mean age in months, then the mean (expectation) $\bar{y}$ of all these m means, $\bar{y}_m$, equals the mean μ of the population. That is, we have an unbiased estimator as well: the mean of the estimates tends for $m \to \infty$ toward the parameter μ – independent of the size n.

The reader can empirically understand this unproven phenomenon of unbiasedness by, for instance, throwing a die six times each, repeated in approximately 30 runs, and recording the integer on the top side. The mean of the means of the observed integers will be almost 3.5, which is equal to the theoretical mean of realized integers.

Master

An estimator $\boldsymbol{S}$ for the parameter θ is called an *efficient estimator*, if its variance $V(\boldsymbol{S})$ under all unbiased estimators for that parameter θ is kept to a minimum. It can be shown that the mean of a random variable is consistent, unbiased, and efficient with regard to the mean μ – independent of which distribution is given.

As concerns the sample variance as defined in Chapter 5 (that is, with the term $n - 1$ in the denominator), its respective estimator is actually an unbiased one. However, the possible estimator $\tilde{\boldsymbol{\sigma}}^2 = \frac{1}{n}\sum_{v=1}^{n}(\boldsymbol{y}_v - \bar{\boldsymbol{y}})^2$ is biased, because it can be shown that

$$E(\tilde{\boldsymbol{\sigma}}^2) = \frac{(n-1)\cdot\sigma^2}{n} = \sigma^2 - \frac{1}{n}\cdot\sigma^2$$

Doctor

An estimator $\boldsymbol{S}$ for the parameter θ is called *sufficient* if the total (relevant) information of the random sampled outcomes is contained in this estimator.

Master

In contrast to the least squares method, there is another approach to gain an estimate $\hat{\theta}$ of an unknown parameter θ. It is the *maximum likelihood method* (MLM). Here any parameter is estimated from the outcomes in such a way that the latter have maximum likelihood (in other words: maximum plausibility) of resulting. Starting from the outcomes y_v, $v = 1, 2, \ldots, n$, the *likelihood function* $L(y \mid \theta)$ is of interest. This is a function of θ, given all y_v. 'Likelihood' does not mean probability, because the function $L(y \mid \theta)$ does not refer to a random variable but to fixed values.

Doctor

The application of the maximum likelihood method requires knowledge of the kind of distribution of the random variable which is modeling the character – we call a random variable's probability function or density function a likelihood function if it is considered as a function of the unknown parameter(s). That parameter value which maximizes this likelihood function for the given outcomes is called the *maximum likelihood estimate* – the corresponding random variable is the *maximum likelihood estimator*. By the way, the maximum likelihood estimator of the expectation of the normal distribution is identical to the least squares estimator.

Summary

Analogous to *statistics* in samples, *parameters* serve for the characterization of the *population*. A character is modeled by a *random variable*, which follows, with respect to the population, some *theoretical distribution*. This is determined by such parameters. Several methods of statistics look for *estimators for these parameters,* i.e. the unknown parameters of the population will be *estimated* by the given sample.

References

Kubinger, K. D. (2009b). *Psychologische Diagnostik – Theorie und Praxis psychologischen Diagnostizierens* (2nd edn) [Psychological Assessment – Theory and Practice of Psychological Consulting]. Göttingen: Hogrefe.

Rasch, D. & Guiard, V. (2004). The robustness of parametric statistical methods. *Psychology Science*, *46*, 175–208.

7

Assumptions – random sampling and randomization

In this chapter we show that the application of statistical methods in planning and analysis of empirical research makes it necessary that the research units are taken from the respective population randomly. Different options for random sampling and randomization of research units in experimental designs are discussed. We will particularly consider complete and incomplete block designs for the elimination of known noise factors.

In responding to an exactly formulated research question (in the area of psychology), a census of the underlying population obviously is not often considered. We have to get by with samples. On the other hand we like to make a 'virtue' of 'necessity': if a census is not essential, but a sample is sufficient for verifying the (psychological) phenomenon in question with given precision, then it would be completely uneconomic to not make a research principle out of this. We will thus look for a sample that is no greater than required in order to describe the population with the desired precision; we will discuss later what is meant by 'precision'.

In order to follow this principle of research, i.e. to conduct, instead of a census, an investigation on a sample of the population, we need a scientific *hypothesis*. That is, a certain assumption underlying the exactly formulated research question – otherwise we would probably not conduct a research study. Aside from a few cases, such an assumption is always of the form: 'there are differences between certain groups of research units' or 'there is an association between two characters'. We call this assumption (purely technically) the *alternative hypothesis* in contrast to the *null hypothesis*, which always states: 'there are no differences; there are no associations', or more generally 'there are no effects'.

Example 3.1 – continued
The psychological effects of a hysterectomy are to be investigated. Summarized, our first reflection on this question leads to the following considerations and

Statistics in Psychology Using R and SPSS, First Edition. Dieter Rasch, Klaus D. Kubinger and Takuya Yanagida.

approaches to planning a study. We focused on the 'psychological effect' on 'self esteem' of a woman or her 'psychological stability', which we suppose to be measurable by means of a psychological assessment tool, Diagnosticum *Y*. Then, we recognized that an arbitrary selection of patients would be critical. As a minimum we have to assess 'healthy' women by *Y* as well (that is women who have not (yet) been positively diagnosed with respect to the given disease), or even better a group of patients with a surgery of comparable severity (gall bladder surgery) as control group. It was also considered critical to provide any arbitrary number of investigated women. The critic's concern would be that nobody would be able to conclude that the results (such as any differences established between the arbitrarily selected groups) were generalizable, i.e. applicable to all women – so that from these results could be deduced a mandatory psycho-hygienic need for action for coping with and prevention of psychological effects. But of course this is the actual aim of every study, namely to generalize: nobody would be interested in differences related only to those two specific groups. Finally, we recognized that it was about an *ex post facto* study; that is, that assignment of the patients to the two groups was not done, as in an experiment, in a random way before the exposure to different treatments, but after the appearance of the disease and hence independent from the researcher. In the end, the exactly formulated research question is: 'Is the psychological effect of a hysterectomy more serious than that of another surgery of comparable severity?' From this, the following alternative hypothesis is deduced: 'The psychological effect of a hysterectomy is more serious than that of another surgery of comparable severity.' In contrast, the null hypothesis is: 'The psychological effects of a hysterectomy are equal to the effect of any other surgery of comparable severity'!

As regards the subject matter, the content of a null hypothesis might often seem vacant, but it is required formally: hypotheses in the empirical sciences concerning stated phenomena can never be proven (in contrast to mathematics); it is only possible to reject a hypothesis based on the evidence of its incorrectness which has been empirically observed (at least once) (see *Karl Popper's* principle of falsification, e.g. Popper, 1959). The statistical strategy in this is to conceptualize a (statistical) test for each research question, in order to examine the null hypotheses with the help of the empirically derived data. The data either (clearly) argues against the null hypothesis or does not (clearly) argue against it; in the latter case, by formal logic we have to maintain it, but the alternative hypothesis, which is in fact of interest, has to be rejected.

For testing a null hypothesis we can revert to the statistics as offered by descriptive statistics. However, these are not sufficient. If two different samples differ, for instance, with respect to the mean, it is not possible to reject or maintain a hypothesis based merely on this. It has to be taken into account that each sample represents only a part of the underlying population. It is thus possible that great distortions have occurred, so that the concretely observed means do not concord sufficiently with the respective parameters in the population in question. Therefore, not every (minor) deviation between two samples is to be interpreted in the way that the null hypothesis has to be rejected.

The crucial element of statistical inference in this context is: if we have a random sample – solely chance decides whether an element of the population is included in the sample or not – calculus of probabilities can be applied to solve our problem. Using it, we can calculate how probable it is to randomly obtain the data which we observed (or data which intuitively argues even more against the null hypothesis) even though in reality the null hypothesis is true. The exact value of this probability will finally decide if we maintain the null hypothesis or reject it (and, in consequence, accept the alternative hypothesis).

We thus do not proceed without random sampling in statistics for answering research questions in scientific psychology; specifically we cannot calculate the error of our decision between null and alternative hypothesis (by means of calculus of probabilities) otherwise, and thus cannot make a judgement about the possible deviation between sample and population.

7.1 Random sampling in surveys

We generally distinguish between surveys and experiments (see Chapter 4), because the latter allow, from a theoretical scientific point of view, more meaningful, that is causal, conclusions. Nevertheless, it has to be assumed for experiments as well that the research units (in psychology mostly persons), who are later assigned to the different experimental treatments, have been selected randomly before; so, the following explanations apply just as well for experiments.

Besides the fact that random samples are required in order to eventually be able to take into account deviations from the population for the interpretation of the research results, the following is a chief concern: *non-random sampling* has the disadvantage that there is always the risk of systematical, not just random, errors. Deviations between sample and population thus cannot be interpreted either concerning their direction or their extent.

Master **Example 7.1** Survey for an election forecast

Although nowadays surveys are not really conducted in such a naive way anymore, this example, in which a forecast concerning the result of a political election will be made by means of the (non-random) sampling method of *haphazard sampling*, is convenient from a didactic point of view. If we situated ourselves, as the researcher, at a bustling place in the center of the capital at 12 noon on a workday in order to interview the first 200 persons who pass by with regard to their political opinion (election preference), we would make the mistake of systematically selecting certain groups of persons in an over- or under-represented way compared to the population (of all eligible voters). As it is to be expected that these groups of persons typically show a different election behavior, we have to anticipate systematical error in the result of the research. Namely, would

1. persons working at the periphery of the city, and other correspondingly locally 'segregated' persons (such as university students or housewives with infants) be under-represented;
2. sales agents with contact addresses in the center be over-represented;
3. pupils eligible to vote and other persons who are bound to certain working hours (teachers, public officers) be under-represented;
4. elderly people in rest homes and poorly mobile elderly people, as well as chronically ill persons in clinical treatment be under-represented;
5. persons from certain social classes be over- or under-represented, respectively.

Here, the following problems are not under consideration at all: certain electors do not announce their true preferences; some electors completely refuse to give a response, other electors give the answer which seems to be socially desired in the given interaction with the interviewer; and often electors change their preference as a consequence of the publication of a survey.

Master

Example 7.2 Telephone survey for an election forecast

Currently election surveys are conducted by means of telephone survey, mostly using a 'well-proven' sample, as the forecast gained in this way in former election surveys has corresponded quite well with the later result of the election. If such a well-proven sample is not to hand, we would have to anticipate the following biases compared to the population:

1. Persons who do not own a fixed line network but only a non-registered mobile phone would not be reached.
2. Persons who are often away from the home would be under-represented.
3. In shared flats with more than one eligible voter, the ones who routinely do not answer the phone first would be under-represented.
4. Persons who do not answer the phone, when they do not know the number which is calling, would be under-represented.

A random sample can be realized theoretically easily, when all elements of the population are known and registered in a data file. Then who will be included in the sample is decided by means of a random number generator. Practically, this approach often is not applicable, because not all elements of the population are known and registered in a data file.

For Lecturers:

A case concerning a presidential election in the US, which is cited as a classical case in the literature, shows that samples using haphazard sampling can be entirely useless (from Wallis & Roberts, 1965, p. 102):

> In 1936, the *Literary Digest*, a magazine that ceased publication in 1937, mailed 10,000,000 ballots on the presidential election. It received 2,300,000 returns, on the basis of which it confidently predicted that Alfred M. Landon would be elected. Actually, Franklin D. Roosevelt received 60 percent of the votes cast, one of the largest majorities in American presidential history. One difficulty was that those to whom the *Literary Digest's* ballots were mailed were not properly selected. They over-represented people with high incomes, and in the 1936 election there was a strong relation between income and party preference. In the preceding four elections, ballots obtained in the same way had correctly predicted the winners, but in those elections there was much less relation between income and party preference.

7.2 Principles of random sampling and randomization

Randomization in experiments means that a random assignment of the research units to the treatments is performed, at which the research units themselves have previously been taken randomly from the population.[1]

7.2.1 Sampling methods

If one talks about 'sample', this is mostly meant as an abbreviation for random sample. A random sample is the result of a *random sampling method*. *Sampling method* describes the rules by which elements of a given (finite) population are selected for the sample. If these rules depend on chance, it is thus a random sampling method; this means that each element in the population can be part of the sample with a given probability greater than zero.

Bachelor As soon as a sample is given, one cannot determine if it is a random sample or not, without knowing the exact sampling method. Thus, with regard to a sample's randomness, it is not decisive which elements of the population are included in the sample, but only how the sample was taken. If, for instance, in lottery the numbers 1, 2, 3, 4, 5, 6 win, this does not generally argue against a random sample of the numbers 1 to 49. Samples which have been taken using random sampling can definitely appear to be extreme and seem not to be free from systematical influences. In lotteries, it is demonstrated, by drawing the numbers while a camera is filming the procedure, that the appearance of seemingly extreme numbers is nevertheless random. However, certainly the principle of chance leads to the fact that extreme sample results occur with a very low probability.

Master One often finds a tighter definition of random sampling; namely that each element of the population has got the same probability of being part of the sample. However, the definition given above, according to which each element has got only a 'given probability', applies as well if we take a random sample from *strata* from a *stratified population* – this happens with probabilities which are proportional to the size of the strata.

Random sampling can be conducted according to one of two different principles:

1. *Random sampling with replacement*: after an element of the population has been taken, it is registered (and the desired information is gathered, e.g. the measurement values of the characters in question are observed) and then put back into the population; in that way it can possibly be taken again. Obviously, this is only possible if the gathering of the information does not influence the elements (this occurs very rarely in psychological treatment or psychological testing) or even destroys them.

2. *Random sampling without replacement*: after an element has been taken from the population, it is not available any more for re-selection.

---

[1] Rarely, the term randomization is used as a generic term and then includes the random sampling in surveys as well.

Master Doctor

Sampling with replacement leads to the fact that the population can be modeled well by an infinite theoretical population even if it is small itself. In sampling without replacement this is only the case when the population is big (approximately $N > 1000$ and $\frac{n}{N} < 0.1$). In sampling with replacement, the sample size n can be bigger than the size of the population N. Naturally, this is not possible in sampling without replacement.

Doctor

If, from a population of size N, a sample without replacement of the size $n < N$ is to be taken, $\binom{N}{n}$possibilities exist for this (see Section 6.2.1). Each of these possibilities has the same probability, which equals $1/\binom{N}{n}$ according to Formula (6.1): the one sample actually taken gives $g = 1$; all possible samples make up $e = \binom{N}{n}$. In this case we say it is an *unrestricted random sampling* (without replacement) in which each possible sample occurs with the same probability.

Different technical options exist for the practical realization of unrestricted random sampling. In the worst case, the method of drawing lots from a box can still be applied. In science research work, this is nowadays done with the help of random number generators, which are used in simulation studies as well (for more detailed information see Section 14.4).

For Lecturers:

The pool from which the random numbers are generated depends on the special random number generator. This can be, for instance, the set of all 32-bit numbers or the set of the real numbers in the interval [0, 1]. Often, it is desirable that the random number generator creates uniformly distributed values; but for some statistical simulations random number generators which produce a given distribution (e.g. a normal distribution or a binomial distribution) are of interest. Most of the random number generators of computer programs produce (only) pseudo-random numbers and therefore are called *pseudo-random number generators*. They serve for a sequence of numbers which seems to be but is not random, because it is calculated by means of any deterministic algorithm: every time when the random number generating process is started with the same starting value (that is the so-called *seed*), the same pseudo-random sequence of numbers is produced. This can be avoided by first determining the 'seed' randomly. For simulation studies and for random sampling, pseudo-random number generators are completely sufficient and thus are used in most cases.

Master

Example 7.3 A random sample without replacement, of the size $n = 5$, will be taken from a population with $N = 100$ elements

In an unrestricted random sample, each element is taken with a probability of 0.05. Element 8, for instance, will be taken with a probability of $\frac{1}{100}$ in the first step, or will not be taken with a probability of $\frac{99}{100}$, respectively. The probability that element 8 is taken (only) in the second step equals $\frac{99}{100} \cdot \frac{1}{99}$; that is, 0.01 as well. Equally, the probability for element 8 up to the fifth step equals 0.01 each time (e.g. $\frac{99}{100} \cdot \frac{98}{99} \cdot \frac{97}{98} \cdot \frac{96}{97} \cdot \frac{1}{96}$). Thus, the sum of these probabilities results as an overall probability of 0.05 that element 8 comes into the sample.

In **R**, we randomly draw 5 elements out of 100 (numbers 1, 2, ..., 100) by typing

```
> sample(1:100, size = 5, replace = FALSE)
```

i.e. we apply the function `sample()` and set the number-set `1:100`, from which we randomly draw elements, as the first argument. With the second argument, `size = 5`, we define the number of elements to be drawn, and specify by setting `replace = FALSE` to draw the sample without replacement. As a result we get a random sequence of 5 numbers from 1 to 100. Since we draw a set of 5 numbers, 1 out of $\binom{100}{5} = 75\,287\,520$ possible sets (permutations) – whereas each one has $5! = 120$ different orders – we hardly expect to get the same result twice. For instance we get, after adding a start value for the random number generator (only for didactic reasons – for the purpose of trying out at home)

```
> set.seed(321)
```

as the first result

```
[1] 96 93 24 25 38
```

and the next (without using `set.seed()`)

```
[1] 35 45 29 44 78
```

However, we can conduct a *systematic sampling* as well. Therefore, we randomly choose a number between 1 and $\frac{100}{5} = 20$ and get, for instance, 12 (*random start sampling*). Starting from this element, we choose every following 20th element and obtain the elements 12, 32, 52, 72, and 92 for the sample. Certainly, the elements 32, 52, 72, and 92 have been chosen systematically depending on element 12, but because 12 is random they are random too. Obviously, this procedure can, however, be realized more easily/quickly than an unrestricted random sampling.

Sometimes the population, of size N, is divided, in a way which is relevant regarding the content, into s sub-populations of size $N_1, N_2, \ldots, N_i, \ldots, N_s$. In particular, the population can occasionally be divided into sub-populations according to the levels of an assumed noise factor. Then, the sub-populations are called strata. If the researcher wants to take samples of size n from the population, he/she has to be concerned that, in the case of an unrestricted random sampling, not all strata are included in the sample at all, or at least not in an appropriate relation. In this case, it is preferable to conduct a *stratified random sampling*. In this, sub-samples of the size $n_i (i = 1, 2, \ldots, s)$ are taken from the ith stratum. The sub-samples are taken from the respective stratum using unrestricted random sampling. Apart from random sampling, this equates to the so-called *quota sampling*, if n_i/n is chosen proportionally to N_i/N.

Master Doctor

Example 7.4 An intelligence test for 12-year-olds will be standardized

Standardization (see also Section 2.3) in the subject of psychological assessment means the creation of a frame of reference, in relation to which the individual test score can be put. In this, the respective standardization tables have to rely on a sample which is representative for the population (see e.g. Kubinger, 2009b).

We suppose that, as an exception, this is about the very tightly defined population of 12-year-olds in Vienna (Austria). It is to be assumed that the social structure of the different districts of Vienna is different and also influences the test score. If we now want to avoid that exclusively children from school classes from only 1 of the 23 districts are observed – which would be quite unlikely, but not impossible when using unrestricted random sampling – then we should use the information about the structure of the urban districts.

Master Doctor

Example 3.1 – continued

In this example, the psychological effects of hysterectomy were to be determined.

If we were to investigate just the first group of patients who come along, we would obviously be applying haphazard sampling. However, actually we wanted to gain insight into all hysterectomy patients compared to other female patients with gall bladder surgery in Western civilization – or at least the English-speaking countries. Our insight will be applicable for the typical age of such patients, their other typical psychosocial characteristics, but particularly also for patients in the conceivable future.

However, a stratification of the population and a corresponding selection according to some relevant psychosocial characteristics is practically impossible, as the underlying frequencies are not known or cannot be made public easily. Stratification according to the hospitals (within the English-speaking area) is possible. If we realize this, it is a multi-center study.

Master Doctor

Random sampling is called *multi-stage* if sequences of random sampling are done one after the other. In the first stage, a random sample of so-called primary units is taken from the population. Each primary unit is then to be regarded as a separate population of so-called secondary units, which are selected using random sampling again, and which already constitute the elements of the sample themselves in a procedure with (only) two stages. From each of the primary units, which have been selected in the first stage, the secondary units are selected in the second stage. In a procedure with three stages, we call the elements of the sample tertiary units and so on.

Master Doctor

Example 7.4 – continued

In Table 7.1 we find the exact information concerning the population and its strata. Accordingly, we can apply unrestricted random sampling per stratum. Assuming that the sample size should equal $n = 100$, we would have to include, for instance, 3 children from the Hernals district. However, we could also apply two-stage random sampling proportional to the size, where in the first stage (preferably without replacement) firstly some of the 23 urban districts (that are the primary units) are selected randomly. Then, unrestricted random sampling

Table 7.1 Number of inhabitants overall and the number of 12-year-olds (estimated from the number of children) in the urban districts of Vienna, 2009. The last column indicates the relative frequency of 12-year-olds in % (from Statistik Austria, 2009)

Urban district	Number of inhabitants overall	Number of 12-year-olds N_i	$100 \cdot f_i = 100 \cdot \frac{N_i}{16\,111}$
Innere Stadt	16 958	97	0.60
Leopoldstadt	94 595	835	5.18
Landstrasse	83 737	684	4.25
Wieden	30 587	228	1.42
Margarethen	52 548	431	2.68
Mariahilf	29 371	214	1.33
Neubau	30 056	190	1.18
Josefstadt	23 912	160	0.99
Alsergrund	39 422	275	1.71
Favoriten	173 623	1 738	10.79
Simmering	88 102	948	5.88
Meidling	87 285	852	5.29
Hietzing	51 147	441	2.74
Penzing	84 187	748	4.64
Rudolfsheim	70 902	658	4.08
Ottakring	94 735	828	5.14
Hernals	52 701	493	3.06
Währing	47 861	412	2.56
Döbling	68 277	590	3.66
Brigittenau	82 369	783	4.86
Floridsdorf	139 729	1 626	10.09
Donaustadt	153 408	1 926	11.95
Liesing	91 759	954	5.92
Total	1 687 271	16 111	100.00

is performed with respect to all of the 12-year-olds in the chosen districts. A three-stage sampling would be meaningful as well, in which a certain number of schools would be selected randomly from each of the districts as secondary units and, after that, from these schools the pupils would be selected, randomly as well. Bear in mind that the sampling proportional to the size can be done according to the number of inhabitants, the area of the districts (less recommendable in this case), the number of schools, or the number of (12-year-old) pupils in the urban districts. According to Table 7.1 it is clear that a higher number of inhabitants does not necessarily mean that the number of 12-year-olds is higher as well. For instance, the Favoriten district has more inhabitants than the Donaustadt district, but more children live in the latter than in Favoriten. It is thus appropriate for our research to choose the districts proportionately with regard to the number of children. If it had been difficult to determine the number of 12-year-olds, we would have had to use the number of inhabitants.

It is also possible in multi-stage sampling that in the last stage all elements are considered; thus here, for instance, all of the 12-year-olds in the selected schools. However, in this case not the children, but the schools are the research units – in this case we say it is a *cluster sampling*.

Master
Doctor

Stratified sampling must not be mixed up with quota sampling which is common nowadays within surveys. Indeed, in this case also, sub-populations are determined with regard to some characters and factors which are considered particularly important, and the relative proportion of these sub-populations in the overall population is taken into account using respective quotas for the sampling. It is thus assured that the sample is in fact representative for the population with regard to these characters and factors. For instance, by doing so, the percentage of women in the sample concords with that of the population, as well as the percentage of, for instance, certain levels of education or age groups. However, the concrete selection is left to the interviewers, so that the selection does actually not depend only on chance.

Master

Example 7.5 Survey for an elective forecast using quota sampling

If the selection of the persons who are to be included in the sample is left to the interviewers, they will naturally try to minimize their effort. They just have to fulfill certain quotas, such as: 'Interview two women of medium education level, living in a city, 50 to 60 years old, as well as five men from the lower educational class from a rural area, of between 18 and 22 years', and so on. Typical mistakes, which could lead to a bias in the sample compared to the population, are:

1. Successfully interviewed persons are asked to arrange contact with other persons from their circle of acquaintances who are also willing to take part in the survey ('pyramid scheme'). However, circles of acquaintances are often characterized by similar political opinions.
2. In telephone surveys the interviewer possibly stays on the same page of the directory, so that often several persons from one family living in different households are addressed – in this case often similar political opinions are given as well.

Summary

If one wants to use a sample for characterizing a population, the sample has to be taken using a *random sampling method*; in this, only chance decides whether a research unit is included into the sample or not. In this case, calculus of probabilities can be applied; with its help we can calculate to what extent the observed data argues for or against a certain *null hypothesis*. *Non-random sampling methods* have the disadvantage that they always run the risk of producing not only random error, but systematical error. Deviances between the sample and the population thus cannot be estimated either with respect to their direction or with respect to their size.

7.2.2 Experimental designs

It is not always easy to determine how the assignment of research units to the treatments/conditions is to be performed. The schedule according to which this is done is called the *experimental design*. A simple or *completely randomized design* is a design in which each treatment/condition can be assigned to each research unit with the same probability. The assignment itself is to be done randomly. And it has to be assumed that the available research units do represent a random sample of the underlying population.

Master Doctor

Example 7.6 The influence of different item response formats in personality questionnaires on the tendency to social desirability is to be investigated

We want to assign (for reasons which are not explained here in detail: exactly) $n = 104$ persons randomly to $v = 4$ treatments/conditions: (1) dichotomous response format (the testee can answer here by 'is true' or 'is not true' to questions like 'I am often despondent'); (2) five-categorical response format ('yes, mostly', 'yes, from time to time', 'I do not know', 'no, rarely', 'no, almost never'); (3) analogue scale-response format (see Section 2.4.3), (4) Q-sort as response format (see Section 2.4.4). Each condition is to be applied on 26 persons.

In **R**, we create a matrix consisting of 4 columns and 26 rows, in which $n = 104$ persons are randomly assigned. We type

```
> matrix(sample(104, size = 104, replace = FALSE), ncol = 4)
```

i.e. we again apply the function `sample()` – see Example 7.3 for the meaning of the arguments – and set the result as the first argument in the function `matrix()`. We add the argument `ncol = 4` to get a matrix with 4 columns.

As a result we get (shortened output):

```
        [,1]  [,2] [,3] [,4]
[1,]     50    58    6   47
[2,]     17    76   64   60
[3,]     36    24   15   95
...
...
...
[24,]    19    42   53   62
[25,]    23    92    2   70
[26,]    65    46   78   79
```

Each column represents one of $v = 4$ treatments, and the numbers stand for the randomly assigned persons, who were serially numbered from 1 to 104.

Sometimes, restrictions due to the content exist concerning the assignment of the research units to the treatments. This is mostly the case when one or more noise factors come into consideration.

Doctor

An experimental design is called a *block design* with b blocks, if a certain noise factor is taken into account by means of the creation of blocks. The number $k_j (j = 1, 2, \ldots, b)$ of the research units in the jth block is called the *block size* of block j; the sum of the k_j equals the sample size n. The number of research units which occur with the ith treatment is called the replication r_i of the treatment i. The sum of the r_i equals the sample size n as well.

For Lecturers:

Many of the notations in statistics have a long tradition. Numerous statistical procedures can be traced back to the British statistician *Ronald A. Fisher*, who worked in an agricultural research station near London for a long time. He often chose symbols from this domain. The treatments often were varieties, and the observed character the yield. The letter v for the number of the treatment comes from the word *variety*; the denotation of y for the observations from the word *yield*, and the denotation of r for the number of replications from the word *replication*.

Doctor

Example 7.6 – continued
One possible noise factor is sex – in psychology a noise factor is often called a *moderator*. In this case we would have $b = 2$ blocks, which are $j = 1$ (female) and $j = 2$ (male). It has to be guaranteed that, in all of the 4 treatments, both categories of this moderator have to be realized equally frequently; thus $k_1 = k_2 = 13$.

Doctor

When constructing a block design, the procedure has to be as follows. First the number, b, of blocks (number, b, of levels of the noise factor) is to be determined and defined by the contents. Then, the research units are to be assigned to the blocks, if they are not already assigned in some natural way, as for instance twins or siblings are. This assignment has to be done in an *exhaustive* and *disjunctive* way. We assume here that all b blocks contain the same number of research units k. The randomization is now to be performed in the following way: the research units in each block are to be assigned randomly to the treatments. A randomization is conducted separately for each block.

Doctor

Example 7.7 Altogether, nine clients will be randomly assigned to three psychotherapists

The three psychotherapists are to be understood as blocks; their possible influence is thus to be eliminated using the creation of blocks. Naturally, the assignment will be done randomly.

In **R**, we type

```
> matrix(sample(9, size = 9, replace = FALSE), ncol = 3)
```

– see Example 7.3 or 7.6 for the workings of the functions and arguments.

As a result we get:

```
[1,]    8    3    4
[2,]    6    7    1
[3,]    5    9    2
```

Doctor In the case of $k < v$ the situation is different. In these cases, we say that we have incomplete blocks in an *incomplete block design*.

Doctor **Example 7.8** Planning a study with five treatments in grouped research units

If we want to test $v = 5$ postoperative treatments in patients in triple bedrooms and to eliminate possible 'room effects', the rooms are our blocks and the block size of each room is $k = 3$. So in every block there are less research units (patients) than treatments, which are to be investigated. Thus it is not possible that all treatments are applied simultaneously in one room.

Doctor As can be shown, the goal to determine all treatment effects with the same precision (and also all possible differences between treatment effects with the same precision) requires that all treatments occur with the same frequency (r times) in the design. The precision of a difference of two treatments depends here on how many times, that is λ times, the respective pair of observations is in the experimental design given. From this goal a certain experimental design results: an incomplete block design (with equal block size k) is called a *balanced block design*, if each of the v treatments occurs with the same frequency (r times) in the experiment, and the number of occurrences of each of the possible treatment pairs always equals λ.

In a balanced incomplete block design we have $n = vr = bk$. On one hand, the total number of research units n results from the product of the number v of treatments and the number r of the applications of these treatments in the experiment; on the other hand, n results as the number b of blocks times the block size k. It also has to be true that $\lambda(v - 1) = r(k - 1)$, as in v treatments $\frac{v(v-1)}{2}$ pairs of treatments exist (see Section 6.2.1), each of which occurs λ times, so that the overall number of pairs in the experiment equals $\lambda\frac{v(v-1)}{2}$. However, this number can also be calculated by multiplying the number b of blocks by the number $\frac{k(k-1)}{2}$ of pairs which appear in each of the blocks; that is $\lambda\frac{v(v-1)}{2} = b\frac{k(k-1)}{2}$; replacing bk by vr, we obtain $\lambda\frac{v(v-1)}{2} = v\frac{r(k-1)}{2}$. There are thus two necessary conditions for the existence of a balanced incomplete block design. This reduces the number of possible quintuples of natural numbers v, b, r, k, *and* λ considerably. If we determine a balanced incomplete block design for reasons of contents by three of these parameters, for instance by v, k, and λ, then the remaining parameters can be calculated with the help of the two formulas. Bear in mind that these *necessary* conditions are not always *sufficient*[2] for the existence of a balanced incomplete block design. For example, the values $v = 16$, $r = 3$, $b = 8$, $k = 6$, $\lambda = 1$ do

[2]For an event A it may be necessary that B has occurred, but the fact that B has occurred may not be sufficient for A to occur; for instance, because C is necessary as well for A.

fulfill the necessary conditions, as $16 \cdot 3 = 8 \cdot 6$ and $1 \cdot 15 = 3 \cdot 5$; anyway such a design does not exist.

It is essential in incomplete block designs that they are *connected block designs*; *disconnected block designs* cannot be analyzed as a whole, but have to be treated like two or more independent experimental designs. An incomplete block design is connected, if for each pair (A_k, A_l) of treatments $A_1, A_2, \ldots, A_v$, a sequence exists which starts with A_k and ends with A_l such that succeeding treatments in this sequence both occur in at least one block.

The formal representation of block designs can be done by creating a so-called incidence or assignment matrix. This type of matrix has v rows and b columns, and it contains the n_{ij}, which describe how often the ith treatment (the ith row) occurs in the jth block (the jth column). Generally in incomplete block designs it is convenient to use, instead of the incidence matrix, a compact notation for its characterization: in this, each block corresponds to a bracket term in which the numbers of the treatments included in the block are written.

Doctor

Example 7.9 Formal example of an incomplete block design without relation to any content

A block design with $v = 4$ treatments and $b = 6$ blocks is defined by the following incidence matrix:

$$\begin{pmatrix} 1 & 0 & 1 & 0 & 0 & 0 \\ 0 & 1 & 0 & 1 & 1 & 1 \\ 1 & 0 & 1 & 0 & 0 & 0 \\ 0 & 1 & 0 & 1 & 1 & 1 \end{pmatrix}$$

As this matrix contains the number zero, it represents an incomplete block design. In the compact notation, it can be written in the form (1, 3), (2, 4), (1, 3), (2, 4), (2, 4), (2, 4). As written, the first bracket, for instance, represents block 1, in which the treatments 1 and 3 occur; for example, the first column, which defines the first block, contains the number 1 in rows 1 and 3, which corresponds to treatments 1 and 3.

Here, the first and the second treatment do not occur together in any of the six blocks: no continuous path of vertical or horizontal steps between the matrix content '1' and '1' can be found between these two treatments; it follows that the design is an unconnected block design. The consequence of this fact becomes clear when we change the numeration of the blocks and the treatments or interchange the columns and rows of the incidence matrix in an appropriate way. Doing so, we do not change anything in the structure of the design. So we interchange the blocks 2 and 3 and the treatments 1 and 4, thus the columns 2 and 3 and the rows 1 and 4 in the incidence matrix. The following matrix results:

$$\begin{pmatrix} 0 & 0 & 1 & 1 & 1 & 1 \\ 0 & 0 & 1 & 1 & 1 & 1 \\ 1 & 1 & 0 & 0 & 0 & 0 \\ 1 & 1 & 0 & 0 & 0 & 0 \end{pmatrix}$$

We can see now that the experimental design consists of two designs with separated partial sets of treatments. In one design we have, in the rearranged notation, two treatments (1 and 2) in the last four blocks. In the other design we have two more treatments (3 and 4) in the first two blocks.

Doctor

Example 7.8 – continued

This concerns five psychological postoperative treatments, which are to be investigated in patients in triple bedrooms. We want to construct a balanced incomplete block design. The triple bedrooms represent blocks of the size $k = 3$; we have $v = 5$ psychological postoperative treatments. We suppose that the patients of seven triple bedrooms are available for the experiment; that is $n = 21$. As we remember, it has to be true that $vr = bk = n$; that is $5r = 21$, as well as $\lambda(v - 1) = r(k - 1)$; that is $4\lambda = 2r$ or $r = 2\lambda$, and consequently $5r = 10\lambda = 3b = 21$. As r and b have to be integers, we cannot construct a balanced incomplete block design under the given circumstances.

Thus it is better to plan our experimental design using relevant software programs – here it will turn out that we cannot incorporate 7 triple bedrooms when testing 5 treatments.

In **R**, we apply the package `OPDOE`, which we load after its installation (see in Chapter 1) using the function `library()`. Next we type

```
> bibd(v = 5, k = 3)
```

i.e. we apply the function `bibd()` and set the number of treatments with `v = 5` and the block size with `k = 3`. The program selects an appropriate method out of 21 construction methods available (which can be found in Rasch, Pilz, Verdooren, & Gebhardt, 2011).

As a result we get:

```
block treatments
1    (1, 2, 3)
2    (1, 2, 4)
3    (1, 2, 5)
4    (1, 3, 4)
5    (1, 3, 5)
6    (1, 4, 5)
7    (2, 3, 4)
8    (2, 3, 5)
9    (2, 4, 5)
10   (3, 4, 5)

v = 5 k = 3 b = 10 r = 6 lambda = 3
```

In column `block` you see the number of the respective block and, beside it, in column `treatments`, the three respectively assigned psychological postoperative treatments. At the end, the five characteristics of the block design are specified.

First, we write the resulting design in compact notation; each of the bracket terms corresponds to a triple room; the numbers included are the numbers of the postoperative treatments numbered consecutively from 1 to 5: (1, 2, 3); (1, 2, 4); (1, 2, 5); (1, 3, 4); (1, 3, 5); (1, 4, 5); (2, 3, 4); (2, 3, 5); (2, 4, 5); (3, 4, 5). If we consider treatment 1, for instance, it is applied in the rooms 1 to 6 for one patient each; treatment 4, for instance, is applied 6 times also – once in each of the rooms 2, 4, 6, 7, 9, 10. If we pick out, in addition, the pair (2, 5) in order to examine the frequency of treatment pairs, we find this pair in the rooms 3, 8, and 9, and thus in fact exactly $\lambda = 3$ times. We do not have to verify that this block design is connected, because each treatment is connected to each other by at least one of the pairs.

Doctor

Example 7.9 – continued

It is difficult to prove that an incomplete, unbalanced block design is connected. We add an additional observation to our design for the treatment 1 in block 4:

$$\begin{pmatrix} 1 & 0 & 1 & 1 & 0 & 0 \\ 0 & 1 & 0 & 1 & 1 & 1 \\ 1 & 0 & 1 & 0 & 0 & 0 \\ 0 & 1 & 0 & 1 & 1 & 1 \end{pmatrix}$$

With this, we obtain a connected block design. In contrast to the case before, a sequence of treatment pairs exists now, and such pairs occur in one of the blocks together. For instance, the sequence: (2, 4) in column 2; (1, 4) = (4, 1) in column 4; and (1, 3) in column 1 sets up the connection between the treatments 2 and 3, because treatment 2 is also found in column 4, and treatment 3 in column 1. That such a sequence can be found for each treatment pair can be recognized from the following: each of the pairs (1, 2), (1, 3), (1, 4), and (2, 4) occurs together in one block, and therefore each pair is connected directly; the sole remaining pair, (3, 4), is, for instance, connected by the sequence (1, 3) and (1, 4).

Doctor

In psychological assessment, within the area of test construction and so-called *Large Scale Assessments*[3], respectively, incomplete connected block designs are often required. This is due to the fact that the number of experimental conditions (test items) is far too big to be able to present them to all testees. Therefore, groups of test items are assembled in so-called test *booklets*, which are presented as a package to the different groups of testees. These test booklets thus are the different blocks of the experimental design. In the simplest case a balanced incomplete block design is feasible. However, as, generally, the population of the persons also has to be stratified into sub-populations and also the set of test items is stratified, a balanced incomplete block design becomes practically impossible. Besides the sex of the testees, their age and their nationality have to be taken into

[3]This means a survey of (all) persons of a defined population with the help of (psychological or) psychological educational tests. Mostly they are about educational contents.

account, and, as regards the test items, their difficulty and the topic of content, and mainly the number of applications in former investigations. Therefore, it is important to insure at least a connected block design, because otherwise the results from the different test booklets and various test items, respectively, would not be comparable.

References

Kubinger, K. D. (2009b). *Psychologische Diagnostik – Theorie und Praxis psychologischen Diagnostizierens* (2nd edn) [Psychological Assessment – Theory and Practice of Psychological Consulting]. Göttingen: Hogrefe.

Popper, K. R. (1959). *The Logic of Scientific Discovery* [Trans.]. New York: Harper.

Rasch, D., Pilz, J., Verdooren, R. L., & Gebhardt, A. (2011). *Optimal Experimental Design with R.* Boca Raton: Chapman & Hall/CRC.

Statistik Austria (2009). *Bevölkerungsstand inklusive Revision seit 1.1.2002* [Demography inclusive Revision since 1.1.2002]. Vienna: Statistik Austria.

Wallis, W. A. & Roberts, H. V. (1965). *The Nature of Statistics*. New York: Free Press.

8

One sample from one population

In this chapter we consider point estimation, the construction of confidence intervals, and the testing of hypotheses, all of these concerning only a single population. The principle of statistical tests, which is based on probability theory, is explained in detail using a typical method as an example. We show, by fixing certain precision requirements, how the planning of a study is performed, i.e. the necessary sample size is calculated. However, this fixing of a sample size in advance is not needed if sequential testing is applied.

We assume that we are investigating a single population concerning a single character and we draw just one random sample.

For most of the following methods we further assume that the character investigated is normally distributed – more exactly that the character can be modeled sufficiently by a normally distributed random variable. In this case the mean and the variance completely characterize the population: the density of the normal distribution is uniquely fixed, as long as the mean μ and the variance σ^2 are known (see Formula (6.5)). As we have already explained in Section 6.2.2, assuming a normal distribution is important mostly for the mathematical derivation of the method, and seldom, however, for the practical application.

8.1 Introduction

Inferential statistics is directed at the estimation of an unknown parameter, the testing of hypotheses – that is to say *hypothesis testing* – regarding this parameter, and the construction of *confidence intervals* for this parameter.

To describe these three subjects we consider only the parameter mean μ of a normally distributed random variable, which is used as a model for the character to be observed. By doing this the presentation becomes less abstract but, nevertheless, remains valid for the general case. We draw a random sample from a population of size n. We denote the random variable by $\boldsymbol{y}$, the n observed values with $y_1, y_2, \ldots, y_v, \ldots, y_n$.

Statistics in Psychology Using R and SPSS, First Edition. Dieter Rasch, Klaus D. Kubinger and Takuya Yanagida.

8.2 The parameter μ of a character modeled by a normally distributed random variable

The assumption of the random variable $\boldsymbol{y}$ being normally distributed is theoretically important for the testing of hypotheses and for the construction of confidence intervals, because the methods discussed below can only be derived under this condition. From a practical point of view this assumption is often not needed, because the differences between results are negligible if the distribution deviates from the normal distribution. For point estimations by the least squares method (contrary to maximum likelihood estimation; see Section 6.5) this assumption is not even theoretically necessary.

8.2.1 Estimation of the unknown parameter μ

We speak about a *point estimation* if we ascertain an unknown parameter of the population by means of a statistic based on a sample – thus leading to a certain value.

Fundamentally, the unknown parameter μ – as well as the outcomes – lies between $-\infty$ and ∞. It was shown (see Section 6.5), that the estimator for μ is $\hat{\boldsymbol{\mu}} = \bar{\boldsymbol{y}} = \frac{1}{n}\sum_{v=1}^{n} \boldsymbol{y}_v$. The mean $\bar{\boldsymbol{y}}$ of the sample is a proper estimator for the mean μ in the population.

Master

A single mean in an observed sample is just one of theoretically infinite possible means in infinite possible samples. All these possible means are possible realizations of a random variable: $\bar{\boldsymbol{y}}$. Derivations in mathematical statistics show that if $\boldsymbol{y}$ is normally distributed then $\bar{\boldsymbol{y}}$ is normally distributed too, with expectation/mean μ and variance $\frac{\sigma^2}{n}$ (or with the standard deviation $\frac{\sigma}{\sqrt{n}}$) – where σ^2 is the variance of $\boldsymbol{y}$. The standard deviation $\frac{\sigma}{\sqrt{n}}$ of the estimator $\bar{\boldsymbol{y}}$ is called the *standard error*.[1]

It is plausible that the differences between the means of several samples become smaller the larger the sample size n, on which the calculation of the mean is based, becomes. But primarily the variability of the means from several samples depends on how much the observed outcomes of the character in question are spread within the population.

Master

Example 8.1 Simulation to demonstrate the sample theory without reference to any content

From a hypothetical population of a normally distributed random variable $\boldsymbol{y}$ with mean $\mu = 50$ and standard deviation $\sigma = 10$ we would like to draw $k = 10$ random samples.[2] Each sample has size $n = 10$. From each of these samples we now calculate the sample mean $\bar{y}_l$ $(l = 1, 2, \ldots, k)$ as well as the estimated standard error, $\frac{s_l}{\sqrt{n}}$ (instead of σ_l we use its estimate; see Section 6.5).

[1] This is different from the standard measurement error in psychological assessment; the latter is based on the reliability of a (psychological) test.

[2] Statistical simulation means repeated sampling from a population that is well defined in advance, without actually gaining the data empirically (more precisely see in Section 14.4).

In **R**, we define a new function by typing

```
> sim.se <- function(k, n, mu = 50, sig = 10) {
+   me <- numeric(length = k)
+   se <- numeric(length = k)
+     for(i in 1:k) {
+       sample <- rnorm(n, mean = mu, sd = sig)
+       me[i] <- round(mean(sample), digits = 2)
+       se[i] <- round(sd(sample)/sqrt(n), digits = 2)
+     }
+   return(cbind("mean" = me, "standard error" = se))
+ }
```

i.e. we apply the function `function()` and use the number of samples `k` as the first argument and the sample size `n` as the second. The mean `mu = 50` and the standard deviation `sig = 10` are used as the third and fourth arguments. The sequence of commands in the braces specifies the simulation procedure and will not be explained in detail here. We assign the function to the object `sim.se`.

Next, we type

```
> set.seed(123)
> sim.se(k = 10, n = 10)
```

i.e. we set the arbitrary starting number `123` using the function `set.seed()` in order to enable repeatability of the results. We do so only for didactic reasons, so that the reader yields the same results. In practical use, the line `> set.seed(123)` is omitted. Now we calculate the sample mean and the standard error of $k = 10$ random samples of size $n = 10$, using the previously created function `sim.se`.

The results are presented in Table 8.1.

Table 8.1 Means and standard errors for sampling/simulation of 10 samples of size 10 out of a population with $\mu = 50$ and $\sigma = 10$.

Sample number	Sample mean	Standard error of the mean
1	50.75	3.02
2	52.09	3.28
3	45.75	2.94
4	53.22	1.67
5	49.91	3.42
6	52.22	2.71
7	51.23	2.96
8	46.37	3.15
9	53.13	1.73
10	54.37	3.38

In the samples we find estimates which sometimes deviate considerably from the parameter $\mu = 50$. This is not surprising, because we have drawn relatively small samples, and with this the chance of big deviations is greater. The reader can try further simulations and will see that, for larger n, the deviations from parameter $\mu = 50$ generally become smaller.

8.2.2 A confidence interval for the unknown parameter μ

Point estimation on its own is not compelling, because, in some cases, the estimator can result in an estimate far from the parameter. Additionally, it is better to estimate a region, in which the parameter is most likely located. Such a region is called a confidence interval.

Master A confidence interval is a function of the random sample; that is to say it is also random in the sense of depending on chance; hence we speak of a random interval. However, once calculated based on observations, the interval has of course non-random bounds. Sometimes only one of the bounds is of interest; the other is then fixed – this concerns the estimator as well as the estimate itself. Contrary to a *two-sided confidence interval*, such a confidence interval is called a *one-sided confidence interval.*

A confidence interval for the mean μ is an interval with at least one random bound, which covers the parameter with probability $1 - \alpha$. This probability – or better to say the *confidence coefficient* – is the probability for the correctness of the conclusion 'μ lies in the confidence interval'. Often we choose $\alpha = 0.05$ or $\alpha = 0.01$; hence we often speak about a '$100 \cdot (1 - \alpha)\%$' interval, namely a 95% or 99% confidence interval.

It can be shown that – given that the normally distributed character's variance is known – the lower (**L**) and the upper (**U**) bound of the confidence interval for the mean is given by:

$$\boldsymbol{L} = \bar{\boldsymbol{y}} - z\left(1 - \frac{\alpha}{2}\right) \cdot \frac{\sigma}{\sqrt{n}}; \quad \boldsymbol{U} = \bar{\boldsymbol{y}} + z\left(1 - \frac{\alpha}{2}\right) \cdot \frac{\sigma}{\sqrt{n}} \tag{8.1}$$

Here $z(1 - \frac{\alpha}{2})$ is the $(1 - \frac{\alpha}{2})$-quantile of the standard normal distribution (see Section 7.1). We find the quantiles in Appendix B, Table B2; as an experienced user of statistics, alternatively we remember the values $z(0.975) = 1.96$ and $z(0.995) = 2.58$. For example, with **R** the *P*-quantiles for the most common distributions can easily be calculated; we demonstrate this in the following examples.

If, as is usually the case, the variance in the population is not known, then instead of the bounds in Formula (8.1), the following are applicable – with $\boldsymbol{s}$ being the standard deviation of the sample according to Formula (5.4) as an estimator $\hat{\boldsymbol{\sigma}}$ of σ:

$$\boldsymbol{L} = \bar{\boldsymbol{y}} - t\left(n - 1;\ 1 - \frac{\alpha}{2}\right) \cdot \frac{\boldsymbol{s}}{\sqrt{n}}; \quad \boldsymbol{U} = \bar{\boldsymbol{y}} + t\left(n - 1; 1 - \frac{\alpha}{2}\right) \cdot \frac{\boldsymbol{s}}{\sqrt{n}} \tag{8.2}$$

Here $t(n-1, 1-\frac{\alpha}{2})$ is the $(1-\frac{\alpha}{2})$-quantile of *Student's (central) distribution* or, for short, *t-distribution* with $df = n - 1$ *degrees of freedom.* This distribution is symmetrical (centered at the value 0) and takes a flatter course than the standard normal distribution. However, with growing sample size $n \to \infty$, it tends towards the standard normal distribution. The quantiles can be found in Appendix B, Table B2.

Master

The derivation of the bounds in (8.1) and (8.2) starts with the probabilities

$$P\left(z\left(\frac{\alpha}{2}\right) \leq \frac{\bar{y}-\mu}{\sigma}\sqrt{n} \leq z\left(1-\frac{\alpha}{2}\right)\right) = 1-\alpha$$

and

$$P\left(t\left(n-1;\frac{\alpha}{2}\right) \leq \frac{\bar{y}-\mu}{s}\sqrt{n} \leq t\left(n-1;1-\frac{\alpha}{2}\right)\right) = 1-\alpha \qquad (8.3)$$

respectively. From the symmetry of the normal distribution and the t-distribution, centered at 0, we have $z(\frac{\alpha}{2}) = -z(1-\frac{\alpha}{2})$ and $t(n-1;\frac{\alpha}{2}) = -t(n-1;1-\frac{\alpha}{2})$. Now the bounds $\boldsymbol{L}$ and $\boldsymbol{U}$ have to be found so that any deviation of estimator and parameter lies in between them with the probability $1-\alpha$. It can be shown that $\frac{\bar{y}-\mu}{\sigma}\sqrt{n}$ is standard normally distributed, and $\frac{\bar{y}-\mu}{s}\sqrt{n}$ is t-distributed, so that the corresponding inequalities result. Algebraic rearrangements lead to Formulas (8.1) and (8.2).

For Lecturers:

The t-distribution is also called the *Student distribution* due to the pseudonym of its 'discoverer' (Student, 1908), who was *William Sealy Gosset*. The distribution he established contradicted the doctrine of his time, and the brewery *Guinness*, where Gosset was employed, did not allow him to publish under his actual name.

For Lecturers:

Obviously, the term degrees of freedom is a function of the sample size. But the following example clarifies the chosen wording exactly. Concerning the variance, the sum of squared deviations $\sum_{v=1}^{n}(y_v - \bar{y})^2$ (the so-called sum of squares) is of relevance. If, once the sample mean is calculated, we may imagine that all n outcomes, y_v, apart from one, i.e. $n-1$, may be changed without changing the mean as well as the sum of squares $\sum_{v=1}^{n}(y_v - \bar{y})^2$, then only the nth value is consequently unequivocally determined: we only have the 'freedom' to choose $n-1$ outcomes arbitrarily.

Master

Example 8.2 An 'infinite' (∞) sample size in statistics by the example of the t-distribution

From Table B2 we take the values $t(10, 0.975)$, $t(30, 0.975)$, and $t(100, 0.975)$. In the corresponding row for *df* and in the column for 0.975 we find the values 2.228, 2.042, and 1.984. If the t-distribution with growing sample size $n \to \infty$ now tends to the standard normal distribution, then $t(\infty, 0.975)$ must be equal to $z(0.975) = 1.96$ – which indeed is the case. We learn that it is not so important numerically whether the sample size is $n = \infty$ or 'only' $n = 101$ ($df = 100$), because the two quantiles are very near to each other.

Master

Often we need only a one-sided confidence interval. Then Formulas (8.1) and (8.2) have to be changed so that one side of the inequality equals $+\infty$ or $-\infty$, respectively, and $\frac{\alpha}{2}$ is replaced by α. For example Formula (8.2) changes to

$$L = \bar{y} - t(n-1, 1-\alpha)\frac{s}{\sqrt{n}}; \quad U = \bar{y} + t(n-1, 1-\alpha)\frac{s}{\sqrt{n}} \tag{8.4}$$

The equation for the lower bound in Formula (8.4) refers to a *left-hand-sided confidence interval*; its upper bound is ∞. The equation for the upper bound in Formula (8.4) refers to a *right-hand-sided confidence interval*; its lower bound is $-\infty$. Whether we use one-sided or two-sided confidence intervals depends on the corresponding research question.

Master
Doctor

As already indicated, once a confidence interval is calculated based on the observations, the interval has of course non-random bounds. Hence it would be better to call these intervals 'realized confidence intervals', but this is seldom the case. Nevertheless, bear in mind that such intervals either include or do not include the unknown parameter: there is no longer a probability statement to give. That is, the statement 'this interval includes the unknown parameter' can be wrong, but in the span of a researcher's life it will be correct in approximately $100 \cdot \alpha\%$ of all cases.

Bachelor

Example 8.3 Confidence interval for the mean of the character *Immediately Reproducing – numerical, 1st test date* for children with German as their native language (Example 1.1)

We choose $\alpha = 0.05$.

In **R**, we first create a data set containing exclusively children with German as a native language, by typing

```
> Example_1.1.g <- subset(Example_1.1,
+                          subset = native_language == "German")
> attach(Example_1.1.g)
```

i.e. we apply the function `subset()` to select those children from data set `Example_1.1` who meet the condition `native_language == "German"`, and assign their data to the object `Example_1.1.g`. Next, we enable access to this database using the function `attach()`.
Now we calculate the confidence interval for `sub5_t1`. We type

```
> t.test(sub5_t1, conf.level = 0.95)
```

i.e. we apply the function `t.test()` to the character *Immediately Reproducing – numerical, 1st test date* (`sub5_t1`). With `conf.level = 0.95` we stipulate the 95% confidence interval. As a result, we get (shortened output):

```
95 percent confidence interval:
 47.45432 52.74568
sample estimates:
mean of x
     50.1
```

In SPSS, we use the following sequence of commands (after having limited our sample to children with German as a native language, analogously to Example 5.5)

Analyze
    Descriptive Statistics
        Explore...

and in the resulting window (see Figure 8.1) we drag and drop Immediately Reproducing – numerical, 1st test date to the panel Dependent List:. Next, we open the window shown in Figure 8.2 by clicking Statistics... The default setting for the Confidence Interval for Mean is set to 95%. A click on Continue and OK yields the results shown in Table 8.2.

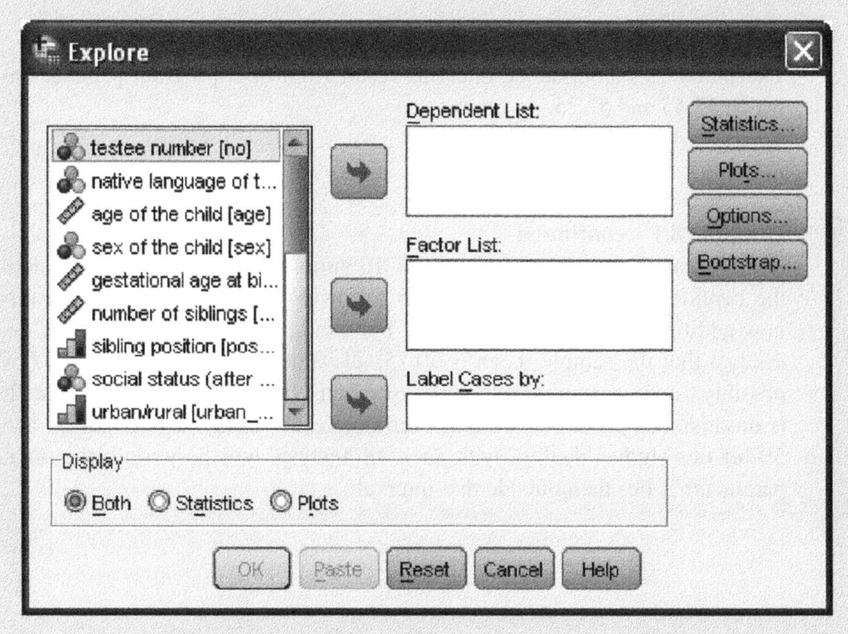

Figure 8.1 SPSS-window for description of data.

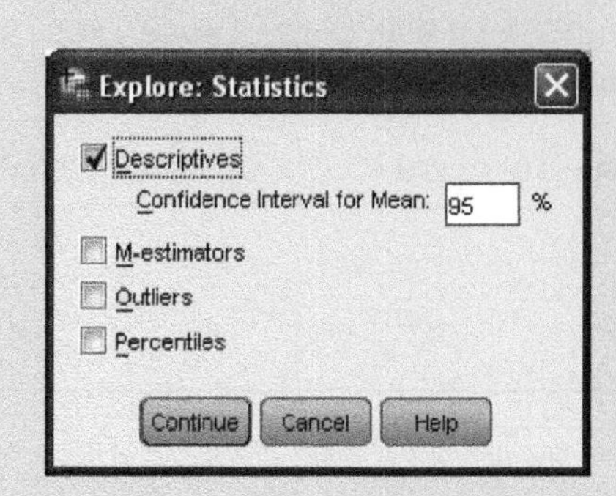

Figure 8.2 SPSS-window for calculation of a confidence interval.

Table 8.2 SPSS-output of the confidence interval for Example 8.3 (shortened output).

Descriptives

			Statistic	Std. Error
Immediately Reproducing - numerical, 1st test date (T-Scores)	Mean		50.10	1.317
	95% Confidence Interval for Mean	Lower Bound	47.45	
		Upper Bound	52.75	

Two-sided 95% confidence intervals result from both program packages as between 47.45 and 52.75.

Master

Example 8.1 – continued

Let us assume that from each of our 10 simulated samples we had calculated the confidence interval for $\alpha = 0.05$ following Formula (8.2). The results are now additionally given in Table 8.3. For example, for the first sample, despite the fact that the sample mean is 50.75, all values between 43.92 and 57.57 are possibly true as concerns the population mean. Naturally such an inaccurate result is unsatisfactory, but beware that even this result might be wrong – on average 5% of our studies dealing with such an analysis will be wrong: the unknown parameter μ lies then outside that interval.

Table 8.3 Confidence intervals for the mean of a population with $\mu = 50$ and $\sigma = 10$ for 10 simulated samples of size $n = 10$.

No. of the sample	Sample mean	Estimated standard error of the mean	Two-sided confidence interval: Lower bound	Two-sided confidence interval: Upper bound
1	50.75	3.02	43.92	57.57
2	52.09	3.28	44.66	59.51
3	45.75	2.94	39.10	52.41
4	53.22	1.67	49.45	56.99
5	49.91	3.42	42.17	57.66
6	52.22	2.71	46.09	58.34
7	51.23	2.96	44.53	57.93
8	46.37	3.15	39.25	53.49
9	53.13	1.73	49.21	57.05
10	54.37	3.38	46.72	62.02

Psychologists sometimes utilize a totally incorrect interpretation of statistical results. Fundamentally, statistics aims for a conclusion which is based on the conceptualization that it is either correct with a certain probability or wrong with 1 minus this probability. However, concerning confidence intervals, researchers frequently conclude 'With 95% certainty the unknown parameter lies between x and z.' This is wrong, because the parameter either lies between x and z or not, but 'the parameter cannot choose to be within this interval today and tomorrow to be outside of the same interval'; the parameter is not random.[3] That is, for a certain calculated result, no probability statement is possible (anymore); any probability statement is only valid for the applied method as a principle, but not at all for each individual result. The point is, that we hazard for each of our studies the conclusion: 'The mean lies in the calculated interval', because the method of calculation has a high probability, $1 - \alpha$, of being correct. In any individual case this statement is either correct or wrong. If, during our lives as researchers, we often make such analyses and conclusions, then in appropriately $100 \cdot (1-\alpha)\%$ of all the cases we will be right.

This statistical approach can be compared to the situation in a hospital in which difficult surgeries often have to be performed. If in the past this surgery was successful in 95% of the cases, the physician may say to the patient 'All will be good'; but he does not know in advance what may happen in a certain given case. Another parallel for this statistical approach would be weather forecasting.

[3] Take into account that random parameters actually occur in the Bayesian statistics approach, which however is not dealt with in this book.

8.2.3 Hypothesis testing concerning the unknown parameter μ

Often we have a certain, fact-based assumption about the unknown parameter μ of a character, which is, as indicated, modeled by a normally distributed random variable. The research question for a study therefore could be whether the unknown parameter μ, after estimation from a sample, is compatible with a certain parameter value μ_0. Our null hypothesis is H_0: $\mu = \mu_0$.

Either we have – in the case that H_0 is wrong – the conjecture that $\mu > \mu_0$ or $\mu < \mu_0$, respectively, or we have no specific conjecture and state $\mu \neq \mu_0$. In the first case our alternative hypothesis is H_A: $\mu = \mu_1 > \mu_0$ and H_A: $\mu = \mu_1 < \mu_0$, respectively; in the second case H_A: $\mu = \mu_1 \neq \mu_0$. In the first case the alternative hypothesis is called *one-sided*, and in the second case *two-sided*. We also speak of *one-sided* and *two-sided problems*. The decision between a one-sided or a two-sided alternative hypothesis, as well as the determination of some other precision requirements, must be made prior to data sampling (data analysis), and founded with regard to content.

Bachelor **Example 8.3 – continued**

Because the test is standardized, i.e. the mean and the standard deviation for the population are fixed with $\mu_0 = 50$ (T-Scores) and $\sigma = 10$, we are now interested in whether our sample stems from a population with just this mean or not. There actually is some risk that our sample is biased – the results from Vienna may systematically differ from the population of German speaking children in Germany, Austria (whole country), Switzerland, or South Tyrol. Furthermore, a biased sample could also result from an arbitrary selection of schools in terms of, for instance, the socio-economic milieu. The null hypothesis therefore is H_0: $\mu = \mu_0 = 50$; the alternative hypothesis is H_A: $\mu \neq 50$.

The matter at hand, that is the principle of *statistical tests*, is to determine whether we reject the null hypothesis due to empirical data or, alternatively, accept it due to these data. In both cases we can be wrong, i.e. make an error. On the one hand, we can reject a null hypothesis which in reality is correct. This is called a *type-I error*. On the other hand the null hypothesis can be accepted despite it being wrong (i.e. in reality the alternative hypothesis is valid). This is called a *type-II error.* All possible decisions in statistical testing are shown in Table 8.4.

We now know that empirical research means making conclusions under uncertainty, with the risk of making a wrong conclusion. We have to assess exactly which risk we are willing to accept: how likely or unlikely each of the possible errors should be. By convention the probability of a type-I error, that is the *type-I risk*, very often is chosen as $\alpha = 0.05$ or

Table 8.4 Decisions in statistical testing.

Decision based on the observations	**True situation**	
	H_0 correct, i.e. $\mu = \mu_0$	H_A correct, i.e. $\mu \neq \mu_0$
H_0 rejected	*type-I error*	no error
H_0 accepted	no error	*type-II error*

sometimes also as $\alpha = 0.01$. In testing any null hypothesis, it is our goal that in all comparable studies the null hypothesis is on average erroneously rejected in only 5% (or 1%) of the cases.

The principle of statistical testing is to split all possible sample results into two parts. The one part consists of all results supporting the null hypothesis; the other part consists of all results supporting the alternative hypothesis, i.e. arguing against the null hypothesis. We then calculate the probability for all results in the second part, i.e. for those arguing against the null hypothesis, given that in reality the null hypothesis is valid. If this probability is small and in fact smaller than the type-I risk, then the observed data (or even more extreme data) are, under the assumption of the null hypothesis, too unlikely to be explained by chance alone: therefore we reject the null hypothesis, because it does not sound at all rational to believe we only were dogged by bad luck.

It is important to know that this principle of statistical testing always applies, whatever the null hypothesis claims, how many observations are given, what scale type the interesting character has, and so on. Nevertheless, the principle of statistical testing can be most easily understood using a binomially distributed character.

Master

Example 8.4 The principle of testing explained by an example of a character which is modeled by a binomially distributed random variable

The question is, whether a certain coin that we happen to have on hand is a 'fair' coin or not.

Hence, the null hypothesis is: 'In half the cases, the coin falls showing heads; in the other half the coin shows tails.' The alternative hypothesis is: 'The coin falls on one of the two sides more often than on the other.' That is, under the null hypothesis, the probability for a head, following Formula (6.1) is *P*(head) = $^1/_2$. The null hypothesis therefore formally reads: H_0: *P*(head): $= p =$ *P*(tails): $= 1 - p = {}^1/_2$. The alternative hypothesis is: $H_A: p \neq {}^1/_2$. For the final conclusion we select – for the decision between null and alternative hypothesis – a type-I risk of $\alpha = 0.05$.

Before data are sampled, we first have to split all possible sample results into two parts: those supporting the null hypothesis on the one hand (*acceptance region*), and those results supporting the alternative hypothesis, i.e. arguing against the null hypothesis (*rejection region*). For this we first have to fix the sample size, which means the number of times n the coin will be tossed. Here we have a population which can never be completely observed, as it is infinite (all possible results of tossing that coin). We decide to toss the coin $n = 10$ times. Now we have to define the acceptance region and the rejection region in such a way that the probability of erroneously rejecting the null hypothesis – the type-I risk – is 0.05.

We first arbitrarily assume that if we obtained heads $k = 2$ times in $n = 10$ turns, then this is against the null hypothesis: heads occurred too seldom. Then, of course, this is the case even more for $k < 2$. However there also exists another extreme case, namely that heads occurred 'too often', i.e. $k = 8$ times (or even more); bear in mind that our alternative hypothesis has actually been formulated as two-sided. Therefore we now have the rejection region defined: the results for k in the set $\{0, 1, 2, 8, 9, 10\}$. All other results, i.e. k from the set $\{3, 4, 5, 6, 7\}$, define the acceptance region. In order to take the type-I risk, α, into account, we must calculate as the next step the probability that a result belonging to the rejection region is observed – although the null hypothesis is true. For this, the number k (times heads) is to be modeled as a binomially distributed random variable $\boldsymbol{k}$ (as above in Example 6.5). Using Formula (6.4) the required probabilities $P(\boldsymbol{k} \leq 2)$ and $P(\boldsymbol{k} \geq 8)$ can be calculated:

$$P(\boldsymbol{k} \leq 2) = \sum_{i=0}^{2} \binom{10}{i} \left[\frac{1}{2}\right]^{i} \left[\frac{1}{2}\right]^{10-i} = (1 + 10 + 45) \left(\frac{1}{2}\right)^{10} = 0.0547$$

and

$$P(\boldsymbol{k} \geq 8) = \sum_{i=8}^{10} \binom{10}{i} \left[\frac{1}{2}\right]^{i} \left[\frac{1}{2}\right]^{10-i} = (45 + 10 + 1) \left(\frac{1}{2}\right)^{10} = 0.0547$$

Because the two cases are mutually exclusive, the total probability, the actual type-I risk for the splitting used for the acceptance region and rejection region is: $0.0547 + 0.0547 = 0.1094 > 0.05 = \alpha$ (addition rule; see Formula (6.2)). That is, our arbitrary splitting of the whole into an acceptance and a rejection region is not appropriate for $\alpha = 0.05$.

Now we analogously calculate the probability of at most 1 time or at least 9 times tossing heads, which means that the creation of completely different acceptance and rejection regions is necessary. The result is $(1 + 10)(^1/_2)^{10} + (10 + 1)(^1/_2)^{10} = 0.0214 < 0.05$ and is also not appropriate. Nevertheless once we have data observed, we can make a unique statement. The case $k \leq 2$ or $k \geq 8$ with $n = 10$ is, under the null hypothesis (a fair coin), very unlikely. But it is more likely than the critical probability that we, in advance, decided to risk; only if the calculated probability were smaller than 0.05 would we have rejected the null hypothesis. But according to 0.1094, we have to accept it. On the other hand, in the case where $k \leq 1$ or $k \geq 9$, the null hypothesis has to be rejected. As a matter of fact, in such a study we can, in the long run, not constrain the actual type-I risk; if we always only reject the null hypothesis in the case where $k \leq 1$ or $k \geq 9$, then we will be wrong in just approximately 2 of 100 studies – given that the null hypothesis is true and the coin fair.

Master

Example 8.4 – continued

The reader may perform a simulation for example with 10, 20, and with 100 runs (tossing a coin).

In **R**, we again create a new function. We type

```
> sim.bin2 <- function(rep, p1 = 2, p2 = 8, n = 10, prob = 0.5) {
+     hit <- table(rbinom(rep, n, prob))
+     rel <- hit/rep
+     p1.e <- sum(rel[which(names(rel) <= p1)])
+     p2.e <- sum(rel[which(names(rel) >= p2)])
```

```
+       cat(paste("\n", "   P(k <= ", p1,") =",sep = ""),
+                round(p1.e, 5),
+          paste("\n", "   P(k >= ", p2,") =",sep = ""),
+                round(p2.e, 5),
+                "\n", "\n")
+   return(invisible(list(hit, rel, c(p1.e, p2.e))))
+ }
```

i.e. we apply the function `function()`, using as an argument the number of runs, given for the time being, by a variable `rep`. Additionally, we use the number of cases in question as arguments; that is `p1` with `2` for $k \leq 2$, and `p2` with `8` for $k \geq 8$. Furthermore, we use as arguments the number of coin tosses per run, `n`, and probability of a tail with `prob = 0.5`. The sequence of commands inside the braces specifies the procedure of the simulation and will not be explained in detail here. Finally, we assign this function to the object `sim.bin2`. Now, we type

```
> set.seed(143)
> sim.bin2(rep = 10)
```

i.e. we set the arbitrary starting number `143` using the function `set.seed()`.We do this only for didactic reasons, so that the reader yields the same results. In practical application, this line is to be omitted. We apply the function `sim.bin2()`, and with `rep = 10` we simulate 10 runs.

This yields:

```
P(k <= 2) = 0.1
P(k >= 8) = 0.1
```

The estimate for 10 runs is 0.1 for the probability $P(\boldsymbol{k} \leq 2)$; and $P(\boldsymbol{k} \geq 8) = 0.1$ as well. The reader may proceed to perform the simulation with 20 (and with 100) runs; if he/she uses the computer program, even 200 runs would pay off. The result will be that the estimate of the probability $P(\boldsymbol{k} \leq 2) + P(\boldsymbol{k} \geq 8)$ comes closer and closer to 0.1094 when the number of runs increases.

We do this now for only 50 runs. Using the seed of 155 we get $P(\boldsymbol{k} \leq 2) + P(\boldsymbol{k} \geq 8) = 0.04$.

For Lecturers:

With discrete distributions like the binomial distribution, a given value of α can seldom be maintained, in contrast to continuous distributions. If in the case of a discrete distribution the value of α must be adhered to, then so-called *randomized tests* have to be used. The strategy is, in the case of $n = 10$ and $p = \frac{1}{2}$ as follows:

'Reject the null hypothesis if $k = 0, 1, 9, 10$; accept it if $k = 3, \ldots, 7$. If $k = 2$ or $k = 8$ then do a simulation; that is to say generate a random number within the interval of 0 and 1 (see Section 14.4); if this random number is smaller than or equal to $\frac{0.05-0.0241}{0.0439} = 0.59$ (see Example 8.4) then we reject the null hypothesis, otherwise we accept it. In doing so we actually realize a type-I risk of $\alpha = 0.05$. Of course, it is hard to understand for a beginner in statistics that he/she sometimes has to use a die so that the type-I risk holds. So, it sounds better to advise him/her to only apply the test with either a type-I risk of $\alpha = 0.0214$ or a type-I risk of $\alpha = 0.1094$ – or even to enlarge the sample size.

It is important to choose the type-I risk before planning the study and sampling (analyzing) the data. Otherwise, if the type-I risk is not chosen until the analysis has been finished and the actual result is known, then all studies of this kind would obviously always use the highest of the available alternative risks. For instance, if an empirical result was obtained that would not reject the null hypothesis at the type-I risk of 0.01, but would allow it to be rejected at the type-I risk of 0.05, then the researcher will always choose the latter type-I risk – though sometimes a level of 0.01 suffices.

Now the statistical concept of *significance* is the central issue. We call a result of a statistical test *significant* when this result is statistically meaningful. More precisely this means that an observed deviation between the corresponding parameter's estimate and the parameter as hypothesized under the null hypothesis (for instance $\hat{\mu} = \bar{y} \neq \mu_0$) is larger than can be explained by chance. Each rejected null hypothesis is therefore based on a significant result. We can say that the parameter of the population from which the sample is drawn differs significantly from the value which is hypothesized by the null hypothesis. On the other hand, a non-significant result of a statistical test means that the empirically given deviation from the null hypothesis (for instance $\hat{\mu} = \bar{y} \neq \mu_0$) must be interpreted as random error. Statistical tests are also called *tests of significance*; the type-I risk, α, is then called the *significance level*.

Analogously to the case of calculating a confidence interval for the unknown parameter μ, testing of a hypothesis concerning μ needs to differ, depending on whether the population's variance is known or not. It can be shown that the following *test statistic* for the case of a normally distributed random variable and known variance has a standard normal distribution:

$$z = \frac{\bar{y} - \mu_0}{\sigma}\sqrt{n} \tag{8.5}$$

For the (usual) case of an unknown variance, the following test statistic is t-distributed with $df = n - 1$ degrees of freedom: $\boldsymbol{s}$ is the standard deviation of the sample from Formula (5.4), as an estimator $\hat{\boldsymbol{\sigma}}$ for σ:

$$\boldsymbol{t} = \frac{\bar{\boldsymbol{y}} - \mu_0}{\boldsymbol{s}}\sqrt{n} \tag{8.6}$$

The test using the test statistic in Formula (8.6) is called the *one-sample t-test*.

Master The change from standard normal distribution to t-distribution is logical, when the variance has to be estimated from the sample. As the t-distribution always takes a flatter course than the standard normal distribution, the *critical region* of the density function, i.e. the rejection region as concerns the null hypothesis, falls in a more extreme area (see Figure 8.3). And of course, if an additional unknown parameter has to be estimated, the sample data must be more extreme as concerns the null hypothesis; that is, they must lead to a larger test statistic or, in

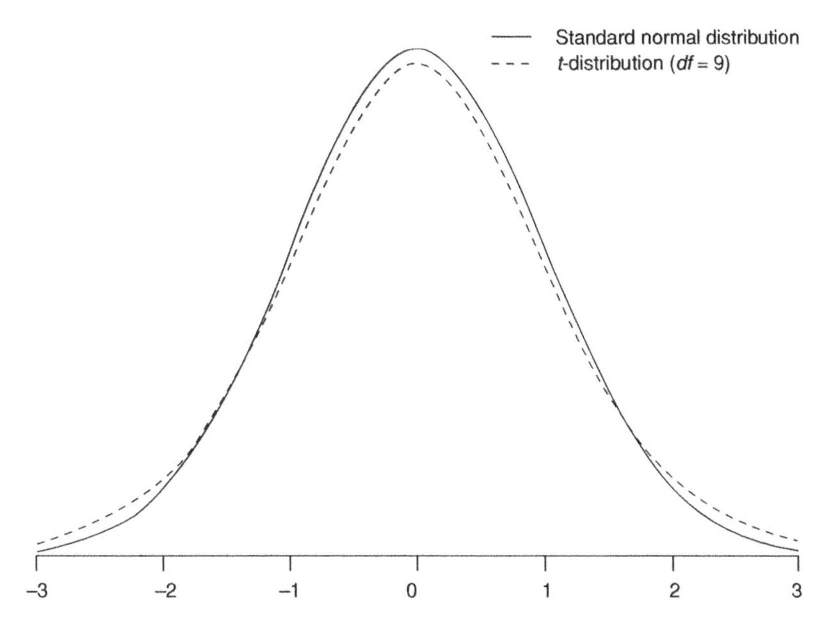

Figure 8.3 Density function of the standard normal distribution and the *t*-distribution.

other words, the deviation $\bar{y} - \mu_0$ must be larger to become significant. Because the *t*-distribution converges for $n \to \infty$ to the standard normal distribution, the smaller *n* is the larger the resulting deviation must be.

When applying both statistical tests to the data of the given sample there is only the test statistic *z* and *t* to calculate, respectively, as well as their respective absolute values $|z|$ and $|t|$; then we must compare the resulting value with its appropriate *P*-quantile in Table B1 and B2 of Appendix B, respectively. In the case of a two-sided alternative hypothesis, we have to use the column from Table B2 with 1 minus half of the type-I risk; that is $1 - \frac{\alpha}{2}$. This *P*-quantile is also called the critical value. If the calculated (absolute) value of the test statistic in question is larger than the critical one, then the observed data (and more extreme values) are, under the null hypothesis, less likely than the type-I risk accounts for; the null hypothesis must be rejected.

If there are content-related reasons to expect a deviation from the hypothesized parameter only in one direction, then we use a one-sided alternative hypothesis. Then $|z|$ and $|t|$, respectively, have to be compared with the *P*-quantile from Tables B1 and B2 corresponding to the (overall) type-I risk α; however first we check whether the values of *z* and *t*, respectively, correspond with the direction of the alternative hypothesis at all.

Bachelor **Example 8.5** Contrasting empirical *t*-values with the critical values (*P*-quantiles) in Table B2 of Appendix B

Without regard to content we consider the (one-sided) alternative hypothesis H_A: $\mu < \mu_0$; H_0: $\mu = \mu_0$, with type-I risk $\alpha = 0.05$. We assume that we come, according to Formula (8.6) for a certain $n = 20$ observations, to a test statistic value of $t = -1.85$. Then we proceed as follows: The *P*-quantile of the *t*-distribution for

$1-\alpha = 0.95$ and $df = 19$ equals $t(19, 0.95) = 1.7291$; because $|-1.85| = 1.85 > 1.7291 = t(19, 0.95)$, the result of our analysis is significant: the null hypothesis must be rejected – the sign is appropriate.

In **R**, we calculate the *P*-quantile in question by typing

```
> qt(0.95, df = 19)
```

i.e. we use the function `qt()` and set `0.95` as the first argument and add, with `df = 19`, the degrees of freedom.

As a result, we get:

```
[1] 1.729133
```

To ascertain the *p*-value for the *P*-quantile 1.85 we type

```
> pt(-1.85, df = 19)
```

i.e. we set the test statistic with the value of `-1.85` as the first argument in the function `pt()`, and the degrees of freedom with `df = 19` as the second argument.

This yields:

```
[1] 0.03996467
```

In the case of a two-sided alternative hypothesis and the same type-I risk $\alpha = 0.05$, we would have to use the *P*-quantile with the value $1 - \frac{\alpha}{2} = 0.975$.

Pertinent software packages usually do this laborious work for the researcher. They almost always report a value '*p*', which is the probability of obtaining any result out of those contradictory to the null hypotheses even though it is true; sometimes this *p*-value is only given in some 'Significance' column. Then we only have to check whether $p < \alpha$. If yes, the null hypothesis must be rejected; if no, the null hypothesis must be accepted. In SPSS, in the case of a one-sided alternative hypothesis, the given *p*-value must be divided by two, because there the *p*-value is always calculated for two-sided alternative hypotheses.

Bachelor **Example 8.3 – continued**

We consider the character *y*, *Immediately Reproducing – numerical, 1st test date* for children with German as their native language; the null hypothesis is H_0: $\mu = \mu_0 = 50$; the alternative hypothesis is H_A: $\mu \neq 50$. We decide on a type-I risk of $\alpha = 0.05$.

In **R**, we have already created the data set `Example_1.1.g`, which contains exclusively children with German as a native language, and we enabled access to the database using the function `attach()`. Now we type

```
> t.test(sub5_t1, mu = 50)
```

i.e. we apply the function `t.test()` and use the character `sub5_t1` as the first argument and set the mean `mu = 50`, which we expect if the null hypothesis is true, as the second argument.

As a result, we get:

```
    One Sample t-test

data:  sub5_t1
t = 0.076, df = 49, p-value = 0.9398
alternative hypothesis: true mean is not equal to 50
95 percent confidence interval:
 47.45432 52.74568
sample estimates:
mean of x
     50.1
```

In SPSS, we select

Analyze
  Compare Means
    One-Sample T Test...

In the resulting window shown in Figure 8.4, we drag and drop Immediately Reproducing – numerical, 1st test date to Test Variable(s):. Since SPSS tests for the null hypothesis $\mu_0 = 0$ according to the default settings, we have to change the Test Value to the value of the parameter stated in the null hypothesis, if it does not equal zero. In the window shown in Figure 8.4 we have already set the Test Value to 50. By clicking OK, we obtain the result shown in Table 8.5. In the column Sig. (2-tailed) the corresponding p-value can be found.

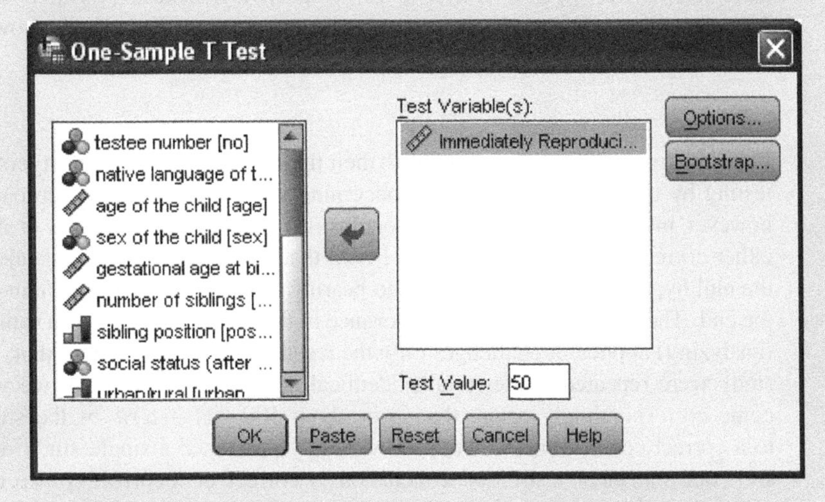

Figure 8.4 SPSS-window for calculating a one sample t-test.

Table 8.5 SPSS-output of one sample t-test in Example 8.3 (shortened output).

One-Sample Test

	Test Value = 50		
	t	df	Sig. (2-tailed)
Immediately Reproducing - numerical, 1st test date (T-Scores)	.076	49	.940

Because of a calculated p-value of 0.9398 and 0.940, respectively, the result of our analysis is not statistically significant ($p \geq \alpha$), so that we accept the null hypothesis H_0: $\mu = \mu_0 = 50$.

Master The advantage of one-sided alternative hypotheses is that the total type-I risk is concentrated in the region of interest. For example, in the case of H_A: $\mu > \mu_0$, if it is not at all possible for the parameter μ to be smaller than μ_0, it would be an unnecessary risk to also take values $\mu \leq \mu_0$ into account. For instance, in Table B2 in Appendix B we realize that $t(n-1, 1-\frac{\alpha}{2}) > t(n-1, 1-\alpha)$, which means – according to Figure 8.3 – that even smaller deviations of μ and $\hat{\mu}_0$ become significant.

Master To be more exact, in the case of a one-sided problem, the null hypothesis is generally H_0: $\mu \leq \mu_0$ or H_0: $\mu \geq \mu_0$, respectively, instead of H_0: $\mu = \mu_0$. Therefore, the null hypothesis is a *composite hypothesis*. Sometimes, however, values $\mu < \mu_0$ cannot occur because of content reasons.

Bachelor Researchers may misstep and establish their final conclusion regarding hypothesis testing by using some probability concerning its validity. Such an approach is, however, incorrect. The established conclusion is not more or less likely at all, but either correct or incorrect. The type-I risk, that is the probability α of rejecting the null hypothesis erroneously, has no bearing on the concrete decision made in the end. The type-I risk is only of relevance to the situation before data sampling (analyzing) applies. It challenges only the researcher's understanding that, if the study were repeated (under totally identical conditions) very often, we would come correspondingly often, that is in about $100 \cdot (1-\alpha)\%$ of the studies, to a correct conclusion. As we, however, only processed a single study of this sort, our statistical test-based conclusion is correct or incorrect. If we could obtain knowledge about the true situation, and hence would transcendentally know whether the null hypothesis holds or not, then our statistical test-based conclusion has no other choice than to be definitely correct or incorrect, but has of course no chance-dependent probability of being correct or wrong. This situation is comparable with drawing a playing card. Whenever a card is drawn from a randomly ordered blind deck, our conclusion, or rather suggestion, that this card is hearts no longer has any probability of being right or wrong but factually is either correct or incorrect. That is, any statistical, test-based conclusion is only to establish that the null hypothesis holds or the null hypothesis does not hold. Type-I risk concerns the method of analysis but not a single concrete result of any study.

8.2.4 Test of a hypothesis regarding the unknown parameter μ in the case of primarily mutually assigned observations

Often we deal with just a single sample (of research units; in psychology mostly persons), but nevertheless there are, per unit, two or more observations for content-related traits, aptitudes or the like. In psychology we speak about *matched samples* as if we had more than only a single sample, but in fact we have only one sample of research units with two or more observations; statistically these observations stem from two or more specially modeled random variables. Therefore, we come to the case of several characters, though from the content point of view it is often the same character. When there are only two observations per unit, the easiest way is to reduce the data to a single character; for example if we are interested in the effect of a treatment in a '*pre* and *post*' design. Formally, we observe, for every subject, two characters ($x = pre$, $y = post$); from a content point of view we observe the same trait/aptitude at two different times. The sample consists of pairs of observations. Given that both characters are interval scaled, we can calculate, for subject v, the difference $d_v = x_v - y_v$; thus we only have a single character.

Example 8.6 As concerns children with German as their native language and the subtest *Immediately Reproducing – numerical,* are there any learning effects due to testing twice? (Example 1.1)

Because we consider an intelligence test, that is fundamental abilities are measured which are stable for a relatively long time, repeated testing of a testee should result in almost the same test scores. As a consequence, we expect, for the variable $\boldsymbol{d} = \boldsymbol{x} - \boldsymbol{y}$, rather small outcomes $d_v = x_v - y_v$; at least values such that their sum comes close to zero – differences from zero should occur only due to the testees' (mental) constitution on that particular day. The null hypothesis is therefore $\mu = \mathrm{E}(\boldsymbol{d}) = 0$. On the other hand, because of learning effects, the performance in the second testing could be better, so that for most of the children we have: $d_v = x_v - y_v < 0$. The alternative hypothesis is therefore H_A: $\mathrm{E}(\boldsymbol{d}) < 0$. The null hypothesis is exactly: H_0: $\mathrm{E}\,(\boldsymbol{d}) \geq 0$.

In **R**, there is an easier way than to compute the differences $d_v = x_v - y_v$ at first. We will show both options here. In doing so, we will continue using the data set `Example_1.1.g`. We have already enabled access to this database by applying the function `attach()` in Example 8.3.

Applying the first method, we type

```
> diff <- sub5_t1 - sub5_t2
```

i.e. we compute the differences between the characters *Immediately Reproducing – numerical, 1st test date* (`sub5_t1`) and *Immediately Reproducing – numerical, 2nd test date* (`sub5_t2`) and assign the result to the object `diff`. Next, we type

```
> t.test(diff, alternative = "less")
```

i.e. we again apply the function `t.test()`. We now use the new variable `diff` as the first argument and `alternative = "less"` as the second argument, as we want to test the one-sided alternative hypothesis H_A: $E(\boldsymbol{d}) < 0$.

As a result, we get (using the first method):

```
        One Sample t-test

data:  diff
t = -1.3014, df = 49, p-value = 0.0996
alternative hypothesis: true mean is less than 0
95 percent confidence interval:
      -Inf 0.1960075
sample estimates:
mean of x
    -0.68
```

We obtain the same result, more easily, if we proceed immediately in the following way. We type

```
> t.test(sub5_t1, sub5_t2, paired = TRUE, alternative = "less")
```

i.e. we specify the two characters in question and request the calculation of pair-wise differences by using `paired = TRUE` as an argument in the function. Additionally, we specify the one-sided alternative hypothesis by using the argument `alternative = "less"`.

This yields:

```
        Paired t-test

data:  sub5_t1 and sub5_t2
t = -1.3014, df = 49, p-value = 0.0996
```

```
alternative hypothesis: true difference in means is less than 0
95 percent confidence interval:
      -Inf 0.1960075
sample estimates:
mean of the differences
                  -0.68
```

In SPSS, there is also an easier way to obtain the result than to compute the differences $d_v = x_v - y_v$ at first. We will again show both options here.

In the first case, we actually apply the first sequence of commands (Transform – Compute Variable...) described in Example 5.3 to calculate a new variable. We type diff as the Target Variable and compute the difference between the observed values of Immediately Reproducing – numerical, 1st test date (sub5_t1) and Immediately Reproducing – numerical, 2nd test date (sub5_t2). To do this, we type sub5_t1 - sub5_t2 in the text field Numeric Expression:. Next we click If..., tick Include if case satisfies condition:, and type native_language = 1. By clicking Continue, we return to the window Compute Variable. Now we click OK. The new variable diff is calculated for each person and can be found in the SPSS Data View in the last column on the right (the first column which was vacant before). Next we proceed analogously to Example 8.3. In the window shown in Figure 8.4, we drag and drop diff to Test Variable(s): and set the Test Value to 0. By clicking OK, we get the result shown in Table 8.6.

Table 8.6 Output (shortened) for Example 8.6.

One-Sample Test

	Test Value = 0		
	t	df	Sig. (2-tailed)
diff	-1.301	49	.199

We can obtain the same result more easily, if we proceed in the following way (from Example 8.3 the sample is already limited to children with German as native language). We apply

Analyze
    Compare Means
        Paired-Samples T Test...

and in the resulting window (see Figure 8.5), we drag and drop Immediately Reproducing – numerical, 1st test date to the column Variable1 in the panel Paired Variables: and Immediately Reproducing – numerical, 2nd test date to the column Variable2. We click OK and obtain the result shown in Table 8.7. However, we have to halve the resulting p-value shown in Sig.

(2-tailed), because testing of one-sided alternative hypotheses is not possible using SPSS in most cases, and this applies here as well.

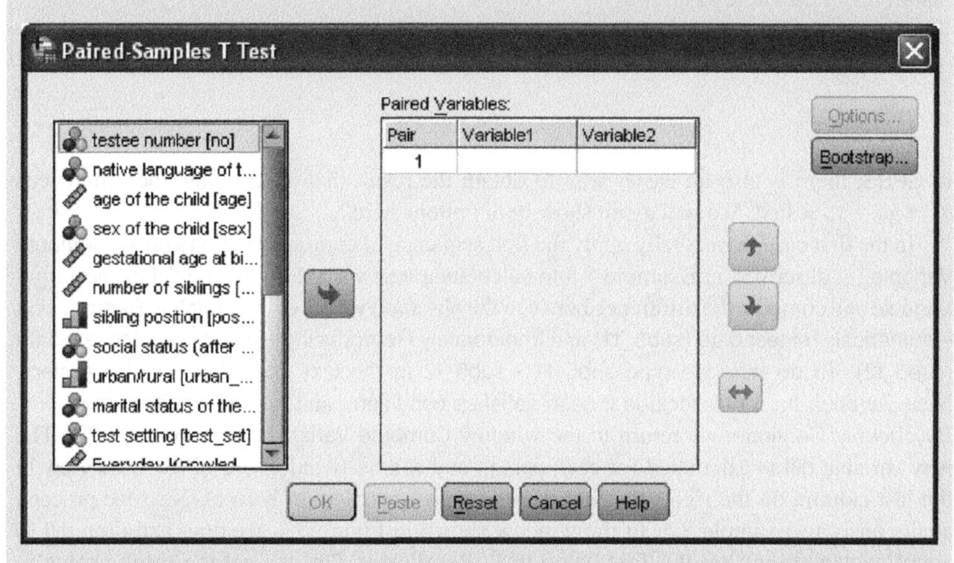

Figure 8.5 SPSS-window for conducting a paired sample t-test.

Table 8.7 SPSS-output showing the result of the paired sample t-test in Example 8.6 (shortened output).

Paired Samples Test

		Paired Differences			
		Mean	t	df	Sig. (2-tailed)
Pair 1	**Immediately Reproducing** – numerical, 1st test date (T-Scores) - **Immediately Reproducing** – numerical, 2nd test date (T-Scores)	-.680	-1.301	49	.199

According to the p-value of 0.0996 and $0.199 / 2 = 0.0995$, respectively, there is a non-significant result. Our research question is to be answered as follows: for children with German as their native language, no learning effects have been found in the subtest *Immediately Reproducing – numerical.*

The alternative approach of testing using the original data instead of the differences $d_v = x_v - y_v$ is known as the *paired sample t-test* (see Example 8.6).

8.3 Planning a study for hypothesis testing with respect to μ

Remember, Table 8.4 summarizes all errors, which potentially result when conclusions are based on a statistical test. Up until now we exclusively have discussed the type-I error. So far, our considerations might induce/encourage always testing with a very small type-I risk. However, type-I risk and the probability of a type-II error, that is the *type-II risk*, β, do directly depend on each other.

In the first instance, we of course try to minimize both sources of erroneous conclusions: of course, we neither like to reject the null hypothesis erroneously (type-I error), nor do we like to accept it erroneously – the latter means that we fail to detect a true alternative hypothesis (type-II error). However, as can be shown, these risks are mutually dependent, meaning that a practical compromise between the two must be found.

Master Doctor

We now illustrate in Figure 8.6 the situation of testing a hypothesis for the case of H_0: $\mu \leq \mu_0$ and H_A: $\mu > \mu_0$. Given that the variance of the population is known, Formula (8.5) comes into play. The test statistic therefore is z. We consider, on the one hand, the case that $\mu = \mu_0$ (the null hypothesis) is true, and on the other hand the case that $\mu = \mu_1 = \mu_0 + 2$ (the alternative hypothesis) is true. The area denoted by α corresponds with the probability of rejecting the null hypothesis – despite it being true – because of a certain sample of observations with mean $\bar{y}$. The area α hence shows the type-I risk. The corresponding $(1 - \alpha)$-quantile now separates the real line, at the point $z(1 - \alpha)$, into the acceptance region of the null hypothesis (left side of $z(1 - \alpha)$) and the rejection or critical region (equal to and right side of $z(1 - \alpha)$). If an observed value, z, falls into the critical region, then we argue as follows: it is not plausible that the null hypothesis is true while we have obtained such an extreme (or rare) result. However, the given specific $\bar{y}$, leading to a specific value z, has as a matter of fact actually been observed; hence it is rational to conclude that our hypothesis, the null hypothesis, is or was wrong. The area denoted by β corresponds with the probability of accepting our null hypothesis, despite it being wrong (i.e. in the case $\mu_1 = \mu_0 + 2$ is true). The area β hence shows the type-II risk. We realize the following:

1. If the researcher draws a sample with a certain size n as discussed thus far, he/she cannot determine β, the type-II risk, but it just results according to μ_1 – the researcher has no control.

2. If the type-I risk, α, were chosen smaller, then the type-II risk, β, becomes larger. It is therefore nonsense to select a type-I risk that is 'too small', because then it is apparent that it could happen that the alternative hypothesis is almost never to be accepted, though it is valid: for $\alpha \to 0$ we find $\beta \to 1$.

3. The type-II risk depends on the real value of $\mu_1 = \mu_0 + \delta$; the more μ_1 differs from μ_0, the smaller thc type-II risk.

4. In the case of an unknown population variance, i.e. where the t-distribution becomes relevant, then – because the t-distribution takes a flatter course than the standard normal distribution (the smaller n, the flatter the course

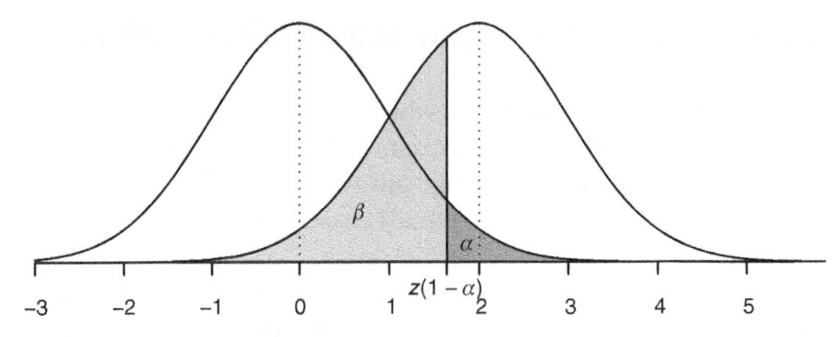

Figure 8.6 The graphs of the density function of the test statistic z in Formula (8.5) for $\mu = \mu_0$ (left) and $\mu = \mu_1 = \mu_0 + 2$ (right).

becomes) – for the same values of $\bar{y}$ and μ_1, the area denoted by α moves to the right; as a consequence the area denoted by β becomes larger. Conversely, for a fixed value $\mu_1 = \mu_0 + \delta$, the type-II risk, β, becomes smaller with increasing sample size $n \to \infty$. Hence, to make both the risks smaller, it is necessary to enlarge the sample size.

5. Again considering the t-distribution, fundamentally every difference between μ_0 and μ_1 might result in significance, even an extremely small one; all we have to do is select a sufficiently large sample size, because then the two density functions become steeper and steeper and tend to no longer overlap.

Doctor By the way, Figure 8.6 alters to Figure 8.7 in the case of an unknown variance of the variable $\boldsymbol{y}$, so that the t-test according to Formula (8.6) has to be applied. Given the null hypothesis, the test statistic's distribution is the (central) t-distribution with corresponding degrees of freedom. However, given the alternative hypothesis, the test statistic's distribution is the non-central t-distribution (with respective degrees of freedom) and the so-called *non-centrality parameter* $\lambda = \frac{\mu_1 - \mu_0}{\sigma}\sqrt{n}$. This distribution is clearly not symmetric.

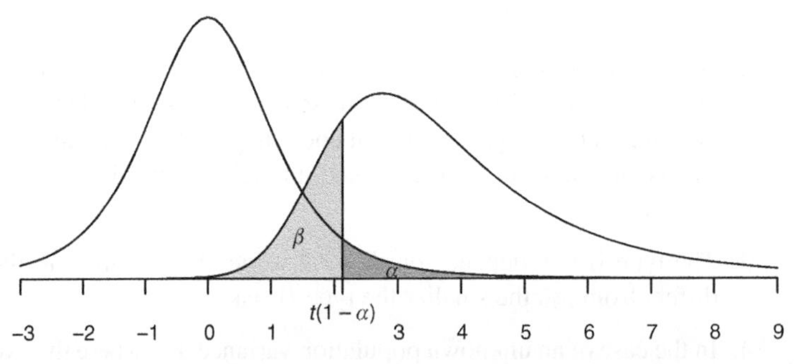

Figure 8.7 Graphs of the density function of the test statistic $\boldsymbol{t}$ in Formula (8.6) for $\mu = \mu_0$ (left) and $\mu = \mu_1 = \mu_0 + 2$ (right).

In psychological research, the type-I risk is almost always fixed as either $\alpha = 0.05$ or 0.01. For the choice of the type-II risk β there are hardly any conventions; this means the researcher can determine it adequately for the given problem.

Master Doctor

If the type-II risk, β, is controlled at all – i.e. the study is planned so that β is fixed from the very beginning of data sampling (see details in the following) – then some text books recommend using $\beta \cong 4\,\alpha$. This is, however, completely unjustified; on the contrary, this recommendation often is made totally in error. In the case, namely, in which a certain research question with any type-II error entails big negative consequences, these being bigger than any type-I error, it must be tested with $\beta \leq \alpha$.

Master Doctor

Example 8.7 The different severities of type-I and type-II error for different research questions

Some research question might lead to the following null hypothesis: 'A certain cognitive head-start program for preschool-age children has no negative consequences on their development of social competence.' If this hypothesis is erroneously accepted, children, parents, and other important persons in the surrounding environment (for instance future peers and teachers) suffer the consequences: promoted children would be striking in their social behavior. If, on the other hand, the hypothesis is erroneously rejected, children would suffer from the fact that they cannot fully realize their achievement potential. Hence, the consequences of an erroneous conclusion must be taken into account when deciding the two kinds of risk. This is of the utmost importance for a researcher, who is willing to plan, perform, and analyze a study. Here, we have to balance whether the burden for the child from social competence deficits is graver than an incomplete use of his/her achievement potential as early as possible – particularly the findings of developmental psychology showing that an uninfluenced cognitive development process sooner or later leads to the same achievement evolvement. For this reason we could choose here $\beta = 0.01$ and $\alpha = 0.05$.

Doctor

If there are several statistical tests (for a one-sided alternative hypothesis), certainly we will prefer the one that reaches, for each type-I risk α for a certain parameter value, the largest value of the power function – which means the smallest type-II risk. Then we have the so-called *most powerful α-test* (at the parameter value in question). If an α-test is the most powerful for all possible parameter values it is called a *uniformly most powerful α-test.* A uniformly most powerful α-test needs, in comparison with all other α-tests, the smallest sample size for a given type-II risk. In the case of a normal distribution, the one-sided t-test according to Formula (8.6) is such a most powerful α-test.

In the case of a two-sided alternative hypothesis, the t-test is no longer a uniformly most powerful α-test, but it is one if only *unbiased tests* are under discussion. Unbiased tests never realize power function values smaller than α. As can be shown, the t-test is a *uniformly most powerful unbiased α-test,* meaning that it is uniformly most powerful for all possible parameter values, but not only for selective ones (see for details Lehmann & Romano, 2005).

Theoretically any difference between the hypothesized μ_0 and the hypothesized μ_1 can result in significance, as long as the sample size is sufficiently large; thus a significant result of its own is not necessarily meaningful from a content point of view. In principle such a result could be meaningless, because the resulting difference might be, for instance, $\mu_1 - \mu_0 = 0.00001$. For example, a difference of 1 T-Score in intelligence tests, in practice, does not mean anything. Such a small difference is without any practical consequences. Therefore studies have to be planned in such a way that any significant differences from the hypothesized value in the null hypothesis (μ_0) are *practically relevant* – we also say: they are of practical interest.

As concerns Figure 8.6 and Figure 8.7 and the type-II risk β, we have behaved until now as if only a single value μ_1 for the alternative hypothesis is possible. Of course, in most of the analyses all the values unequal μ_0 (in the case of a two-sided alternative hypothesis) and all the values smaller or larger than μ_0, respectively (for one-sided alternative hypotheses), are possible. But for each value of μ_1, another value of the type-II risk, β, results. As already seen in Figure 8.7, β becomes increasingly smaller the more $\mu_1 - \mu_0$ differs from zero. Coming to the point, we are interested in the practically relevant minimal difference; for this, the expression $E = (\mu_1 - \mu_0)/\sigma$, that is the relative or in other words standardized practically relevant difference, is called the *(relative) effect size*.

An important step in planning a study is therefore the determination of the practically interesting minimal deviation $\delta = \mu_1 - \mu_0$. Once δ is determined, given a certain type-I risk α and a certain type-II risk β, one can calculate the necessary sample size.

Master Doctor

Once α, β, and δ are determined, the precision requirements are fixed (see Section 3.2). In fact the matter at hand is the requirement to get knowledge about erroneous conclusions as concerns the null hypothesis in the case that a practically relevant difference is given.

Master Doctor

The exact meaning of a relevant difference $\delta = \mu_1 - \mu_0$ is the following. All differences $\mu_1 - \mu_0$ which are equal or even larger than the fixed δ should preferably not be missed by the statistical test in question. In other words, not detecting those differences should happen only with the probability β or an even smaller probability.

Master Doctor

The sample size we are looking for, given certain precision requirements, can (graphically) be found using the *power function*. The power function quantifies, per sample size for all possible values of δ, the respective *power*; that is, the probability to reject the null hypothesis. If the alternative hypothesis is false, the power amounts generally to $1 - \beta$, or more exactly to $1 - \beta(\delta)$ (obviously, the power depends on δ). If the null hypothesis is true, the power function equals α.

Of course, it would be unfair to compare the power of a test at $\alpha = 0.01$ with the power of a test at $\alpha = 0.05$, because a larger α means that the power is also larger (often at all values of the alternative hypothesis) – in Figure 8.8, the reader

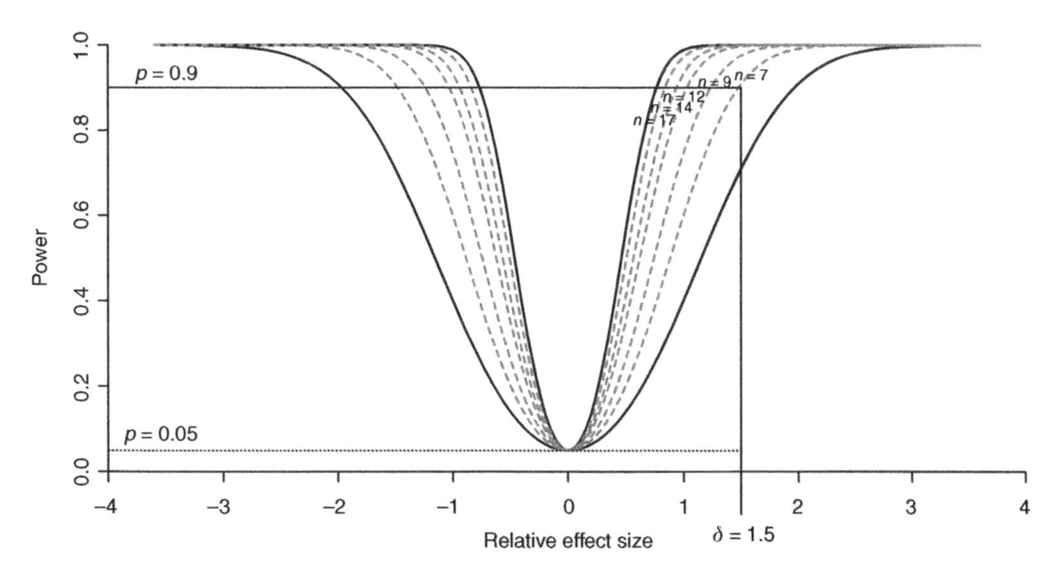

Figure 8.8 The power function of the t-test in Formula (8.6) for testing the null hypothesis $H_0 : \mu = \mu_0$ against H_A: $\mu \neq \mu_0$, using the type-I risk $\alpha = 0.05$. Shown are the cases for $n = 5$ (lowest curve) and some other sample sizes, up to $n = 20$ (topmost curve). For instance, trying a power of 0.9, we find on the abscissa the relative effect size; this is 1.5 in the case of $n = 7$.

may imagine that for an $\alpha > 0.05$ the curves will be moved upwards along the ordinate. For this, only tests with the same α are comparable.

For the calculations of the (necessary) sample size (given a one-sided alternative hypothesis), we look first for all power functions of all possible sample sizes which equal α at μ_0; this being the parameter hypothesized under the null hypothesis. Then we look for the practically relevant (minimal) difference δ. We decide on that power function which results, at this point, in the probability $1 - \beta$ (the probability of rejecting the null hypothesis). As a consequence, at this point the probability of not rejecting the null hypothesis (the probability of making a type-II error) is β. The respective power function fixes the sample size n. Given a two-sided alternative hypothesis, we use the points $-\delta$ and $+\delta$. Figure 8.8 shows that differences larger than δ and $|\delta|$, respectively, occur with an even smaller probability than β.

A special problem arises in the case of an unknown variance σ^2. Though this parameter can be replaced in Formula (8.6) by the sample variance, planning a study already requires intrinsic observations in order to estimate σ^2. However, we can proceed as follows: the anticipated range of the character in question (the difference of maximal and minimal outcome; see Section 5.3.2) has to be divided by six, and the result may be used as an estimate for σ.

Master The estimation of σ for a normally distributed random variable by $\frac{1}{6}$ of the range is due to the fact that, with a probability of more than 0.99 (exactly 0.9974), such a variable has nearly all its values between $\mu - 3\sigma$ and $\mu + 3\sigma$. Hence, the range is approximately 6σ (see in Table B1 in Appendix B for $\sigma = 1$ the values $P(z = -3)$ and $P(z = 3)$.

Another option to estimate σ^2 is to take the variance from an earlier study. Sometimes there is a 'least favorable' value for σ^2, which may be used. If nothing is known about the distribution of the character in question, the researcher may first take a small sample of the size $10 \leq n_0 \leq 30$ and calculate a first estimate of the variance; the result can be taken for the calculation of n. If this so-calculated $n > n_0$, then $n - n_0$ further observations have to be sampled – otherwise data sampling can be stopped (see in detail Rasch, Pilz, Verdooren, & Gebhardt, 2011; size $10 \leq n_0 \leq 30$ for the pre-sample suits, according to simulation studies).

In practice, one uses pertinent computer packages in order to calculate the sample size needed for a given α, β, and δ (for how to do it by hand, see Rasch, Pilz, Verdooren, & Gebhardt, 2011).

Example 8.8 A psychological test for measuring stress resistance will be extreme-group validated (see, for validation of psychological tests, in particular, the method of 'extreme-group validation', for instance Kubinger, 2009b)

The test is standardized on a random sample from the population 'non-academic earners between 25 and 35 years of age': $N(\mu = 0, \sigma^2 = 1)$. As an extreme group, an 'approved' field staff (door-to-door salesmen) is available. We are now interested in the hypothesis that the mean of test scores is at most $\mu_0 = 0.67$ in the field-staff group, against the one-sided alternative hypothesis that this is not the case. We decide on $\delta = \frac{2}{3}\sigma$, $\alpha = 0.01$ and $\beta = 0.05$; the hypotheses are

$$H_0 : \mu \leq 0.67$$
$$H_A : \mu > 0.67$$

The hypothesized $\mu_0 = 0.67$ is due to the fact that this is a difference from the mean 0 of the whole population to an extent of two-thirds of the standard deviation. Interpreted as a p-quantile of the distribution under the null hypothesis, this value corresponds to the probability $1-p$ of nearly 0.75. If the test were valid, then persons who have appeared to be apparently stress resistant should attain on average at least such a percentile rank (in comparison with the population as a whole). Because then the validity is dubious, from the content point of view a deviation (to below) of two-third of a standard deviation is – according to our opinion – relevant; for smaller deviations, however, measurement errors might just be responsible, as far as our experience proves. The variance of the field staff we suppose is also $\sigma^2 = 1$.

We now calculate the sample size n.

In **R**, we use the package `OPDOE`, which we have already installed in Example 7.8. Now we type

```
> size.t.test(delta = 0.67, sd = 1, sig.level = 0.01, power = 0.95,
+             type = "one.sample", alternative = "one.sided")
```

i.e. we apply the function `size.t.test()` and use `delta = 0.67` as the first argument to specify the practically relevant (minimum) difference δ. The standard deviation with `sd = 1` is used as the as the second argument, and the type-I error with `sig.level = 0.05` as the third one. The required power (`power = 0.95`), thus $1 - \beta$, is used as the fourth argument. The fifth argument, `type = "one.sample"`, specifies the number of samples. Finally, `alternative = "one.sided"` specifies the type of the alternative hypothesis.

As a result we get

```
[1]  38
```

Thus, we need a sample size of $n = 38$.

Now suppose we have the test scores of $n = 38$ field-staff members, given in the file *Example_8.8* (see Chapter 1 for its availability).

In **R**, we enable access to the database *Example_8.8* by using the function `attach()` (see Chapter 1). To conduct the *t*-test we type

```
> t.test(stress, mu = 0.67, alternative = "greater")
```

– the meaning of the arguments is explained in Example 8.3.

As a result, we get:

```
          One Sample t-test

data:  stress
t = 2.0356, df = 37, p-value = 0.0245
alternative hypothesis: true mean is greater than 0.67
95 percent confidence interval:
 0.7247016       Inf
sample estimates:
mean of x
0.9895362
```

In SPSS, we apply the same sequence of commands (Analyze – Compare Means – One-Sample T Test...) as in Example 8.3 to conduct a one-sample *t*-test after having loaded the database. In the resulting window (shown in Figure 8.4), we drag and drop the character stress resistance to the panel Test Variable(s): and type the value stated in the null hypothesis (0.67) in the field Test Value:. By clicking OK, we obtain the results shown in Table 8.8 and Table 8.9.

Table 8.8 SPSS-output showing mean and standard deviation in Example 8.8 (shortened output).

One-Sample Test

	Test Value = 0.67		
	t	df	Sig. (2-tailed)
stress resistence	2.036	37	.049

Table 8.9 SPSS-output showing the result of the one-sample t-test in Example 8.8 (shortened output).

One-Sample Statistics

	N	Mean	Std. Deviation
stress resistence	38	.9895	.96767

Because the p-values are 0.0245 and $0.049 / 2 = 0.0245$, respectively, the result is significant; that is to say the null hypothesis 'the mean of test scores in the population of field staff is (at most) $\mu_0 = 0.67$' must be rejected. However, as Table 8.9 indicates, the mean of the sample is 0.9895, so we must conclude that the population mean of field-staff members is clearly larger than 0.67: the extreme-group validation was therefore successful.

Usefulness of planning a study with respect to the sample size is proven by the following: if a study is done accordingly, then it is guaranteed that, with given type-I risk, the relevant effect $(\mu_1 - \mu_0)$ will actually be discovered by the statistical test with a known, high probability of $1 - \beta$.

Master Doctor

In psychology, however, the sample size for answering any research question is almost always determined 'intuitively', or better to say 'easy-going pragmatically'. 'Easy going' insofar as no (pretended) exaggerated effort is put forth in the study. 'Pragmatically' insofar as there have been some practices established for the determination of the sample size. In textbooks there sometimes is the recommendation to choose a sample size of $n > 30$. This is comprehensible, because in these cases most likely the central limit theorem applies (i.e. the assumption of normal distribution of a character's modeled random variable is beside the point; see Section 6.2.2). However, that argument is no longer needed, because simulation studies about the robustness show that, even for large deviations from the normal distribution, the test statistic's distribution hardly alters (Rasch & Guiard, 2004).

The 'intuitive' determination of the sample size is not at all in accordance with statistical foundations. The sample size n is to be derived exclusively from α, β, δ (and σ).

If no planning applies, then in the case of a significant result nothing is known, as concerns the (relative) effect size. Therefore, it should at least be estimated from the data. Even less is known in the case of a non-significant result about the type-II risk, given that the true effect is as large as the estimated effect size. Therefore, the researcher may be interested in this risk; we will call it 'result-based type-II risk'. In the given case we estimate $E = (\mu_1 - \mu_0)/\sigma$ by $\hat{E} = (\bar{y}_1 - \mu_0)/\sigma$ and $\hat{E} = (\bar{y}_1 - \mu_0)/s$, respectively. The calculation of the 'result-based type-II risk' β^* can be done by program packages very easily, for instance by **R**; in psychological research the program package *G*Power 3* is often used, and like **R** this is available as freeware (www.psycho.uni-duesseldorf.de/abteilungen/aap/gpower3/).

Example 8.8 – continued

The estimated relative effect size is $\hat{E} = 0.9895$ and $\hat{E} = 0.9895/0.9677 = 1.0225$, respectively.

Although the study has been planned, and we therefore know that at a given type-I risk of 0.01 a practically relevant (relative) effect of at least 0.67 can be detected by the test with the probability of 0.95, we now calculate just for a tutorial the type-II risk β^* based on observations, with probability of 0.95.

In **R**, we type

```
> power.t.test(n = 38, delta = 1.0225, sig.level = 0.01,
+              type = "one.sample", alternative = "one.sided")
```

i.e. we again apply the function `power.t.test()`. We set the sample size with `n = 38`, and the estimated relative effect size with `delta = 1.0225`. The meaning of the remaining arguments is explained in the application of the function `size.t.test()` in this example, above.

As a result, we get:

```
     One-sample t test power calculation

              n = 38
          delta = 1.0225
             sd = 1
      sig.level = 0.01
          power = 0.999907
    alternative = one.sided
```

The result of $1 - \beta^* \approx 1.00$ differs from 0.95, upwards.

If the true relative effect were really 1.0225 then we nearly do not have a type-II risk at all.

Example 8.9 Example to estimate the 'result-based type-II risk' β^* without relation to any content

We assume, without planning the study, that we have got a random sample of $n = 12$ persons and have applied the t-test according to Formula (8.6); there is a one-sided research question ($\alpha = 0.01$). The analysis resulted in no significance. The estimated (relative) effect size may be $\hat{E} = 0.5$. This is considered relevant with regard to content. We now calculate the 'result-based type-II risk' β^*.

In **R**, we type, analogously to Example 8.8 – continued,

```
> power.t.test(n = 12, delta = 0.5, sig.level = 0.05,
+              type = "one.sample", alternative = "one.sided")
```

i.e. we again apply the function `power.t.test()` and use the arguments explained above.

As a result, we get:

```
     One-sample t test power calculation

              n = 12
          delta = 0.5
             sd = 1
      sig.level = 0.05
          power = 0.4918845
    alternative = one.sided
```

The result of $\beta^* = 1 - 0.4919 = 0.5081$ shows – given the true effect equals the estimated one, which we consider as practically relevant – that in comparable studies with $n = 12$, a false null hypothesis would more often be accepted than rejected. The results of this study should not be published, because it indicates inadequate or absent planning of the study.

Determination of sample sizes is also possible for the topics of point estimation and confidence intervals. As the first one regularly results in calculations of a confidence interval, and the latter can be used for testing a hypothesis, we do not deal with planning a study as concerns these topics in this textbook (but see Rasch, Pilz, Verdooren, & Gebhardt, 2011).

Master Choosing the type-I risk α of statistical tests equal to that α used for a confidence interval, the following connection exists: if the parameter value hypothesized in the null hypothesis lies inside the corresponding confidence interval, the null hypothesis is to be accepted. Otherwise, if a statistical test results in significance, then we know that the confidence interval does not cover the hypothesized value of the null hypothesis.

Summary

In *testing hypotheses,* the researcher has to decide which probability for a *type-I error* and which probability for a *type-II error* he/she can tolerate. The corresponding risks are denoted by α and β, whereby most of the time $\alpha = 0.05$ (sometimes 0.01) is used. In relation to α, the probability β has to be decided; this depends on which of the two kinds of error is more important for the given research question. The *statistical test* is based on the idea of calculating the probability that the observed data or (from the null hypothesis' point of view) even more extreme data may occur – though the null hypothesis is actually true. If this probability is smaller than α, the null hypothesis is to be rejected; we say: the result is *significant*. Probability theory, however, can only be applied if the data stems from a random sample. For planning a study as concerns the determination of the sample size, a practically relevant minimal difference has to be determined for which the null hypothesis is not rejected with probability β (type-II risk); that is for which that difference is not detected by the test. Such a relative or standardized practically relevant difference is called (*relative*) *effect size.*

8.4 Sequential tests for the unknown parameter μ

Once accepted that the scientific gain of findings (in psychology) is only possible by planning a study in order to come to unequivocal conclusions, by known risks, then the demonstrated strategy can be optimized. The method is *sequential testing*. This is a specific technique of statistical inference.

Sequential testing can be applied, if the observations of a character can or must be done in a study one after the other. Typical examples are laboratory analyses, psychological assessment by individual testing, or consulting of patients and clients in hospitals or institutions alike. The basic idea is to analyze existing observed data before the next research unit is sampled.

For instance, for a test of the hypothesis $H_0 : \mu = \mu_0$ versus H_A: $\mu \neq \mu_0$ after each step of analysis, after each research unit, there are the following three options:

- accept H_0
- reject H_0
- continue the study and data sampling, respectively.

The advantage of sequential testing compared with the 'classical' approach is that, with the average of many studies, many fewer observations are needed. However, only if α, β, and δ are fixed in advance does sequential testing work; any decision for the given three options is possible.

Bachelor Sequential testing is still rarely applied in psychology. However, where each research unit causes high costs, it is already a part of everyday usage, for example in pharmacological research.

There are two different approaches to sequential testing. In the original approach by Wald (1947), at the very beginning of a study it is totally unknown how many research units will be needed. Such tests therefore are called *open sequential*

tests. Nevertheless, the average sample size for these sequential tests is smaller than for the corresponding tests with a planned and therefore fixed sample size. But there is another approach: testing and data sampling ends at the latest with a maximal number of research units; these tests are called *closed sequential tests*. Their disadvantage is that their average sample size is a little bit larger than that of the open sequential tests; but it is still smaller than that of the corresponding tests with a planned and therefore fixed sample size.

A special group of closed sequential tests are the *sequential triangular tests*. Their principle goes back to Whitehead (1992) and Schneider (1992) and is as follows. Using the observed values y_v, some cumulative ascertained ancillary values Z_v and V_v are calculated. In the case of a test with respect to the mean μ that is after the mth step, i.e. after we have observed the values $y_1, y_2, \ldots, y_m$, it is to calculate

$$D_m = \sqrt{\frac{1}{m}\sum_{v=1}^{m} y_v^2},\ Z_m = \frac{\sum_{v=1}^{m} y_v}{D_m},\ \text{and } V_m = m - \frac{Z_m^2}{2m}$$

Here D_m is obviously essentially a function of the estimated standard deviation of $\boldsymbol{y}$, and Z_m is a function of the standardized estimated mean of $\boldsymbol{y}$. The values Z_m can be marked on the abscissa of a rectangular coordinate system, and values V_m on the ordinate. For a one-sided alternative hypothesis, two straight lines are defined in dependence of type-I and type-II risk and of the practically relevant minimal difference δ. They form a triangle open to the left side, and intersect each other at a point $Z_{\max}$; this is the value corresponding to the maximal sample size $n_{\max}$ of the sequential triangular test. As long as the sequence of the Z_m values is within this triangle, then sequential testing and data sampling, respectively, are to continue. If one of the two borderlines is met or even exceeded, then the procedure stops. Depending on which line is concerned, the null hypothesis is either accepted or rejected. In the case of a two-sided alternative hypothesis, this is split into two one-sided alternative hypotheses in such a way that the two-sided alternative hypothesis is to be rejected when one of the two one-sided alternative hypotheses is rejected. Using $\frac{\alpha}{2}$ instead of α for each of the one-sided alternative hypotheses, a triangle as described above is constructed. Both are open towards the left side and end at the same point of the abscissa on the right side. As long as the sequence of the Z_m values is within any of these triangles, sequential testing and data sampling, respectively, are to continue. If one of the outside borderlines of the triangles is met or even exceeded, then the null hypothesis is to be rejected. If the sequence of the Z_m values stays in the region between the two triangles, the null hypothesis is to be accepted.

For the sequential triangular tests there are program packages like PEST (see Whitehead, 1992) and CADEMO-TRIQ (www.biomath.de). SPSS offers no sequential testing, but in **R** we find sequential triangular tests in the package `OPDOE`. Because we are not dealing with commercial software apart from SPSS, we will now introduce those sequential triangular tests available in **R**.

Example 8.10 Is the sample of children with German as their native language in Example 1.1 a representative sample as concerns the personality trait *conscientiousness*?

At the end of the study of Example 1.1, suspicion rises that the children with German as their native language were negatively biased as concerns personality trait *conscientiousness*. Therefore in a follow-up study with the aid of a personality questionnaire[4], this character has additionally been tested. *Conscientiousness* is standardized to T-Scores, i.e. modeled as a normally distributed random variable with mean $\mu_0 = 50$ and standard deviation $\sigma = 10$. The null hypothesis is therefore H_0: $\mu = \mu_0 = 50$ and the alternative hypothesis is H_A: $\mu < 50$: if the sample does not stem from the population in question, then we expect the children to achieve lower test scores. We choose $\alpha = 0.05$, $\beta = 0.20$, and $\delta = 0.67$.

Because we prefer to not test, if possible, all 50 children with German as their native language with the personality questionnaire, we proceed sequentially. We start with the first five children of our data set, who have been tested with the intelligence test-battery.

Their results are as follows: 50, 52, 53, 40, and 48.

In **R**, we type

```
> conscient.tt <- triangular.test.norm(x = c(50, 52), mu0 = 50,
+                                      mu1 = 43.3, sigma = 10,
+                                      alpha = 0.05, beta = 0.2)
```

i.e. we apply the function `triangular.test.norm()`. We use the vector of the first and second observation value `c(50, 52)` of the character *conscientiousness* as the first argument, the mean $\mu_0 = 50$ (`mu0 = 50`) as the second, and with `mu1 = 43.3` indirectly δ as third argument. Furthermore, we specify the standard deviation of $\sigma = 10$ (`sigma = 10`) and the respective precision requirements by `alpha = 0.05` and `beta = 0.2`. Now we assign the function to the object `conscient.tt`.

As a result, we get:

```
Triangular Test for normal distribution

Sigma known: 10

H0: mu = 50  versus H1: mu < 43.3
alpha: 0.05  beta: 0.2

Test not finished, continue by adding single data via update()
current sample size for x:  2
```

[4]For example in NEO PI-R (*NEO Personality Inventory-Revised*; Costa & McCrae, 1992), *conscientiousness* is measured by questions like the statement '*I'm known for my prudence and common sense.*'

A final decision based on the first two observation values is not possible. Thus, we type

```
> conscient.tt <- update(conscient.tt, x = 53)
```

i.e. we apply the function `update()` to amend its first argument, object `conscient.tt`, with the third observation value `x = 53` which is used as the second argument. The result is similar to the one above; that is, we are informed that further observation values have to be added.

Because after $n = 5$ children no terminal decision is possible, either for accepting or for rejecting the null hypothesis, we test a further child. As a result, after sequentially testing one child after the other up to altogether $n = 13$ children, no decision is possible there, either: even these test scores (38, 45, 56, 45, 50, 53, 68, 44) force us to continue testing. After three more children (with test scores 55, 54, 51) we come to a terminal conclusion.

In **R**, we add now the 16th observation value by typing

```
> conscient.tt <- update(conscient.tt, x = 51)
```

i.e. we apply the function `update()` to amend the object `conscient.tt` with `x = 51`.

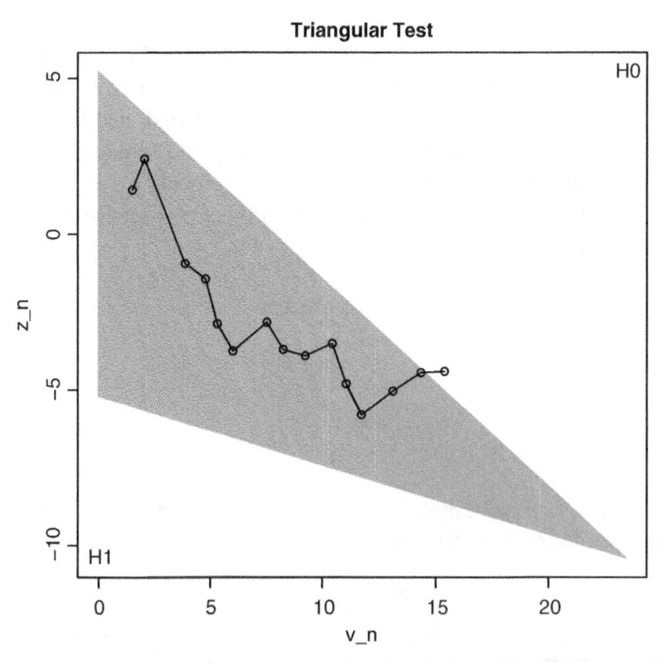

Figure 8.9 **R**-output showing the result of the sequential triangular test in Example 8.10 after 16 observation values have been entered.

As a result, we get (see also Figure 8.9):

```
Triangular Test for normal distribution

Sigma known: 10

H0: mu = 50   versus H1: mu < 43.3
alpha: 0.05   beta: 0.2

Test finished: accept H0
Sample size for x:   16
```

As a matter of fact, the conclusion is that the null hypothesis has to be accepted; the sample of children with German as their native language in Example 1.1 is representative with respect to the personality trait *conscientiousness*.

With SPSS no sequential testing is possible.

8.5 Estimation, hypothesis testing, planning the study, and sequential testing concerning other parameters

8.5.1 The unknown parameter σ^2

Since Section 5.3.1 we know that an appropriate (that is, unbiased) estimator of the population variance σ^2 is the sample variance according to Formula (5.4): $\hat{\boldsymbol{\sigma}}^2 = s^2 = \frac{1}{n-1}\sum_{v=1}^{n}(\boldsymbol{y}_v - \bar{\boldsymbol{y}})^2$.

In psychology there are hardly any research questions leading to a confidence interval for the variance of a character modeled as a normally distributed random variable. The same is true as concerns hypothesis testing pairs like H_0: $\sigma^2 = \sigma_0^2$, H_A: $\sigma^2 \neq \sigma_0^2$ (and H_A: $\sigma^2 < \sigma_0^2$ or H_A: $\sigma^2 > \sigma_0^2$). In this respect the planning of a study and sequential testing are also of no relevance (see, if needed, Rasch, Pilz, Verdooren, & Gebhardt, 2011).

Doctor As can be shown, given a normally distributed random variable, the test statistic

$$\chi^2 = \frac{\hat{\boldsymbol{\sigma}}^2}{\sigma^2}(n-1)$$

is χ^2-distributed with $df = n - 1$ degrees of freedom (see in Table B3 in Appendix B the α- and $(1-\alpha)$-quantiles $\chi^2(n-1, \alpha)$ and $\chi^2(n-1, 1-\alpha)$; values missing in this table can of course be determined by **R** and SPSS). As a consequence, bounds of the confidence interval for σ^2 are deducible given:

$$\boldsymbol{L} = \frac{(n-1)\cdot \boldsymbol{s}^2}{\chi^2\left(n-1, 1-\dfrac{\alpha}{2}\right)}; \quad \boldsymbol{U} = \frac{(n-1)\cdot \boldsymbol{s}^2}{\chi^2\left(n-1, \dfrac{\alpha}{2}\right)} \tag{8.7}$$

A test immediately follows – the null hypothesis is accepted with a type-I risk α if the hypothesized value σ_0^2 lies within the bounds of the confidence interval.

Doctor **Example 8.11** The question is, whether students functioning as examiner of IQ did their job fairly

Skepticism always arises that hired helpers are lazy and produce testees' answers in psychological tests and questionnaires on their own instead of actually testing the testees. Clever, but dilatory student examiners will try to hit approximately the average IQ in doing so. But what is hard to anticipate for them and even harder to control is to 'generate' the variance of the *IQ* values sufficiently precisely over all testees. For this reason, for each student examiner, the variance of the *IQ* values he/she delivers for analysis shall be estimated.

More precisely, for each student examiner a confidence interval ($\alpha = 0.05$) for the variance of the *IQ* values is calculated.

Let us assume that for a certain student examiner the sample variance is $\hat{\sigma}^2 = 144$, due to $n = 21$ testees. We remember that the intelligence tests are standardized to $\sigma = 15$. From Table B3 in Appendix B we obtain $\chi^2(20, 0.025) = 9.591$ and $\chi^2(20, 0.975) = 34.17$, and using this, from Formula (8.7) the 95% confidence interval results as $[20 \cdot 144/34.17 = 84.28; 20 \cdot 144/9.591 = 300.28]$. We use this confidence interval for testing the pair of hypotheses H_0: $\sigma = 15$; H_A: $\sigma \neq 15$. Because the observed value 225 lies within the interval, we accept the null hypothesis and trust our test administrator. On the other hand, we could have tested the given null hypothesis directly by $\chi^2 = (144/225) \cdot 20 = 12.8 > 9.591 = \chi^2(20, 0.025)$; the probability of this data, or more extreme data, given the null hypothesis, is not small enough (is not less than 0.05) to reject it.

8.5.2 The unknown parameter p of a dichotomous character

Without any derivation it is plausible that the probability p of one of the two possible outcomes of a dichotomous character is best estimated by $\hat{p} = \frac{h}{n}$ with h (again) the absolute frequency of that value, given n observations and research units, respectively.

The exact formula for a confidence interval is complex and hardly used in psychology. This is because, for even relatively small sample sizes n, the distribution function of the binomial distribution is well approximated by a normal distribution. And therefore the bounds of the confidence intervals are:

$$L = \hat{p} - z\left(1 - \frac{\alpha}{2}\right)\sqrt{\frac{\hat{p}(1-\hat{p})}{n-1}}; \quad U = \hat{p} + z\left(1 - \frac{\alpha}{2}\right)\sqrt{\frac{\hat{p}(1-\hat{p})}{n-1}} \tag{8.8}$$

Apart from the fact that from a confidence interval a test also results (the null hypothesis is to be accepted, with the type-I risk α, if the hypothesized value p_0 lies within the bounds of the confidence interval), a test of this null hypothesis is also possible by an approximately standard normally distributed test statistic (with a two-sided or one-sided alternative hypothesis):

$$z = \frac{\hat{p} - p_0}{\sqrt{\frac{\hat{p}(1-\hat{p})}{n-1}}} \tag{8.9}$$

One has just to compare this resulting value z with the respective standard normal distribution's $\frac{\alpha}{2}$- and $(1-\frac{\alpha}{2})$-quantile or α- and $(1-\alpha)$-quantile, respectively, from Table B2 in Appendix B.

A rule of thumb is: if $n \cdot p \geq 5$ and $n \cdot (1-p) \geq 5$, the approximation is sufficient.

Example 8.12 From the study of Example 1.1, the relative proportion of children with exactly two siblings will be estimated

We look for a point estimate as well as for a confidence interval ($\alpha = 0.05$). First of all we count how often the value '2' has been realized in the character *number of siblings*.

In **R**, we type (after having attached the original data set of Example 1.1)

```
> addmargins(table(no_siblings))
```

i.e. we request the output of the frequency distribution of the character *number of siblings* by applying the function `table()`. The function `addmargins()` ascertains the sum of all frequencies. As a result, we get:

```
no_siblings
 0   1   2   3   4   5 Sum
17  30  24  16  10   3 100
```

In SPSS, we select

Analyze
  Descriptive Statistics
    Frequencies...

and drag and drop number of siblings to Variable(s): and click OK. As a result, we get Table 8.10.

Table 8.10 SPSS-output showing the frequencies in Example 8.12.

number of siblings

		Frequency	Percent	Valid Percent	Cumulative Percent
Valid	0	17	17.0	17.0	17.0
	1	30	30.0	30.0	47.0
	2	24	24.0	24.0	71.0
	3	16	16.0	16.0	87.0
	4	10	10.0	10.0	97.0
	5	3	3.0	3.0	100.0
	Total	100	100.0	100.0	

Table 8.10 shows that $h = 24$ and $n = 100$. The relative proportion we are looking for is to estimate $\hat{p} = \frac{24}{100} = 0.24$. The confidence interval results, according to Formula (8.8), as $[0.24 - 1.96\sqrt{\frac{0.24 \cdot 0.76}{99}}; 0.24 + 1.96\sqrt{\frac{0.24 \cdot 0.76}{99}}]$; that is [0.1971; 0.2829].

Bachelor **Example 8.13** Will party A win the next election with an absolute majority, according to an opinion poll?

In an opinion poll, $h(\text{A}) = 842$ of $n = 2000$ interviewed randomly sampled persons will vote for party A. The question is whether, from these results, at the next election (which will be held in the very near future), party A can expect to win with an absolute majority (that is, simplified: $P(\text{A}) = p = 0.5$; exactly: $P(\text{A}) = p = 0.500 \cdots 01$). The null hypothesis is thus H_0: $p \geq 0.5$; the alternative hypothesis is H_A: $p < 0.5$. Type-I risk is chosen with $\alpha = 0.05$. As $\hat{p} = f(\text{A}) = 842/2000 = 0.421$, we get:

$$z = \frac{0.421 - 0.5}{\sqrt{\frac{0.421 \cdot 0.579}{1999}}} = -7.15$$

From Table B2 in Appendix B we find $-z(1-0.05) = -2.33$; the result is significant, and the null hypothesis must be rejected. Party A cannot reckon on absolute majority.

Until now we have discussed only the (relative) effect size for comparing means. Of course, estimates of an effect size from the result of a statistical test are always available; in which way an effect size for a standard normally distributed test statistic can generally be estimated will be dealt with, however, later in Section 11.3.5.

Several methods exist for planning a study, but there is no single method that leads, for all values of p, systematically to the smallest sample size. By the way, each of them needs some *a priori* information of roughly the size of p.

Master Doctor

The least favorable case for the sample size n is that with $p = 0.5$; the nearer p is to 0 or 1, the smaller the required sample size. We apply a method based on the normal distribution approximation; the alternative hypothesis may be one-sided. If p^* is that value which roughly is to be expected, then n is to be determined approximately as

$$\frac{\left[z_{1-\beta}\sqrt{(p^*+\delta)(1-p^*-\delta)} + z_{1-\alpha}\sqrt{p^*(1-p^*)}\right]^2}{\delta^2}$$

or the next larger integer, respectively: δ is the practically relevant difference between the (by the null hypothesis) hypothesized parameter p_0 and the actual true parameter p. This difference will not result in significance in cases with a probability of $\beta \cdot 100\%$. The formula holds for $p^* \leq \frac{1}{2}$ only; if $p^* > \frac{1}{2}$, replace p^* by $1-p^*$. If the alternative hypothesis is two-sided, $1 - \frac{\alpha}{2}$ is to be used instead of $1-\alpha$.

Master Doctor

Example 8.14 The number of applicants in a psychological study in Vienna that come from Germany in the calendar year X + 1, in comparison with those in calendar year X

We assume that in year X the proportion of applicants from Germany is 0.4. In order to predict the respective proportion in the year X + 1 – for instance to gain insight into future enrollment fees – the sample of the first n applicants should be used. The null hypothesis is H_0: $p_{X+1} = p$; the alternative hypothesis is H_A: $p_{X+1} \neq p$. We choose $\alpha = 0.05$, $\beta = 0.05$, and $\delta = 0.2$.

In **R**, we compute the sample size with the help of the package `OPDOE` by typing

```
> size.prop_test.two_sample(p1 = 0.4, p2 = 0.6, alpha = 0.05,
+                           power = 0.95, alt = "two.sided")
```

i.e. we use the probability stated in the null hypothesis, `p1 = 0.4`, as the first argument and the probability resulting from $\delta = 0.2$, that is `p2 = 0.6`, as the second argument in the function `size.prop_test.two_sample()`. Moreover, we use the type-I risk (`alpha = 0.05`) as the third, the type-II risk subtracted from 1 (`power = 0.95`) as the fourth, and the type of the alternative hypothesis (`alt = "two.sided"`) as the fifth argument in the function.

As a result, we get:

```
[1] 170
```

We have to wait until $n = 170$ applicants are there; we then check their nationality. Note: in this example we have to assume that the order of applicants is random.

Master Doctor

Testing of hypotheses concerning the parameter p of a dichotomous character is also possible with a sequential triangular test.

Example 8.14 – continued

Assume that we had already asked 88 applicants for their nationality and $\hat{p}_{X+1} = 0.5114$.

In **R**, we obtain the preliminary result shown in Figure 8.10 (given that the last observation was 'Germany') by typing

```
> nation.tt <- triangular.test.prop(nation, p0 = 0.4,
+                                   p1 = 0.2, p2 = 0.6,
+                                   alpha = 0.05, beta = 0.05)
```

i.e. we apply the function `triangular.test.prop()` and use all 88 observation values in the object `nation` as the first argument in the function. With `p0 = 0.4` we specify the null hypothesis and with `p1 = 0.2` and `p2 = 0.6` we indirectly define δ. Finally, we use the arguments `alpha = 0.05` and `beta = 0.05` to specify the respective precision requirements. We assign the result to the object `nation.tt`.

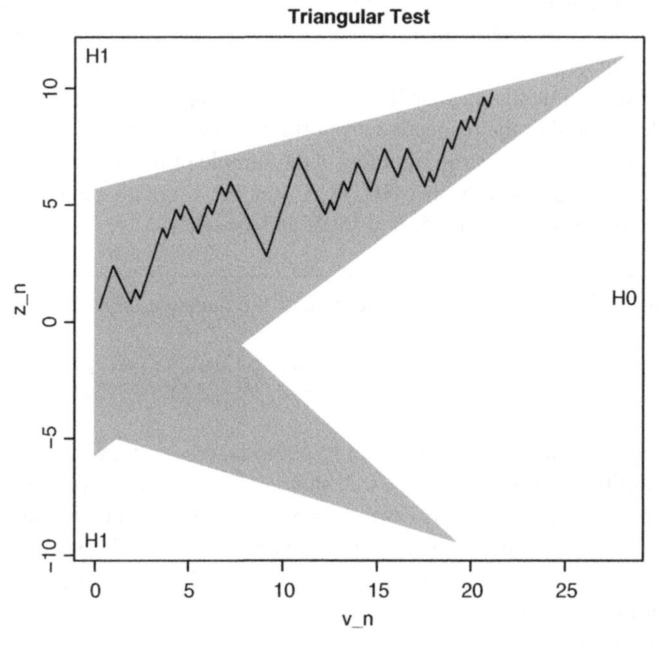

Figure 8.10 **R**-output showing the result of the sequential triangular test in Example 8.14 after 88 observation values have been entered.

As a result we get:

```
Triangular Test for bernoulli distribution R-
H0: p = 0.4  versus H1: p > 0.6  or p < 0.2
alpha: 0.05  beta: 0.5

Test not finished, continue by adding single data via update()
current sample size for x:   88
```

Analyzing the 89th observation – again an applicant from Germany – we get the terminal decision: in the year X + 1 we expect a larger proportion of German applicants than in the year X. The resulting sample size of $n = 89$ with sequential testing is much smaller than when planning a fixed sample size for the study.

8.5.3 The unknown parameter p of a dichotomous character which is the result of paired observations

Often there is just a single sample (of research units; in psychology mostly persons); but nevertheless, content-wise, two or more values for the same character are observed. In such a case, given a quantitative character, we have already dealt with two values per research unit in Section 8.2.4. In the case of a qualitative character, the methodical approach differs somewhat from that in Section 8.2.4, because there can be no differences ascertained. In the following we consider a dichotomous character and once again only two observations per unit. Again, for instance a certain treatment's effect is of interest.

In this context, the primary aim is testing a null hypothesis. If we denote the two possible values with '+' and '–', then we expect under the null hypothesis that the relative frequency of, for instance, '+' within the population in question (i.e. the population's probability) does not change from the first observation time to the second. That is H_0: $p_1 = p_2$, H_A: $p_1 \neq p_2$. As with quantitative characters, the matter at hand is measuring the change per research unit. There are four possibilities of two outcomes per unit: (+, +), (+, –), (–, +), and (–, –). The corresponding relative frequencies and probabilities are $f(+, +), \ldots, f(-, -)$ and $p(+, +), \ldots, p(-, -)$, respectively. The only data which count contra-null hypothesis and pro-alternative hypothesis are $f(+, -)$ and $f(-, +)$. Given the null hypothesis, $f(+, -) > 0$ and/or $f(-, +) > 0$ occur just by chance; principally only (+, +) or (–, –) would be realized. As a consequence, the null hypothesis is to be reformulated: H_0: $p(+, -) = p(-, +)$; H_A: $p(+, -) \neq p(-, +)$.

As can be shown, some test statistic is (given the null hypothesis) *asymptotically* χ^2-distributed ('chi-square') with $df = 1$ degree of freedom. This means that the actual distribution of that test statistic can – for sufficiently large sample sizes – be approximated with the *χ^2-distribution*. The rule of thumb is: $h(+, -) + h(-, +) \geq 25$. The test based on this test statistic is called the *McNemar test*. Because nowadays almost nobody will calculate it manually, we refrain from presenting the formula.

How the effect size for tests with a χ^2-distributed test statistic can be estimated is shown in Section 11.3.5.

Apart from the fact that any planning of a study would require prior (rough) knowledge, calculation of the (overall) sample size is completely impossible.

Example 8.15 Is the advanced training in a seminar 'aptness in negotiations' efficient?

In advanced training in a seminar within a company, several co-workers are trained during a week to use proper strategies in negotiations. Before starting, as well as about a month after the end of the seminar, the boss of each of the 25 participants judges their behavior in sales conversations either as positive ('1') or as negative ('0'). The results are (each column corresponds to a participant):

Before the seminar	1	0	1	1	1	0	0	0	0	0	0	0	0	0	0	0	0	0	0	0	0	0	0	0	0
After the seminar	1	0	0	0	0	0	1	0	1	1	1	0	1	1	1	0	1	1	1	0	0	1	1	1	1

More clearly arranged, this data is:

		After the seminar +	After the seminar −	Total
Before the seminar	+	1	3	4
	−	14	7	21
	Total	15	10	25

Now we apply the McNemar test (see Chapter 1 for the availability of the data).

In **R**, we apply the package `exact2x2`, which we load after its installation (see Chapter 1) using the function `library()`. First, we create two vectors according to the values shown above. To do this, we type

```
> before <- c(1, 0, 1, 1, 1, 0, 0, 0, 0, 0, 0, 0, 0, 0, 0, 0, 0, 0,
+             0, 0, 0, 0, 0, 0, 0)
> after <- c(1, 0, 0, 0, 0, 0, 1, 0, 1, 1, 1, 0, 1, 1, 1, 0, 1, 1,
+            1, 0, 0, 1, 1, 1, 1)
```

i.e. we apply the function `c()` to concatenate all of its arguments into a vector. We assign the resulting vectors to the objects `before` and `after`, respectively. Now we conduct a McNemar test by typing

```
> mcnemar.exact(before, after)
```

i.e. we use the two vectors `before` and `after` as arguments in the function `mcnemar.exact()`.

As a result, we get (shortened output):

```
        Exact McNemar test (with central confidence intervals)

data:  before and after
b = 14, c = 3, p-value = 0.01273
alternative hypothesis: true odds ratio is not equal to 1
95 percent confidence interval:
1.302461 25.326134
```

In SPSS, we first type in the variables before and after according to the values shown above. To conduct a McNemar test we proceed in the following way. We use the sequence of commands from Example 5.13 (Analyze - Descriptive Statistics - Crosstabs...) to create a two-dimensional frequency table. In the resulting window (see Figure 5.28), we drag and drop before to the field Row(s): and after to the field Column(s):. Next we click Statistics... and tick McNemar in the following pop-up window (see Figure 8.11). After clicking Continue and OK we get Table 8.11 as a result.

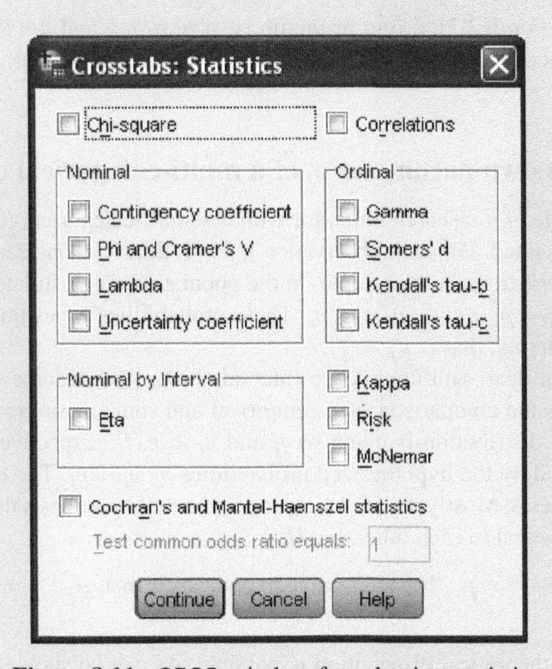

Figure 8.11 SPSS-window for selecting statistics.

Table 8.11 SPSS-output showing the result of the McNemar test in Example 8.15.

Chi-Square Tests

	Value	Exact Sig. (2-sided)
McNemar Test		.013[a]
N of Valid Cases	25	

a. Binomial distribution used.

The null hypothesis is to be rejected. As $h(+, -) < h(-, +)$, the seminar was successful.

Master The McNemar test can be replaced by the so-called *binomial test* (see Example 8.5): we look for the conditional probability $P[\boldsymbol{k} = h(+, -)|n = h(+, -) + h(-, +); p = {}^1\!/\!_2)]$ – given that $h(+, -) > h(-, +)$; as indicated we have to interpret, for instance, $h(+, -)$ as k. This probability can be approximated by

$$P\left(z \leq \frac{k - np}{\sqrt{np(1-p)}}\right) \tag{8.10}$$

(see Section 6.2.2); a rule of thumb is: if $n \cdot p \geq 5$ and $n \cdot (1 - p) \geq 5$, then the binomial distribution is sufficiently approximated by the normal distribution.

8.5.4 The unknown parameter p_j of a multi-categorical character

We consider a sample of n research units, for which a multi-categorical (c-categorical) character has been ascertained. For each observation j, $j = 1, 2, \ldots, c$, one can count the number h_j of respective observations in the sample. In the population we postulate the respective parameters (probabilities) $p_1, p_2 \ldots, p_j, \ldots, p_c$. These probabilities are estimated by the relative frequencies in the sample; that is $\hat{p}_j = f_j$.

Again we do not deal with confidence intervals here. As concerns hypothesis testing, the matter at hand is the comparison of an empirical and some theoretical distribution. The empirical frequency distribution is given by f_j and $h_j = n \cdot f_j$, respectively. The theoretical distribution is defined by the hypothesized probabilities p_j and np_j. The test depends on the concrete null hypothesis. Mostly we are interested in the null hypothesis that the probabilities of all categories are equal to each other, i.e. H_0:

$p_1 = p_2 = \cdots = p_j = \cdots = p_c$. The alternative hypothesis is then H_A: $p_j \neq p_g$, for at least one pair $j \neq g$.

However, any other theoretical distribution is possible. As can be shown, the following test statistic is (given the null hypothesis is true) asymptotically χ^2-distributed; that is:

$$\chi^2 = \sum_{j=1}^{c} \frac{(\boldsymbol{h}_j - np_j)^2}{np_j} \quad \text{is approximately } \chi^2\text{–distributed, with } df = (c-1) \text{ degrees of freedom} \tag{8.11}$$

Though, these days, nobody would compute this test statistic by hand, Formula (8.11) shows that with the help of the $(1-\alpha)$-quantiles of the χ^2-distribution, $\chi^2((c-1), 1-\alpha)$, a decision for or against the null hypothesis is feasible (for instance according to Table B3). If the resulting value in Formula (8.11) exceeds this quantile, the null hypothesis is to be rejected, and the result is significant; as concerns the empirical distribution with its frequencies, some of them (or even all) differ from those which were hypothesized by the null hypothesis. The observed differences between the relative frequencies and the probabilities under H_0 cannot be explained by chance alone.

Master The test statistic in Formula (8.11) is based on a squared mathematical term; therefore take into account that the direction of the differences is not considered with the alternative hypothesis. Hence, no matter which content and which direction of deviation is of interest, this test always refers to a two-sided alternative hypothesis. Nevertheless, from the formal point of view we are interested only in one side of the χ^2-distribution, the right one – this is the side corresponding to large (squared) differences. If one considers also the left side – this is the side tending towards 0 – then the test examines (in addition) whether the (squared) differences are less than expected by chance.

A rule of thumb for the sample sizes in order to get a sufficient approximation of this χ^2-tests' distribution is: $np_j \geq 5$ for all j.

Example 8.16 Do suicides cumulate in certain calendar months?

Unsystematic observations from co-workers in mental health professions show that in central Europe in the winter, more precisely between November and February, relatively more suicides are observed than in the other months. A psychologist therefore in a certain political region records over some years the frequencies of suicides per calendar month (January = 1, ..., December = 12):

Calendar month	1	2	3	4	5	6	7	8	9	10	11	12
Number of suicides observed	28	32	16	20	12	14	5	8	12	20	24	35

The null hypothesis is that there are no differences between the calendar months; hence H_0: $p_1 = p_2 = \cdots = p_{12} = \frac{1}{12}$; the alternative hypothesis is H_A: $p_j \neq p_g$, for at least one pair $j \neq g$; i.e. at least two probabilities differ from $\frac{1}{12}$. We decide on the type-I risk $\alpha = 0.01$.

In **R**, we start by creating a vector containing the number of documented suicides. To do this, we type

```
> number <- c(28, 32, 16, 20, 12, 14, 5, 8, 12, 20, 24, 35)
```

i.e. we apply the function `c()` to concatenate all its arguments into a vector, which we assign to the object `number`. Next, we conduct a χ^2-test by typing

```
> chisq.test(number)
```

i.e. we submit the vector `number` containing the number of documented suicides to the function `chisq.test()`. According to the default settings, the null hypothesis states that the probability of all categories is the same.

As a result we get:

```
Chi-squared test for given probabilities
data:   number
X-squared = 52.1239, df = 11, p-value = 2.589e-07
```

In SPSS, we create a data file with two columns: in the first one, we type in the character *calendar month* (month) with values from 1 to 12 and in the second column the corresponding frequencies (f) according to the list shown above.

In SPSS Variable View we set character *calendar month* (month) in the column Measure to Nominal and the corresponding frequencies (f) to Scale. Next, we weight the character month using the corresponding frequencies. To do this, we select the sequence of commands

Data
 Weight Cases...

In the resulting pop-up window (not shown) we select Weight cases by and drag and drop the character f to the field Frequency Variable:. We confirm the weighting by clicking OK. To start the analysis, we apply

Analyze
 Nonparametric Tests
  One Sample...

In the resulting window (not shown), we activate the tab Fields in the control panel at the top. This gets us to the window shown in Figure 8.12, where calendar month has already been automatically moved into the panel Test Fields:.

Now we switch to the tab Settings (see Figure 8.13), where we select Customize tests and tick Compare observed probabilities to hypothesized (Chi-Square test). The null hypothesis

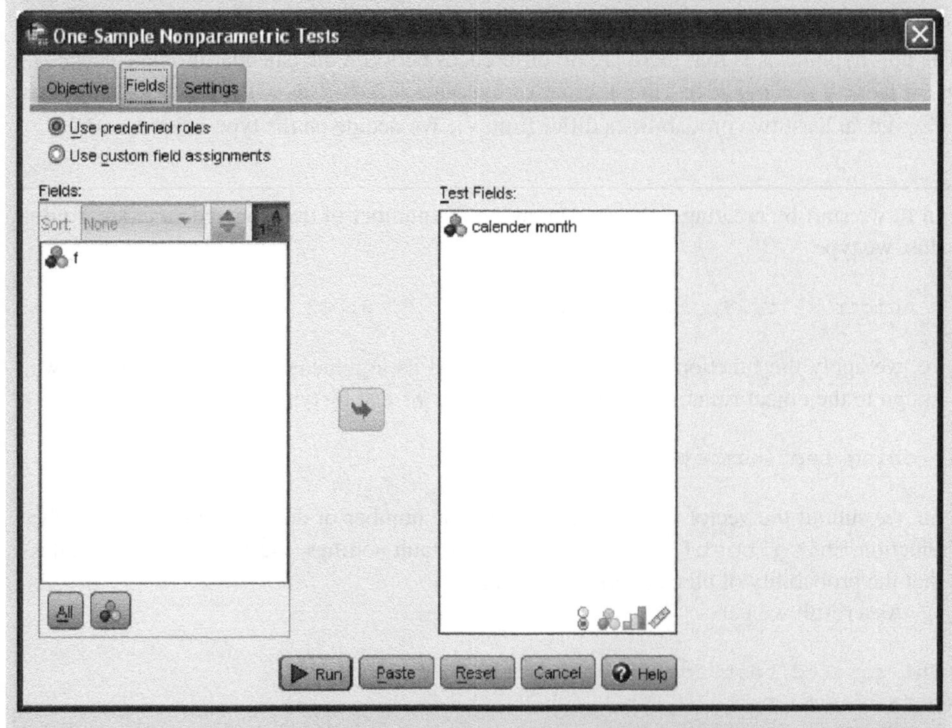

Figure 8.12 SPSS-window I for calculating a χ^2-test.

assumes that the probabilities for all categories are equal, according to the default setting. After clicking Run, the output window shows the table Hypothesis Test Summary (not shown), which we double-click. As a consequence, a window containing a table with the essential results pops up. The p-value, which can be found in Asymptotic Sig. (2-sided test):, equals 0.000.

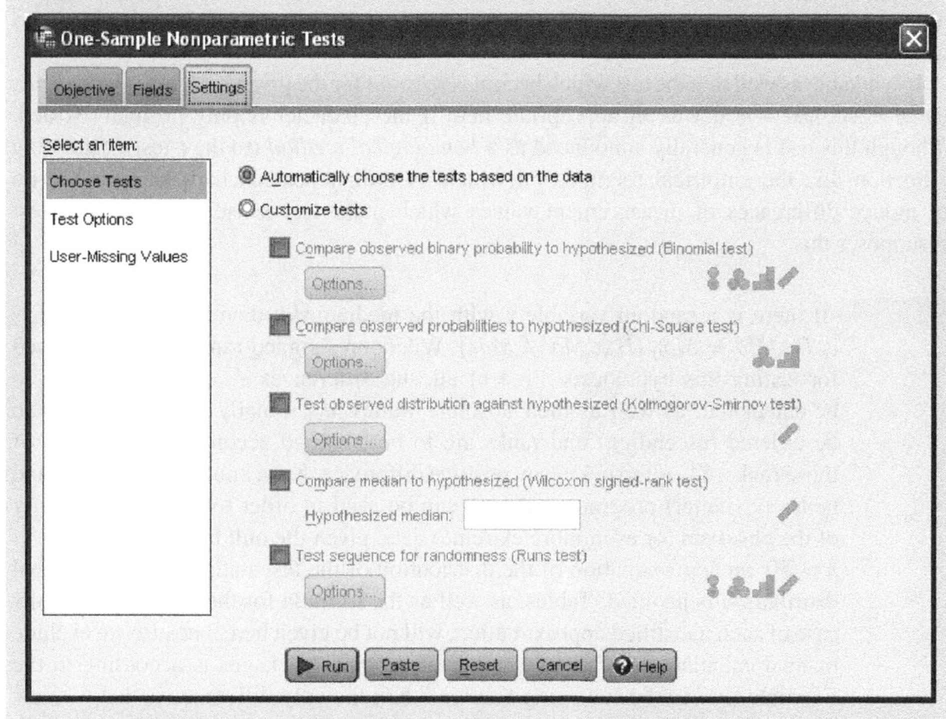

Figure 8.13 SPSS-window II for calculating a χ^2-test.

The result is significant. The frequencies of suicides in all calendar months are, in contrast to the null hypothesis, not equal; the null hypothesis is to be rejected. At least in December we found many more suicides than in July.

Concerning the estimation of the effect size, see Section 11.3.5.

Planning of a study concerning n is, for this χ^2-test, not possible, though it is concerning c.

8.5.5 Test of a hypothesis about the median of a quantitative character

Very seldom in psychology it is of interest to test a hypothesis with respect to a quantitative character's median instead of its mean. However, for example if the character and its modeled variable, respectively, have an extremely skewed distribution – so that the median better characterizes the data than the mean – then we might access *non-parametric tests/distribution-free tests*. These are tests without a specific assumption about the distribution.

It is notable that such an approach is not intrinsically justified by a suspected deviation from the normal distribution of the character in question *per se*. Particularly as Rasch and Guiard (2004) showed that all tests based on the t-distribution or its generalizations are extraordinarily robust against violations from the normal distribution. That is, even if the normal random variable is not a good model for the given character, the type-I risk actually holds almost accurately. Therefore, if a test statistic results on some occasion in significance, we can be quite sure that an erroneous decision concerning the rejection of the null hypothesis has no larger risk than in the case of a normal distribution.

It is notable as well that the test which becomes relevant for the given problem – *Wilcoxon's signed-ranks test* – is not at all appropriate here if the character is only ordinally scaled. Although this test is generally announced as a *homological method* (to the t-test): according to Section 4.2, the empirical relations on which an ordinal-scaled character is based do not induce differences of measurement values which make any sense; however, this test presupposes this.

Doctor

If there is a random variable $\boldsymbol{y}$ with the median Md, then the null hypothesis is H_0: $Md = Md_0$ (H_A: $Md \neq Md_0$). Wilcoxon's signed-ranks test can be used for testing this hypothesis. First of all, the differences $d_v = y_v - Md_0$ are to be calculated, as well as their absolute values $|d_v|$. Finally these values are to be ordered (ascending) and ranks are to be assigned accordingly. The sum of those ranks, S^+, referring to any positive difference, has a known distribution, and tables or (better) program packages can be used in order to get the probability of the observed (or even more extreme) data, given the null hypothesis is true. If $n > 50$, an approximation of the distribution of the test statistic S^+ by a normal distribution is justified. Tables, as well as the formula for the test statistic in the case of such a justified approximation, will not be given here, because we exclude manual calculations. The application of program packages is according to the Example 8.17 in the following Section 8.5.6; the only difference is that we now have to use, in addition to the outcomes y_v of the character y, the values $x_v = Md_0$ for a virtual character x.

Planning a study is generally difficult for non-parametric tests, because it is hard to formulate the alternative hypothesis (however see, for instance, Brunner & Munzel, 2002).

8.5.6 Test of a hypothesis about the median of a quantitative character which is the result of paired observations

As already discussed in Section 8.2.4, we often have a single sample, but two observations (x_v, y_v) per research unit for the same character from a content point of view. If now the median is more appropriate than the mean for the characterization of the differences $d_v = x_v - y_v$ (given quantitative data), then again Wilcoxon's signed-ranks test can be used. Once the differences $d_v = x_v - y_v$ have been calculated the procedure is the same as described above. The null hypothesis concerning the modeled random variable $\boldsymbol{d}$ is: H_0: $Md = 0$.

Again a possible given deviation from a normal distribution does not necessarily entail the application of this procedure. And of course it is not at all appropriate for ordinal-scaled characters.

Doctor

Example 8.17 Evaluation of a campaign

An opinion research institute tries to evaluate the political party *A* during a campaign. 250 persons are sampled using stratified sampling (see in Section 7.2.1) and are interviewed before and after the campaign about their 'Sympathy for party A'. The persons could take a position using an analogue scale-response format (see Section 2.4.3). In the analysis, the 7-inch line of the analogue scale-response format was split into 140 small intervals of equal length; therefore each answer corresponds with one value $0 \leq y_v \leq 141$ (0 inch 'very little sympathy' up to 7 inches 'much sympathy'). Usually in this context the empirical distribution of the observations follows a '*U-shaped*' distribution; i.e. either a person finds the party very sympathetic or the person has a strong antipathy. To judge the effect of the campaign we calculate the differences of the outcomes after and before the campaign, and test for significance (i.e. H_0: $Md = 0$). We apply Wilcoxon's signed-ranks test (type-I risk $\alpha = 0.05$). We use the data *Example_8.17* (see Chapter 1 for their availability). Because such a campaign may also have negative effects on the sympathy for the party A, we decide on a two-sided alternative hypothesis.

In **R**, we enable access to the database *Example_8.17* by using the function `attach()` (see Chapter 1). Then we apply Wilcoxon's signed-ranks test by typing

```
> wilcox.test(sym1, sym2, paired = TRUE, correct = FALSE)
```

i.e. we apply the function `wilcox.test()` and use the characters `sym1` and `sym2` as arguments in it. With `paired = TRUE`, we request the calculation of pair-wise differences, and with `correct = FALSE` we prevent the continuity correction, which otherwise is conducted automatically according to the default settings.

As a result, we get:

```
       Wilcoxon signed rank test

data:  sym1 and sym2
V = 13117, p-value = 0.04012
alternative hypothesis: true location shift is not equal to 0
```

In SPSS, we apply

Analyze
 Nonparametric Tests
  Related Samples...

after having loaded the data. In the resulting window (not shown), we select the tab Fields. This gets us to a window very similar to the one in Figure 8.12. Now, we drag and drop both Sympathy2 and Sympathy1 to Test Fields:. Generally, any order of the variables can be chosen, though interchanging the variables alters the sign of the resulting test statistic.

Next, we select the tab Settings. In the resulting window (see Figure 8.14), we activate Customize tests and tick Wilcoxon matched-pair signed-rank (2 samples). Finally, we Run the analysis and, as a result, we get an output window showing the table Hypothesis Test Summary (not shown here), which we double-click. The resulting window shows a table containing the relevant results. Here we can find the Standardized Test Statistic with a value of −2.053 and the p-value, which equals 0.040, in Asymptotic Sig. (2-sided test).

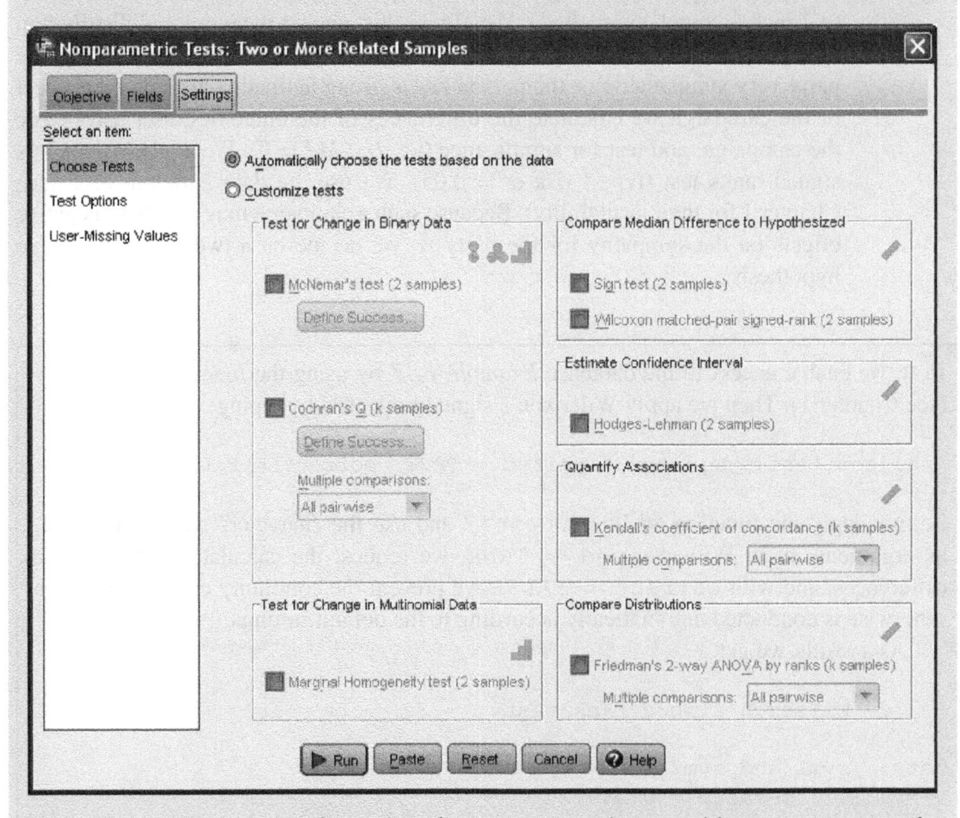

Figure 8.14 SPSS-window for settings for non-parametric tests with two or more samples.

Because the p-value is smaller than 0.05, the campaign had a significant effect; we reject the null hypothesis. The test statistic (in SPSS) is negative, which means negative differences have higher ranks; that is, changes happen towards antipathy.

Because the test statistic of Wilcoxon's signed-ranks test is asymptotically standard normal distributed, the estimation of the effect size can be done as described in Section 11.3.5.

Summary

For hypothesis testing there are different statistical tests depending on the scale type of the character of interest, and depending on the kind of parameter which is hypothesized. Their application with computer packages is generally very easy. For *sequential testing* the research units are sampled one after the other and analysis is always applied subsequently, that is until a terminal decision for or against the null hypothesis can be made.

References

Brunner, E., & Munzel, U. (2002). *Nicht-parametrische Datenanalyse* [Non-parametric Data Analysis]. Heidelberg: Springer.

Costa, P. T. & McCrae, R. R. (1992). *Revised NEO Personality Inventory (NEO-PI-R) and NEO Five-Factor Inventory (NEO-FFI). Professional Manual.* Odessa, FL: Psychological Assessment Resources.

Kubinger, K. D. (2009b). *Psychologische Diagnostik – Theorie und Praxis psychologischen Diagnostizierens* (2nd edn) [Psychological Assessment – Theory and Practice of Psychological Consulting]. Göttingen: Hogrefe.

Lehmann, E. L. & Romano, J. P. (2005). *Testing Statistical Hypotheses* (3rd edn). New York: Springer.

Rasch, D. & Guiard, V. (2004). The robustness of parametric statistical methods. *Psychology Science, 46*, 175–208.

Rasch, D., Pilz, J., Verdooren, R. L., & Gebhardt, A. (2011). *Optimal Experimental Design with R.* Boca Raton: Chapman & Hall/CRC.

Schneider, B. (1992). An interactive computer program for design and monitoring of sequential clinical trials. In *Proceedings of the XVIth International Biometric Conference* (pp. 237–250). Hamilton, New Zealand.

Student (Gosset, W. S.) (1908). The probable error of a mean. *Biometrika, 6*, 1–25.

Wald, A. (1947). *Sequential Analysis.* New York: John Wiley & Sons, Inc.

Whitehead, J. (1992). *The Design and Analysis of Sequential Clinical Trials* (2nd edn). Chichester: Ellis Horwood.

9

Two samples from two populations

This chapter is about statistical tests for comparing parameters of two populations, from each of which samples have been drawn. Wherever possible (and where practically applicable), it will be shown how planning of a study is carried out by specifying certain precision requirements. Furthermore, it will be demonstrated how sequential testing works for certain tests.

We assume that two independent samples from each of two populations are of interest; independent, as per its definition in Chapter 6, means that each outcome in one sample, as a particular event, will be observed independently of all other outcomes in the other sample. Once the outcomes of the particular character are given, (point) estimates can be calculated for each sample. However the main question is: do the two samples stem from the same population or from two different populations?

Bachelor **Example 9.1** Do, at the first test date, children with German as a native language differ from those with Turkish as a native language in the subtest *Everyday Knowledge* (Example 1.1)

This example deals with children whose native language is German and children whose native language is Turkish as two independent (random) samples. The question is now whether these two samples stem from two different populations in terms of the character *Everyday Knowledge, 1st test date*; of course they stem, by definition, from different populations in relation to other characters, in particular with regard to their ancestral nationality. Nevertheless, it is conceivable, perhaps not very plausible, that these two samples as representatives of Viennese students (in terms of the character *Everyday Knowledge, 1st test date*) stem from a 'uniform' single population.

Statistics in Psychology Using R and SPSS, First Edition. Dieter Rasch, Klaus D. Kubinger and Takuya Yanagida.

9.1 Hypothesis testing, study planning, and sequential testing regarding the unknown parameters μ_1 and μ_2

As it is the most illustrative, we will first consider again the parameter μ; that is, the two means of the two populations underlying the samples, which are considered as different for the time being. The character of interest is modeled by a normally distributed random variable. That is to say, a random sample of size n_1 and n_2, respectively, will be drawn from the two populations 1 and 2. The observations of the random variables $\boldsymbol{y}_{11}, \boldsymbol{y}_{12}, \ldots, \boldsymbol{y}_{1n_1}$ on one hand and $\boldsymbol{y}_{21}, \boldsymbol{y}_{22}, \ldots, \boldsymbol{y}_{2n_2}$ on the other hand will be $y_{11}, y_{12}, \ldots, y_{1n_1}$ and $y_{21}, y_{22}, \ldots, y_{2n_2}$. We term the underlying parameters μ_1 and μ_2, respectively, and σ_1^2 and σ_2^2, respectively. The unbiased estimators are then $\hat{\boldsymbol{\mu}}_1 = \bar{\boldsymbol{y}}_1$ and $\hat{\boldsymbol{\mu}}_2 = \bar{\boldsymbol{y}}_2$, respectively, and $\hat{\boldsymbol{\sigma}}_1^2 = \boldsymbol{s}_1^2$ and $\hat{\boldsymbol{\sigma}}_2^2 = \boldsymbol{s}_2^2$, respectively (according to Section 8.5.1 and Formula (5.4)).

The null hypothesis is H_0: $\mu_1 = \mu_2 = \mu$; the alternative hypotheses is H_A: $\mu_1 \neq \mu_2$, and in the case of a one-sided hypothesis either H_A: $\mu_1 > \mu_2$ or H_A: $\mu_1 < \mu_2$.

The solution to the problem is simple according to recent findings from simulation studies (Rasch, Kubinger, & Moder, 2011), while the traditional approach has almost always been unsatisfactory because it wasn't theoretically regulated for all possibilities. Nevertheless, an introduction to the traditional approach must be given at this point: the current literature on applications of statistics (in psychology) refers to it, and also methods for more complex issues are built on it. Lastly, sequential testing is currently not possible without the corresponding traditional test.

The traditional approach is based upon the test statistic in Formula (9.1). It examines the given null hypothesis for the case that the variances of σ_1^2 and σ_2^2, respectively, of the relevant variables in the two populations, are not known; this is the usual case. However, it assumes further that these variances are equal in both populations; yet experience has shown that this is often not the case. The test statistic

$$\boldsymbol{t} = \frac{\bar{\boldsymbol{y}}_1 - \bar{\boldsymbol{y}}_2}{\sqrt{\dfrac{(n_1 - 1) \cdot \boldsymbol{s}_1^2 + (n_2 - 1) \cdot \boldsymbol{s}_2^2}{n_1 + n_2 - 2}}} \cdot \sqrt{\frac{n_1 n_2}{n_1 + n_2}} \tag{9.1}$$

is t-distributed with $df = n_1 + n_2 - 2$ degrees of freedom. This means (despite the fact that today hardly anyone calculates by hand), that we can decide in favor of, or against, the null hypothesis, analogous to Section 8.2.3, by using the quantile of the (central) t-distribution, $t(n_1 + n_2 - 2, 1 - \frac{\alpha}{2})$ and $t(n_1 + n_2 - 2, 1 - \alpha)$, respectively, depending on whether it is a two- or one-sided problem. It will therefore be rejected if $|\,t\,| > t(n_1 + n_2 - 2, 1 - \frac{\alpha}{2})$ and if $|\,t\,| > t(n_1 + n_2 - 2, 1 - \alpha)$, respectively, provided the sign of $\bar{y}_1 - \bar{y}_2$ is in the given direction of the alternative hypothesis. If the null hypothesis has to be rejected, it is again called a significant result: as concerns the two populations, the two means are different; the observed differences between $\bar{y}_1$ and $\bar{y}_2$ cannot be explained by chance (alone).

Master Doctor

The given condition for the test statistic in Formula (9.1), namely the assumption of equal variances in both populations, strictly speaking limits the null and alternative hypothesis to the following: H_0: $\mu_1 = \mu_2 = \mu$, $\sigma_1^2 = \sigma_2^2 = \sigma^2$; H_A: $\mu_1 \neq \mu_2$, $\sigma_1^2 = \sigma_2^2 = \sigma^2$; or in the case of a one-sided question either H_A: $\mu_1 > \mu_2$, $\sigma_1^2 = \sigma_2^2 = \sigma^2$ or H_A: $\mu_1 < \mu_2$, $\sigma_1^2 = \sigma_2^2 = \sigma^2$.

Master

Formula (9.1) can be understood more precisely in that the test statistic used is centrally t-distributed under the null hypothesis – i.e. with mean 0. However, under the alternative hypothesis, this test statistic is non-centrally t-distributed, with the *non-centrality parameter*

$$\frac{\mu_1 - \mu_2}{\sigma} \cdot \sqrt{\frac{n_1 n_2}{n_1 + n_2}}$$

The calculation of this test, the *two-sample t-test*, is particularly simple with statistical computer programs; however the theory-based justification is not:

- First, it has to be considered that the application of the test is only justified if there is a quantitative character, which is at least interval scaled; computer programs can, of course, be used for ordinally scaled characters without any justification as concerns the scaling, and can also be used for multi-categorical, nominally scaled characters, which however would be illogical (see also Section 5.3.2).
- So far we have assumed that the character of interest can be modeled by a normally distributed random variable. But it remains unclear how to proceed if this assumption is not plausible.
- The two variances in the two underlying populations were certainly not assumed to be known, but were estimated from the respective samples. However, it was assumed that the two populations have the same variance even though they perhaps have different means. It still remains unclear how to proceed if this assumption is not plausible.
- A further stringent requirement is that, before the data collection and data analysis applies, the study is planned; given a chosen type-I risk α, a difference $\delta = \mu_1 - \mu_2$, which is considered to be relevant, will not be detected by the test only with a satisfactorily low type-II risk, β. That is, the sample sizes n_1 and n_2 need to be accordingly determined in advance.

Master
Doctor

As for the previously made assumption of a normally distributed modeled variable in the t-test, a number of things have to be considered. Firstly, according to our explanation of the central limit theorem, the only necessary assumption is that both samples should be of an agreeable size; however, rules of thumb are controversial and not necessarily sufficiently empirically researched (see again in Section 6.2.2). Secondly, if such rules of thumb are ignored, another test or 'pre-test' of normality of the variable $\boldsymbol{y}$ in both populations is not appropriate because the available inventory of methods is of very low power (i.e. $1 - \beta$ is relatively low). Thirdly, the assumption of a normal distribution is practically negligible because the t-test is, even in the case of small sample sizes, *robust* against any deviation from this assumption (see Rasch & Guiard, 2004).

Master

Rasch and Guiard (2004) simulated 10 000 pairs of samples each (from two populations) where the null hypothesis was correct, and performed the t-test. Different sample sizes, equal and unequal variances, and distributions other than the normal distribution were used in these simulations. At the end, each of the specified type-I risks α was compared with the relative frequency of (false) rejections among the 10 000 cases. All deviations were (with the exception of the t-test in the case of unequal variances) less than 20% of the given α, which phenomenon is called '20% robust' against non-normality.

Master Doctor

There is also more to be said regarding the assumption of equal variances in the two populations. Firstly, the t-test basically gives incorrect results and thus often leads to false interpretations and conclusions if the assumption of equality of variances is not met in reality. Secondly, there indeed is an appropriate 'pre-test', i.e. a test that checks whether the hypothesis of equal variances can be accepted or has to be rejected (see in Section 9.2.2). But then we would have to consider all combinations of two possible type-I errors and two possible type-II errors in our investigation, which becomes very complicated (Rasch, Kubinger, & Moder, 2011). Thirdly, planning the study with such a pre-test would hardly be possible; to determine the two sample sizes needed, one would need data supporting the hypothesis that both variances of both populations are equal.

For planning a study (and subsequent analyzing), it is much more practical to strive for a test which is not based on such stringent, namely actually not verifiable, assumptions. It remains essential that there is an (at least) interval-scaled character. Although the test statistic that is recommended in the following also derives from the assumption of a normally distributed modeled character, actual deviations from the normal distribution, however, prove to be practically meaningless (Rasch & Guiard, 2004; Rasch, Kubinger, & Moder, 2011). That is to say, the stated null hypothesis can be better tested with a test-statistic that is slightly different from the test statistic in Formula (9.1) – regardless of whether the two variances in both populations are equal or not. The test statistic itself is, again, t-distributed or approximately so; the test is called the *two-sample Welch test*. The procedure described above as the traditional approach, namely the application of the (two-sample) t-test, including testing the equality of variances with a special (pre-) test is, however, strictly to be rejected based on the new level of knowledge. If it were to be adhered to, it would result in unknown (large) type-I and type-II risks in many cases.

Master

Since the Welch test also permits unequal variances in the populations, the degrees of freedom of its test statistic also depend on the corresponding variance estimates, due to its derivation:

$$t = \frac{\bar{y}_1 - \bar{y}_2}{\sqrt{\frac{s_1^2}{n_1} + \frac{s_2^2}{n_2}}} \tag{9.2}$$

is approximately t-distributed with

$$df = \frac{\left(\frac{s_1^2}{n_1} + \frac{s_2^2}{n_2}\right)^2}{\frac{s_1^4}{(n_1 - 1)\,n_1^2} + \frac{s_2^4}{(n_2 - 1)\,n_2^2}}$$

degrees of freedom. The computation of the Welch test is very simple using statistical computer programs.

Bachelor **Example 9.1 – continued**

We fix $\alpha = 0.05$ and decide on a one-sided alternative hypothesis; that is H_A: $\mu_G > \mu_T$, because we do not assume that children with Turkish as a native language perform better in the test than children with German as a native language.

In **R**, we first enable access to the data set *Example_1.1* (see Chapter 1) by using the function `attach()`. Then we type

```
> t.test(sub1_t1 ~ native_language, alternative = "greater",
+         var.equal = FALSE)
```

i.e. we use the command to analyze the character *Everyday Knowledge, 1st test date* (`sub1_t1`) in reference to the factor *native language of the child* as the first argument in the function `t.test()`; as second argument, we state, with `alternative = "greater"`, that the alternative hypothesis is one-sided, and, with the argument `var.equal` set to `FALSE`, we adjust for the (possible) case of unequal variances.

As a result, we get:

```
        Welch Two Sample t-test

data:  sub1_t1 by native language
t = 0.6006, df = 97.369, p-value = 0.2747
alternative hypothesis: true difference in means is greater than 0
95 percent confidence interval:
 -2.188438       Inf
sample estimates:
 mean in group German mean in group Turkish
                53.16                 51.92
```

In SPSS, we select

Analyze
    Compare Means
        Independent-Samples T Test...

to obtain the window shown in Figure 9.1. There we move the character Everyday Knowledge, 1st test date into the panel Test Variable(s):. Next we move the character native language of the child into the panel Grouping Variable:. Afterwards, we click the button Define Groups... in order to set how the groups are labeled. In this case, in the window shown in Figure 9.2 we type the digit 1 (corresponding to children with German as a native language) in the field Group 1:, and in the field Group 2: we type the digit 2 (this corresponds to children with Turkish as a native language); by clicking Continue, we return to Figure 9.1, from which OK yields the result in Table 9.1.

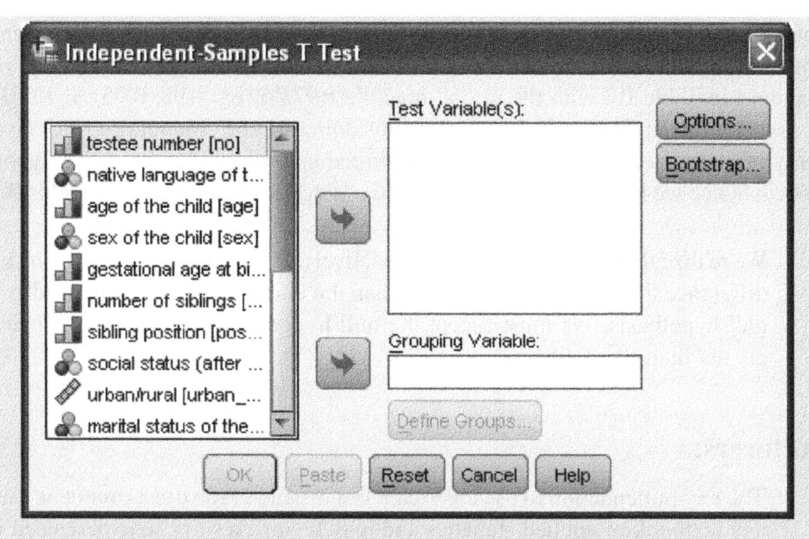

Figure 9.1 SPSS-window for comparing two sample means.

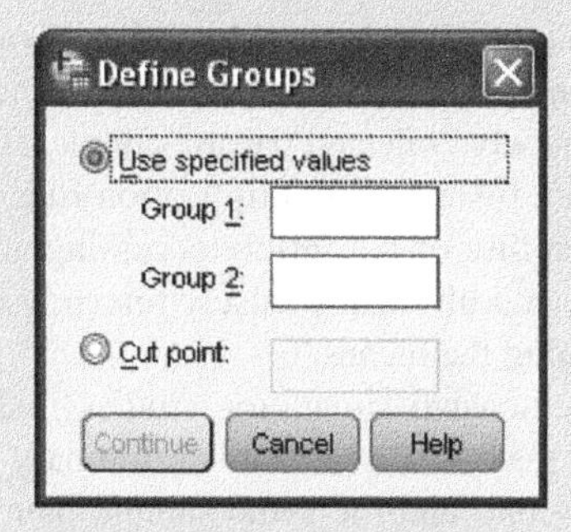

Figure 9.2 SPSS-window for selecting the samples for a comparison of two sample means.

Table 9.1 SPSS-output for the Welch test in Example 9.1 (shortened output).

Independent Samples Test

		t-test for Equality of Means		
		t	df	Sig. (2-tailed)
Everyday Knowledge, 1st test date (T-Scores)	Equal variances assumed	.601	98	.549
	Equal variances not assumed	.601	97.369	.549

Table 9.1 makes it immediately plain that it is necessary to distinguish two cases: the case that both populations have equal variances, and the case that the variances are distinct. Since we have no advance information regarding this, we have to assume the more general second case; this corresponds to the Welch test. In column t we read the value of the according

t-distributed test statistic. Because it is positive, in our case $\bar{y}_1 = \bar{y}_G > \bar{y}_T = \bar{y}_2$ applies, which means that the difference has the expected tendency. We could compare the value t from column t in Table B2 with the $(1-\alpha)$-quantile $t(97.369) \approx t(98, 0.95) = 1.661$. The result is non-significant. However, SPSS spares us doing so: the column (2-tailed) gives the probability of getting a value $|t| \geq |t|$. Note that in the given case the alternative hypothesis is one-sided; hence we have to bisect (mentally) the stated probability, which yields 0.2745.

We realize that 0.2747 or 0.2745, respectively, is the probability of getting a mean difference that is equal to or greater than the one observed here by validity of the null hypothesis. We must accept the null hypothesis because this probability is greater than $\alpha = 0.05$.

For Lecturers:

The recommendation to use the Welch test instead of the t-test cannot be found in any textbook of applied statistics and may therefore surprise experienced users. This recommendation is instead based on the results of current research from Rasch, Kubinger, & Moder (2011).

Strictly speaking, each step, i.e. testing the null hypothesis for normal distribution per sample, testing the null hypothesis for equality of variances, and, finally, testing the null hypothesis for equality of means, should each have to be carried out with a newly collected pair of samples. Only in this way would it be possible to calculate (using the multiplication rule for independent events), for example, the probability of not falsely accepting any of the null hypotheses of the three pre-tests, and ultimately falsely rejecting or falsely accepting the null hypothesis concerning the means.

However, the extensive simulation studies in Rasch, Kubinger, & Moder (2011) concern the realistic case that all preliminary tests for the t-test are carried out on the basis of one and the same sample pair. Various sample sizes from systematically varied distributions with both equal and also unequal variances, and both equal and also different means, were used to randomly generate 100 000 sample pairs each. In the case of equal means, the relative frequency of the (falsely) rejected null hypothesis of interest ($H_0 : \mu_1 = \mu_2$) is a good estimate of the actual type-I risk α_{act} for the nominal $\alpha = 0.05$.

The results show quite clearly that the actual type-I risk, i.e. the risk estimated using simulation, is only in the case of the Welch test near the chosen value $\alpha = 0.05$ – it is exceeded by more than 20% in only a few cases. By contrast, the t-test rejects the null hypothesis to an unacceptably large extent. Moreover, the Wilcoxon U-test (see Section 9.2.1) does not maintain the type-I risk in any way: it is much too large.

Referring to the t-test, the desired relative effect size is, for equal sample sizes, according to Section 8.3

$$E = \frac{\mu_1 - \mu_2}{\sigma}$$

For the Welch test it is then:

$$E = \frac{\mu_1 - \mu_2}{\sqrt{\frac{\sigma_1^2 + \sigma_2^2}{2}}}$$

If it is only supposed to be estimated for illustration purposes after an analysis, then it is possible to do this precisely with

$$\hat{E} = \frac{\bar{y}_1 - \bar{y}_2}{\sqrt{\frac{s_1^2 + s_2^2}{2}}}$$

or to simplify matters – without overestimating the effect – with

$$\hat{E} = \frac{\bar{y}_1 - \bar{y}_2}{s_{\max}}$$

with $s_{\max}$ being the larger of the two standard deviations.

Planning a study for the Welch test is carried out in a rather similar manner to that in Chapter 8 using pertinent computer programs. A problem arises, however: in most cases it is not known whether the two variances in the relevant population are equal or not, and if not, to what extent they are unequal. If one knew this, then it would be possible to calculate the necessary sample size exactly. In the other case, one must be content with determining the largest appropriate size, which results from equal, realistically maximum expected variances.

Example 9.2 How large should the sample sizes in Example 1.1 have been for testing, with given precision requirements, whether children with German as a native language differ from those with Turkish as a native language in the subtest *Everyday Knowledge*, at the second test date? (See Example 9.1)

Since children with Turkish as a native language may at best show an equally high performance compared to children with German as a native language regarding the character *Everyday Knowledge, 2nd test date* (due to possible language-related handicaps), we have to use the one-sided alternative hypothesis H_A: $\mu_1 > \mu_2$, while the null hypothesis is H_0: μ_1 $(= \mu_G) = \mu_2$ $(= \mu_T)$; precisely, H_0: $\mu_1 \leq \mu_2$. We take a type-I risk of $\alpha = 0.05$ and a type-II risk of $\beta = 0.05$; as a (minimum) relevant difference $\delta = \mu_1 - \mu_2$, we fix $\delta = 6.67$ (T-Scores). Since T-Scores are calibrated to a mean $\mu = 50$ and a standard deviation $\sigma = 10$ (see Example 5.5), this means that a difference of two-thirds of the standard deviation of the standardization sample is considered to be relevant. In the case that a group of children (with German as a native language) on average matches the standardization sample ($\mu = 50$ T-Scores, corresponding to a percentile rank of 50), it would be considered as relevant if the other (with Turkish as a native language) shows on average only 43.33 T-Scores in the underlying population; that would mean a percentile rank of 25 compared to the standardization sample.

We do not know whether the two variances in the relevant population are equal or not. Thus, we can only determine the largest appropriate size for both samples.

In **R**, we use the package OPDOE, which we load (see Chapter 1) using the function library(), to ascertain the sample size. We type

```
> size.t.test(delta = 6.67, sd = 10, sig.level = 0.05, power = 0.95,
+              type = "two.sample", alternative = "one.sided")
```

i.e. we submit the appropriate precision requirements (see Example 8.9), along with the information that the standard deviation (sd) is 10, to the function size.t.test(), specify with type = "two.sample" that the two samples are from two populations, and finally set alternative = "one.sided", because we plan to conduct a one-sided test.

As a result, we get:

```
[1] 50
```

For the given precision requirements, a one-sided research question and equal variances, the result is therefore (rounded to) n_1 $(= n_G) = n_2$ $(= n_T) = 50$. With these sample sizes, when using the significance level 0.05, a mean difference of at least 6.67 will not be discovered with a maximum probability of 0.05.

In Psychology, often even in the case of two samples, *homogeneous variances* are spoken of, though it is generally used only for more than two samples, where the respective variances are equal. Similarly there are *heterogeneous variances* if the variances are unequal. Accordingly, the Welch test is often referred to as the '*t*-test for homogeneous variances' and the so-called '*t*-test for heterogeneous variances', instead of the Welch test.

Master Doctor

As stated in Section 8.4, the technique of sequential testing offers the advantage that, given many studies, on average far fewer research units need to be sampled as compared to the 'classic' approach of hypothesis testing with sample sizes fixed beforehand. Nevertheless, we need also the precision requirements α, β, and δ. In the case of two samples from two populations and testing the null hypothesis H_0: $\mu_1 = \mu_2$, there is again a sequential triangular test. Its application is completely analogous to Section 8.4. The only difference is that we use $\delta = 0 - (\mu_1 - \mu_2) = \mu_2 - \mu_1$ for the relevant (minimum difference) between the null and the alternative hypothesis, instead of $\delta = \mu_A - \mu_0$. Again we keep on sampling data, i.e. outcomes y_{iv}, $i = 1, 2$, until we are able to make a terminal decision: namely to accept or to reject the null hypothesis.

Master Doctor

Example 9.3 Validation of a psychological test

The validity of a psychological test refers to the fact that it actually measures the particular trait/aptitude which it claims to measure (see, for example, Anastasi & Urbina, 1997).

As an example we will take the validation of a psychological aptitude test for programmers. The prognostic validity criterion is the professional success achieved at a large computer centre after a probationary period of six months. The aptitude test was given at the beginning of the probationary period; no candidate was excluded from employment on probation on the basis of his or her test score in the aptitude test. Up until the start of the first evaluation, 25 candidates were employed on a probationary basis. After the six-month probationary period was up, 3 of them (group 1) had been dismissed due to lack of qualification and 2 of them (group 2) had been hired on a permanent basis; the other 20 candidates are still on probation. In the following, the test scores of the first 5 candidates are quoted:

Group 1	Group 2
48	54
51	55
47	

The study is designed in such a way that $\alpha = 0.05$, $\beta = 0.2$, and $\delta = 10$ T-Scores. The deviation of $\delta = 10$ T-Scores corresponds, for example, to the case that the average test score matches the percentile rank of 50% in one group and the percentile rank of 83.3% in the other group. The alternative hypothesis is one-sided: H_A: $\mu_1 < \mu_2$; hence, the null hypothesis is H_0: $\mu_1 \geq \mu_2$.

In **R** we use again the package OPDOE, and ascertain the sample size by typing

```
> size.t.test(delta = 10, sd = 10, sig.level = 0.05, power = 0.8,
+             type = "two.sample", alternative = "one.sided")
```

i.e. we apply the function `size.t.test()`, in which we use the appropriate precision requirements along with the specification of the standard deviation (`sd`), as well as the number of samples and the type of alternative hypothesis `alternative = "one.sided"` as arguments.

As a result, we get:

```
[1] 14
```

Thus, one would have to plan a sample size per group of $n_1 = n_2 = 14$ for the worst case, i.e. equal variance in both populations. In this way, the researcher becomes aware of the fact that the available sample size of $n_1 + n_2 = 25$ is basically too small to meet the given precision requirements.

With $n_1 = 3$ and $n_2 = 2$ at the beginning, sequential testing shows, not surprisingly, that more data must be sampled (see Figure 9.3).

In **R**, for now a sequential triangular test corresponding to the Welch test is not available. Hence we have to, rather inappropriately, use the one for the t-test, staying well aware that this leads, in the case of distinct variances in both populations, to falsely rejecting the null hypothesis more often than nominal for the assumed α.

```
> valid.tt <- triangular.test.norm(x = 48, y = 54,
+                                  mu1 = 50, mu2 = 60,
+                                  sigma = 10, alpha = 0.05,
+                                  beta = 0.2)
```

i.e. we apply the function `triangular.test.norm()` and use the first observation value of group 1 (`x = 48`) and group 2 (`y = 54`) as the first two arguments; furthermore we set μ_0 (`mu1 = 50`) and μ_1 (`mu2 = 60`), and `sigma = 10` for σ. Finally, we set the appropriate precision requirements with `alpha = 0.05` and `beta = 0.2`. All of this we assign to object `valid.tt`.

As a result, we get:

```
Triangular Test for normal distribution

Sigma known: 10

H0: mu1=mu2= 50  versus H1: mu1= 50  mu2>= 60
alpha: 0.05  beta: 0.2
Test not finished, continue by adding single data via update()
```

After the first round (with altogether two observations), no final decision is achievable and for that reason we enter further observations by typing

```
> valid.tt <- update(valid.tt, x = 51)
```

i.e. we apply the function `update()` and use the object `valid.tt` as the first argument; as second argument we add the second observation value of group 1 (`x = 51`). We again assign the result to the object `valid.tt`.

Now, this yields the result:

```
Triangular Test for normal distribution

Sigma known: 10

H0: mu1=mu2= 50  versus H1: mu1= 50  mu2>= 60
alpha: 0.05  beta: 0.2

Test not finished, continue by adding single data via update()
```

Even after the second round (with altogether three observations) no final decision is achievable; hence we type

```
> valid.tt <- update(grp.tt, y = 55)
```

i.e. we submit to the function `update()` the object `valid.tt` again as first argument and, as second, with `y = 55`, the second observation value of group 2. We again assign the result to the object `valid.tt`.

This yields the result:

```
Triangular Test for normal distribution

Sigma known: 10

H0: mu1=mu2= 50  versus H1: mu1= 50  mu2>= 60
alpha: 0.05  beta: 0.2

Test not finished, continue by adding single data via update()
```

Even after analyzing altogether four observations no final decision is possible; such being the case we proceed by typing

```
> valid.tt <- update(valid.tt, x = 47)
```

i.e. we add the third observation value of group 1, `x = 47`.

The result is:

```
Triangular Test for normal distribution

Sigma known: 10

H0: mu1=mu2= 50  versus H1: mu1= 50  mu2>= 60
alpha: 0.05  beta: 0.2

Test not finished, continue by adding single data via update()
```

After including altogether five observations, still no final decision is attainable (additionally see Figure 9.3).

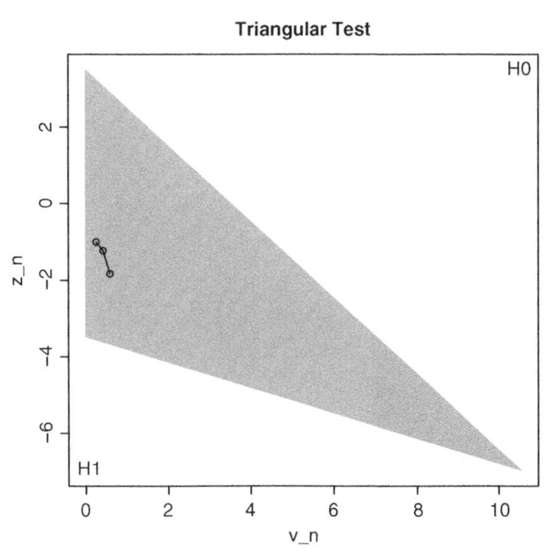

Figure 9.3 **R**-output of the sequential triangular test in Example 9.3 after altogether five observations (starting from the second, observations are entered as points into the triangle).

There are similar results if one analyzes the subsequent incoming data. For the following candidates, namely candidate 6 with test score 49, who turned out to be qualified and was hired on a permanent basis after the probationary period and therefore belongs to group 2, and for the candidates 7 to 15 (with: 54, group 1; 51, group 2; 58, group 2; 46, group 2; 52, group 1; 44, group 1; 57, group 2; 56, group 1; 49, group 1), the reader may recalculate this him/herself. The result after candidate number 16 with test score 62, group 2, is provided in the following.

Hence, in **R**, we type

```
> valid.tt <- update(valid.tt, y = 62)
```

i.e. we use, in the function `update()`, the object `valid.tt` as the first argument and, as second, with `y = 62`, the 16th observation value. The result is assigned to the object `grp.tt` and as output we get:

```
Triangular Test for normal distribution

Sigma known: 10

H0: mu1=mu2= 50   versus H1: mu1= 50   mu2>= 60
alpha: 0.05   beta: 0.2

Test not finished, continue by adding single data via update()
```

Still no final decision can be achieved (see Figure 9.4.)

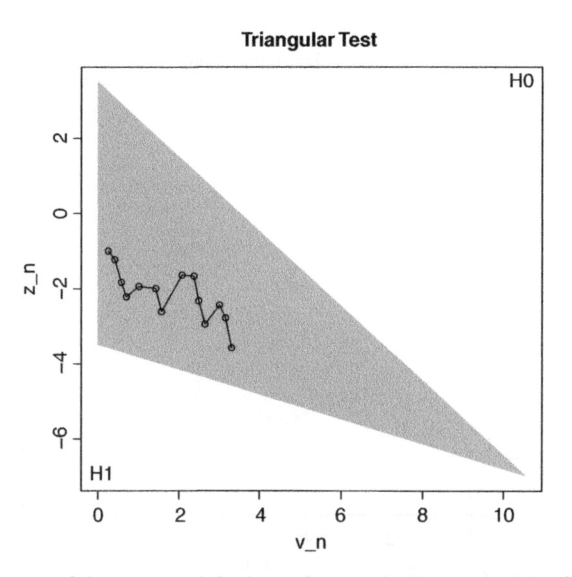

Figure 9.4 **R**-output of the sequential triangular test in Example 9.3 after altogether 16 observations.

A terminal decision is, therefore, still not possible. It is still the same case after observations 17 to 21 (51, group 1; 53, group 2; 49, group 1; 56, group 2; 49, group 1). However, after observation 22 (54, group 2), sequential testing terminates.

In **R**, after looking at 21 observations altogether, we type

```
> valid.tt <- update(valid.tt, y = 54)
```

i.e. we use, in the function `update()`, the 11th observation value of group 2, `y = 54`.

As a result, we get:

```
Triangular Test for normal distribution

Sigma known: 10

H0: mu1=mu2= 50  versus H1: mu1= 50  mu2>= 60
alpha: 0.05  beta: 0.2

Test finished: accept H1
```

And Figure 9.5 appears.

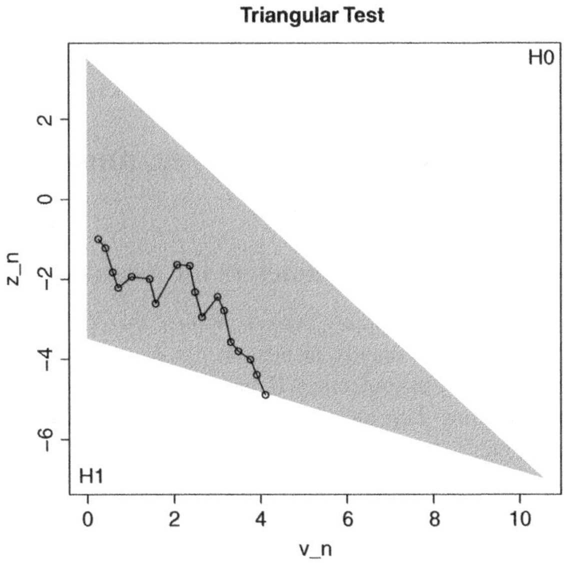

Figure 9.5 **R**-output of the sequential triangular test in Example 9.3 after altogether 22 observations.

Sequential testing ends after observation 22; the decision is made in favor of the alternative hypothesis. As the average test score from the aptitude test for the programmers who were later hired is significantly greater, the test principally meets the quality criterion of validity.

Master Doctor

Usually, sequential testing starts with one or two outcomes from each of the two groups and then new outcomes are observed, alternating between group 1 and 2. Thus, the allocation rate between the groups is selected as $1:1$. Of course the allocation rate does not have to be set to $1:1$, if there are economical or ethical reasons (especially in the clinical area when comparing two treatments). One could also simultaneously sample more than one observation in one step, even at a later time point and not only at the beginning. However, in this case 'overshooting' may occur; that is, a terminal decision might be missed and hence the selected risks would no longer hold. Finally, it is not always necessary to register each observation alternating between the groups (see Example 9.3).

As discussed in Section 8.2.3, we often have two outcomes of content-related traits or the like from each research unit, but statistically these outcomes stem from two specially modeled random variables (we gave the example of being interested in the effect of a treatment in a *pre* and *post* design). Then it is simplest, also in case of two samples, to trace the data to a single character. In other words, analogous to the case of only a single sample in Chapter 8, we calculate the difference $d_v = y_{1v} - y_{2v}$ of the respective pair of outcomes per research unit and use them as new $y_{11}, y_{12}, \ldots, y_{1n_1}$ and $y_{21}, y_{22}, \ldots, y_{2n_2}$– given the original characters are interval scaled.

9.2 Hypothesis testing, study planning, and sequential testing for other parameters

9.2.1 The unknown location parameters for a rank-scaled character

If the character of interest y is ordinally scaled, neither the Welch test nor the t-test is admissible: as explained most accurately in Section 5.3.2, empirical differences would be implied that are not expressed by the outcomes. If one, however, wants to compare two samples from two populations regarding their location, the *Wilcoxon rank-sum test* (Wilcoxon, 1945) would be adequate. We refer to the test here as the *U-test*, according to the common term for the test statistic, $\boldsymbol{U}$.

However, the U-test is derived for hypothesis testing with regard to the equality of two arbitrary continuous distributions. In essence this means that not only the location of the distribution of the outcomes is compared, but also the distribution as a whole. Most applications of the U-test (in psychology) falsely interpret the results 'automatically' as though the means of the populations in question differ in the case that the null hypothesis has been rejected. Yet, the U-test, for example, also becomes significant (when the sample size is appropriate) if the variance (in the case of a quantitative character) or the variability of the outcomes (in the case of an ordinal-scaled character) is different in both populations. For the purpose of location comparison, the U-test therefore only makes sense if it is plausible that the populations do not differ with regard to variance or variability, respectively.

Doctor

The test goes back to Wilcoxon, 1945 who derived tables of critical values of the test statistic W for equal sizes in both samples. Later Mann & Whitney (1947) published tables for different sizes for a transformed test statistic U. The relationship between the test statistic W of Wilcoxon and the test statistic U of Mann and Whitney is as follows. At first we renumber (if needed) the samples so that the size n_1 of the first sample is not larger than the size n_2 of the second one. Then all $n_1 + n_2$ observations are ordered by size, and ranks are given to them. Then W is just the sum of the ranks of those observations belonging to the first sample. The test statistic of the Mann–Whitney test is then

$$U = W - \frac{n_1(n_1 + 1)}{2}$$

Master

The presupposition for the derivation of the U-test is that the character is continuous. This is not self-evidently met for the intended case of application, namely an ordinal-scaled character. For example, grades of some grade scales are obviously not continuous: a student receives the grade '2' or '3', but e.g. not 2.34590912... On the one hand, psychology raises the argument that the respective ordinal scales or the ordinal-scaled outcomes, respectively, are due to some theory of measurement abstraction (the consequence of a factual unalterable measurement imprecision), but actually an interval-scaled character is under consideration. On the other hand, there are corrections for the test statistic which take the case of ties, i.e. identical observations, into account.

The U-test thus examines the null hypothesis that the two relevant populations with continuous distribution functions $F(z)$ and $G(z)$, respectively, are equal, i.e. H_0: $F(z) = G(z)$, for all values of z. In contrast, the two-sided alternative hypothesis claims H_A: $F(z) \neq G(z)$ for at least one value z. The one-sided alternative hypothesis is H_A: $F(z) > G(z)$ or H_A: $F(z) < G(z)$ for at least one value z.

Master
Doctor

The derivation of the test statistic of the U-test for testing the null-hypotheses is based on the assignment of ranks (see Section 5.3.2) for the outcomes of both samples. Then, basically, the test from Formula (9.1) is applied, whereby the variance for the numbers 1 to $n_1 + n_2$ is always equal and therefore known.

The U-test is very simple to calculate with statistical computer programs.

Master

Example 9.4 Calculation example for the U-test without relation to any content

Consider the outcomes $y_{11}, y_{12}, \ldots, y_{1i}, \ldots, y_{1n_1}$ in sample 1 and the outcomes $y_{21}, y_{22}, \ldots, y_{2j}, \ldots, y_{2n_2}$ in sample 2 as a starting point. Let's say $n_1 = n_2 = 5$. When we rank all observations altogether, we will use 'a' instead of y_{1i} and 'b' instead of y_{2j}, for all i and j, in order to simplify matters. The alternative hypothesis may be H_A: $F(z) < G(z)$ for at least one value z, with F for population 1 and G for population 2; the null hypothesis is therefore H_0: $F(z) \geq G(z)$. We choose $\alpha = 0.01$.

If all y_{1i} differ among themselves and all y_{2j} differ among themselves, and finally all y_{1i} differ from all y_{2j}, there results a unique sequence, for example: *b-b-b-a-a-b-b-a-a-a*. This result intuitively contradicts H_0. That is to say, under H_0 we would expect something like: *b-a-b-a-b-a-b-a-b-a* or *a-b-a-b-a-b-a-b-a-b* or the like, or even *b-b-a-a-a-a-a-b-b-b* or *a-a-a-b-b-b-b-b-a-a*. Our data therefore tend to speak in favor of the alternative hypothesis of the type: *b-b-b-b-b-a-a-a-a-a*. Now let us introduce the concept of inversion, which means that any *b* precedes an *a*; then we recognize that in our example the number of inversions is $U = 1 \cdot 5 + 1 \cdot 5 + 1 \cdot 5 + 1 \cdot 3 + 1 \cdot 3 = 21$; in the case of the alternative hypothesis given above it would be $U_{\max} = 5 \cdot 5 = 25$.

We can now easily calculate the probability that the number of inversions is equal to or greater than 21 though H_0 is correct. For this purpose we calculate the number of possible arrangements to distribute 5 *a*'s and 5 *b*'s over a total of 10 positions in accordance with Section 6.2.1; i.e.

$$\binom{10}{5} = \frac{10 \cdot 9 \cdot 8 \cdot 7 \cdot 6}{1 \cdot 2 \cdot 3 \cdot 4 \cdot 5} = 252$$

We are interested in all arrangements with $U \geq 21$; $U = 25$ precisely fulfils one of the 252 arrangements, as does $U = 24$. $U = 23$ fulfils two of the arrangements because one *b* can precede two *a*'s, or two *b*'s can precede one *a*. As a provisional result we obtain $P(U \geq 23) = \frac{4}{252} = 0.01587$; and this is larger than $\alpha = 0.01$, so that $P(U \geq 21)$ is also larger than $\alpha = 0.01$. Therefore, the probability of our result (or of one that is more in favor of the alternative hypothesis) is certainly small if we were to assume H_0, but not so small that we would reject H_0.

It can be demonstrated that H_0 can be rewritten as a function of rank positions for *a* and *b*. For $n = n_1 + n_2$ rank positions, the sum is $1 + 2 + \ldots + n = \frac{n(n+1)}{2}$, so that, for example for $n_1 = n_2$, under H_0 the same ranking sum $\frac{n(n+1)}{4}$ is expected for both samples (for $n_1 \neq n_2$ accordingly we expect that the average rank position is the same in both samples). In the case of tied ranks in principle nothing would change in this approach.

As the test statistic of the U-test is asymptotically standard normally distributed, the estimate of the effect size can be carried out accordingly (see Section 11.4).

Planning the study as concerns non-parametric methods generally results in problems because the alternative hypothesis is hard to quantify.

Bachelor **Example 9.5** The *social status* will be compared between children with German as a native language and children with Turkish as a native language (Example 1.1)

For this purpose we want to exclude children with the measurement value 'single mother in household' from the analysis – this changes the character *social status* from a nominal-scaled character to an ordinal-scaled one. The null hypothesis H_0 is 'The distribution of children in the measurement values of *social status* is equal for both populations in question'; the alternative hypothesis H_A is 'On the whole, the *social status* of children with German as a native language is higher'. We choose $\alpha = 0.01$.

In **R**, we use character `socialnew`, defined in Section 5.2.2. To conduct the U-test we type

```
> wilcox.test(as.numeric(socialnew) ~ native_language,
+              alternative = "less", correct = FALSE)
```

i.e. we apply the function `wilcox.test()`, in which we use the command to analyze the character `socialnew` in reference to the character *native language of the child* (`native_language`) as first argument; with the function `as.numeric()` we transform the character `socialnew` to numeric values. With the second argument, `alternative = "less"`, we specify the alternative hypothesis as one-sided, and with the third argument we specify that the analysis will be conducted without continuity correction.

As a result, we get (shortened output):

```
        Wilcoxon rank sum test

data:  as.numeric(socialnew) by native language
W = 770, p-value = 0.04467
alternative hypothesis: true location shift is less than 0
```

In SPSS, for this purpose we use character socialnew, already defined in Section 5.2.2, for which we need – factually incorrect, but according to an SPSS peculiarity – to redefine the type of scale from Ordinal to Scale (see Example 5.3). Next we select the sequence of commands

Analyze
  Nonparametric Tests
    Independent Samples...

and select in the resulting window (not shown) the Fields tab and get to a window very similar to Figure 8.12; we drag and drop socialnew into Test Fields: and native language of the child into Groups:. Next we switch to the Settings tab and get to the window in Figure 9.6. Here, we select Customize tests and tick Mann-Whitney U (2 samples). After clicking Run, we get in the resulting output window the table Hypothesis Test Summary (not shown), which we double-click; subsequently a window pops up, including a table from which we learn the essential result – the p-value in Asymptotic Sig. (2-sided test) is 0.089. Due to the one-sided alternative hypothesis we have to bisect it, and get 0.0455.

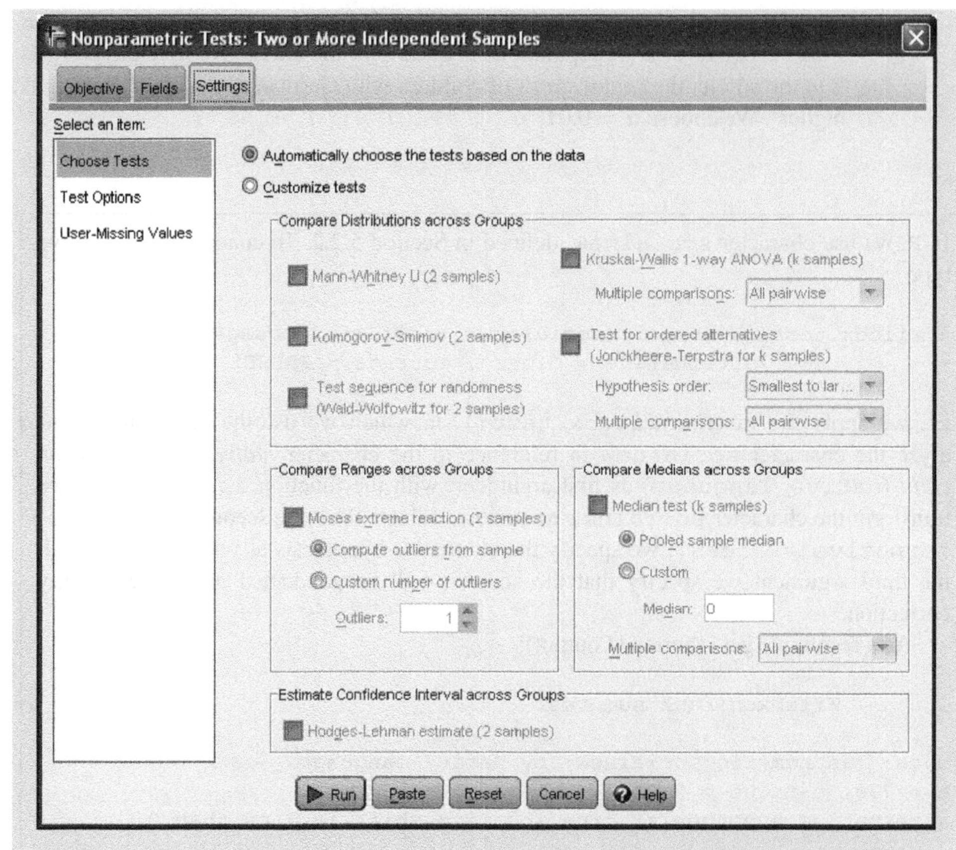

Figure 9.6 SPSS-window for choosing among a collection of non-parametric tests.

As the p-value is larger than 0.01, the null hypothesis is to be accepted. Viennese students with Turkish as a native language do not come from different social classes in comparison to students with German as a native language.

Sequential testing for the U-test is possible in principle, even by using a sequential triangular test. Besides the fact that, regarding the alternative hypothesis, certain input requirements have to be established, which is hard to do before research, the program package **R** currently does not provide such a routine.

9.2.2 The unknown parameters σ_1^2 and σ_2^2

Hypothesis testing regarding the variance of a normally distributed modeled random variable ***y*** from two independent samples is at most important for the traditional approach as concerns the t-test and hence regarding hypothesis testing with respect to the mean – as explained above in more detail. Apart from this, hypothesis testing regarding the variance is relatively unusual in psychology. However, it basically is a matter of the null hypothesis H_0: $\sigma_1^2 = \sigma_2^2$ versus the two-sided alternative hypothesis $H_A : \sigma_1^2 \neq \sigma_2^2$ (one-sided problems are quiet uncommon in practice). Only one test statistic is indicated below, however we will not address planning a study, which generally is possible.

The method of choice is *Levene's test*. Its test statistic is based on the absolute values of the difference between the outcomes and their mean per sample, i.e. $d_{1v} = |y_{1v} - \bar{y}_1|$ or $d_{2v} = |y_{2v} - \bar{y}_2|$, respectively. The (two-sample) t-test is then applied to the resulting values of the new random variable $\boldsymbol{d}$. Of course there are computer programs available for direct computation of the test statistic in question; however, in SPSS only, for instance, in connection with the computation of the t-test or the Welch test.

Master Doctor

Example 9.6 Calculation example for Levene's test without relation to any content

Let us take the outcomes 2, 4, 5, 6, and 8 in the character y for sample 1, and the outcomes 1, 2, 5, 8, and 9 for sample 2. Then obviously $\bar{y}_1 = \bar{y}_2 = 5$, and, therefore, 3, 1, 0, 1, and 3 results for $d_{1v,}$ as well as 4, 3, 0, 3, and 4 for d_{2v} of the character d. We apply the t-test and calculate Levene's test with **R** and SPSS directly.

In **R**, we can conduct Levene's test in two ways. For both we start by typing

```
> x <- gl(2, k = 5)
> y <- c(2, 4, 5, 6, 8, 1, 2, 5, 8, 9)
```

i.e. we apply the function `gl()` and create a factor with two (2) levels and `k = 5` observations each, and assign it to the object `x`. Applying the function `c()`, we concatenate all observation values into a vector and assign it to the object `y`.

The first option to conduct Levene's test is via t-test. We type

```
> tapply(y, x, mean)
```

i.e. we ascertain the `mean` of the vector `y` for both levels of `x`.

As a result, we get:

```
1 2
5 5
```

Thus in both groups (`1` and `2`) the mean is 5. Next we type

```
> d <- abs(y - 5)
```

i.e. we subtract the mean `5` from each observation value and ascertain the absolute value with the function `abs()`. The result is assigned to object `d`. Last, we type

```
> t.test(d ~ x, var.equal = TRUE)
```

and conduct a t-test, analyzes `d` in reference to `x`.

As a result, we get:

```
        Two Sample t-test

data:  d by x
t = -1.2649, df = 8, p-value = 0.2415
alternative hypothesis: true difference in means is not equal to 0
95 percent confidence interval:
 -3.3876676  0.9876676
sample estimates:
mean in group 1 mean in group 2
            1.6             2.8
```

For the other option, we use the function `leveneTest()` of package `car`, which we load after its installation (see Chapter 1) using the function `library()`. Next, we type

```
> leveneTest(lm(y ~ x), center = "mean")
```

i.e. we apply the function `leveneTest()` and use the function `lm()` to analyze the character `y` in reference to character `x` as the first argument; as the second argument, we request the original Levene test with `center = "mean"`.

As a result, we get:

```
Levene's Test for Homogeneity of Variance (center = "mean")
      Df F value Pr(>F)
group  1     1.6 0.2415
       8
```

In SPSS, we use the same sequence of commands as in Example 9.1 (Analyze – Compare Means – Independent-Samples T Test...) two times (once for the character y and the other time for the character d) and each time get to the window in Figure 9.1. There, we move the respective character to Test Variable(s):. The test statistic of the t-test in the table t-test for Equality of Means (that is with reference to character d) leads to the p-value of 0.242 in Sig. (2-tailed). The test statistic of the (directly conducted) Levene's test in the table Levene's Test for Equality of Variances (that is with reference to the character y) also leads to $p = 0.242$.

Master Doctor

In many textbooks the so-called *F-test* is still recommended in order to test the null hypothesis H_0: $\sigma_1^2 = \sigma_2^2$. Simulation studies (see Rasch & Guiard, 2004) have shown, however, that this test is very sensitive to even small deviations from the postulated normal distribution of the character of interest; it then does not maintain the type-I risk. It is a different case with Levene's test; this test is very robust with respect to deviations from the normal distribution. Therefore SPSS uses Levene's test exclusively; however the symbol F, which is used for the respective test statistic, is confusing.

9.2.3 The unknown parameters p_1 and p_2 of a dichotomous character

Given two independent samples with the sizes n_1 and n_2 respectively, we consider the dichotomous character $\boldsymbol{y}$. In each of the two samples, we can determine the relative frequency with which one of the two measurement values has been ascertained. From Section 8.5.2, we know that these two relative frequencies $f_1 = \hat{p}_1$ and $f_2 = \hat{p}_2$ are estimates for the two unknown parameters p_1 and p_2. The null hypothesis is H_0: $p_1 = p_2$; the alternative hypothesis is H_1: $p_1 \neq p_2$ or H_1: $p_1 < p_2$ or H_1: $p_1 > p_2$.

Now it can be shown that the following test statistic is asymptotically χ^2-distributed under the null hypothesis; i.e.

$$\chi^2 = \sum_{i=1}^{r} \sum_{j=1}^{c} \frac{(o_{ij} - e_{ij})^2}{e_{ij}} \tag{9.3}$$

is asymptotically χ^2-distributed, with $df = (r - 1)(c - 1)$ degrees of freedom. In this case, o_{ij} are the absolute frequencies of the measurement values in categories $i = 1, 2, \ldots, r$ and sample $j = 1, 2, \ldots, c$ (see Table 9.2); e_{ij} are the expected values under the null hypothesis, calculated as:

$$e_{ij} = \frac{\sum_{i=1}^{r} o_{ij} \cdot \sum_{j=1}^{c} o_{ij}}{\sum_{i=1}^{r} \sum_{j=1}^{c} o_{ij}}$$

In our case it is $r = c = 2$ and thus $df = 1$; the more general illustration chosen indicates that the test statistic in Formula (9.3) applies to more than two samples and or for a multi-categorical qualitative character (apart from that, Formula (8.11) is a special case of (9.3)). This means (despite the fact that these days hardly anyone is going to calculate by hand) that one can use the $(1 - \alpha)$-quantile of the χ^2-distribution $\chi^2((r - 1)(c - 1), 1 - \alpha)$ to decide for or against the null hypothesis. If the null hypothesis has to be rejected, we again speak of a significant result: for the two populations, the distributions concerning the two measurement values are different; the observed differences in the relative frequencies between the two samples cannot be explained by chance (alone).

Table 9.2 2×2 table.

		Measurement values	
		A	$\bar{A}$
Sample from	Population 1	o_{11}	o_{21}
	Population 2	o_{12}	o_{22}

Master

Since the test statistic in Formula (9.3) is essentially a quadratic term, the direction of deviation between the observed and the expected frequencies under the null hypothesis (that is, as concerning the measurement values, the frequencies are divided proportionally across both samples) cannot be adopted in the alternative hypothesis. Therefore, the test is always two-sided with regard to content, although formally only one side of the χ^2-distribution is of relevance: namely the right side – i.e. the one which corresponds to large (squared) deviations. If we also took the left side of the χ^2-distribution into account, i.e. the side that approaches 0, then we would test whether the total (squared) deviation is smaller than would be expected by chance.

The most common rule of thumb with regard to what constitutes sufficient sample sizes for the approximation of the χ^2*-test* is: $e_{ij} \geq 5$, for all i and j (see however e.g. Kubinger, 1990). Computer programs often refer to what extent such rules of thumb are satisfied or violated in a particular case. The *Yates' continuity correction*, which aims to improve the adjustment of distribution, is best recommended for $r = c = 2$ and a sample size of $o_{11} + o_{12} + o_{21} + o_{22} < 20$.

In Section 11.3.5 we will additionally provide a measure to estimate the effect size. There is no way of planning a study as regards the sizes n_1 and n_2 for the χ^2-test.

Master
Doctor

As concerns planning a study for the χ^2-test, it can be said that it is certainly possible to determine the appropriate number of degrees of freedom with a given type-I and type-II risk, as well as a given relevant difference (of probabilities); however the size of sample(s) is unobtainable. Thus, in the sense used here, planning a study with this test, or with all χ^2-distributed test statistics, is not possible.

Master
Doctor

Example 9.7 Is the frequency of *only children* the same in families where German is the child's native language and in families where Turkish is the child's native language (Example 1.1)?

We deal with two populations and indicate the probability of an only child with p_1 for children with German as a native language and p_2 for children with Turkish as a native language. In fact, awareness of cultural differences leads us to expect more children in families with Turkish as a native language than in families with German as a native language; therefore $p_1 > p_2$; but, as an alternative hypothesis to the null hypothesis H_0: $p_1 = p_2$, the χ^2-test tests H_1: $p_1 \neq p_2$. We choose a type-I risk of $\alpha = 0.05$, a type-II risk of $\beta = 0.2$, and we determine $\delta = 0.2$ as a minimum relevant difference $\delta = p_1 - p_2$.

In **R**, we use, analogously to Example 9.6, the package `car`, which we have already installed and loaded with the function `library()`. To begin with, we transform the character `no_siblings` into the new character `only_child` by typing

```
> only_child <- recode(no_siblings, "0 = 1; else = 2")
```

i.e. we apply the function `recode()` to the character `no_siblings` and define the coding rules of the values in quotation marks; namely `0 = 1` and `else = 2`; the thusly recoded values are assigned to the object `only_child`. Next we create a cross-tabulation, by typing

```
> addmargins(table(native_language, only_child))
```

i.e. we use the characters `native_language` and `only_child` as arguments for the function `table()`; the function `addmargins()` specifies the inclusion of row and column totals in the resulting table.

As a result, we get:

```
                  only_child
native_language    1    2 Sum
         German   14   36  50
        Turkish    3   47  50
            Sum   17   83 100
```

Next we conduct a χ^2-test. Hence we type

```
> chisq.test(native_language, only_child, correct = FALSE)
```

i.e. we apply the function `chisq.test()` and use `native_language` and `only_child` as arguments; due to `correct = FALSE` the analysis is conducted without Yates' correction.

As a result, we get:

```
          Pearson's Chi-squared test

data:  native_language and only_child
X-squared = 8.5755, df = 1, p-value = 0.003407
```

In SPSS, we start with

Transform
    Recode into Different Variables...

to transform, in the resulting window (Figure 9.7), the character number of siblings into the new character only_child. In order to do this, we select the character number of siblings and move it to the field Input Variable -> Output Variable:. In the panel Output Variable we type only_child in the field Name: and click Change. The button Old and New Values... gets us to the next window (Figure 9.8). For Value:, in the panel Old Value we type 0, and 1 in the panel New Value. Next we press Add, and select the option All other values in panel Old Value, and type 2 into the field Value in panel New Value. A click on Add followed by Continue returns us to the previous window, and after pressing OK the transformation is performed.

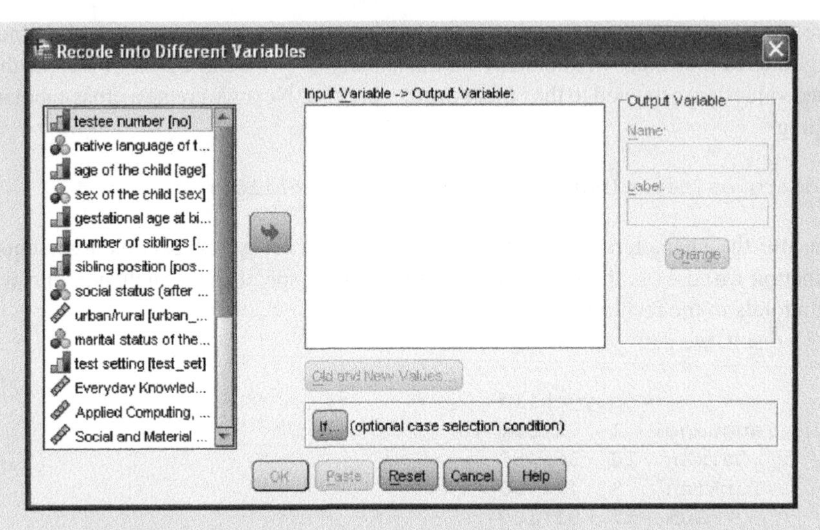

Figure 9.7 SPSS-window for recoding into a new character.

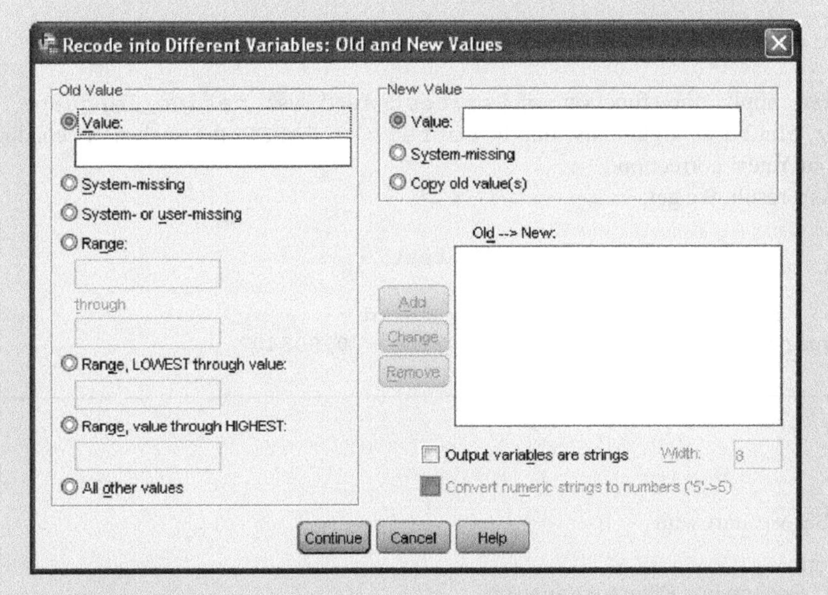

Figure 9.8 SPSS-window for setting the values of a new character.

To create a two-dimensional frequency table we use the steps described in Example 5.13 (Analyze - Descriptive Statistics - Crosstabs...) and drag and drop the character only_child into the field Row(s): and the variable native language of the child into the field Column(s):. A click on Statistics... gets us to the next window, where we tick Chi-square, click Continue and complete our query with OK. Table 9.3 shows the cross-tabulation including the respective frequencies; in Table 9.4 among others is the result of the χ^2-test (as Pearson Chi-Square).

Table 9.3 SPSS-output of the cross-tabulation in Example 9.7.

only child * native language of the child Crosstabulation

Count

		native language of the child		
		German	Turkish	Total
only child	yes	14	3	17
	no	36	47	83
Total		50	50	100

Table 9.4 SPSS-output of the χ^2-test in Example 9.7 (shortened output).

Chi-Square Tests

	Value	df	Asymp. Sig. (2-sided)
Pearson Chi-Square	8.575[a]	1	.003
Continuity Correction[b]	7.087	1	.008
N of Valid Cases	100		

a. 0 cells (.0%) have expected count less than 5. The minimum expected count is 8.50.
b. Computed only for a 2x2 table

It proves to be a significant result. The null hypothesis should be rejected. We can calculate by hand from Table 9.3: $f_1 = \hat{p}_1 = \frac{14}{50} = 0.28$ and $f_2 = \hat{p}_2 = \frac{3}{50} = 0.06$; these are the best (point) estimates of the probability of an only child in the two populations.

Another test statistic can also be derived for the given problem; this also allows study planning and sequential testing. The test statistic

$$\boldsymbol{u} = \frac{|f_1 - f_2| - \dfrac{\min(n_1, n_2)}{2n_1 n_2}}{\sqrt{\hat{\boldsymbol{p}}(1 - \hat{\boldsymbol{p}})}} \cdot \sqrt{\frac{n_1 n_2}{n_1 + n_2}} \tag{9.4}$$

is asymptotically $N(0, 1)$-distributed, with

$$\hat{p} = \frac{n_1 f_1 + n_2 f_2}{n_1 + n_2}$$

– the term subtracted in the numerator of the test statistic reduces the difference between the two observed relative frequencies and corresponds to Yates' continuity correction. Thus, the procedure of hypothesis testing is as usual; in the case of a one-sided problem, it is important to make sure that the sign of $f_1 - f_2$ is in the hypothesized direction.

Planning a study basically takes place in a similar manner to that of cases already discussed by using relevant computer programs. The planning and analysis, however, is not possible in SPSS.

Master Doctor

Example 9.7 – continued

For the test statistic in Formula (9.4), planning the study is possible as follows. We already decided, above, on $\alpha = 0.05$ and $\beta = 0.2$, as well as $\delta = 0.2$, but now we need to determine the values of p_1 and p_2 more precisely: realistically, it may be $p_1 = 0.3$ and $p_2 = 0.1$. Of course, here we use the possibility of a one-sided alternative hypothesis H_A: $p_1 > p_2$.

In **R**, we ascertain the sample size[1] by typing

```
> size.prop_test.two_sample(p1 = 0.3, p2 = 0.1, alpha = 0.05,
+                           power = 0.8, alt = "one.sided")
```

i.e. we apply the function `size.prop_test.two_sample()` and use the value of p_1 (`p1 = 0.3`) as the first argument and the value of p_2 (`p2 = 0.1`) as the second. We set `alpha = 0.05` as the third argument and `power = 0.8`, which means $1 - \beta$, as the fourth; and as fifth we set the type of the alternative hypothesis to one-sided (`alt = "one.sided"`).

As a result we get:

```
[1] 59
```

And,

```
> power.prop.test(p1 = 0.3, p2 = 0.1, sig.level = 0.05, power = 0.8,
+                 alternative = "one.sided")
```

leads to $n = 49$.

It follows that 59 or 49 persons have to be included in each of the samples. In Example 1.1, we had 50 children each, and this we regard as sufficient.

We obtain

$$u = \frac{|0.28 - 0.6| - \dfrac{50}{2 \cdot 50 \cdot 50}}{\sqrt{0.17 \cdot 0.83}} \cdot \sqrt{\frac{50 \cdot 50}{100}} = 2.80 > 1.645 = u(1 - 0.05)$$

[1] There exist several formulas but no one is uniformly the best. Due to this fact there are two R-programs for calculating n: `size.prop_test.two_sample()` of OPDOE which is based on a corrected formula (Formula (2.51), p. 49 in Rasch, Pilz, Verdooren, & Gebhardt, 2011), and `power.prop.test()` which is based on the original formula (Formula (2.50), p. 49 in Rasch, Pilz, Verdooren, & Gebhardt, 2011). Both of them may be used. Which one is more appropriate depends on the values of the probabilities and their difference.

In **R**, we can compute the test statistic from Formula (9.4), by defining an appropriate function; we call it `p.test()`; so we type

```
> p.test <- function(f1, f2, n1, n2) {
+   p <- (n1*f1 + n2*f2)/(n1 + n2)
+   numer <- abs(f1 - f2) - (min(c(n1, n2))/(2*n1*n2))
+   denom <- sqrt(p*(1-p))
+   nn <- sqrt((n1*n2)/(n1+n2))
+   u <- numer/denom*nn
+ return(round(u, digits = 3))
+ }
```

i.e. we use `function()` and declare the relative frequencies f_1 and f_2 as the first two arguments and n_1 and n_2 as third and fourth. The sequence of commands between the braces corresponds to Formula 9.4 and won't be further explained at this point.

Next, we conduct the analysis by typing

```
> p.test(f1 = 0.28, f2 = 0.06, n1 = 50, n2 = 50)
```

i.e. we use the previously created function `p.test()` and set the determinants according to Example 9.7 as arguments.

As a result, we get:

```
[1] 2.795
```

The result is equivalent to the χ^2-test result.

Sequential testing is possible using appropriate computer programs.

Example 9.8 Calculation example for sequential testing for the comparison of two unknown parameters of a dichotomous character without relation to any content. The null hypothesis is H_0: $p_1 = p_2 = 0.5$. The alternative hypothesis is $H_A : |p_1 - p_2| = 0.4$.

The reader may recalculate that after the following successively sampled outcomes '1' or '0' for the groups 1 and 2, a terminal decision has to be made in favor of the alternative hypothesis: group 1 '1', group 2 '0', group 1 '1', group 2 '1', group 1 '0', group 2 '0', group 1 '1', group 2 '1', group 1 '0', group 2 '0', group 1 '1', group 2 '0', group 1 '1', group 2 '0', group 1 '1' ($\alpha = 0.05$; $\beta = 0.20$; $\delta = 0.4$). Without the last outcome, sequential testing cannot be completed under the given precision requirements.

In **R**, using the package `OPDOE` again, we do this by typing

```
> Example.tt <- triangular.test.prop(x = 1, y = 0,
+                                    p0 = 0.1, p1 = 0.5, p2 = 0.9,
+                                    alpha = 0.05, beta = 0.2)
```

Here we wrote (according to the **R**-notation) `p0 = p1 - 0.4 = 0.1`, `p1 = 0.5`, `p2 = p1 + 0.4 = 0.9`; we use the function `triangular.test.prop()`, for which we apply, as the first two arguments, the first observation value of group 1 (`x = 1`) and group 2 (`y = 0`), respectively. Additionally, we set `p0 = 0.1, p1 = 0.5`, and `p2 = 0.9` for H_0: $p_1 = p_2 = 0.5$ and H_A: $p_2 = 0.1$ or $p_2 = 0.9$, respectively. Furthermore we set the appropriate precision requirements with `alpha = 0.05` and `beta = 0.2`. All this we assign to the object `Example.tt`.

As a result, we get:

```
Triangular Test for bernoulli distribution

Sigma unknown, estimated as  0.4315642314
H0: p1=p2= 0.5  versus H1: p1= 0.5  and p2>= 0.9  or p2< 0.1
alpha: 0.05  beta: 0.2

Test not finished, continue by adding single data via update()
```

Since, with this as well as with the following observations, no final decision is obtainable, we proceed analogously until we, after the 15th observation value, can stop the sequential testing (see Figure 9.9).

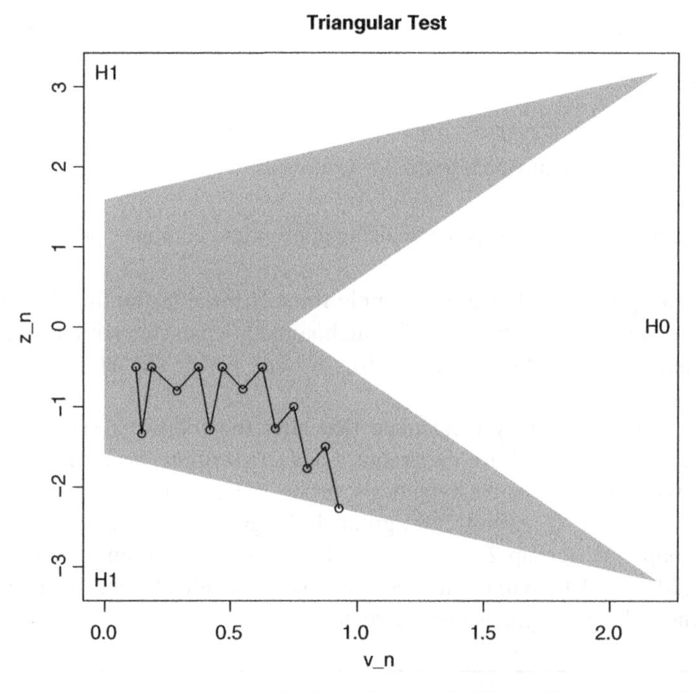

Figure 9.9 **R**-output of the sequential triangular test in Example 9.8 after altogether 15 observations.

Therefore, the hypothesis of $p_1 > p_2$ is accepted.

9.2.4 The unknown parameters p_i of a multi-categorical nominal-scaled character

Again, there are two independent samples with sizes n_1 and n_2, respectively, in which we consider a multi-categorical, nominally scaled character y. The character should have c different measurement values. Per value $1, 2, \ldots, j, \ldots, c$, we now have the absolute and relative frequencies h_{ij} and f_{ij}, with $i = 1, 2$ (in general: $i = 1, 2, \ldots, r$ samples). The underlying probabilities should be p_{ij}. All considerations will then remain similar to the case of a dichotomous character. Formula (9.3) as a test statistic can be applied in order to test the null hypothesis H_0: $p_{1j} = p_{2j}$ for all j; the alternative hypothesis is H_1: $p_{1j} \neq p_{2j}$, for at least one j.

Bachelor **Example 9.9** The difference in *marital status of the mother* between children with German as a native language and children with Turkish as a native language will be investigated (Example 1.1)

We decide on a type-I risk of $\alpha = 0.05$.

In **R**, we conduct the χ^2-test analogously to Example 9.7; hence we type

```
> chisq.test(marital_mother, native_language, correct = FALSE)
```

i.e. we use the function `chisq.test()` and apply both characters, *marital status of the mother* and *native language of the child*, that is, `marital_mother` and `native_language`, and conduct the analysis with `correct = FALSE`, hence without Yates' correction.

As a result, we get:

```
            Pearson's Chi-squared test

data:  native_language and marital_mother
X-squared = 16.4629, df = 3, p-value = 0.0009112
```

In SPSS, we follow the steps described in Example 9.7, except we apply the character marital_mother instead of the character only_child. First we get the frequencies of the character combinations, arranged as in Table 9.5. The result of the χ^2-test is, with a p-value of 0.001, significant.

Table 9.5 SPSS-output of the two-dimensional frequency table in Example 9.9.

marital status of the mother * native language of the child Crosstabulation

Count

		native language of the child		
		German	Turkish	Total
marital status of the mother	never married	6	3	9
	married	23	42	65
	divorced	18	4	22
	widowed	3	1	4
Total		50	50	100

As the p-value is 0.001, the null hypothesis has to be rejected: there are significant differences between mothers of children with German as a native language and those of children with Turkish as a native language, concerning the marital status. These relate, at least, to the marital status 'divorced', which is more apparent in mothers of children with German as a native language than in mothers of children with Turkish as a native language.

Summary

When it comes to hypothesis testing regarding the comparison of two populations, different statistical tests can be applied, depending on the type of scale of the character of interest and also on the desired parameter. For quantitative characters, in terms of mean differences, it is best to apply the *Welch test*; for ordinal-scaled characters, in terms of distribution differences, the *U-test,* and for nominal-scaled characters, in terms of different probabilities of the measurement values, the *χ^2-test*. In the case of the Welch test, both planning of the study and sequential testing is possible in accordance with certain given precision requirements. On the other hand, planning a study for the *U*-test is problematic because it will be hard to quantify the alternative hypothesis; it is not usual to apply sequential testing with this test. Planning a study with respect to the sample sizes needed is not possible in the case of the χ^2-test.

9.3 Equivalence testing

In recent times, the use of so-called *equivalence tests* has come about. For example, in the case of two kinds of therapy or support programs, one of which is favored for some reason (for example due to economic reasons), then the question would be raised as to whether both are equal in terms of the desired effect, i.e. are equivalent. Due to the fact that there are probably also certain irrelevant differences, it is necessary to examine more precisely whether the differences in terms of the desired effect exceed a certain relevance threshold or not. Thus, it is not about 'total equality' as is the case with significance testing, but rather about 'approximate equality', where we have to specify exactly what degree of inequality still conforms to 'approximate equality' or 'approximate equivalence'.

Doctor

As far as the mean difference $\mu_1 - \mu_2$ is concerned, regarding the two normally distributed random variables $\boldsymbol{y}_1$ and $\boldsymbol{y}_2$, the null hypothesis as well as the alternative hypothesis is defined quite differently in comparison to before. First, we denote the tolerable extent of a difference with ε. Thus, there is an equivalence interval between $-\varepsilon$ and ε for the difference $\mu_1 - \mu_2$. For any differences within this interval, mean differences between the two populations are interpreted as practically meaningless, and values outside the interval as meaningful. Although we could formulate the null and alternative hypothesis as follows: H_0: $|\mu_1 - \mu_2| \leq \varepsilon$; H_A: $|\mu_1 - \mu_2| > \varepsilon$, there are, however, mathematical reasons which argue against this. In this case, the test statistic (9.1) would no longer be centrally t-distributed, even under the null hypothesis, and consequently the test would not be applicable in the form presented here. However, if we swap the null and alternative hypothesis, then the problem is relatively easy to solve. Thus, the hypotheses are now $H_0 : |\mu_1 - \mu_2| > \varepsilon$ and $H_A : |\mu_1 - \mu_2| \leq \varepsilon$. That is to say, the null hypothesis claims that the therapies or support programs are not equivalent regarding the two means of the character y; but the alternative hypothesis states that they are ('approximately') equivalent.

If we now decide for ('approximate') equivalence on the basis of the empirical data, this should be correct with a probability of $1 - \alpha$. This is synonymous with the fact that both the hypothesis '$\mu_1 - \mu_2$ is greater than or equal to ε' and the hypothesis '$\mu_1 - \mu_2$ is less than or equal to $-\varepsilon$' will be rejected with the risk α, respectively, although one of them is true. Unlike significance testing (in this case the t-test), in equivalence tests, both hypotheses $\mu_1 - \mu_2 > \varepsilon$ and $\mu_1 - \mu_2 < -\varepsilon$ are to be tested collectively with the risk α. Hence, exact equivalence means that both these hypotheses are false.

Compared to the significance testing, we are only dealing with one risk here. Although we can also calculate a power, Wellek (2003) shows that this is often only possible with simulation studies.

In principal, all tests for significance in this chapter can also be rewritten as equivalence tests (see further details in Rasch, Herrendörfer, Bock, Victor, & Guiard, 2008).

In our case of an equivalence test for the difference of means, $\mu_1 - \mu_2$, the calculation amounts to two one-sided confidence intervals. It can be deduced that these – if the variances of the two populations are equal – can be found as follows:

$$\boldsymbol{L} = \bar{\boldsymbol{y}}_1 - \bar{\boldsymbol{y}}_2 - t(n_1 + n_2 - 2,\ 1 - \alpha)\boldsymbol{s} \cdot \sqrt{\frac{n_1 + n_2}{n_1 n_2}} \quad \text{and}$$

$$\boldsymbol{U} = \bar{\boldsymbol{y}}_1 - \bar{\boldsymbol{y}}_2 + t(n_1 + n_2 - 2,\ 1 - \alpha)\boldsymbol{s} \cdot \sqrt{\frac{n_1 + n_2}{n_1 n_2}} \tag{9.5}$$

If the calculated limits U and L are now both within the range of the alternative hypothesis $H_A : -\varepsilon < \mu_1 - \mu_2 \leq \varepsilon$, then H_0 will be rejected and the alternative hypothesis will be accepted. Planning a study is carried out by specifying a size δ

as the upper limit for the average (expected) distance $\delta \geq (\mu_1 - \mu_2) - E(\boldsymbol{L})$ of the interval limit $\boldsymbol{L}$ and/or as the lower limit for the average (expected) distance $\delta \geq E(\boldsymbol{U}) - (\mu_1 - \mu_2)$ of the interval limit $\boldsymbol{U}$, and by determining the value of α. The closer the limit(s) of the interval (9.5) is/are to the difference $(\mu_1 - \mu_2)$ on average, the more accurate the result is. It is best to choose the same size n for both samples. Then we can solve the equation

$$N = 4\frac{\sigma^2}{\delta^2}t^2(N-2, 1-\alpha)$$

with $N = n_1 + n_2$. Both n_1 and n_2 then have to be rounded up to integers.

Doctor

Example 9.10 Comparison of two rehabilitation programs in terms of their effectiveness

Two different programs A and B, for the rehabilitation of different memory functions after traumatic brain injury, differ insofar as A has to be carried out by the occupational therapist, whereas B can be carried out on a computer, without the necessity for staff to be present. It is assumed that the personal attention from a therapist has an additional positive or negative effect that is, however, practically negligible. As the affected population has a mean of 35 T-Scores with a standard deviation of 10 in a pertinent memory test, one hopes that the application of any of the two programs leads to an increase of 20 T-Scores in the long term, where differences between the two programs of up to $\varepsilon = 5$ T-Scores would indicate ('approximate') equivalence. That is to say, based on the rehabilitation success (T-Score in the memory test after applying the respective rehabilitation program minus T-Score in the memory test before its application), the null and alternative hypothesis are stated as follows: $H_0 : |\mu_A - \mu_B| > 5$ and $H_A : |\mu_A - \mu_B| \leq 5$. The risk α will be 0.05.

Planning the study leads therefore to $N = 4\frac{10^2}{5^2}t^2(120, 0.95) = 16 \cdot 1.66^2 = 44.09$ (rounded up i.e. $N = 45$), if one sets $2n - 2 = 120$ in a first iteration step. Correcting this in the next step to $16 \cdot t^2(45,0.95) = 16 \cdot 1.68^2 = 45.15$ hardly changes the result; thus 23 people in each group are sufficient.

In **R**, we use the function `power.t.test` – in doing so we have to submit `power = 0.5` as argument. The reason for this is that the formula for the sample size of the confidence interval only relates to the p-quantile of $1 - \frac{\alpha}{2}$, whereas the function `power.t.test` requires a p-quantile of $1 - \beta$, too, which we, in our case, have to set to null, which is achieved by $\beta = 0.5$. Hence we type

```
> power.t.test(delta = 0.5, sd = 1, sig.level = 0.05, power = 0.5,
+              type = "two.sample", alternative = "one.sided")
```

i.e. we submit the precision requirements as separate arguments to the function `power.t.test()`(instead of $\sigma = 10$ and $\delta = 5$ we use `sd = 1` and `delta = 0.5`; more simply $\sigma = 1$ and $\delta = 0.5$, though), as well as the further arguments `type =`

"two.sample", since we have two samples, and alternative = "one.sided", because of the one-sided alternative hypothesis.

As a result, we get:

```
     Two-sample t test power calculation

              n = 22.34895
          delta = 0.5
             sd = 1
      sig.level = 0.05
          power = 0.5
    alternative = one.sided

NOTE: n is number in *each* group
```

By rounding up, according to the **R**-program, 23 persons can be chosen from each group; that is, therefore, $N = 46$.

In the appropriately planned study, we might obtain $\bar{y}_A = 18.18$ and $\bar{y}_B = 15.51$;

$$s^2 = \frac{s_1^2 \cdot 23 + s_2^2 \cdot 23}{46} = 100$$

From this follows

$$U = 2.67 - 1.68 \cdot 10 \cdot \sqrt{\frac{46}{23 \cdot 23}} = -2.28$$

as well as $O = 7.62$.

As we see, not both limits of the confidence interval are in the equivalence range –5 to 5; therefore the null hypothesis, which states lack of equivalence, has to be accepted.

References

Anastasi, A. & Urbina, S. (1997). *Psychological Testing* (7th edn). Upper Saddle River, NJ: Prentice Hall.

Kubinger, K. D. (1990). Übersicht und Interpretation der verschiedenen Assoziationsmaße [Review of measures of association]. *Psychologische Beiträge*, *32*, 290–346.

Mann, H. B. & Whitney, D. R. (1947). On a test whether one of two random variables is stochastically larger than the other. *Annals of Mathematical Statistics*, *18*, 50–60.

Rasch, D. & Guiard, V. (2004). The robustness of parametric statistical methods. *Psychology Science*, *46*, 175–208.

Rasch, D., Herrendörfer, G., Bock, J., Victor, N., & Guiard, V. (2008). *Verfahrensbibliothek Versuchsplanung und -auswertung. Elektronisches Buch.* [Collection of Procedures in Design and Analysis of Experiments. Electronic Book]. Munich: Oldenbourg.

Rasch, D., Kubinger, K. D., & Moder, K. (2011). The two-sample *t*-test: pre-testing its assumptions does not pay off. *Statistical Papers, 52*, 219–231.

Rasch, D., Pilz, J., Verdooren, R., & Gebhardt, A. (2011). *Optimal Experimental Design with R*. Boca Raton: Chapman & Hall/CRC.

Wellek, S. (2003). *Testing Statistical Hypotheses of Equivalence*. Boca Raton: Chapman & Hall/CRC.

Wilcoxon, F. (1945). Individual comparisons by ranking methods. *Biometrics Bulletin, 1*, 80–83.

10

Samples from more than two populations

This chapter is about selection procedures and statistical tests for comparing parameters from more than two populations from which samples have been drawn. The latter especially include the numerous forms of analysis of variance, but also multiple comparisons of means. Regarding analysis of variance, we distinguish between one-way analysis of variance and multi-way analysis of variance; in the latter between cross classification and nested classification. Of interest are always the differences between the factor levels; the factors can have fixed or random (caused by chance) levels. Depending on whether there are fixed or random factors, this dictates which particular tests have to be performed. Given multi-way analysis of variance, besides the main effects of each factor, the interaction effects of factors (and factor levels, respectively) will be dealt with thoroughly. Also, methods for ordinally- and nominal-scaled characters will be discussed.

We start from at least three independent samples from each of the respective number of populations of interest – independent in the sense defined in Section 6.1, which states that each outcome in one sample will be observed independent of all other outcomes in the other samples. So, this definitely precludes the existence of research units in one sample that are related to specific other research units of the other sample (e.g. siblings). Once the outcomes of the particular character are given, (point) estimates can be calculated in each sample. However, the main question is: do the samples stem from the same population or from different populations? The latter answer is still given even when a single sample does not stem from the same population as the others.

Statistics in Psychology Using R and SPSS, First Edition. Dieter Rasch, Klaus D. Kubinger and Takuya Yanagida.

Bachelor **Example 10.1** Do children differ regarding test scores on *Everyday Knowledge, 1st test date*, depending on their sibling position? (Example 1.1)

The example is about several independent (random) samples of first-borns, second-borns, etc. The question is, if these samples stem from different populations in terms of the character *Everyday Knowledge, 1st test date*; of course they stem, by definition, from different populations regarding other characters. In the psychological literature, there are theories claiming that intelligence decreases with later position in the ranking of siblings.

10.1 The various problem situations

Again, we try to introduce the problem on the basis of the statistic: mean, and thus on the basis of a quantitative character.

To simplify matters, we assume that the variances in the a populations, from which a independent samples are drawn, are all equal. The particular character should also be modeled sufficiently well by a normally distributed random variable $\boldsymbol{y}_i$, $i = 1, 2, \ldots, a$ in each population. There are now two fundamentally different problem situations.

Assume that a researcher investigates several psychological tests regarding their quality to measure a particular ability. In this case, he/she wants to choose one of these tests for practical use, similar to the way somebody may want to choose one drug to combat a specific disease from several drugs on the basis of a clinical study. This illustrates problem-situation 1: from a populations, on the basis of sample results, the most appropriate one for a clearly defined purpose has to be selected. This brings us to *selection procedures*.

It is a completely different situation if we want to find out, considering the data from Example 1.1, whether there are differences between children from places with $a = 3$ different measurement values (character: *urban/rural* with the categories 'city', 'town', and 'rural') regarding the subtest *Everyday Knowledge*. Regardless of which consequences would be possible in the medium or long term, it is not about the selection of the 'best' population for a particular purpose. This represents problem-situation 2: the question is whether there are differences between the means of the a populations. This leads to so-called *overall tests*, also called *omnibus tests*, namely in the form of *analysis of variance*. So, we are not interested which population or populations differ(s) from which other(s), but only in the existence of such differences at all. However, if we are interested in more details, regarding which populations differ from each other, i.e. the comparison of each and every population with one another, then this represents problem-situation 3: the need to examine between which of the a populations there are differing means. This leads to *multiple comparisons of means*.

We start with the presentation of selection procedures and then deal with multiple comparisons of means. Analysis of variance will be discussed afterwards, though it certainly provides less information initially in comparison to multiple comparisons of means, but it is usually preferable in terms of meeting precision requirements.

Essentially, this chapter is actually about the comparison of means, and 'homologous' methods, i.e. those that do not result in conclusions about the arithmetic mean, but nevertheless all give evidence about the location of a distribution of a character. We will point out where access to particular methods is given for other statistical measures, such as relative frequency, as for example is the case in selection procedures.

10.2 Selection procedures

If the aim of a study actually consists of determining which population has the smallest or largest mean of a populations, then researchers should not perform hypothesis testing, but instead a selection procedure should be done. In this case, the very simple selection rule is: once a sample has been drawn from each population, declare the 'best' population as the one which has the largest – or smallest – sample mean. Such an approach is always reasonable, regardless of whether this one mean is significantly different from (any or all) other means or not. In contrast, multiple comparisons of means are particularly inappropriate for the given question. They unnecessarily increase the required total sample size.

Below, we consider the special case of looking for that population with the largest mean μ_i $(i = 1, 2, \ldots, a)$.

Unlike hypothesis testing, planning a study only requires the determination of the probability of a false selection – it is traditionally also called β, although this probability has nothing to do with a type-II risk – as well as a minimum relative effect size δ, which is considered to be relevant. In the particular case of means, δ specifies the minimum difference between the 'best' (in the situation-specific defined sense of the word) and second-best population. Since actual deviations smaller than δ are practically irrelevant, it does not matter if we select, not the best, but instead, erroneously, the second best etc. However, we do not want to make a false selection in the case of a deviation exceeding δ. That is, providing that the difference between the largest and the second-largest mean exceeds the value δ, then the probability of a false selection should not be larger than the predetermined β.

The required sample size for the given values of δ and β also depends on the number a of the populations to be investigated. It has to be determined for the worst case. This occurs if there is just a difference of δ between the largest mean, $\mu_{\max}$, and the second largest, μ_{II}, and if all other means are equal, μ_{II}. If the sample size is calculated for this worst-case situation and with respect to any false decision's probability equal to β, then one is just on the side of caution: the probability of a false selection is smaller than β in all other cases. It is optimal, in the sense of an as-small-as-possible total sample size, to draw samples of the same size n in all populations.

The analysis is, as already stated, quite simple: the researcher calculates the sample mean from each of a samples, and determines that population with the largest sample mean to be the best (Bechhofer, 1954).

Example 10.2 One of six training programs for reading and spelling difficulties has to be selected, namely the one with the greatest treatment success.

Given that all $a = 6$ training programs are equally time-consuming, expensive, and can be used equally well and willingly by any psychologist, then each of the 6 training programs will be carried out on the same number of randomly selected children (e.g. six- to seven-year-olds). The character of interest is the difference in the test scores in an appropriate reading and spelling test before and after the application of the training program. We decide on a probability of a false selection with $\beta = 0.05$, and the relative minimum difference with $\delta = 0.33\sigma$.

In **R**, we first load the package `OPDOE`, which we have already used in Chapter 8, applying the function `library()`. For calculating the sample size, we type

```
> size.selection.bechhofer(a = 6, beta = 0.05, delta = 0.333,
+                          sigma = 1)
```

i.e. we use the number of populations, a, as the first, and the probability of an erroneous selection, β, as the second argument. The relative minimum difference, δ, is used as the third, and the standard deviation, σ, as the fourth argument in the function. As a result, we get:

```
[1] 90
```

Consequently, each training program has to be conducted with at least 90 children.

For Lecturers:

The reasons why selection procedures are rarely used (not only) in psychology are as follows:

- Selection procedures were suggested about 50 years after the other statistical methods were developed.
- Selection procedures are 'unexciting'; with their application one can hardly impress.
- Selection procedures are not included in statistical packages, such as SPSS, due to their simplicity.

If one is interested in the probability p of a dichotomous character, the aim of the study will be to find, from a populations, that one with the largest or smallest probability, respectively, with regard to one of the two measurement values of this dichotomous character. Then, as above, the probability β of a false selection and a minimum effect size δ, which is considered to be relevant, has to be determined, where δ describes the minimum difference between the 'best' (in the particular defined sense) and second-best population. More details can be found in Rasch, Pilz, Verdooren, & Gebhardt (2011).

10.3 Multiple comparisons of means

The starting point is a independent random samples from the same number of populations; we assume that we do not (yet) know, from the application of an analysis of variance, if there are any differences at all between the means of these populations. That is, the question: 'Which populations are different regarding their means?' still allows for the non-existence of differences. In the sense of the first step of empirical research as stated in Section 3.2, this question has yet to be formulated more precisely. It must be clarified: does one only want to test the null hypothesis H_0: $\mu_i = \mu_l$, $i \neq l$ for some (particular) pair of means, or for all pair-wise differences of $a > 2$ means. All the methods by which one can do this are called multiple comparisons of means.

Essentially new, in comparison to what has already been said, is that we now have to distinguish between two alternatives for each of the risks; namely between two different type-I risks and two different type-II risks. The *comparison-wise risk* refers to each individual comparison. Therefore, one takes a certain (given) risk in each comparison, i.e. the risk of falsely rejecting the null hypothesis on one hand and of falsely accepting the null hypothesis on the other hand. The consequence of such an approach is that the greater the number of pair-wise comparisons, the greater the probability of a wrong decision in at least one of these comparisons within the same study. This is exactly the reason why one usually will not take a comparison-wise risk. Then one decides for a *study-wise risk*. So, if one uses a study-wise risk, then one has a certain (given) risk for the entire study; i.e., for all performed comparisons, a certain (given) accumulated risk of falsely rejecting the null hypothesis on the one hand, and a certain (given) accumulated risk of falsely accepting the null hypothesis on the other hand. Then, it does not matter whether there was only a wrong decision regarding a single comparison or, in extreme cases, regarding all comparisons. For example, if we actually perform all $\binom{a}{2}$ possible comparisons and test the corresponding null hypothesis H_0: $\mu_i = \mu_l$, then the study-wise type-I risk α indicates the probability that at least one of the null hypotheses is falsely rejected. The study-wise type-II risk β indicates the probability of falsely accepting at least one of the null hypotheses.

Now, we have to distinguish between methods that determine both risks either comparison-wise or study-wise; besides that, there are also methods that take the type-I risk study-wise and the type-II risk comparison-wise into account.

Statistical tests that determine both risks study-wise are the tests of analysis of variance. For these, study-wise risks can even be taken into consideration as precision requirements. But first we consider the case of comparison-wise risks in the following.

It is almost obvious why a comparison-wise type-I risk is usually undesirable. If one actually performs all $\binom{a}{2}$ comparisons and every time takes the risk of a type-I error with a probability of α, then the overall type-I risk tends, given a is quite high, quickly to 1. However, there are sometimes research questions for which a comparison-wise risk is quite appropriate. In such a case it is recommended to perform comparisons only between each pair of adjacent means, when all means are ranked by size, instead of performing all comparisons. Then there are only $a - 1$ comparisons instead of $\binom{a}{2}$. Of course it is also conceivable to only compare each of the (estimated) means from $a - 1$ populations with a standard (reference) population.

Example 10.3 Validation of a psychological test via extreme groups

For the evaluation of a psychological test's validity, it is recommendable to compare the (mean) test scores from two groups of testees that show obvious extreme values of the character that it is intended to measure by the test. If this is about a test to measure the ability to cope with stress, we might have the idea to compare the following three groups (populations): I counter service, II staff nurse, and III kindergarten teacher. We are especially interested in the comparison of I and II as well as the comparison of I and III. We do not need a study-wise (type-I) risk, because we absolutely want to validate the test twice: for both the groups II and III, we need proof of validity; whereas a type-I error here does not change anything regarding the evidence of quality there (and vice versa). So in this case, one will take a comparison-wise risk.

Of course, the researcher has already to decide on a study-wise or a comparison-wise risk while planning the study. Since the study-wise risk represents a stricter (precision) requirement than a comparison-wise risk, it requires a larger sample size.

An appropriate method for comparison-wise type-I and type-II risks is the so-called *least significant difference (LSD) test*. It is simply the pair-wise application of the t-test, essentially in accordance with Formula (9.1); though, because there are more than two samples, there is a different estimation of the common variance σ^2. According to the explanations in Section 9.1, it is again preferable to perform a pair-wise applied Welch test – for which planning of the study has already been dealt with.

Master
Doctor

In most cases (in psychology), the LSD-test (that is the multiply-applied t-test according to Formula (9.1)) is performed without any thought put into whether a comparison-wise type-I risk is at all appropriate for the particular research question.

If one wants to perform the LSD-test in SPSS without calculating all $\binom{a}{2} = \frac{a(a-1)}{2}$ pair-wise t-tests separately one after another, as is demonstrated in Example 9.1, but instead more conveniently (and using the more accurate estimation of σ), simultaneously with a specific program command, then one has to perform the analysis of variance (see Section 10.4.1) beforehand (to call up the option LSD in Post Hoc...). In this case the user should be aware that he/she first tests study-wise, but continues testing comparison-wise.

If the researcher has – according to a carefully considered research question – decided on a comparison-wise type-I risk, then the pair-wise applied Welch test has to be preferred over the pair-wise t-test in any case, as the Welch test better meets the nominal type-I risk in the case of violations of the presumptions of the t-test.

In **R**, we can conduct a pair-wise Welch test by typing

```
> pairwise.t.test(y, g = group, p.adj = "none", pool.sd = FALSE)
```

i.e. we apply the function `pairwise.t.test()`, using the character of interest `y` as the first argument. As the second argument we use `g = group`, which determines to which of the respective samples an observation unit belongs. With `p.adj = "none"` we omit the so-called α correction (see Section 15.2.1.1) and, using `pool.sd = FALSE`, we specify the Welch test.

As a result, we obtain the p-values of all pair-wise tests, arranged in a matrix.

As concerns ordinal-scaled characters, one can proceed analogously to the method of multiple comparisons of means, if interested in distribution differences; in terms of comparison-wise risks, the U-test is applicable.

10.4 Analysis of variance

It is indeed correct that variances (precisely: sums of squares) are decomposed by the method of analysis of variance; nevertheless, this method is exclusively about tests for comparing means. Basically, it is about establishing the effect of (qualitative) factors on a quantitative character y. Thereby, each nominal- or ordinal-scaled factor, in the sense of Section 4.5, has $a \geq 2$ measurement values, or that is to say categories. In the simplest case, we only have a single such factor; then, the *one-way analysis of variance* or *one-way layout*, respectively, is the direct generalization of the situation underlying the t-test. However, it is often the case that more than just one factor exists, generally p factors. This is called p-way analysis of variance or, generally, a *multi-way analysis of variance* (two-way layout etc.).

10.4.1 One-way analysis of variance

In general, we denote the one single factor by A, and its a levels by A_i, i.e. $A_1, A_2, \ldots, A_a$.

Master The statistical model of analysis of variance is a special case of the so-called general linear models (see more details in Section 13.1). As a matter of fact, the character of interest, y, is modeled as a random variable $\boldsymbol{y}$ of some linear function. That is, besides the presumption of a normally distributed variable with population mean $E(\boldsymbol{y}) = \mu$ and population variance $\sigma_y^2 = \sigma^2$, it is assumed that every (theoretically possible) observation of $\boldsymbol{y}$ differs from μ by an error component $\boldsymbol{e}$. From the definition of the random variable in question, $\boldsymbol{y} = \mu + \boldsymbol{e}$, it is obvious that the error variance, i.e. the variance of $\boldsymbol{e}$, is also equal to σ^2. However, this model is too simple and is only appropriate for the case in which we only consider a single sample and in which the null hypothesis (that all samples stem from the same population) is actually true. (See also the similarity of this model to the basic equation of the so-called *classical test theory*; e.g. in Lord & Novick, 1968.) If we also consider the case of the alternative hypothesis, $\boldsymbol{y} = \mu + \boldsymbol{e}$ therefore has to be supplemented by a further term (parameter) which takes the diversity of means in different populations into account. By doing so, we have defined a formalism that basically allows us to test several hypotheses.

In analysis of variance, we always assume that the variability of the research units in $\boldsymbol{y}$ depends linearly on certain (new) model parameters. Thus, several models of analysis of variance differ by type and number of such parameters.

The method is about the means of populations, which underlie the given samples and initially are considered to be dissimilar. The character of interest will be modeled as a normally distributed random variable. That is, a random sample of the size n_i is collected from each of the a populations G_i, $i = 1, 2, \ldots, a$. The outcomes of the random variables $\boldsymbol{y}_{i1}$, $\boldsymbol{y}_{i2}, \ldots, \boldsymbol{y}_{in_i}$ are $y_{i1}, y_{i2}, \ldots, y_{in_i}$. The underlying parameters are μ_i and σ_i^2. The corresponding estimators are then (according to Section 8.5.1 and Formula (5.4)): $\hat{\boldsymbol{\mu}}_i = \bar{\boldsymbol{y}}_i$ and $\hat{\boldsymbol{\sigma}}_i^2 = \mathbf{s}_i^2$.

In analysis of variance, it is typical to differentiate between different (basic) models. Certainly, the type of the model has, in the case of one-way analysis of variance (i.e. with only one factor), no effect at all as concerns hypothesis testing; but the effect is, however, serious in the case of more than one factor.

Case 1: exactly the a levels (measurement values) of the factor being included in the analysis are of interest. In such a case, model I of analysis of variance is on hand. Since the levels of the factor are fixed from the outset, we talk about a *fixed factor*.

Bachelor **Example 10.1 – continued**

In designing such a study, the researcher would probably decide to record the birth order: first-, second-, third-, and fourth-born; because the population of fifth- or even later-born persons is relatively small, these sibling positions can be ignored to some extent. It might be discussed whether the fourth- and all later-born should be taken as a common factor level, or all those later than fourth-born should not be recorded at all and therefore be excluded from consideration. In any case, it is just these four factor levels in which we are interested. The factor *sibling position* is fixed.

Case 2: the factor has many measurement values (levels); their number theoretically has to be considered infinite. The a levels, which have to be included in the study, are randomly selected by drawing a random sample from all levels of the population. Thus, the factor has to be modeled as a random variable itself. This case represents model II of the analysis of variance. Seeing that the levels of the factor are obtained by chance and are therefore not fixed, we talk about a *random factor*.

Bachelor **Example 10.1 – continued**

Although unrealistic, the following case still has great illustrative qualities. If we want to investigate exactly $a = 3$ different sibling positions, but randomly obtain the sibling positions of third-, fourth-, and seventh-born from all practically possible sibling positions 1 to 9, it can be demonstrated that the number of measurement values (levels) of the factor in this example is too small for model II to really be applicable.

Example 10.4 Do the test scores of students from different schools differ in test T?

For example, for calibrating a psychological test, a sample of schools could be randomly drawn (in a multi-stage random sampling procedure); these schools would then represent the levels of the factor *school* – in these schools, the children could be selected randomly again. The number of schools, for example in Boston, would therefore be large enough that we do not have to use a model with finite populations of levels. Bear in mind, the question was: are there differences in the test scores between children from all Boston schools? So, we are not only interested in the randomly selected schools.

Case 3: there is a finite number, namely S_A levels, of factor A, where the number S_A is not large enough to be modeled by case 2. On the other hand, there is no possibility or desire to include all these levels in the investigation. We are not dealing with this case here (but see, for example, Rasch, 1995).

Bachelor **Example 10.5** Do the test scores of students from different municipal districts differ in test T?

For example, for calibrating a psychological test, a sample of municipal districts could be randomly drawn (using a multi-stage random sampling procedure); these would then represent the levels of the factor *municipal districts* – in these municipal districts, schools, and within that, children could be selected randomly again. The population of the levels of the factor *municipal districts* in Boston consists of $S_A = 21$ elements; the number 21 is so small that model II should not be chosen.

10.4.1.1 Model I

For case 1, we first model the character of interest more precisely than with the random variables $\boldsymbol{y}_{iv}$, $i = 1, \ldots, a$; $v = 1, \ldots, n_i$. Model I of one-way analysis of variance is then:

$$\boldsymbol{y}_{iv} = E(\boldsymbol{y}_{iv}) + \boldsymbol{e}_{iv} = \mu_i + \boldsymbol{e}_{iv} = \mu + a_i + \boldsymbol{e}_{iv} \quad (i = 1, 2, \ldots, a; v = 1, 2, \ldots, n_i) \tag{10.1}$$

In Formula (10.1), the terms a_i are fixed,[1] i.e. parameters; $\boldsymbol{e}_{iv}$ describe the errors of the random variables $\boldsymbol{y}_{iv}$. The terms a_i are the effects of the factor levels A_i; they are called *(main) effects* in statistics. We recognize that the variance of $\boldsymbol{y}_{iv}$ is therefore equal to the variance of $\boldsymbol{e}_{iv}$; i.e. $\sigma^2(\boldsymbol{y}_{iv}) = \sigma^2 = \sigma^2(\boldsymbol{e}_{iv})$. The total size of the research is defined as N as a matter of simplification; i.e. $N = \sum_{i=1}^{a} n_i$.

If all $n_i = n$, we talk about equal cell frequencies, otherwise they are termed unequal cell frequencies. The term *cell* arises from the fact that the outcomes y_{iv} can be systematically arranged in different cells of a table (see Table 10.1).

Table 10.1 Data structure of a one-way analysis of variance.

Level A_1	Level A_2	...	Level A_a
y_{11}	y_{21}	...	y_{a1}
y_{12}	y_{22}	...	y_{a2}
⋮	⋮	⋮	⋮
y_{1n_1}	y_{2n_2}	...	y_{an_a}

We consider the null hypothesis H_0: $\mu_i = \mu_l = \mu$ for all i and all l, and the alternative hypothesis H_A: $\mu_i \neq \mu_l$ for at least one $i \neq l$. A one-sided alternative hypothesis of the type H_A: $\mu_i > \mu_l$ is not treatable with elementary methods of statistics.

[1] a_i is an (unknown) parameter, like (until now) μ and σ; basically, we always denote parameters by Greek letters – a previous exception was parameter p for the probability. But we deviate from that in the context of analysis of variance. The reason is less that this is usual in the international literature of statistics and not because the corresponding Greek letter α has a fixed, completely different meaning; the reason is rather that, in the formulaic representation, instead of this parameter, i.e. a fixed value, also random variables are defined, for which we uniformly use a bold notation in Latin in this book.

For derivation, especially of hypothesis testing, the following must be assumed:

- The e_{ij} are normally distributed according to $N(0, \sigma^2)$, i.e. with the same expected value 0 and equal variance σ^2.
- The e_{ij} are independent of each other.
- $a \geq 2$ and $n_i \geq 2$ for all i.

Doctor The $a+1$ parameters μ, a_i $(i = 1, 2, \ldots, a)$ of Formula (10.1) are estimated by the least squares method (see Section 6.5). It follows, by setting $\sum_{i=1}^{a} a_i = 0$, that the estimates are $\hat{\mu} = \bar{y} = \frac{1}{N}\sum_{i=1}^{a}\sum_{v=1}^{n_i} y_{iv}$ and $\hat{a}_i = \bar{y} - \bar{y}_i$, with $\bar{y}_i = \sum_{v=1}^{n_i} y_{iv}$; $i = 1, 2, \ldots, a$.

Master To better understand the method, but also because computer programs represent it as follows, we look at the so-called *analysis of variance table* more precisely in the following. We distinguish between a realized analysis of variance table, i.e. referring to empirical data, and a theoretical table – which includes certain expected values.

Fundamentally, analysis of variance is about decomposing the total sum of certain squared differences into several partial sums; we call them all, for short, *sums of squares* (*SS*). Amongst others, these are clearly summarized in the analysis of variance table (see Table 10.2).

Table 10.2 Empirical analysis of variance table of one-way analysis of variance for model I $(n_i = n)$.

Source of variation	*SS*	*df*	*MS*
Factor *A*	$SS_A = \sum_{i=1}^{a}\sum_{v=1}^{n}(\bar{y}_i - \bar{y})^2$	$a-1$	$MS_A = \frac{SS_A}{a-1}$
Residual	$SS_{res} = \sum_{i=1}^{a}\sum_{v=1}^{n}(y_{iv} - \bar{y}_i)^2$	$N-a$	$s^2 = \hat{\sigma}^2 = MS_{res} = \frac{SS_{res}}{N-a}$
Total	$SS_t = \sum_{i=1}^{a}\sum_{v=1}^{n}(y_{iv} - \bar{y})^2$	$N-1$	

The sum of squares SS_t is the sum of the squared differences of all outcomes y_{iv} from the total mean $\bar{y} = \frac{1}{N}\sum_{i=1}^{a}\sum_{v=1}^{n} y_{iv}$; it expresses the total variability of outcomes in the study. SS_A is the sum of squares of the level means $\bar{y}_i$ from the total mean $\bar{y}$; it expresses the variability of the level means. And SS_{res} is the sum of squares of the outcomes y_{iv} from the respective level mean; it expresses the (average) variability of outcomes per level.

It can be shown that $SS_t = SS_A + SS_{res}$. Analogously, it is obvious that $df_t = df_A + df_{res}$ (see Table 10.2); the derivation of the degrees of freedom will not be given here. By dividing the individual sums of squares by their corresponding degrees of freedom we obtain the *mean squares* (*MS*); there is no further interest in the conceivable statistic MS_t.

As can be shown, the *error variance* or *residual variance* σ^2 can be estimated by $s^2 = \hat{\sigma}^2 = MS_{\text{res}}$; the corresponding estimator $\boldsymbol{s}^2 = \boldsymbol{MS}_{\text{res}}$ is unbiased. Hypothesis testing arises from the theoretical analysis of variance table in which the expected values of the mean squares are recorded (Table 10.3). Thereby it is assumed that the following arbitrary condition of reparameterization is met: $\sum_{i=1}^{a} a_i = 0$.

Table 10.3 Theoretical analysis of variance table of one-way analysis of variance for model I ($n_i = n$).

Source of variation	SS	df	MS	$E(MS)$
Factor A	$\boldsymbol{SS}_A = \sum_{i=1}^{a}\sum_{v=1}^{n}(\bar{\boldsymbol{y}}_i - \bar{\boldsymbol{y}})^2$	$a-1$	$\boldsymbol{MS}_A = \dfrac{\boldsymbol{SS}_A}{a-1}$	$\sigma^2 + \dfrac{n}{a-1}\sum a_i^2$
Residual	$\boldsymbol{SS}_{\text{res}} = \sum_{i=1}^{a}\sum_{v=1}^{n}(\boldsymbol{y}_{iv} - \bar{\boldsymbol{y}}_i)^2$	$N-a = a(n-1)$	$\boldsymbol{MS}_{\text{res}} = \dfrac{\boldsymbol{SS}_{\text{res}}}{N-a}$	σ^2
Total	$\boldsymbol{SS}_{\text{t}} = \sum_{i=1}^{a}\sum_{v=1}^{n}(\boldsymbol{y}_{iv} - \bar{\boldsymbol{y}})^2$	$N-1 = an-1$		

The above, given null hypothesis H_0: $\mu_i = \mu_l = \mu$ for all i and all l, and the alternative hypothesis H_{A}: $\mu_i \neq \mu_l$ for at least one $i \neq l$, now have to be rephrased as follows: H_0: $a_i = a_l = a = 0$ for all i and all l, and H_{A}: $a_i \neq a_l$ for at least one $i \neq l$. Now, a test can be derived from Table 10.3: it involves comparing two mean squares that both have the same expected value under the null hypothesis. If the test provides a significant result, both variances have to be interpreted as unequal in the population, which means not all $a_i = 0$; and therefore the means are different.

It can be shown that the null hypothesis can be tested with the following test statistic;

$$\boldsymbol{F} = \frac{\boldsymbol{MS}_A}{\boldsymbol{s}^2} \tag{10.2}$$

is F-distributed with $df_1 = a - 1$ and $df_2 = N - a$ degrees of freedom. Given that the null hypothesis is true, $\boldsymbol{MS}_A$ as well as $\boldsymbol{s}^2$ estimate the variance of $\boldsymbol{y}$ and $\boldsymbol{e}$, respectively; the former from the variance of the means $\bar{\boldsymbol{y}}_i = \hat{\boldsymbol{\mu}}_i$. So, this *F-test* examines if the means vary more than is explainable by the variability of the variable itself. If this is the case, then the means differ systematically.

This means, the null hypothesis can be tested using the $(1 - \alpha)$-quantile of the F-distribution. It will be rejected if the calculated value of $F > F(a-1, N-a : 1-\alpha)$; otherwise it will be accepted. If the null hypothesis has to be rejected, we again talk about a significant result. However, this does not yet state which means differ in which direction.

Doctor Incidentally, for $a = 2$ we have the case of Section 9.1; namely testing a hypothesis regarding the comparison of two means from two populations. The test statistic in Formula (10.2) is just the square of the test statistic in Formula (9.1). For the quantiles, it is therefore $F(1, N-2 : 1-\alpha) = t^2\left(N-2, 1-\frac{\alpha}{2}\right)$, as the reader can easily verify by comparing corresponding values in Tables B2 and B4 in Appendix B.

Considerations regarding the estimation of the effect size will be made in Section 11.3.4.

Planning a study, i.e. the determination of the necessary sample size, is done analogously to Chapters 8 and 9 by specifying certain precision requirements. Type-I risk, type-II risk, α and β, are chosen again analogously; but the parameter δ has to be redefined: δ will now be the distance from the largest of all a means to the smallest. However, in analysis of variance, the sample size also depends on the location of the remaining $a-2$ means. The available computer programs determine the sample size for both the worst and the best case. Basically, the worst case should be assumed; i.e. the case that leads to the largest (minimum) size for given precision requirements. This case occurs if all other $a-2$ means are located exactly in the middle of both extreme means (see Figure 10.1). In that case, it is – as generally applies – optimal in terms of a smallest possible sample size, if the size is equal in all samples, i.e. $n_i = n$.

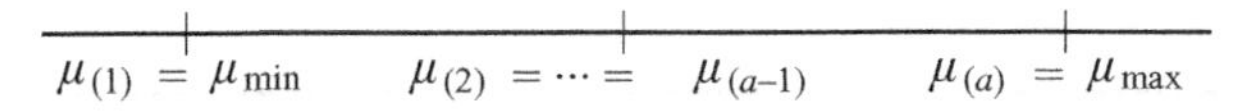

Figure 10.1 The (worst) case for the location of the a means $\mu_{(i)}$ ($i = 1, \ldots, a$) ordered by size, which leads to the largest sample size.

Example 10.6 Planning a study for one-way analysis of variance without relation to any content

Assume $a = 4$. We decide on $\alpha = 0.05$; $\beta = 0.2$; $\delta = \mu_{\max} - \mu_{\min} = 0.67\sigma$; so it will follow that $\delta = 0.67$ by assuming $\sigma = 1$.

In **R**, we calculate the sample size again using the package `OPDOE`; we type

```
> size_n.one_way.model_1(alpha = 0.05, beta = 0.2, delta = 0.67,
+                        a = 4, cases = "maximin")
```

i.e. we apply the function `size_n.one_way.model_1()` and use as arguments the precision requirements $\alpha = 0.05$, $\beta = 0.2$, and $\delta = \mu_{\max} - \mu_{\min} = 0.67$ (with `delta = 0.67`). With `a = 4` we fix the number of populations, and with `cases = "maximin"` we request the sample size for the worst case. As a result, we get:

```
[1] 50
```

Thus, we have $n = 50$.

In instances where it is known from the outset that unequal sample sizes are unavoidable, we have no algorithm that meets the precision requirements. In any case, the smallest sample size should be equal to the calculated n.

Bachelor **Example 10.1 – continued**

In Example 1.1, there is a total of 100 children. First, we investigate how many children are distributed into each sibling position, but we instantly decide, because of the low relative frequency of fifth- or even later-born sibling positions in the population, to analyze only up to the fourth sibling position. We obtain a corresponding frequency table using **R** or SPSS, as we have done in Example 5.4 by using *sibling position* as the character in question. We note: $n_1 = 44$, $n_2 = 24$, $n_3 = 16$, $n_4 = 8$. The sample size of $n = 50$, which we calculated in advance by planning the study above, is rather too small – bearing in mind that planning of the given study was originally designed for another research question (see Example 9.2). We now analyze with the one-way analysis of variance.

In **R**, we first have to exclude all children whose sibling position is bigger than four from the analysis. To do this, we type

```
> Example_1.1.s <-  subset(Example_1.1,
+                            subset = unclass(pos_sibling) < 5)
```

i.e. we use the database `Example_1.1` as the first argument in the function `subset()`, and the condition `subset = unclass(pos_sibling) < 5` as the second argument. We need the function `unclass()` to make the selection `pos_sibling < 5`. We assign the data which we have selected to `Example_1.1.s`. After enabling access to the new database `Example_1.1.s` by applying the function `attach()` (see Chapter 1), we type, to conduct the analysis of variance

```
> aov.1 <- aov(sub1_t1 ~ pos_sibling)
> summary(aov.1)
```

i.e. we apply the function `aov()`, specifying the analysis of the character *Everyday Knowledge, 1st test date* (`sub1_t1`) with respect to the factor *sibling position* (`pos_sibling`). We assign the result of the analysis to the object `aov.1` and submit this object to the function `summary()`.

As a result, we get:

```
            Df Sum Sq Mean Sq F value Pr(>F)
pos_sibling  3  253.9   84.65  0.7719 0.5128
Residuals   88 9651.0  109.67
```

We consider only the two last columns: The F-test is not significant, because the value shown in `Pr(>F)` is bigger than $\alpha = 0.05$.

In SPSS, we first have to exclude from the analysis all children whose sibling position is bigger than four. To do this, we use the same sequence of commands (Transform – Recode into Different Variables...) as in Example 9.7. This gets us to the window shown in Figure 9.7, where we drag and drop the character sibling position to the panel Input Variable -> Output Variable:. In the section Output Variable, we type pos_sib in the text field Name and click Change. A click on Old and New Values... gets us to the window shown in Figure 9.8, where we tick Range, value through HIGHEST: in the section Old Value and type the value 5 in the text box. Now, we tick System-missing in the section New Value and click Add. Next, we tick All other values in the section Old Value and Copy old value(s) in the section New Value. A click on Add and Continue gets us back to the previous window. We click OK and the transformation is done. To conduct the analysis, we select

Analyze
  Compare Means
    One-Way ANOVA...

In the resulting window, shown in Figure 10.2, we drag and drop the character Everyday Knowledge, 1st test date to the panel Dependent List: and the character pos_sib to the panel Factor: (we have already done this in Figure 10.2). By clicking OK, we get the results shown in Table 10.4.

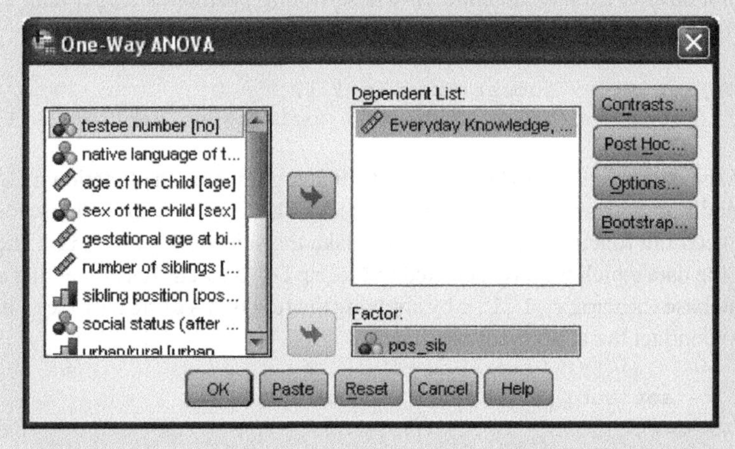

Figure 10.2 SPSS-window for the one-way analysis of variance.

Table 10.4 SPSS-output of the table of variances for one-way analysis of variance (model I) in Example 10.1.

ANOVA

Everyday Knowledge, 1st test date (T-Scores)

	Sum of Squares	df	Mean Square	F	Sig.
Between Groups	253.949	3	84.650	.772	.513
Within Groups	9651.040	88	109.671		
Total	9904.989	91			

In the output shown in Table 10.4, we consider only the two last columns: the F-test is not significant, as the p-value indicated in the column Sig. is bigger than 0.05.

The children in different sibling positions do not differ with respect to the character *Everyday Knowledge, 1st test date.*

It is usually advised to test the presumption that there are homogeneous variances, i.e. that there is *variance homogeneity*, before performing one-way analysis of variance. However, by analogy to the problem of the *t*-test, we suggest not to do so here. Instead, in the case of equal cell frequencies, i.e. $n_i = n$, we recommend not applying the one-way analysis of variance, but to run the analysis in accordance to a special case of *Hotelling's* T^2. This test statistic, which is based on all pair-wise mean differences of the a factor levels, is intended for more than one character and will be discussed in more detail in Section 12.2; it is not influenced by deviations from variance homogeneity (see Moder, 2007, 2010).

Example 10.1 – continued

In the following, we assume that we have actually sampled data according to Example 10.6; that is we have tested each of $n = 50$ first-, second-, third-, and fourth-born male high school students in 6th class, say for instance with the subtest *Social and Material Sequencing* from the intelligence test-battery AID 2. We can now calculate Hotelling's T^2 with the data of *Example 10.1* (see Chapter 1 for its availability) regarding the factor *sibling position* (with the levels first, second, third, and fourth position) and therefore test the null hypothesis H_0: $\mu_1 = \mu_2 = \mu_3 = \mu_4$.

In **R**, we define a new function with the help of `function()` by typing

```
> hotT2.aov <- function(X, nf, nrep) {
+   require(MASS)
+   ncol <- nf*(nf-1)/2
+   D <- matrix(1:(nrep*ncol), nrep, ncol)
+   k = 0
+   for(i in 1:(nf-1))
+     {
+     for(j in (i+1):nf)
+       {
+       k <- k+1
+       D[,k] <- X[,i]-X[,j]
+       }
+     }
+   Xq <- apply(D, 2, mean)
+   V <- var(D)
+   rank <- qr(V)$rank
+   IV <- ginv(V)
+   T2 <- t(Xq)%*%IV%*%Xq*nrep
+   df1 <- rank
+   df2 <- nrep-rank
+   F <- T2*(nrep-rank)/(rank*(nrep-1))
+   probf <- 1-pf(F, df1, df2)
+   cat("      df1      df2              T2              F          ProbF\n",
+       formatC(df1, width = 7), formatC(df2, width = 7),
+       formatC(T2, format = "f", width = 12,
+               digits = 5, flag = " "),
+       formatC(F, format = "f", width = 11,
+               digits = 5, flag = " "),
+       formatC(probf, format = "f", width = 11,
+               digits = 5, flag = " "),"\n")
```

```
+     res <- list(T2 = T2, F = F, df1 = df1, df2 = df2, probf = probf,
+                 Xq = Xq, D = D, V = V, IV = IV)
+ return(invisible(res))
+ }
```

i.e. we apply the function `function()` using the data matrix `X` as the first argument, which contains the observation values per factor level in its columns. As the second argument, we use, with `nf`, the number of factor levels $a = 4$ and, with `nrep`, the number of observation values n per factor level. The sequence of commands within the braces determines the procedure of the function and is not explained in detail here. Finally, we assign this function to the object `hotT2.aov`. Next, we enable access to the database *Example_10.1* (see Chapter 1) by using the function `attach()`. Then we prepare the data for the analysis by typing

```
> pos1 <- sub4[which(unclass(pos_sibling) == 1)]
> pos2 <- sub4[which(unclass(pos_sibling) == 2)]
> pos3 <- sub4[which(unclass(pos_sibling) == 3)]
> pos4 <- sub4[which(unclass(pos_sibling) == 4)]
> pos.dat <- cbind(pos1, pos2, pos3, pos4)
```

i.e. we create a vector of the observation values in the character *Social and Material Sequencing* (`sub4`), separately for each of the factor levels of the factor *sibling position* (`pos_sibling`). The function `which()` serves to make sure that only certain research units are considered. The function `unclass()` is needed to select the observation units which stem from the respective factor level of *sibling position* (`pos_sibling`). We assign each of the vectors consisting of the values selected in this way to a new object (`pos1` to `pos4`). Next, we submit these objects to the function `cbind()` for creation of an adapted data matrix. We assign this data matrix to the object `pos.dat`. Now, we finally conduct the analysis using Hotelling's T^2 by typing

```
> hotT2.aov(pos.dat, nf = 4, nrep = 50)
```

i.e. we apply our function `hotT2.aov()` and submit the data matrix `pos.dat` as the first argument to the function. As the second argument, we use the number of factor levels $a = 4$, with `nf = 4`, and as the third one the number of children per factor level $n = 50$, with `nrep = 50`.

As a result, we get:

```
df1      df2          T2          F       ProbF
  3       47     6.15429    1.96770     0.13170
```

In SPSS, we first have to rearrange the data of data set *Example 10.1*. Therefore, we select

Data
    Restructure...

and, in the resulting window (Figure 10.3), we tick Restructure selected cases into variables. With a click on Next we come to the following window (not shown here), where we drag and drop the character sibling position to the field Identifier Variable(s):. After clicking Next, three more windows appear successively (all without figure here), in which we just confirm everything by clicking Next, and in the last one (not shown here) we click Finish. This brings us to a small window showing a warning, to which we react by clicking OK.

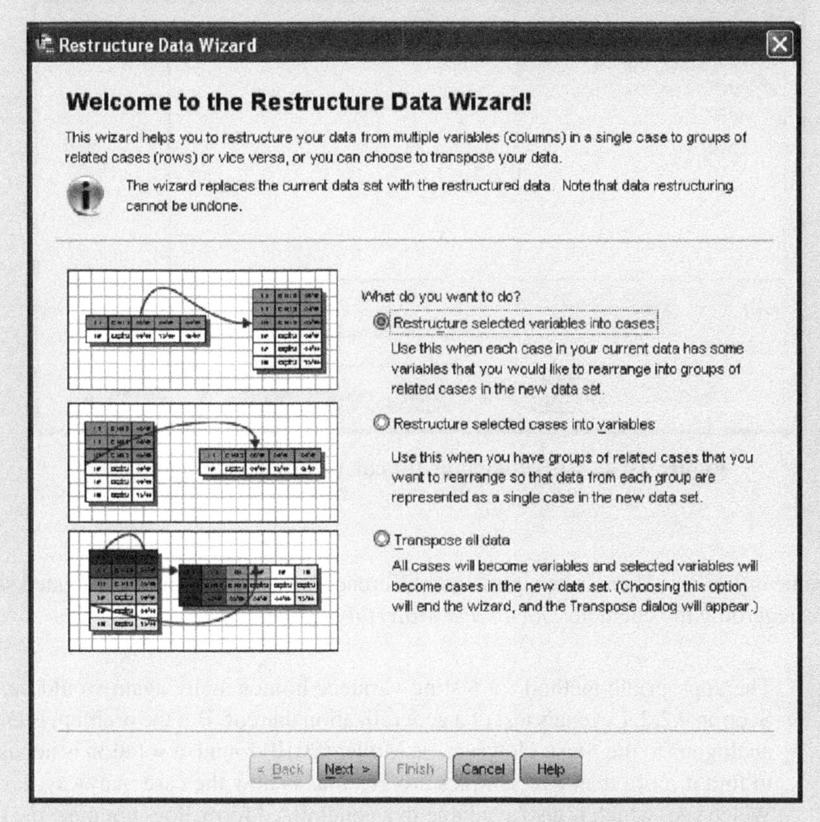

Figure 10.3 SPSS-window for restructuring data.

Now we select

Data

    Transpose...

In the resulting window (not shown), we drag and drop all characters from sub4_t1.1 to sub4_t1.50 to the field Variable(s): and click OK. As a consequence, a small window (not shown) with a warning appears. We click OK. In this way, we obtain a new data sheet with which we can continue working. In order to conduct Hotelling's T^2, we select

Analyze

    Scale

        Reliability Analysis...

In the resulting window (Figure 10.4) we drag and drop the variables var001 to var004 to the field Items:. We click Statistics... and in the resulting window (not shown here) we tick

Hotelling's T-Square. After clicking Continue and OK, we get a table, where we find the main result with regard to Hotelling's T^2: the p-value equals 0.132.

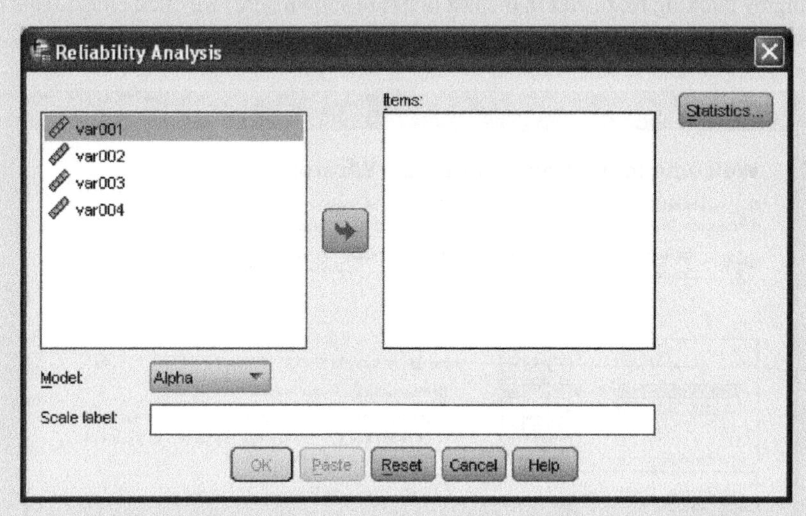

Figure 10.4 SPSS-window for calculating Hotelling's T^2.

The p-value is 0.132, so there is no significant difference between the four investigated sibling positions regarding the character *Social and Material Sequencing*.

Master Doctor

The appropriate method for testing variance homogeneity again would be, as in Section 9.2.2, Levene's test or a generalization thereof. But the problem is entirely analogous to the t-test. However, as Moder (2010) found, a solution is not as easy to find if more than two samples are given as is now the case. Anyway, even the Welch test, which is also available in a generalized form, does not meet the type-I risk if the distributions differ greatly between the samples. However, one-way analysis of variance itself is very sensitive to deviations from the presumption of variance homogeneity. Finally, also the generalization of the U-test, namely the Kruskal–Wallis H-test (see below), does not only test differences in the location of the relevant populations' distributions.

Even when the application of Hotelling's test statistic T^2 will be commonly used in the given context, there is a problem of principle. Namely, that for all different types of analysis of variance, especially in cases of multi-way (see below, starting from Section 10.4.4) and multivariate analyses of variance (see Section 13.3), there are presumptions referring to several parameters of variability of the characters in question; just for example, referring to the variances. Certainly there are tests to examine these; but such tests are neither independent from the test of the actual questions, i.e. analysis of variance, nor are they generally robust with regards to violations of certain distributional assumptions as is the case with analysis of variance itself. But what aggravates the problem is the fact that the researcher, if he/she uses these pre-tests in spite of everything, does not have an alternative to analysis of variance, given that any pre-test results in significance. In particular, non-parametric methods are just as unsatisfactory a solution, because, as we have pointed out, they test an entirely different null hypothesis from that intended by analysis of variance; despite the fact that for multi-way analysis of variance there are barely any adequate non-parametric methods, and for multivariate analysis of variance there are even fewer. In the practice of psychological research, this results in the following unacceptable consequences: either to disclaim the respective presumption or, in the case of the application of such pre-tests, to ignore their possible significant results – without realization that the result of the analysis of variance could be consequently interpreted completely contrary to the actual issue.

In this book, we therefore recommend proceeding as follows:

1. Where possible use procedures that are mostly independent of violations from homogeneity – so far this has been the Welch test and Hotelling's test with the test statistic T^2.
2. Where there are no such methods (yet), identify homogeneity or heterogeneity on a descriptive basis according to certain rules of thumb (especially with multi-way analyses of variance) or apply the appropriate pre-test, albeit without guarantee of a precise total type-I risk, and without accurate knowledge of the actual type-I risk for the primarily interesting test of comparing the means, respectively.
3. If there is heterogeneity, the researcher should either exclude individual factors (in terms of multi-way analyses of variance) and/or individual factor levels and/or individual characters (in terms of multivariate analyses of variance) from the analysis. Of course that is the strategy of *a posterior*[2] model fitting, which can lead to distorted results in the direction of one or the other hypothesis. If such an exclusion is not possible, the actual posed research question cannot be answered. The researcher then may have to change the (null) hypothesis or change from a study-wise to a comparison-wise type-I risk (see also below).

As concerns the abovementioned rule of thumb, we recommend considering differences between the standard deviations of two factor levels of larger than a factor of 1.5 as being too large to apply the methods of analysis of variance. Where, as concerns the one factor level, 68.3% of the population are located between $\mu - 1$ and $\mu + 1$ (see Example 6.8), as concerns the other factor level, only 49.5% of the population are located in this area (given the null hypothesis is true); thus, a comparison of the two samples regarding a measure of location on its own does not seem convincing at all.

As already mentioned in Section 8.2.3 and Section 9.1, we often have two outcomes of each research unit, which are statistically interpreted as each stemming from a separately modeled random variable but, with respect to content, describe the same thing (bear the example in mind that some treatment effects should be ascertained with a *pre* and *post* design). Then it is simplest, also with more than two samples, to trace the data to a single character. That is, given interval-scaled characters, we calculate the difference of the respective pair of outcomes per research unit and use them as values $y_{i1}, y_{i2}, \ldots, y_{in_i}$.

[2] Latin: from those things that follow

10.4.1.2 *Post hoc* tests

Analysis of variance is unsatisfying insofar as, in the case of a significant result, the researcher does not know between which means and samples, respectively, and therefore between which populations, there are differences. That is the price of having a study-wise type-I risk (and also a study-wise type-II risk).

Therefore, the inventory of statistical methods provides so-called *post hoc*[3] *tests*. In the simplest case this is the LSD test, which was discussed in Section 10.3 regarding multiple comparisons of means. However, this test uses comparison-wise risks and is therefore inappropriate. On the other hand, the so-called *Newman–Keuls procedure* also tests pair-wise, but with a study-wise type-I risk (and a comparison-wise type-II risk). Following an analysis of variance, where the researcher has already decided on a study-wise type-I risk, this procedure is more appropriate.

Basically, the Newman–Keuls procedure is only designed for use with equal sample sizes, as a consequence of which the approximations often used in computer programs do not guarantee that the type-I risk will be precisely met. In any case, using computer programs the analysis is easily done; hence, no formulae will be given here.

Computer programs are also available for planning the study; their application is, however, a moot point, because the researcher will likely plan the study in terms of (one-way) analysis of variance.

Bachelor **Example 10.7** Do the mean test scores in *Coding and Associating, 1st test date* differ with regard to the *social status* (Example 1.1)?

On the one hand, the mentioned subtest is non-verbal and thus, no disadvantages have to be expected for members of lower social classes (see, for example, Cattell, 1987); on the other hand, there are school-atypical tasks with particular materials that members of certain social classes may be more familiar with than others. So, we test the null hypothesis H_0: $\mu_i = \mu$ for all i, knowing already, from Example 5.3, that there are $a = 6$ factor levels regarding the factor *social status*. The alternative hypothesis is H_A: $\mu_i \neq \mu_l$ for at least one $i \neq l$. We also know from Example 5.3 that the corresponding n_i have the sizes: 5, 16, 12, 25, 30, 12. This is different from the calculation according to Example 10.6, because the data were originally sampled for another research question – given the same precision requirements we get $n = 59$; nevertheless, we test with a type-I risk of $\alpha = 0.05$. Following the same procedure as in Example 10.1, we obtain a significant result this time (the p-value is 0.006). We now want to locate the differences precisely and therefore apply the Newman–Keuls procedure.

In **R**, we apply the package `agricolae`, which we first install (see Chapter 1) and then load using the function `library()`. Furthermore, we enable access to the original database *Example_1.1* (see Chapter 1) using the function `attach()`. Now, we type

```
> aov.2 <- aov(sub7_t1 ~ social_status)
> SNK.test(aov.2, trt = "social_status")
```

[3] Latin: after the fact

i.e. we first conduct an analysis of variance, applying the function `aov()`. We assign the result of this analysis to the object `aov.2`. Next, we apply the function `SNK.test()`, to which we submit the object `aov.2` as the first argument, and as the second one, with `trt = "social_status"`, the factor of interest *social status*.

As a result, we get (shortened output):

```
                            sub7_t1  std.err replication
lower classes              44.60000 5.045790           5
lower middle class         52.92000 2.391596          25
middle classes             53.23333 2.016987          30
single mother in household 51.83333 2.312133          12
upper classes              62.58333 2.230839          12
upper lower class          47.62500 2.411561          16

alpha: 0.05; Df Error: 94

Critical Range
       2         3         4         5         6
8.480896 10.171840 11.172138 11.880532 12.426881

Harmonic Mean of Cell Sizes  11.94030

Different value for each comparison
Means with the same letter are not significantly different.

Groups, Treatments and means
a       upper classes                   62.58333
 b      middle classes                  53.23333
 b      lower middle class              52.92
 b      single mother in household      51.83333
 b      upper lower class               47.625
 b      lower classes                   44.6
```

We thus obtain two disjunctive groups of factor levels: the ‘upper classes’ on one hand and all other factor levels of *social status* on the other hand.

In SPSS, we have to (preferably after knowing that the analysis of variance yields a significant result) tick Post Hoc... in the window shown in Figure 10.2 and obtain Figure 10.5. There, we click S-N-K (for Student–Newman–Keuls – so called the Newman–Keuls procedure in SPSS) followed by Continue – the type-I risk is already set to $\alpha = 0.05$ according to the default settings. Clicking OK gets us to the result shown in Table 10.5. This table summarizes the groups, which do not differ in their means. Here, two such combined groups result. We can see that the three samples for the categories ‘lower classes’, ‘upper lower class’ and ‘upper classes’ are not contained all together in the same combined group; i.e. the two former groups differ significantly from the latter. Apart from that, there are no significant differences. Figure 10.6 illustrates these results graphically.

One-Way ANOVA: Post Hoc Multiple Comparisons

Equal Variances Assumed

LSD S-N-K Waller-Duncan
Bonferroni Tukey Type I/Type II Error Ratio: 100
Sidak Tukey's-b Dunnett
Scheffe Duncan Control Category: Last
R-E-G-W F Hochberg's GT2 Test
R-E-G-W Q Gabriel 2-sided < Control > Control

Equal Variances Not Assumed

Tamhane's T2 Dunnett's T3 Games-Howell Dunnett's C

Significance level: 0.05

Continue Cancel Help

Figure 10.5 SPSS-window for the calculation of *post hoc* tests.

Table 10.5 SPSS-output of Newman–Keuls procedure in Example 10.7.

Coding and Associating, 1st test date (T-Scores)

Student-Newman-Keuls[a, b]

social status (after Kleining & Moore according to occupation...	N	Subset for alpha = 0.05	
		1	2
lower classes	5	44.60	
upper lower class	16	47.63	
single mother in household	12	51.83	51.83
lower middle class	25	52.92	52.92
middle classes	30	53.23	53.23
upper classes	12		62.58
Sig.		.264	.064

Means for groups in homogeneous subsets are displayed.

a. Uses Harmonic Mean Sample Size = 11.940.

b. The group sizes are unequal. The harmonic mean of the group sizes is used. Type I error levels are not guaranteed.

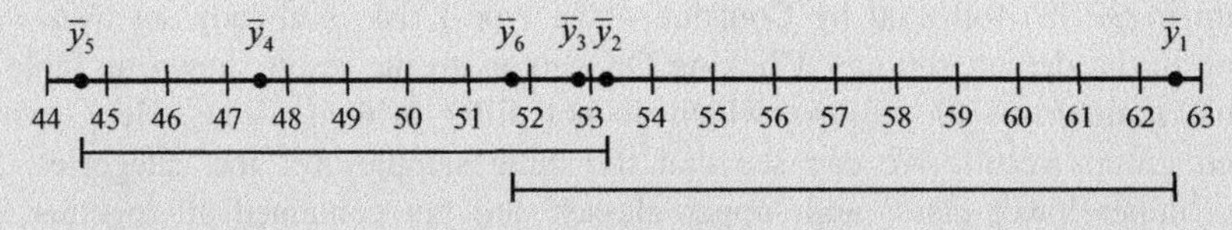

Figure 10.6 Graphical summary of the results from Table 10.5.

R and SPSS provide different results. This is because there are unequal sample sizes and the two program packages obviously adjust the test statistic differently. We recommend interpreting the results from SPSS. With respect to the subtest *Coding and Associating*, children from 'lower classes' and the 'upper lower class' achieve significantly lower mean test scores than children from 'upper classes'.

Master Doctor

As can be seen in Figure 10.5, there are many other *post hoc* tests. We will, however, not deal with them (but see Hsu, 1996). Note the following though: sometimes we want to compare all $a - 1$ means (and populations respectively) with a specific reference/standard/control mean, i.e. the ath mean. In that case, the so-called *Dunnett procedure* is the method of choice; it also tests with a study-wise type-I risk (the type-II risk is comparison-wise). For instance, in Example 10.7, we could determine 'single mother in household' as the reference population. Compared with the Newman–Keuls procedure, there is a difference regarding planning the study, insofar as it is optimal to provide more research units n_s for the standard than for the other factor levels, each having n units; that is $n_s = n\sqrt{a-1}$. It is strongly advised to not apply the *Duncan test*: with this test, it is not known whether the risks are to be interpreted as study- or comparison-wise.

10.4.1.3 Model II

Regarding case 2, i.e. model II, it occurs far less frequently than model I in psychology.

Master

The distinction between model I and model II has no effect on hypothesis testing in terms of one-way analysis of variance and, therefore, it does not matter if model I is always selected due to a lack of knowledge that there is actually also a model II of analysis of variance in existence. Nevertheless, model II should be addressed here for the sake of completeness. In this case, the model is as follows:

$$\boldsymbol{y}_{iv} = E(\boldsymbol{y}_{iv}) + \boldsymbol{e}_{iv} = \mu_i + \boldsymbol{e}_{iv} = \mu + \boldsymbol{a}_i + \boldsymbol{e}_{iv} \qquad (i = 1, 2, \ldots, a; \; v = 1, 2, \ldots, n_i) \tag{10.3}$$

While the (main) effects a_i in model I of Formula (10.1) are fixed statistics, the $\boldsymbol{a}_i$ are corresponding random variables; they have a mean, $E(\boldsymbol{a}_i)$, and a variance $\sigma^2(\boldsymbol{a}_i)$.

In addition to the presumptions of model I, the following must also be provided to derive a test:

- The $\boldsymbol{e}_{iv}$ are independent of the $\boldsymbol{a}_i$.
- $E(\boldsymbol{a}_i) = 0$ for all i.
- The $\boldsymbol{a}_i$ all have the same variance, σ_a^2.

It follows that the random variables $\boldsymbol{y}_{iv}$ do not have the same variance as the errors $\boldsymbol{e}_{iv}$, but have the variance $\sigma_a^2 + \sigma^2$.

The variance of $\boldsymbol{y}_{iv}$ therefore consists of the two components σ_a^2 and σ^2: the so-called *variance components*. It can be shown that the $\boldsymbol{y}_{iv}$ within a cell are not independent as in model I. This dependency is quantified by the so-called intra-class correlation coefficient, which quantifies the proportion of factor A in the variance of the character (see more details in Section 11.3.4).

Master

The column for the expected values of the theoretical analysis of variance table looks different compared to model I (see Table 10.6).

Table 10.6 Theoretical analysis of variance table of one-way analysis of variance for model II ($n_i = n$).

Source of variation	*SS*	*df*	*MS*	*E*(*MS*)
Factor A	$SS_A = \sum_{i=1}^{a} \sum_{v=1}^{n_i} (\bar{y}_i - \bar{y})^2$	$a - 1$	$MS_A = \frac{SS_A}{a-1}$	$\sigma^2 + n\sigma_a^2$
Residual	$SS_{\text{res}} = \sum_{i=1}^{a} \sum_{v=1}^{n_i} (y_{iv} - \bar{y}_i)^2$	$N - a = a(n-1)$	$MS_{\text{res}} = \frac{SS_{\text{res}}}{N-a}$	σ^2
Total	$SS_{\text{t}} = \sum_{i=1}^{a} \sum_{v=1}^{n} (y_{iv} - \bar{y})^2$	$N - 1 = an - 1$		

It is not possible to estimate the (main) effects in model II, because it is about random variables, not about parameters. Instead, we estimate the variance component of factor A, σ_a^2. There are several methods, but the simplest is the *analysis of variance method*. One equates the *MS* calculated from the data with the expected values *E*(***MS***), by replacing the variance components σ^2 and σ_a^2 with their estimates s^2 and s_a^2.

The null hypothesis states again that factor A with its a levels has no influence on the random variable $\boldsymbol{y}$ which models the investigated character y. This means that there is no variability of the factor levels, i.e. that H_0: $\sigma_a^2 = 0$. By contrast, the alternative hypothesis is H_{A}: $\sigma_a^2 > 0$. Accordingly, as given by Table 10.5, an F-test can be performed to test the null hypothesis; $\boldsymbol{F} = \frac{\boldsymbol{MS}_A}{\boldsymbol{MS}_{\text{res}}}$ is F-distributed with $df_1 = a - 1$ and $df_2 = N - a$ degrees of freedom. So, this is again about comparing two mean squares, which have the same expected value given the null hypothesis. If the F-test provides a significant result, the two mean squares are interpreted as being not equal in the population; from this we infer $\sigma_a^2 > 0$.

Unfortunately, there is no getting around taking both the presumption of variance homogeneity and independence of all random variables in Formula (10.3) untested.

Example 10.8 The extent of differences in effectiveness of psychotherapists should be tested

The experimental design requests that clients with the same diagnosis according to ICD-10 (e.g. 'Other anxiety disorders', F41[4]) are treated by different psychotherapists. The effectiveness of the psychotherapy per client will be measured, where effectiveness is operationalized as the difference of the test score in an anxiety questionnaire before and after the six months of therapy. If the psychotherapists were typical representatives of different schools (such as: Psychoanalysis, Behavior Therapy, Gestalt Therapy, Client-centered Therapy, Systemic Therapy), we would have – given that the assignment of the clients to the psychotherapists would happen by chance – a model I, because fixed factor levels would be given. In such a case, we would be interested in the differences between these schools. However, we want to determine the extent of differences regarding the effectiveness of psychotherapists in this particular case, irrespective of school affiliation.

Specifically, clients with the aforementioned diagnosis were randomly assigned to $a = 22$ psychotherapists. These were in turn taken randomly from the sufficiently large population of all the country's psychotherapists. Now, the individual psychotherapists represent the factor levels of a random factor and therefore, a model II situation is at hand.

We use the data of Example 10.8 (see Chapter 1 for its availability), which contains, besides the character *effectiveness* and the factor *therapist*, the characters *sex of the client*, and *sex of the therapist* per client. For reasons that are not related to *effectiveness*, not all psychotherapists had the same number of clients, so the sample sizes n_i ($i = 1, 2, \ldots, 22$) are not all equal; instead the number of clients varies between 20 and 30. The test scores of the anxiety questionnaire are factor scores according to a factor analysis (see Section 15.1) and therefore have decimal places. The data are just simply collected; that is, unfortunately no planning of the study was done beforehand. We decide on a type-I risk of $\alpha = 0.01$; indirectly, compared to the possible type-I risk of 5%, this results in a larger, although unknown type-II risk.

By way of exception we decide to apply the appropriate pre-test to demonstrate its calculation. Therefore, we consider first, using the generalized Levene's test, whether the variances are homogeneous; i.e. whether the null hypothesis holds, H_0: $\sigma_i^2 = \sigma_l^2$ for all $i \neq l$ (H_1: $\sigma_i^2 \neq \sigma_l^2$ for at least one pair i and l, $i \neq l$). We take a type-I risk of $\alpha = 0.01$ here as well.

In **R**, we apply the package `lawstat`, which we load after its installation (see Chapter 1) using the function `library()`. Moreover, we enable access to the database *Example_10.8* using the function `attach()`. For calculating Levene's test, we type

```
> levene.test(effect, group = therapist, location = "mean")
```

i.e. we use the argument `effect` for the character *effectiveness*, the argument `group = therapist` for the character *therapist*, and `location = "mean"` as one of different possible calculation procedures in the function `levene.test()`.

[4] *The ICD-10 Classification of Mental and Behavioural Disorders* (World Health Organization, 1993).

As a result, we get:

```
classical Levene's test based on the absolute deviations from the
mean

data:  effect

Test Statistic = 0.67, p-value = 0.8639
```

As the result of the Levene's test is not significant, we continue calculating the analysis of variance. We type

```
> aov.3 <- aov(effect ~ 1 + Error(therapist))
> summary(aov.3)
```

i.e. we apply the function `aov()`, and determine that the character *effectiveness* (`effect`) should be analyzed with regard to the not fixed (`1`), but random (`Error`) factor *therapist* (`therapist`). We assign the result of this analysis to the object `aov.3`. Applying the function `summary()` with the object `aov.3` as its argument, we request a summary of the results.

This yields:

```
Error: therapist
          Df Sum Sq Mean Sq F value Pr(>F)
Residuals 21 386.39  18.400

Error: Within
           Df Sum Sq Mean Sq F value Pr(>F)
Residuals 566 3247.5  5.7376
```

We calculate the F-value in question manually: $F = 18.400/5.7376 = 3.207$, with $df_1 = 21$ and $df_2 = 566$ degrees of freedom. Now, we can determine the critical F-value $F(21, 566{:}\ 0.99) = 1.89$ by consulting Table B4 in Appendix B. Alternatively, we can calculate the p-value which corresponds to the value $F = 3.207$ in the following way, using **R**:

```
> pf(3.207, df1 = 21, df2 = 566, lower.tail = FALSE)
```

i.e. we apply the function `pf()`, using the F-value as the first argument and the degrees of freedom df_1 and df_2 as the second and the third argument. With `lower.tail = FALSE`, we request the corresponding p-value.

As a result, we get:

```
[1] 2.432629e-06
```

Thus, the p-value equals 0.000.

In SPSS, we can calculate Levene's test only in combination with the analysis of variance as the actual procedure of analysis. This is done by choosing a certain option within this procedure. Selecting

Analyze
General Linear Model
Univariate...

we reach the window shown in Figure 10.7. There, we drag and drop the character effectiveness into the field Dependent Variable:. The character therapist, as the factor, is dragged and dropped into the field Random Factor(s):. Next, we click Options... and in the resulting window (shown in Figure 10.8), we tick Homogeneity tests. If we want to apply only the rule of thumb described above, we can tick Descriptive statistics instead. In this case, the output would show e.g. all of the standard deviations per factor level. Clicking Continue and OK, we obtain the results: a non-significant Levene's test at first, and then the result of the analysis of variance shown in Table 10.7.

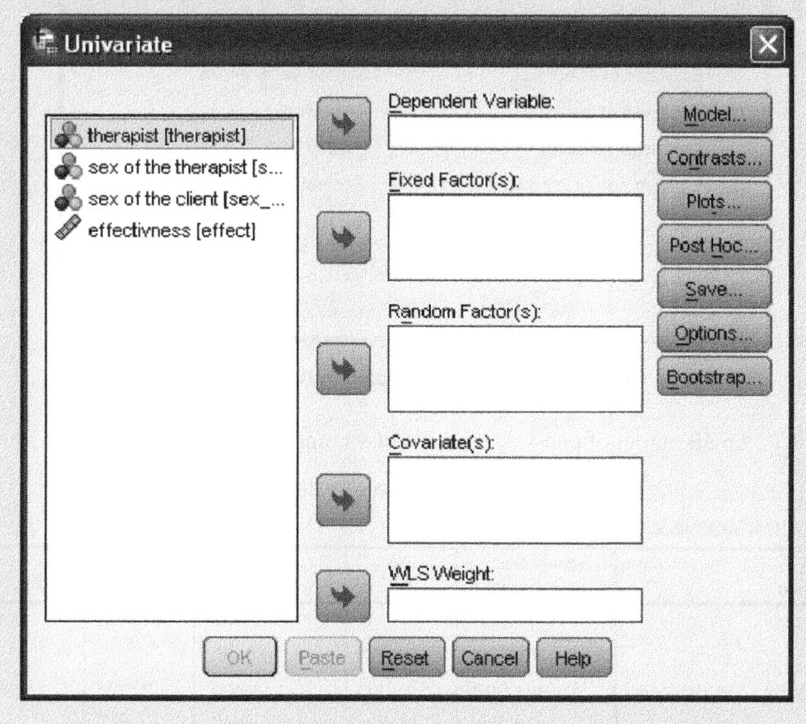

Figure 10.7 SPSS-window for calculating analyses of variance.

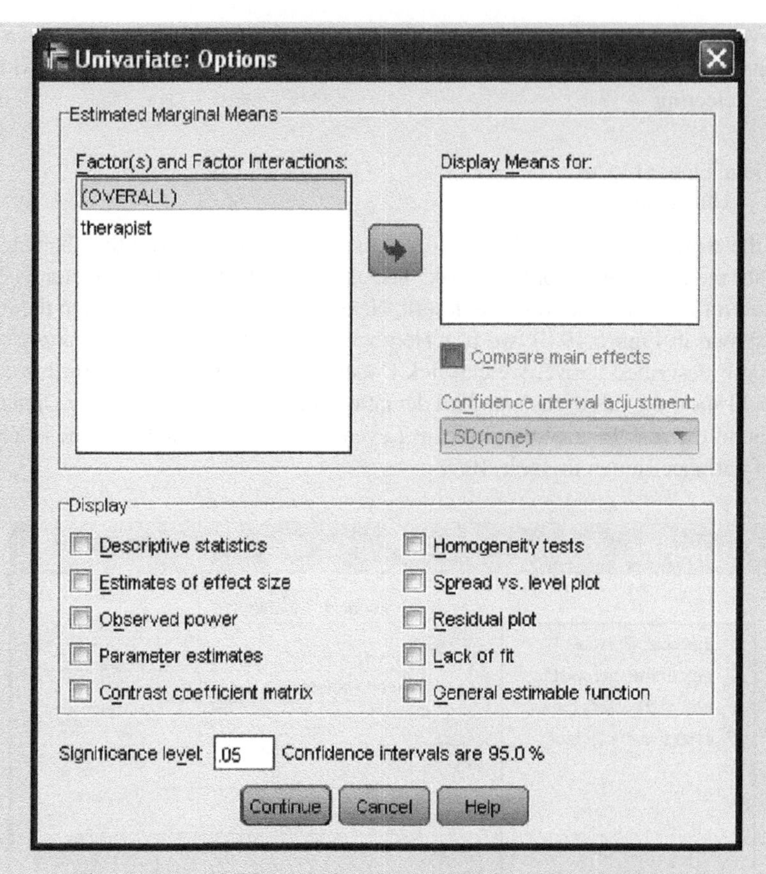

Figure 10.8 SPSS-window for calculation of Levene's test.

Table 10.7 SPSS output of one-way analysis of variance (model II) in Example 10.8.

Tests of Between-Subjects Effects

Dependent Variable: effectiveness

Source		Type III Sum of Squares	df	Mean Square	F	Sig.
Intercept	Hypothesis	14477.693	1	14477.693	802.165	.000
	Error	385.789	21.375	18.048[a]		
therapist	Hypothesis	386.392	21	18.400	3.207	.000
	Error	3247.474	566	5.738[b]		

a. .972 MS(therapist) + .028 MS(Error)
b. MS(Error)

The rows Intercept with Hypothesis and Error are not of interest; the statistic SS_{res} is labeled as Error in SPSS.

Because the F-test is significant (the p-value is smaller than $\alpha = 0.01$), the variance of the psychotherapists with regard to therapy effectiveness is significantly different from zero.

Both planning a study design and the calculation of the result-based type-II risk β^* are more difficult for model II than for model I, and are not provided (yet) by the computer programs used here.

10.4.2 One-way analysis of variance for ordinal-scaled characters

If the examined character is ordinal-scaled, the application of one-way analysis of variances is inappropriate. Instead, a generalization of the U-test, the (*Kruskal–Wallis*) *H-test* is adequate.

It tests the null hypothesis that the populations under discussion, with the distribution functions $F_i(z)$, $i = 1, 2, \ldots, a$, are equal. For the application of the H-test, everything is analogous to the U-test (see Section 9.2.1).

The derivation of the H-test statistic to test this null hypothesis is based on the assignment of ranks for the outcomes of all samples in total. Essentially, the F-test of Formula (10.2) is then applied to these, with the variances for the numbers 1 to $N = \sum n_i$ always being equal and therefore known; this is the reason why there is a χ^2-distributed test statistic instead of an F-distributed one.

The calculation of the H-test is very simple to perform using statistical computer programs.

A measure for the effect size to be estimated is not commonly used for the H-test.

As already stated in Section 9.2.1, planning a study for non-parametric methods is difficult, because the alternative hypothesis can hardly be quantified; we will, therefore, not go into that here (but see e.g. Brunner & Munzel, 2002). *Post hoc* tests are uncommon.

Bachelor

Example 10.9 Three different techniques regarding an opinion poll will be compared

Opinion polls, implemented using appropriate attitude questionnaires, always suffer from the fact that the response rate is much lower than desired. For this reason, certain 'supporting measures' are often used. Given a factor *method* – (1) normal postal sending; (2) previous announcement of the questionnaire by telephone; (3) promise of remuneration in the case of returning the questionnaire – the character in question in a study (by randomization of the addressees) is the number of days until returning the questionnaire. For the case that the questionnaire was returned only after a reminder, the measurement value '900 + number of days after warning' is given; for the case that there is no return at all, the measurement value is 999.

An interval scale for this character is not at all plausible, not only because of the arbitrary encoding of the return period, but also because of the different reactions: returning sooner or later, returning sooner or later only after warning, and not returning at all 'impede' the investigation in a quantifiable way; anyhow, an ordinary ranking is obviously possible. The character is ordinal-scaled.

We use the data of *Example 10.9* (see Chapter 1 for its availability). It contains the character *duration* besides the factor level of the factor *method* for each returned questionnaire.

In **R**, we conduct the H-test by typing

```
> kruskal.test(duration ~ method, data = Example_10.9)
```

i.e. we apply the function `kruskal.test()` and specify that the character `duration` should be analyzed with regard to the factor `method`. With `data = Example_10.9`, we determine the database which is to be used.

As a result, we get:

```
        Kruskal-Wallis rank sum test

data:  duration by method
Kruskal-Wallis chi-squared = 66.6024, df = 2, p-value = 3.447e-15
```

The p-value thus equals almost zero.

Now, we calculate the mean ranks by typing

```
> with(tapply(rank(duration), method, mean), data = Example_10.9)
```

i.e. we apply the function `rank()` to the character `duration`. We use this function as the first argument in the function `tapply()`. With `method` we indicate the factor, and with `mean` we request the mean per factor level. Finally, we determine the database which is to be used by applying the function `with()`.

As a result, we get the mean ranks as shown in Table 10.8.

In SPSS, the H-test can be found in

Analyze
  Nonparametric Tests
    Independent Samples...

In the resulting window (not shown here), we select the tab Fields, which gets us to a window very similar to the one shown in Figure 8.12. We drag and drop duration to Test Fields: and method to Groups:. Now, we select the tab Settings and, as a consequence, we reach the window shown in Figure 9.6. There, we choose Customize tests and tick Kruskal-Wallis 1-way ANOVA (k samples). After a click on Run, the table Hypothesis Test Summary is produced in the output window (not shown here). We just double-click this table, and a window with a table opens. Here, we find the main result in Asymptotic Sig. (2-sided test). The p-value equals 0.000.

For calculating the mean ranks, we select the sequence of commands (Transform – Rank cases...) from Example 5.10. In the resulting window (not shown here) we drag and drop the character duration to the field Variable(s): and click OK. As a consequence, the character Rank of duration is added to the Data view. Now, we select the sequence of commands (Analyze – Compare Means – Means...) from Example 5.1, and reach the window shown in Figure 5.1. There, we drag and drop the character Rank of duration to the field Dependent List: and the character method to the field Independent List:. After a click on OK, we obtain Table 10.8 as a result.

Table 10.8 SPSS-output of the mean ranks in Example 10.9.

Report

Rank of duration

mehod	Mean	N	Std. Deviation
normal	159.92568	74	49.782519
telephone	143.50595	84	56.832797
remuneration	77.63333	90	68.709157
Total	124.50000	248	69.412296

Results show significant differences between the three techniques, with number (3), i.e. financial reward, leading to the shortest return periods: this can be seen from the mean ranks. The mean rank is lowest for (3) and it has to be understood as follows: the greater the *duration* until return, the higher the rank.

10.4.3 Comparing more than two populations with respect to a nominal-scaled character

To generalize from Sections 9.2.2 and 9.2.3, we sometimes are interested in more than two samples – either for a dichotomous or a multi-categorical (nominal)-scaled character. In that case, the null hypothesis is H_0: $p_{ij} = p_{lj}$ for all j in all a samples and in all sample pairs i and l, respectively; the alternative hypothesis is H_1: $p_{ij} \neq p_{lj}$ for at least one j and one sample pair i and l. Then, the test statistic in Formula (9.3) applies again. In the case of a significant result, the same problem arises as in analysis of variance; the χ^2-test is an overall test, too. As concerns interpretation, the best way is to consider the largest (relative) deviations $o_{ij} - e_{ij}$ (or their squared values, respectively).

Bachelor **Example 10.10** Do four different schools of psychotherapy lead to different satisfaction ratings of the clients?

We are interested in the four schools Psychoanalysis, Behavior Therapy, Self-centered Psychotherapy, and Systemic Therapy. Sixty clients with similar complaints ('anxiety') were in turn alternately assigned to one of the four schools of psychotherapy, and were asked about their satisfaction regarding the success of the therapy after eight weeks. The response options were 'very satisfied', 'medium', and 'not satisfied'. Type-I risk is = 0.05.

The analysis of the data of *Example 10.10* (see Chapter 1 for its availability), analogous to Example 9.9, provided a p-value of 0.106; the null hypothesis has to be accepted.

As for all tests with χ^2-distributed test statistics, we will specify a measure for the estimated effect size in Section 11.3.5.

Summary
If there are more than two populations from which independent samples are drawn, the goal may be to select the 'best' of them, such as the one with the largest mean; this leads to *selection procedures*. On the other hand, it can be about comparing all, for example, means; this leads, if it really is about means, to *analysis of variance*, and in exceptional cases to *multiple comparisons of means*. These last methods make it clear that one can basically take a *study-wise* or a *comparison-wise risk* in hypothesis testing. This has to be taken into account when interpreting the results. Analysis of variance always leads to an F-test. For ordinal-scaled characters, there is the H-test; for nominal-scaled characters, again, the χ^2-test.

10.4.4 Two-way analysis of variance

In two-way analysis of variance we consider, in addition to factor A with its levels A_1, $A_2, \ldots, A_a$, a second factor B with its levels $B_1, B_2, \ldots, B_b$. The levels of the two factors can be combined in two ways: either according to a *cross classification* or according to a *nested classification/hierarchical classification.*

In cross classification – we write $A \times B$ – observations do not have to occur in all combinations of A and B, i.e. not in every one of the $a \cdot b$ combinations of factor levels A_iB_j of the a levels A_i of A with the b levels, B_j, of B. If they do, it is a *complete cross classification*; otherwise it is an *incomplete cross classification*. We always assume that incomplete cross classifications are 'connected' in the sense of the block designs given in Section 7.2.2.

In nested classification, the levels of B occur in exactly one level of A; we write $B \prec A$.

Again, the character of interest is modeled as a normally distributed random variable. That is to say, in a cross classification from each of the $a \cdot b$ populations, a random sample of size n_{ij} will be drawn. The outcomes of the random variables $\boldsymbol{y}_{ijv}$ are then y_{ijv}, $i = 1, 2, \ldots, a; j = 1, 2, \ldots, b; v = 1, 2, \ldots, n_{ij}$. The underlying parameters are μ_{ij} and σ^2_{ij}. The corresponding estimators are $\hat{\boldsymbol{\mu}}_{ij} = \bar{\boldsymbol{y}}_{ij}$ and $\hat{\boldsymbol{\sigma}}^2_{ij} = \boldsymbol{s}^2_{ij}$ (according to Section 8.5.1 and Formula (5.4)).

In addition to the distinction of cross and nested classification, we have to differentiate between fixed and random factors. It is possible that both factors are fixed, i.e. model I; or that both factors are random, i.e. model II; or that one factor is fixed and the other random. The last represents mixed-models.

10.4.4.1 Cross classification – model I

For model I of analysis of variance (in which both factors are fixed), the model equation for cross classification is:

$$\boldsymbol{y}_{ijv} = \mu + a_i + b_j + (ab)_{ij} + \boldsymbol{e}_{ijv} \quad (i = 1, \ldots, a; j = 1, \ldots, b; v = 1, \ldots, n_{ij}) \qquad (10.4)$$

In this equation, μ is the total mean, the terms a_i are the effects of the individual factor levels A_i of A, the main effects of A; the terms b_j are the effects of the individual factor levels B_j of B, the main effects of B, and the terms $(ab)_{ij}$ express the extent of the *interaction* between the levels A_i and the levels B_j – the so-called *interaction effects* of $A \times B$. Such an interaction effect is only defined if there are actually observations in the combinations of factor levels A_iB_j. If $n_{ij} = n$, there are equal cell frequencies. In this case, we have the data structure of Table 10.9. For reasons of simplicity we deal in the following formulas most of the time only with equal cell frequencies. Unequal cell frequencies need more complicated

Table 10.9 Data structure of a cross classification of factor A with a levels and factor B with b levels with equal cell frequencies ($n_{ij} = n > 1$).

	Levels of B			
Levels of A	B_1	B_2	...	B_b
A_1	y_{111}	y_{121}	...	y_{1b1}
	y_{112}	y_{122}	...	y_{1b2}
	$\vdots$	$\vdots$	$\vdots$	$\vdots$
	y_{11n}	y_{12n}	...	y_{1bn}
A_2	y_{211}	y_{221}	...	y_{2b1}
	y_{212}	y_{222}	...	y_{2b2}
	$\vdots$	$\vdots$	$\vdots$	$\vdots$
	y_{21n}	y_{22n}	...	y_{2bn}
$\vdots$	$\vdots$	$\vdots$	$\vdots$	$\vdots$
A_a	y_{a11}	y_{a21}	...	y_{ab1}
	y_{a12}	y_{a22}	...	y_{ab2}
	$\vdots$	$\vdots$	$\vdots$	$\vdots$
	y_{a1n}	y_{a2n}	...	y_{abn}

formulas. If such a case applies in an example, we deal with it appropriately using program packages.

The special thing with two-way (generally: multi-way) analysis of variance in the case of a cross classification is that it is possible to examine interactions between factors. If there were no interest in such an interaction, it would of course be possible to calculate a one-way analysis of variance regarding each of the factors A and B. However, if there also is interest in examining the interaction effect between the factors, an additional hypothesis needs to be tested, besides the two null hypotheses regarding the main effects. While the two null hypotheses $H_0(A)$ and $H_0(B)$ examining these main effects independently are stated and tested analogously to the one-way analysis of variance, the third null hypothesis $H_0(A \times B)$ is: there are no mean differences between the combinations of factor levels A_iB_j, which cannot be explained by the two main effects alone.

Example 10.11 Is there a different course of development in athletic stamina as concerns girls and boys in the age range of 12 to 14 years?

It is imaginable that 12- to 13-year-old girls show higher scores in athletic stamina on average than boys of that same age, but that boys show higher scores than girls at the age of 14. If this were indeed the case, we would have an interaction between age (factor A with three levels A_1: 12, A_2: 13, and A_3: 14) and sex (factor B with two levels, B_1: female and B_2: male). This would even be the case if the girls' superiority just weakens with age but never reaches inferiority.

For the derivation of hypothesis testing, it must be assumed:

- The $\boldsymbol{e}_{ijv}$ are normally distributed according to $N(0, \sigma^2)$, i.e. with the same expected value 0 and equal variance σ^2.
- The $\boldsymbol{e}_{ijv}$ are independent of each other.
- The cross classification is connected in the sense of Section 7.2.2.
- $a \geq 2$ and $b \geq 2$.

Depending on the model, these universal presumptions need to be supplemented.

Master As described for one-way analysis of variance, both the total mean squares as well as their degrees of freedom will be decomposed into components. In two-way analysis of variance these will be, in addition to the 'residual', the components which are dedicated to the factors A and B, respectively, and also to the component of interaction. Thus, the analysis of variance table becomes more extensive (Table 10.10). It is, however, identical for all models in the theoretical analysis of variance table, with the exception of column $E(\boldsymbol{MS})$(see Table 10.11). The following new notations are useful: the mean of the levels of factor A is written as $\bar{\boldsymbol{y}}_{i.}$, the mean of the levels of factor B as $\bar{\boldsymbol{y}}_{.j}$, and the mean of the combinations of factor levels A_iB_j as $\bar{\boldsymbol{y}}_{ij}$; $N = \sum_{i=1}^{a}\sum_{j=1}^{b} n_{ij}$.

Table 10.10 Analysis of variance table for two-way cross classification ($n_{ij} = n > 1$).

Source of variation	$\boldsymbol{SS}$	df	$\boldsymbol{MS}$
Factor A	$\boldsymbol{SS}_A = \sum_{i=1}^{a}\sum_{j=1}^{b}\sum_{v=1}^{n}(\bar{\boldsymbol{y}}_{i.} - \bar{\boldsymbol{y}})^2$	$a-1$	$\boldsymbol{MS}_A = \frac{\boldsymbol{SS}_A}{a-1}$
Factor B	$\boldsymbol{SS}_B = \sum_{i=1}^{a}\sum_{j=1}^{b}\sum_{v=1}^{n}(\bar{\boldsymbol{y}}_{.j} - \bar{\boldsymbol{y}})^2$	$b-1$	$\boldsymbol{MS}_B = \frac{\boldsymbol{SS}_B}{b-1}$
Interaction $A \times B$	$\boldsymbol{SS}_{AB} = \sum_{i=1}^{a}\sum_{j=1}^{b}\sum_{v=1}^{n}(\bar{\boldsymbol{y}}_{ij} - \bar{\boldsymbol{y}}_{i.} - \bar{\boldsymbol{y}}_{.j} + \bar{\boldsymbol{y}})^2$	$(a-1)(b-1)$	$\boldsymbol{MS}_{AB} = \frac{\boldsymbol{SS}_{AB}}{(a-1)(b-1)}$
Residual	$\boldsymbol{SS}_{res} = \sum_{i=1}^{a}\sum_{j=1}^{b}\sum_{v=1}^{n}(\boldsymbol{y}_{ijv} - \bar{\boldsymbol{y}}_{ij})^2$	$ab(n-1)$	$\boldsymbol{MS}_{res} = \frac{\boldsymbol{SS}_{res}}{ab(n-1)}$
Total	$\boldsymbol{SS}_t = \sum_{i=1}^{a}\sum_{j=1}^{b}\sum_{v=1}^{n}(\boldsymbol{y}_{ijv} - \bar{\boldsymbol{y}})^2$	$N-1$	

Table 10.11 Expected values of the mean squares *E(MS)* for the models of two-way cross classification ($n_{ij} = n > 1$); σ_a^2, σ_b^2, and σ_{ab}^2 are the variance components.

Source of variation	Model I	Model II	Mixed model (*A* fixed, *B* random)
Factor *A*	$\sigma^2 + \frac{bn}{a-1}\sum_{i=1}^{n} a_i^2$	$\sigma^2 + n\sigma_{ab}^2 + bn\sigma_a^2$	$\sigma^2 + n\sigma_{ab}^2 + \frac{bn}{a-1}\sum_{i=1}^{n} a_i^2$
Factor *B*	$\sigma^2 + \frac{an}{b-1}\sum_{j=1}^{b} b_j^2$	$\sigma^2 + n\sigma_{ab}^2 + an\sigma_b^2$	$\sigma^2 + an\sigma_b^2$
Interaction $A \times B$	$\sigma^2 + \frac{n}{(a-1)(b-1)}\sum_{i,j} (ab)_{ij}^2$	$\sigma^2 + n\sigma_{ab}^2$	$\sigma^2 + n\sigma_{ab}^2$
Residual	σ^2	σ^2	σ^2

Using the mean squares of the analysis of variance table, the null hypotheses can be tested independently for each model.

Master

More specifically, the assumptions for model I are as follows:

$H_0(A)$: $a_i = 0$ for all i; H_A: $a_i \neq a_l$ for at least one $i \neq l$
$H_0(B)$: $b_j = 0$ for all j; H_A: $b_j \neq b_l$ for at least one $j \neq l$
$H_0(A \times B)$: $(ab)_{ij} = 0$ for all i, j; H_A: $(ab)_{ij} \neq (ab)_{lm}$ for at least two pairs i, $j \neq l, m$.

The estimators of the model parameters μ, a_i, b_j, and $(ab)_{ij}$ (in model I, in case of equal cell frequencies) are: $\hat{\boldsymbol{\mu}} = \bar{\mathbf{y}}$, $\hat{\boldsymbol{a}}_i = \bar{\mathbf{y}}_{i.} - \bar{\mathbf{y}}$, $\hat{\boldsymbol{b}}_i = \bar{\mathbf{y}}_{.j} - \bar{\mathbf{y}}$, $\widehat{(\boldsymbol{ab})}_{ij} = \bar{\mathbf{y}}_{ij} - \bar{\mathbf{y}}_{i.} - \bar{\mathbf{y}}_{.j} + \bar{\mathbf{y}}$, given $\sum_{i=1}^{a} a_i = 0, \sum_{j=1}^{b} b_j = 0, \sum_{i=1}^{a} (ab)_{ij} = \sum_{j=1}^{b} (ab)_{ij} = 0$ for all i and all j.

The corresponding F-tests can be taken from Table 10.11: again, these are about comparing mean squares; those which are to be compared having the same expected value under the null hypothesis. If the F-test provides a significant result, the two mean squares are to be interpreted as unequal in the population; i.e. the relevant model parameters are not all zero – the respective means are different.

Post hoc tests, in particular the Newman–Keuls procedure (see Section 10.4.1.2), are provided by pertinent computer programs also for multi-way analyses of variance; however, just for the main effects of each factor, but not for interaction effects. However, it is possible to graphically illustrate the means of the factor levels and combinations of factor levels and to create a graph of the means of the combinations of factor levels, respectively, to describe the interaction effects.

Example 10.11 – continued

We only consider the example theoretically. There might be a mean score of $\mu = 100$ (*athletic stamina*; standardized arbitrarily). The main effects of factor A, *age*, are $a_1 = -10$, $a_2 = 3$, and, because of the model assumption $\sum_{i=1}^{a} a_i = 0$, it follows $a_3 = 7$. Furthermore, the main effects of factor B, *sex*, are $b_1 = 5$ and therefore $b_2 = -5$. We are only free to choose two of the six interaction effects; the others are fixed because of the model assumptions. We think of $(ab)_{11} = 6$ and $(ab)_{21} = 4$; so we have $(ab)_{12} = -6$, and $(ab)_{22} = -4$; also it then has to be $(ab)_{31} = -(6 + 4) = -10$ and therefore $(ab)_{32} = 10$. Thereby, the expected values $\mu_{ij} = \mu + a_i + b_j + (ab)_{ij}$ in the six combinations of factor levels are given (Table 10.12).

Table 10.12 Values of $\mu + a_i + b_j + (ab)_{ij}$ of Example 10.11.

Age (years)	Female	Male
12	$\mu_{11} = 100 - 10 + 5 + 6 = 101$	$\mu_{12} = 100 - 10 - 5 - 6 = 79$
13	$\mu_{21} = 100 + 3 + 5 + 4 = 112$	$\mu_{22} = 100 + 3 - 5 - 4 = 94$
14	$\mu_{31} = 100 + 7 + 5 - 10 = 102$	$\mu_{32} = 100 + 7 - 5 + 10 = 112$

If we took scores from several students, these outcomes in the six combinations of factor levels would disperse around the expected values stated in Table 10.12.

As this example is designed for illustration, we assume three significant F-tests, given a sufficiently large total sample size for both main effects as well as for the interaction effect between *age* and *sex*. It turns-out that younger children/teenagers show less *athletic stamina* than older children, and that girls of the investigated age show more *athletic stamina* than boys, but that the superiority of girls regarding *athletic stamina* changes, with progressing age, to inferiority. 12-year-old boys show the lowest, while 13-year-old girls and 14-year-old boys disclose the highest levels of *athletic stamina*. Figure 10.9 shows the course of the means for all combinations of factor levels in order to illustrate the observed interaction effect. It can be seen that the average *athletic stamina* increases almost in a linear fashion through the age levels of male children, but reaches a maximum in girls at the age of 13 and thereafter decreases.

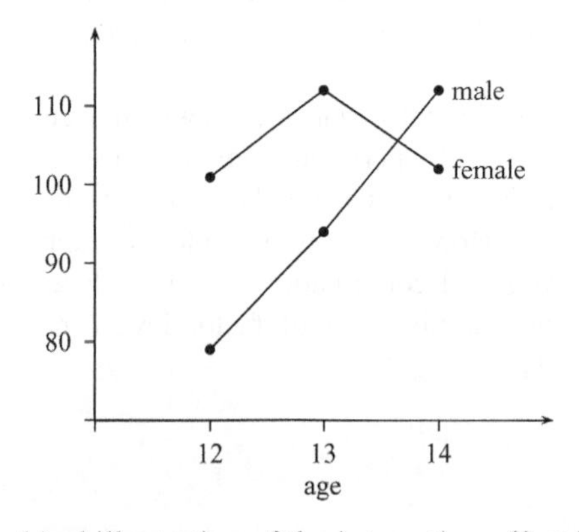

Figure 10.9 Graphical illustration of the interaction effect in Example 10.11.

Doctor Occasionally in two-way (and multi-way) analyses of variance, the special case of having $n = 1$ arises; i.e. only one outcome (of only a single research unit) per combination of factor levels occurs. In such an instance, the interaction between two (or even more) factors can no longer be estimated. Apart from that the analysis can be performed completely analogously. If there is, however, still interest in the interaction, there are special approaches in two-way analysis of variance that will not be described here (see Rasch, Rusch, Šimečková, Kubinger, Moder & Šimeček, 2009).

In model I the null hypothesis regarding the main effect of factor A, $H_0(A)$, has to be tested with the following test statistic:

$$\boldsymbol{F}_A = \frac{\boldsymbol{MS}_A}{\boldsymbol{MS}_{\text{res}}} \tag{10.5}$$

It is F-distributed with $df_1 = a - 1$ and $df_2 = ab(n - 1)$ degrees of freedom. Given the null hypothesis is true, $\boldsymbol{MS}_A$ as well as $\boldsymbol{s}^2 = \hat{\sigma}^2 = MS_{\text{res}}$ estimate the variance of $\boldsymbol{y}$ and $\boldsymbol{e}$, respectively; $\boldsymbol{MS}_A$ depends on the variance of the means of all levels A_i, $i = 1, 2, \ldots, a$.

The null hypothesis regarding the main effect of factor B, $H_0(B)$, has to be tested with the following test statistic:

$$\boldsymbol{F}_B = \frac{\boldsymbol{MS}_B}{\boldsymbol{MS}_{\text{res}}} \tag{10.6}$$

It is F-distributed with $df_1 = b - 1$ and $df_2 = ab(n - 1)$ degrees of freedom. Given the null hypothesis is true, $\boldsymbol{MS}_B$ and $\boldsymbol{MS}_{\text{res}}$ estimate the variance of $\boldsymbol{y}$ and $\boldsymbol{e}$, respectively. $\boldsymbol{MS}_B$ depends on the variance of the means of all levels B_j, $j = 1, 2, \ldots, b$.

The null hypothesis regarding the interaction effect $A \times B$, $H_0(A \times B)$, has to be tested with the following test statistic:

$$\boldsymbol{F}_{AB} = \frac{\boldsymbol{MS}_{AB}}{\boldsymbol{MS}_{\text{res}}} \tag{10.7}$$

It is F-distributed with $df_1 = (a - 1)(b - 1)$ and $df_2 = ab(n - 1)$ degrees of freedom. Given the null hypothesis is true, $\boldsymbol{MS}_{AB}$ as well as $\boldsymbol{MS}_{\text{res}}$ estimate the variance of $\boldsymbol{y}$ and $\boldsymbol{e}$, respectively. $\boldsymbol{MS}_{AB}$ depends on the variance of the means of all combinations of factor levels A_iB_j.

Planning a study, i.e. the determination of the necessary sample size, can basically be done analogously to Section 10.4.1.1, but it has to be decided which of the three null hypotheses is in the limelight and of primary interest. The **R**-package `OPDOE` provides all variants. Again, the sample size is determined primarily for the worst case; i.e. the case which leads to the largest size for given precision requirements. And again, it is optimal in the sense of the smallest possible sample size, if there are equal sizes in all samples, i.e. $n_{ij} = n$.

Example 10.12 The influence of a child's native language and sex on its test score on the subtest *Applied Computing, 1st test date* will be investigated (Example 1.1)

Native language of the child (factor A) and *sex of the child* (factor B) are fixed factors, both with two levels each; i.e. $a = b = 2$. We decide on $\alpha = 0.05$, $\beta = 0.10$, and $\delta = 0.67\sigma$. Of interest is primarily the interaction, for which the sample size will be determined.

In **R**, the package OPDOE is used to calculate the sample size (see Example 10.2 as well). Thus, we type

```
> size_n.two_way_cross.model_1_axb(alpha = 0.05, beta = 0.1,
+                                  delta = 0.67, a = 2, b = 2,
+                                  cases = "maximin")
```

i.e. we apply the function `size_n.two_way_cross.model_1_axb`, using as arguments the precision requirements $\alpha = 0.05$, $\beta = 0.10$, and $\delta = 0.67\sigma$ (with `delta = 0.67`). Furthermore, we put the number of factor levels $a = b = 2$ by using `a = 2` and `b = 2` as arguments, and, with `cases = "maximin"`, we request the calculation of the sample size for the worst case.

```
[1] 48
```

Therefore, 48 persons have to be investigated in each of the four cells; as already mentioned, planning the study of Example 1.1 was done with regard to another research question; thus, in particular equal sample sizes cannot be obtained here.

The analysis is primarily about the null hypothesis $H_0(A \times B)$: 'There is no interaction between the child's native language and sex regarding its test score on the subtest *Applied Computing, 1st test date*' (all $(ab)_{ij}$ are equal to zero). Additionally, there is of course also interest in the null hypotheses $H_0(A)$: 'The mean test scores on the subtest *Applied Computing, 1st test date* do not differ between the levels of the factor *native language of the child* ("German" vs. "Turkish")' (all a_i are equal to zero), as well as $H_0(B)$: 'The mean test scores on the subtest *Applied Computing, 1st test date* do not differ between the levels of the factor *sex of the child* ("male" vs. "female")' (all b_j are equal to zero).

We first have to examine to what extent the standard deviations of all combinations of factor levels differ.

In **R**, we type

```
> tapply(sub3_t1[native_language == "German"],
+        sex[native_language == "German"], sd)
> tapply(sub3_t1[native_language == "Turkish"],
+        sex[native_language == "Turkish"], sd)
```

i.e. we apply the function `tapply()` and use the character *Applied Computing, 1st test date* with `sub3_t1` as the first argument. As the second argument, we use *sex of the child* (`sex`), and as the third one the statistic which is to be calculated separately for the different categories of the character *sex of the child*, thus `sd`. With `[native_language == "German"]` and `[native_language == "Turkish"]`, we specify that the respective analyses are to be conducted for children with German, and children with Turkish, as a native language.

As a result, we get:

```
  female     male
7.916228 7.563729

  female     male
8.013114 8.784456
```

In SPSS, we split the data file based on the characters *native language of the child* and *sex of the child*. To do this, we select the sequence of commands (Data – Split File...) as shown in Example 5.11, and reach the window shown in Figure 5.23. There, we choose Compare groups and drag and drop the characters native language of the child as well as sex of the child to the field Groups Based on:. Now, we click OK and select the next sequence of commands (Analyze - Descriptive Statistics - Frequencies...) analogously to Example 5.2. We reach the window shown in Figure 5.4, where we drag and drop the character Applied Computing, 1st test date to the field Variable(s):. Next, we click Statistics... and tick Std. deviation in the section Dispersion. After a click on Continue and OK, we obtain Table 10.13.

Table 10.13 SPSS-output of the standard deviations in Example 10.12 (shortened output).

Statistics

Applied Computing, 1st test date (T-Score)

German	female	N	Valid	25
			Missing	0
		Std. Deviation		7.916
	male	N	Valid	25
			Missing	0
		Std. Deviation		7.564
Turkish	female	N	Valid	25
			Missing	0
		Std. Deviation		8.013
	male	N	Valid	25
			Missing	0
		Std. Deviation		8.784

Now, we go back to the window shown in Figure 5.23, where we tick Analyze all cases, do not create groups. Clicking OK, we make sure that all further analyses are done for the overall sample again.

The ratio of largest to smallest standard deviation is $8.784/7.564 = 1.161 < 1.5$; i.e. the plausible limit according to the rule of thumb (see above). The application of a two-way analysis of variance seems, therefore, to be justified.

In **R**, we type

```
> aov.4 <- aov(sub3_t1 ~ native_language * sex)
> summary(aov.4)
```

i.e. we use the same function as in Example 10.8. The analysis is now conducted for the character *Applied Computing, 1st test date* (sub3_t1) with regard to the two factors *native language of the child* (native_language) and *sex of the child* (sex). Using the connection '*' between the two characters, both of the main effects and the interaction effects are tested. We assign the result to the object aov.4, which we submit to the function summary().

As a result, we get:

```
                     Df Sum Sq Mean Sq F value Pr(>F)
native_language       1   90.3  90.250  1.3818 0.2427
sex                   1    0.2   0.250  0.0038 0.9508
native_language:sex   1    4.4   4.410  0.0675 0.7955
Residuals            96 6270.1  65.313
```

In SPSS, we select the same sequence of commands (Analyze – General Linear Model – Univariate...) as in Example 10.8. In the resulting window shown in Figure 10.7, we drag and drop the character Applied Computing, 1st test date to the field Dependent Variable:. The factors native language of the child and sex of the child are dragged and dropped to the field Fixed Factor(s):. Next, we click Options... and reach the window shown in Figure 10.8, where we tick Descriptive statistics. After clicking Continue and OK, we get the result shown in Table 10.14 as well as (again) the standard deviations shown in Table 10.13.

Table 10.14 SPSS-output of the table of variances of the two-way analysis of variance (model I) in Example 10.12.

Tests of Between-Subjects Effects

Dependent Variable:Applied Computing, 1st test date (T-Score)

Source	Type III Sum of Squares	df	Mean Square	F	Sig.
Corrected Model	94.910[a]	3	31.637	.484	.694
Intercept	254924.010	1	254924.010	3903.093	.000
native_language	90.250	1	90.250	1.382	.243
sex	.250	1	.250	.004	.951
native_language * sex	4.410	1	4.410	.068	.796
Error	6270.080	96	65.313		
Total	261289.000	100			
Corrected Total	6364.990	99			

a. R Squared = .015 (Adjusted R Squared = -.016)

None of the three null hypotheses have to be rejected.

Master Doctor

The analysis is more complicated if some of the frequencies for the combinations of factor levels in the data structure of Table 10.9 are zero. We will not discuss planning the study here, as it is mostly possible – if there is planning at all – to obtain (equal) cell frequencies. So, the following will be about a case where there was no (appropriate) planning and therefore zero frequencies of individual combinations of factor levels simply 'happen'. However, as long as we have a connected cross classification in the sense of a connected block design (see Section 7.2.2), the analysis can be done rather simply using approximations from computer programs.

In SPSS, we can choose four different ways of calculating the sums of squares; namely from type I to type IV. Type III is appropriate for complete cross classifications; this is in fact the standard setting. Type IV should be used for incomplete cross classifications (with interactions). Type I and II are used for nested and mixed (three-way or multi-way) classifications. Moreover, if there are unequal cell frequencies, SPSS always automatically uses that type which suits best. In **R** (standard package) type I is always used, leading in the case of equal cell frequencies to the same result as type III.

Master Doctor

Example 10.13 Do the test scores in the subtest *Applied Computing, 1st test date* differ depending on the combination of *social status* and *marital status of the mother*? (Example 1.1)

The respective null hypotheses are as follows. $H_0(A)$: 'The mean test scores in the subtest *Applied Computing, 1st test date* do not differ between the levels of the factor *marital status of the mother*' (all a_i are equal to zero). $H_0(B)$: 'The mean test scores in the subtest *Applied Computing, 1st test date* do not differ between the levels of the factor *social status*' (all b_j are equal to zero). $H_0(A \times B)$: 'There is no interaction between *marital status of the mother* and *social status* regarding the mean test scores in the subtest *Applied Computing, 1st test date*' (all $(ab)_{ij}$ are equal to zero). Both factors are again fixed – there is only interest in these factor levels. During the analysis we find out about the cell frequencies of all combinations of factor levels.

In **R**, we create a two-dimensional frequency table by typing

```
> table(social_status, marital_mother)
```

i.e. we submit the characters *social status* (`social_status`) and *marital status of the mother* (`marital_mother`) as arguments to the function `table()`.

As a result, we get a table which is essentially identical with the one shown in Table 10.15.

In SPSS, we create Table 10.15 via Analyze – Descriptive Statistics - Crosstabs..., proceeding analogously to Example 5.13.

Table 10.15 SPSS-output of the two-dimensional frequency table in Example 10.13.

social status (after Kleining & Moore according to occupation of father/alternatively of the single mother)* marital status of the mother Crosstabulation

Count

		marital status of the mother				
		never married	married	divorced	widowed	Total
social status (after Kleining & Moore according to occupation of father/alternatively of the single mother)	upper classes	1	8	2	1	12
	middle classes	4	24	2	0	30
	lower middle class	1	18	5	1	25
	upper lower class	0	13	3	0	16
	lower classes	0	2	3	0	5
	single mother in household	3	0	7	2	12
Total		9	65	22	4	100

As we can see, there are several empty cells. There is no interaction $(ab)_{ij}$ defined for the corresponding combinations of factor levels A_iB_j. However, one can easily prove that the classification is connected in the sense of Section 7.2.2. For example, it is possible without having to skip empty cells to get from 'never married', 'upper classes' (in the row 'upper classes') to 'divorced' (in the column 'divorced') and from there to 'single mother in household', and finally (in the row 'single mother in household') to 'widowed'.

Before we actually apply analysis of variance, we again examine the ratio of the largest to the smallest standard deviation; but we only want to use those combinations of factor levels that are at least populated with 10 observations – conspicuous extreme values in the others could have easily occurred by chance. Both in **R** and SPSS, these standard deviations can be determined completely analogously to Example 10.12. The result is $9.456/7.715 = 1.226 < 1.5$, which is the plausible limit according to the rule of thumb (see above). Therefore the application of the analysis of variance seems to be justified.

In **R**, we type

```
> aov.5 <- aov(sub3_t1 ~ marital_mother * social_status)
> summary(aov.5)
```

i.e. we apply the function `aov()` and use as argument the formula `sub3_t1 ~ marital_mother * social_status`. Thus, the character *Applied Computing, 1st test date* (`sub3_t1`) is to be analyzed with regard to the factors *marital status of the mother* (`marital_mother`) and *social status* (`social_status`). Using the connection '*' between the two factors, all main effects and the interaction effect are tested. We assign the result of the analysis to the object `aov.5`. Finally, we submit this object to the function `summary()` in order to obtain a summary of the results (shortened output):

```
                                Df Sum Sq Mean Sq  F value  Pr(>F)
marital_mother                   3  116.0  38.678  0.6116 0.60936
social_status                    5  759.8 151.963  2.4031 0.04376
marital_mother:social_status     9  303.9  33.762  0.5339 0.84584
Residuals                       82 5185.3  63.235
```

In SPSS, we proceed analogously (Analyze – General Linear Model – Univariate...) to Example 10.12, but now we first click Model... in the window shown in Figure 10.7. In the resulting window (Figure 10.10), we select Type IV in the pull-down menu Sum of squares:, as this is preferable in the case that vacant cells exist (see above). Clicking Continue gets us back to the window shown in Figure 10.7, where we drag and drop the character Applied Computing, 1st test date to the field Dependent Variable:. In addition, we drag and drop the factors social status and marital status of the mother to the field Fixed Factor(s):. After clicking OK, we get the result shown in Table 10.16.

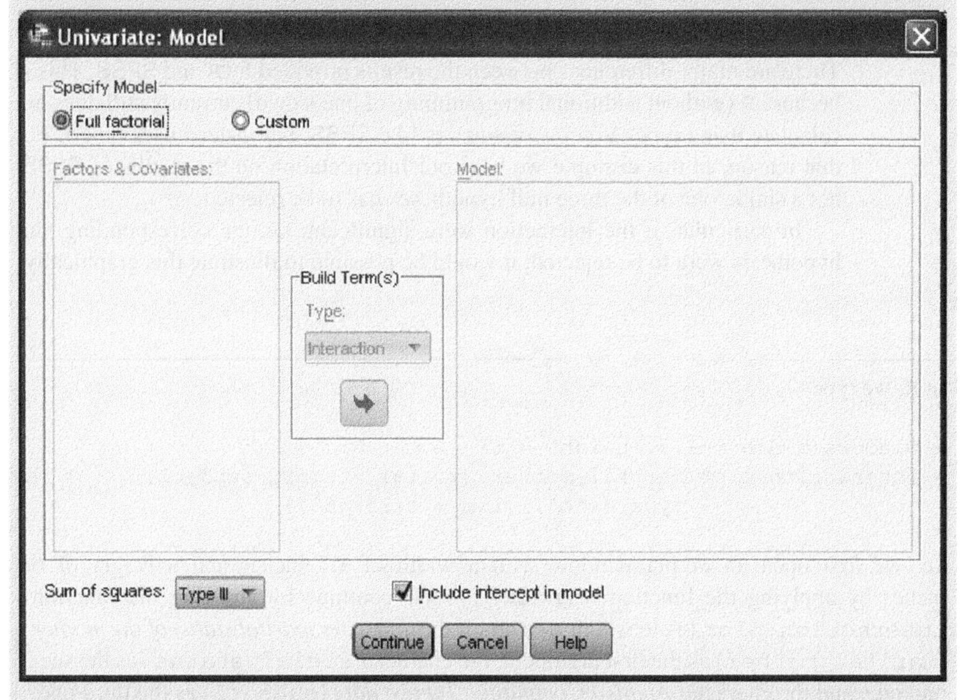

Figure 10.10 SPSS-window for selecting between Type I to IV methods of calculation of the sum of squares in two-way (and multiple) analysis of variance.

Table 10.16 SPSS-output of table of variances in two-way analysis of variance (model I) in Example 10.13.

Tests of Between-Subjects Effects

Dependent Variable:Applied Computing, 1st test date (T-Score)

Source	Type IV Sum of Squares	df	Mean Square	F	Sig.
Corrected Model	1179.703[a]	17	69.394	1.097	.371
Intercept	103803.889	1	103803.889	1641.552	.000
social_status	139.716[b]	5	27.943	.442	.818
marital_mother	105.462[b]	3	35.154	.556	.646
social_status * marital_mother	303.857	9	33.762	.534	.846
Error	5185.287	82	63.235		
Total	261289.000	100			
Corrected Total	6364.990	99			

a. R Squared = .185 (Adjusted R Squared = .016)

b. The Type IV testable hypothesis is not unique.

There are major differences between the results provided by **R** and SPSS. This is because **R** (without additional programming of one's own) currently still does not calculate using type IV sums of squares like SPSS, but instead uses type I. For that reason, in this example we base our interpretation on the results of SPSS: not a single one of the three null hypotheses has to be rejected.

In particular, if the interaction were significant, i.e. the corresponding null hypothesis were to be rejected, it would be possible to illustrate this graphically.

In **R**, we type

```
> windows(width = 8.5, height = 6)
> interaction.plot(marital_mother, social_status, sub3_t1,
+                  type = "b", ylim = c(35, 65))
```

i.e. we first open an output window with a width of 8.5 inches and a height of 6 inches by applying the function `windows()`. We continue by applying the function `interaction.plot()`, to which we submit the character *marital status of the mother* (`marital_mother`) as the first argument, the character `social_status` as the second one, and the character *Applied Computing, 1st test date* (`sub3_t1`) as the third one. With `type = "b"` we request lines as well as labeling (with numbers), and with `ylim = c(35, 65)` we define the range of values to be presented; that is 35 to 65.

As a result, we get Figure 10.11a.

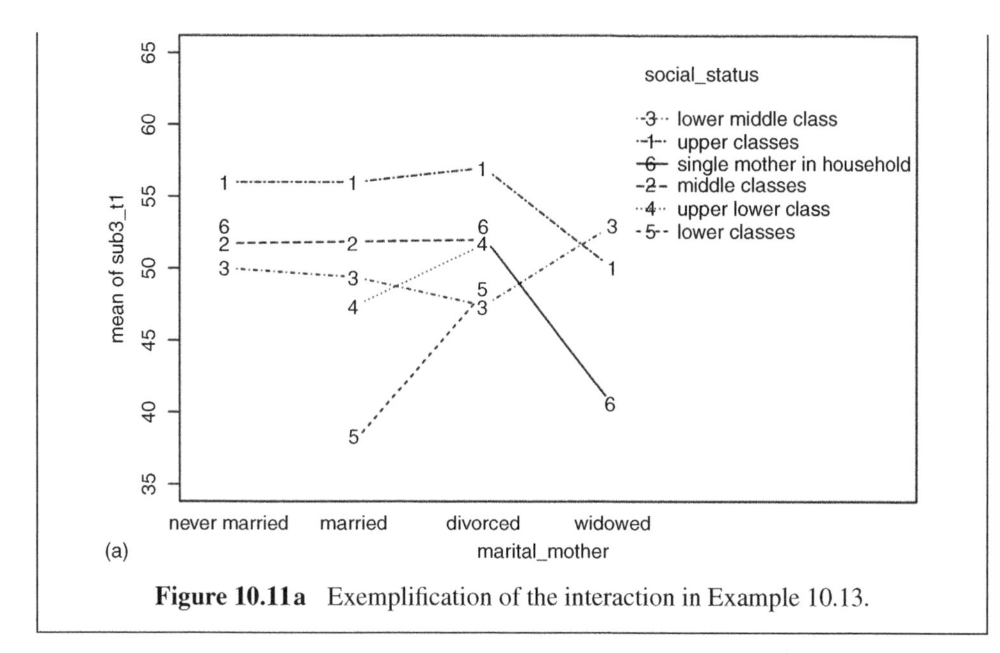

Figure 10.11a Exemplification of the interaction in Example 10.13.

In SPSS, we click Plots... in the window shown in Figure 10.7. As a consequence, the window shown in Figure 10.12 appears. There, we drag and drop the factor marital_mother to the field Horizontal Axis: and the factor social_status to the field Separate Lines:. Then, we click Add and Continue, followed by OK, in order to create the diagram, which is shown in Figure 10.11b. In the given example, the apparent differences are not significant and thus are not to be overvalued.

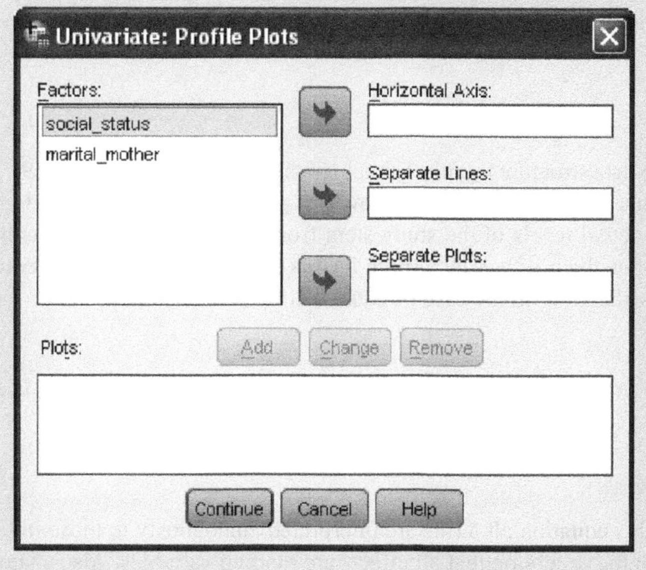

Figure 10.12 SPSS-window to create a graph of interactions.

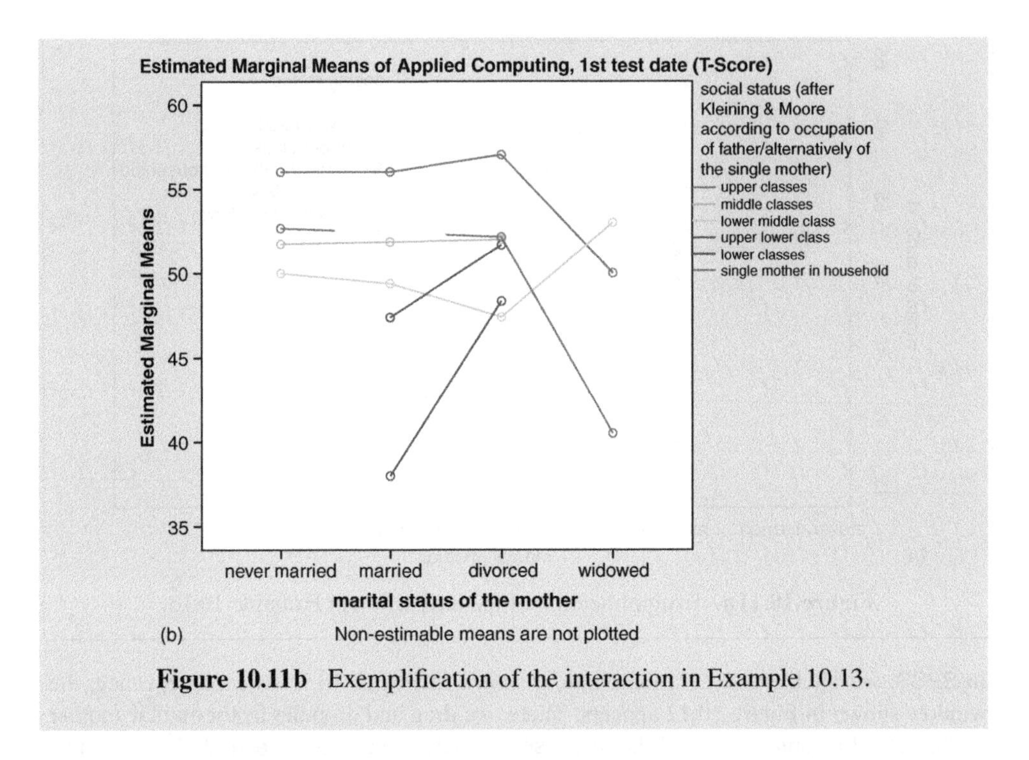

Figure 10.11b Exemplification of the interaction in Example 10.13.

10.4.4.2 Cross classification – model II

Model II rarely plays a role in psychology. But where it is important, the erroneous application of model I would often lead to incorrect results, because the corresponding F-tests would be incorrect.

Master Doctor

The data structure is entirely analogous to model I (see Table 10.9). The difference is that a population of factor levels is given for levels of both A and B, in which the actual levels of the study stem from a random sampling method. For model II of analysis of variance (both factors are random), the model equation for cross classification (in the case of equal cell frequencies) is:

$$\boldsymbol{y}_{ijv} = \mu + \boldsymbol{a}_i + \boldsymbol{b}_j + (\boldsymbol{ab})_{ij} + \boldsymbol{e}_{ijv} \quad (i = 1, \ldots, a; j = 1, \ldots, b; v = 1, \ldots, n) \tag{10.8}$$

In this equation all terms are interpreted analogously to those in Formula (10.4), with the exception that all effects are random variables. Also analogous to model I, an interaction effect is defined only if there is at least one outcome in the combination of factor levels A_iB_j.

For the derivation of hypothesis testing the following is required in addition to the model I assumptions:

- $E(\boldsymbol{a}_i) = E(\boldsymbol{b}_j) = E((\boldsymbol{ab})_{ij}) = E(\boldsymbol{e}_{ijv}) = 0$ for all i, j, v.
- The variances of $\boldsymbol{a}_i$, $\boldsymbol{b}_j$, and $(\boldsymbol{ab}_{ij})$ are each the same: σ_a^2, σ_b^2, and σ_{ab}^2.
- All random variables on the right side of the model equation are independent of each other.

The hypotheses are as follows:

$$H_0(A)\colon \sigma_a^2 = 0;\ H_A\colon \sigma_a^2 > 0$$
$$H_0(B)\colon \sigma_b^2 = 0;\ H_A\colon \sigma_b^2 > 0$$
$$H_0(A \times B)\colon \sigma_{ab}^2 = 0;\ H_A\colon \sigma_{ab}^2 > 0.$$

In the case of equal cell frequencies per combination of factor levels, there exist the following exact tests. They can be derived from Table 10.11; we recognize that, unlike in model I, the F-test does not, for all three null hypotheses, relate the respective mean squares ($\boldsymbol{MS}_A$, $\boldsymbol{MS}_B$, and $\boldsymbol{MS}_{AB}$) to that of the 'residual' ($\boldsymbol{MS}_{\text{res}}$); instead for both main effects the F-test has to be related to the mean square of the interaction effect ($\boldsymbol{MS}_{AB}$).

Given unequal cell frequencies, the *Satterthwaite procedure* is notable, which, however, only leads to an approximately F-distributed test statistic.

There is no program package in **R** for planning the study.

We do not give an example at this point; however, the whole procedure is suitably illustrated by an example of three-way analysis of variance (see Section 10.4.7).

10.4.4.3 Cross classification – mixed model

The mixed model also plays only a minor role in psychology. However, where it is important, the erroneous application of model I would lead to incorrect results, because the respective F-tests would be wrong.

Master Doctor

The data structure is again completely analogous to model I (see Table 10.9). The difference is that, for example, the levels of A are fixed, but those of B are drawn from a population of factor levels; that is specifically for the particular study using a random sampling method. For the mixed model of analysis of variance (factor A is fixed, factor B is random) the model equation for cross classification (in the case of equal cell frequencies) is:

$$\boldsymbol{y}_{ijv} = \mu + a_i + \boldsymbol{b}_j + (\boldsymbol{ab})_{ij} + \boldsymbol{e}_{ijv} \quad (i = 1, \ldots, a;\ j = 1, \ldots, b;\ v = 1, \ldots, n) \tag{10.9}$$

The terms of this equation are equivalent to all the abovementioned. Again, the interaction is only defined for those combinations of factor levels for which outcomes exist.

For derivation of hypothesis testing, in addition to the model I assumptions the following apply:

- $E(\boldsymbol{b}_j) = E((\boldsymbol{ab})_{ij}) = E(\boldsymbol{e}_{ijv}) = 0$ for all i, j, v.
- The variances of $\boldsymbol{b}_j$ and $(\boldsymbol{ab}_{ij})$ are each the same: σ_b^2 and σ_{ab}^2.
- All random variables on the right side of the model equation are independent of each other.

The hypotheses are as follows:

$H_0(A)$: $a_i = 0$ for all i; H_{A}: $a_i \neq a_l$ for at least one $i \neq l$; with $\sum_{i=1}^{a} a_i = 0$

$H_0(B)$: $\sigma_b^2 = 0$; H_{A}: $\sigma_b^2 > 0$

$H_0(A \times B)$: $\sigma_{ab}^2 = 0$; H_{A}: $\sigma_{ab}^2 > 0$.

There are exact tests in the case of equal cell frequencies. These can be derived from Table 10.11; we recognize that, unlike in model I, the F-test does not, for all three null hypotheses, relate the respective mean squares ($\boldsymbol{MS}_A$, $\boldsymbol{MS}_B$, and $\boldsymbol{MS}_{AB}$) to that of the 'residual' ($\boldsymbol{MS}_{\text{res}}$); instead, for the main effect of the fixed factor A the F-test has to be related to the mean square of the interaction effect ($\boldsymbol{MS}_{AB}$).

Given unequal cell frequencies, the Satterthwaite procedure is notable, but only leads to an approximately F-distributed test statistic.

Planning a study design, i.e. the determination of the necessary sample size, can only be done with regard to hypothesis testing of the fixed factor A. This is not to determine the necessary cell frequency – in fact it can be specified as small as possible with $n = 2$ – but instead the number of levels of the random factor which have to be sampled. We do not give an example here (for this see Rasch, Pilz, Verdooren, & Gebhardt, 2011).

10.4.4.4 Nested classification – all models

A fundamentally different data structure than that found in Table 10.9 results from a nested classification; i.e. if there is no cross classification. Table 10.17 illustrates this data structure.

Table 10.17 Data structure of a two-way analysis of variance with nested classification ($n_{ij} = n > 1$).

Levels of A	A_1			A_2			$\cdots$	A_a			
Levels of B	B_{11}	B_{12}	$\cdots$	B_{1b}	B_{21}	B_{22}	$\cdots$	B_{2b}	$\cdots$	B_{a1}	B_{a2}	$\cdots$	B_{ab}
	y_{111}	y_{121}	$\cdots$	y_{1b1}	y_{211}	y_{221}	$\cdots$	y_{2b1}	$\cdots$	y_{a11}	y_{a21}	$\cdots$	y_{ab1}
	y_{112}	y_{122}	$\cdots$	y_{1b2}	y_{212}	y_{222}	$\cdots$	y_{2b2}	$\cdots$	y_{a12}	y_{a22}	$\cdots$	y_{ab2}
	$\vdots$	$\vdots$	$\vdots$	$\vdots$	$\vdots$	$\vdots$	$\vdots$	$\vdots$	$\vdots$	$\vdots$	$\vdots$	$\vdots$	$\vdots$
	y_{11n}	y_{12n}	$\cdots$	y_{1bn}	y_{21n}	y_{22n}	$\cdots$	y_{2bn}	$\cdots$	y_{a1n}	y_{a2n}	$\cdots$	y_{abn}

It is factor B which is nested in factor A (that is, $B \prec A$). The difference between cross classification and nested classification is that in nested classification not all b factor levels of factor B are combined with all a factor levels of factor A, but instead always b other (generally even: b_i other) levels of B are combined with a certain level A_i of A. The random variable $\boldsymbol{y}_{ijv}$ is therefore the vth observation in B_{ij}. An interaction is thereby understandably not defined.

The following model equations are due to the four different models I, II, and (two) mixed models – with the mixed model we must distinguish between the two cases that either A is fixed and B is random or that A is random and B is fixed.

$$\boldsymbol{y}_{ijk} = \mu + a_i + b_{ij} + \boldsymbol{e}_{ijv} (i = 1, \ldots, a; j = 1, \ldots, b_i; v = 1, \ldots, n_{ij}) \quad (10.10)$$

$$\boldsymbol{y}_{ijk} = \mu + \boldsymbol{a}_i + \boldsymbol{b}_{ij} + \boldsymbol{e}_{ijv} (i = 1, \ldots, a; j = 1, \ldots, b_i; v = 1, \ldots, n_{ij}) \quad (10.11)$$

$$\boldsymbol{y}_{ijk} = \mu + a_i + \boldsymbol{b}_{ij} + \boldsymbol{e}_{ijv} (i = 1, \ldots, a; j = 1, \ldots, b_i; v = 1, \ldots, n_{ij}) \quad (10.12)$$

$$\boldsymbol{y}_{ijk} = \mu + \boldsymbol{a}_i + b_{ij} + \boldsymbol{e}_{ijv} (i = 1, \ldots, a; j = 1, \ldots, b_i; v = 1, \ldots, n_{ij}) \quad (10.13)$$

In these equations, μ is the total mean, the terms a_i and $\boldsymbol{a}_i$, respectively, are the effects of the individual factor levels A_i of A, the main effects of A; b_{ij} and $\boldsymbol{b}_{ij}$, respectively, are the effects of factor level B_{ij} of B within A_i. If $n_{ij} = n$ is valid for all B_{ij} and if $b_i = b$, we talk about a *balanced case*. The equations show that no interaction effects are modeled.

For derivation of hypothesis testing, it must generally be assumed:

- The $\boldsymbol{e}_{ijv}$ are normally distributed according to $N(0, \sigma^2)$; i.e. with the same expected value 0 and equal variance σ^2.
- The $\boldsymbol{e}_{ijv}$ are independent of each other.
- $a \geq 2$ and $b \geq 2$.

The analysis of variance table is equal for all models (see Table 10.18); the model-specific column $E(\boldsymbol{MS})$ of the theoretical analysis of variance table is shown in Table 10.19.

Table 10.18 Analysis of variance table for two-way nested classification – balanced case.

Source of variation	*SS*	*df*	*MS*
Factor A	$\boldsymbol{SS}_A = \sum_{i=1}^{a} \sum_{j=1}^{b} \sum_{v=1}^{n} (\bar{\boldsymbol{y}}_i - \bar{\boldsymbol{y}})^2$	$a - 1$	$\boldsymbol{MS}_A = \frac{SS_A}{a-1}$
Factor B within factor A	$\boldsymbol{SS}_{B \text{ in } A} = \sum_{i=1}^{a} \sum_{j=1}^{b} \sum_{v=1}^{n} (\bar{\boldsymbol{y}}_{ij} - \bar{\boldsymbol{y}}_i)^2$	$ab - a$	$\boldsymbol{MS}_{B \text{ in } A} = \frac{SS_{B \text{ in } A}}{ab-a}$
Residual	$\boldsymbol{SS}_{\text{res}} = \sum_{i=1}^{a} \sum_{j=1}^{b} \sum_{v=1}^{n} (\boldsymbol{y}_{ijv} - \bar{\boldsymbol{y}}_{ij})^2$	$N - ab$	$\boldsymbol{MS}_{\text{res}} = \frac{SS_{\text{res}}}{N-ab}$
Total	$\boldsymbol{SS}_{\text{t}} = \sum_{i=1}^{a} \sum_{j=1}^{b} \sum_{v=1}^{n} (\boldsymbol{y}_{ijv} - \bar{\boldsymbol{y}})^2$	$N - 1$	

Table 10.19 Expected values of the mean squares *E(MS)* for the models of two-way nested classification – balanced case; σ_a^2 and $\sigma_{b\ \text{in}\ a}^2$ are the variance components per factor level.

Source of variation	Model I	Model II	Mixed model (*A* fixed, *B* random)	Mixed model (*A* random, *B* fixed)
Factor *A*	$\sigma^2 + \frac{bn}{a-1}\sum_i a_i^2$	$\sigma^2 + n\sigma_{b\ \text{in}\ a}^2 + bn\sigma_a^2$	$\sigma^2 + n\sigma_{b\ \text{in}\ a}^2 + bn\frac{\sum_{i=1}^a a_i^2}{a-1}$	$\sigma^2 + bn\sigma_a^2$
Factor *B* within factor *A*	$\sigma^2 + \frac{n}{a(b-1)}\sum_{i,j} b_{ij}^2$	$\sigma^2 + n\sigma_{b\ \text{in}\ a}^2$	$\sigma^2 + n\sigma_{b\ \text{in}\ a}^2$	$\sigma^2 + n\frac{\sum_{i=1}^a\sum_{j=1}^b b_{ij}^2}{a(b-1)}$
Residual	σ^2	σ^2	σ^2	σ^2

Nothing further has to be presumed for model I for the derivation of hypothesis testing; but we (again) assume the following arbitrary reparameterization: $\sum_{i=1}^a a_i = 0$ and $\sum_{j=1}^b b_{ij} = 0$ for all i. For model II, the following must also be assumed:

- $E(\boldsymbol{a}_i) = E(\boldsymbol{b}_{ij}) = 0$ for all i, j.
- The variances of $\boldsymbol{a}_i$ and $\boldsymbol{b}_{ij}$ are each the same: σ_a^2 and $\sigma_{b\ \text{in}\ a}^2$.
- All the random variables are independent of each other; in particular $\boldsymbol{e}_{ijv}$ are independent of $\boldsymbol{a}_i$ and $\boldsymbol{b}_{ij}$.

For the mixed model (*A* fixed, *B* random), the following must also be assumed in addition to model I:

- $E(\boldsymbol{b}_{ij}) = 0$ for all i, j.
- The variances of $\boldsymbol{b}_{ij}$ are the same: $\sigma_{b\ \text{in}\ a}^2$.
- The random variables are independent of each other; i.e. $\boldsymbol{e}_{ijv}$ independent of $\boldsymbol{b}_{ij}$.

And for the mixed model (*A* random, *B* fixed), it must be assumed additionally to model I:

- $E(\boldsymbol{a}_i) = 0$ for all i.
- The variances of $\boldsymbol{a}_i$ are the same: σ_a^2.
- The random variables are independent of each other; i.e. $\boldsymbol{e}_{ijv}$ independent of $\boldsymbol{a}_i$.

The hypotheses for all four types of models are now clearly summarized in Table 10.20.

Table 10.20 The hypotheses of the four types of models for a two-way analysis of variance with nested classification ($n_{ij} = n > 1$).

	Null hypothesis H_0		Alternative hypothesis H_A	
Factor / Model	A	B	A	B
I	$a_i = 0$ for all i	$b_{ij} = 0$ for all i, j	$a_i \neq a_l$ for at least one $i \neq l$	$b_{ij} \neq b_{il}$ for at least one $j \neq l$
II	$\sigma_a^2 = 0$	$\sigma_{b \text{ in } a}^2 = 0$	$\sigma_a^2 > 0$	$\sigma_{b \text{ in } a}^2 > 0$
Mixed (A fixed, B random)*	$a_i = 0$ for all i	$\sigma_{b \text{ in } a}^2 = 0$	$a_i \neq a_l$ for at least one $i \neq l$	$\sigma_{b \text{ in } a}^2 > 0$
Mixed (A random, B fixed)**	$\sigma_a^2 = 0$	$b_{ij} = 0$ for all i, j	$\sigma_a^2 > 0$	$b_{ij} \neq b_{il}$ for at least one $j \neq l$

* with $\sum_{i=1}^{a} a_i = 0$.

** with $\sum_{j=1}^{b} b_{ij} = 0$.

There are exact tests for the case of equal cell frequencies; they can be derived from Table 10.19. It can be recognized that the F-test does not relate, for all models and both null hypotheses, the respective mean squares to that of the 'residual'. Instead, for model II and for the mixed model (A fixed, B random), the F-test for the main effect of factor A has to use the mean squares with respect to the main effect of B (within A) as the denominator.

Given unequal cell frequencies, the Satterthwaite procedure is notable, but only leads to an approximately F-distributed test statistic.

Planning the study, i.e. the determination of the necessary sample size, can only be done with regards to hypothesis testing of a fixed factor. Planning a study regarding hypothesis testing for a random factor has not (yet) been realized in **R**. For the models of Formula (10.10) and (10.13), the cell frequency n is calculated in accordance with the precision requirements. In contrast, it is not the cell frequency which is determined for the model of Formula (10.12) – it can be chosen as small as possible with $n = 2$ – but instead the number of sampled levels of factor B within each level of factor A. We do not give an example here (for this see Rasch, Pilz, Verdooren, & Gebhardt, 2011).

Master Doctor

Example 10.14 The extent of differences in the *effectiveness* of psychotherapists will be examined in dependence of their sex.

In an extension of the research question in Example 10.8, the extent of differences in the *effectiveness* of psychotherapists will be investigated more accurately in dependence of their sex. The sample of $a = 22$ psychotherapists is stratified; in fact, there were 11 psychotherapists selected from the stratification group 'male' as well as 11 from the stratification group 'female'. Thus, the *sex of the therapist* represents factor A, which is a fixed factor. The different *psychotherapists* represent the 11 levels of the nested factor *therapist*, B, which still is a random factor. Therefore, this is the mixed model (A fixed, B random) of the two-way nested classification. The 'planning', which happens just now, after the fact of given data, provides the following insight. If interest is primarily about the null hypothesis regarding factor A, the *sex of the therapist*, then the sample size n of the clients (per cell) does not occur in the degrees of freedom of the respective F-test ($\boldsymbol{MS}_A$ has to be related to $\boldsymbol{MS}_{B \text{ in } A}$); thus, planning the study is just about determining b; namely the number of psychotherapists per factor level of A. Thus, the sample size has to be only $n > 1$; so, $n = 2$ would be sufficient. We decide on: $\alpha = 0.05$, $\beta = 0.20$, and $\delta = 1\sigma$.

In **R**, we load the package OPDOE and type

```
> size_b.two_way_nested.b_random_a_fixed_a(alpha = 0.05,
+                                          beta = 0.2,
+                                          delta = 1, a = 2,
+                                          cases = "maximin")
```

i.e. we apply the function `size_b.two_way_nested.b_random_a_fixed_a` and use successively the arguments for $\alpha = 0.05$, $\beta = 0.20$, and $\delta = 1\sigma$ (`delta = 1`). With `a = 2`, we determine the number of levels of the factor A, and with `cases = "maximin"` we request the calculation of the sample size for the worst case.

As a result, we get:

```
[1] 17
```

The calculated number of psychotherapists per sex, $b = 17$, comes up short in the given data.

We want to examine the presumption of variance homogeneity in accordance with the abovementioned rule of thumb. We proceed in correspondence with Example 10.8. The proportion of largest to smallest standard deviation is $2.875/2.021 = 1.422 < 1.5$, the plausible limit. Therefore the application of analysis of variance seems to be justified.

In **R**, to conduct the analysis we type

```
> aov.6 <- aov(effect ~ sex_ther + Error(therapist))
> summary(aov.6)
```

i.e. we apply the function `aov()` and request the analysis of the character *effectiveness* (`effect`) with regard to the (fixed) factor *sex of the therapist* (`sex_ther`) and the random (`Error(therapist)`) factor `therapist`. The result of this analysis is assigned to the object `aov.6`, which we submit to the function `summary()`.

As a result, we get:

```
Error: therapist
          Df Sum Sq Mean Sq F value Pr(>F)
```

```
sex_ther    1   0.10   0.0983   0.0051 0.9438
Residuals 20 386.29 19.3147
Error: Within
           Df Sum Sq Mean Sq F value Pr(>F)
Residuals 566 3247.5   5.7376
```

Regarding the main effect of the factor *therapist* (`therapist`), we manually calculate the F-value of interest, namely $F = 19.3147 / 5.7376 = 3.366$, with $df_1 = 20$ and $df_2 = 566$ degree of freedom. Now, we can determine the critical value $F(20, 566: 0.99) = 1.91$ by consulting Table B4 in Appendix B. Alternatively, we can calculate the p-value which corresponds to the value $F = 3.366$ using **R**, in the following way:

```
> pf(3.366, df1 = 20, df2 = 566, lower.tail = FALSE)
```

i.e. we apply the function `pf()` and use as the first argument the observed F value of 3.366. As the second and third argument we use the degrees of freedom df_1 and df_2. With `lower.tail = FALSE`, we request the corresponding p-value.

As a result, we get:

```
[1] 1.347957e-06
```

In SPSS, we proceed in the same way (Analyze – General Linear Model – Univariate...) as in Example 10.8, and in the window shown in Figure 10.7, we additionally drag and drop the factor sex_ther to the field Fixed Factor(s):. After a click on Model..., the window shown in Figure 10.10 appears. In the pull-down menu Sum of squares:, we select Type I, which is appropriate for the calculation of a nested classification. A click on Continue gets us back to the previous window. Next, we click Paste and change the last line in the appearing window, which contains SPSS-syntax, to the last one shown in Figure 10.13, to adjust it for nested classification. In order to conduct the analysis, we select All in the pull-down menu Run and obtain the result shown in Table 10.21.

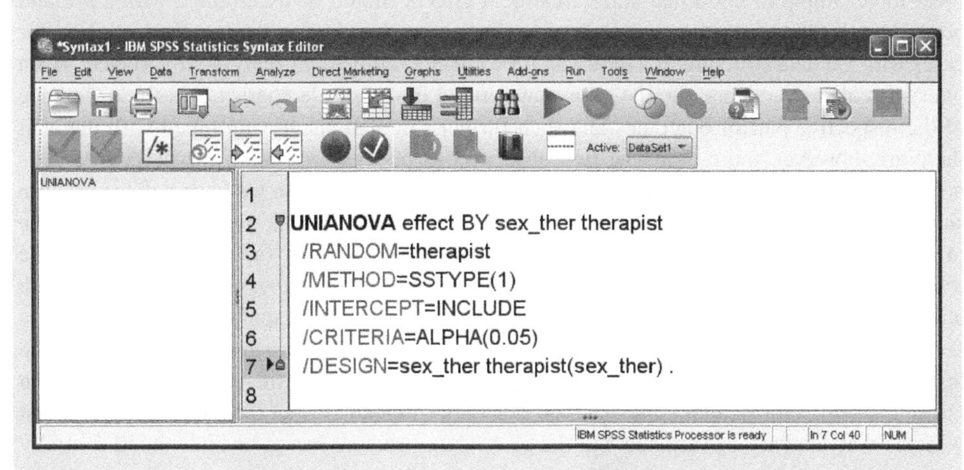

Figure 10.13 SPSS-syntax for Example 10.14.

Table 10.21 SPSS-output of the table of variances for the two-way analysis of variance with nested classification (mixed model) in Example 10.14.

Tests of Between-Subjects Effects

Dependent Variable:effectiveness

Sour Source ce		Type I Sum of Squares	df	Mean Square	F	Sig.
Intercept	Hypothesis	14908.014	1	14908.014	757.342	.000
	Error	387.512	19.686	19.685[a]		
sex_ther	Hypothesis	.098	1	.098	.005	.944
	Error	387.512	19.686	19.685[b]		
therapist(sex_ther)	Hypothesis	386.294	20	19.315	3.366	.000
	Error	3247.474	566	5.738[c]		

a. 1.027 MS(therapist(sex_ther)) - .027 MS(Error)
b. 1.027 MS(therapist(sex_ther)) - .027 MS(Error)
c. MS(Error)

We realize that the sex of the psychotherapist has no significant influence on the effectiveness of the therapy. However, the effect of the psychotherapist is of course significant, as it is in Example 10.8. Certain psychotherapists are, therefore, more effective than others.

As already mentioned several times, most recently in Section 10.4.1.1, we often have two outcomes for each research unit, which are statistically interpreted as each stemming from an individually modeled random variable, but with respect to content describe the same thing (bear the example in mind that some treatment effects should be ascertained with a *pre* and *post* design). Of course, the repeatedly described procedure is also the simplest solution for cases with the data structure of a two-way (or even any multi-way) analysis of variance: given interval-scaled characters, the data are transformed into a single character as the difference of the respective pair of outcomes and then further calculations are continued with them. If there are, however, more than just two observations per research unit, this procedure does not work. Formally, the study design has to be 'redefined'. Suppose we have a data structure such as a one-way analysis of variance (see Table 10.1), except that now the given outcome observations stem, row for row by line, over all factor levels of factor A, from the same research unit (person) v (bear in mind that, up to now, there was always a different research unit v for each factor level). Then these persons $1, 2, \ldots, n$ represent the factor levels of a second factor, B, which is random in any case: we must indeed assume further on that the research units were drawn by a random sample procedure and that there is not just interest in the actual sampled research units. That is, it is only 'pretended' that the data structure is that of a one-way analysis of variance, but the given data structure has to be understood formally as a two-way analysis of variance. The analysis is thereby resolved (but see also Section 13.3).

10.4.5 Two-way analysis of variance for ordinal-scaled characters

Non-parametric analogues of two-way analysis of variance for both cross classification and nested classification can be found in Brunner & Munzel (2002). They are particularly appropriate for ordinal-scaled characters, but they are not yet counted as standard procedures and can only be performed using the computer program SAS (www.sas.com), which is not discussed in this book. So, when there is a data structure of a two-way analysis of variance for an only ordinal-scaled character, we recommend using an H-test for each factor separately and not testing the interaction effect, if the reader does not want to familiarize him/herself with that computer program.

10.4.6 Bivariate comparison of two nominal-scaled factors

If there are two qualitative factors, in particular two nominal-scaled factors, and if one considers the observed frequencies per combination of factor levels (in one sample), then there is basically no other data structure than the one discussed in Section 10.4.3. The hypotheses H_0 and H_A stated there can now be interpreted in a different way; namely H_0: $p_{ij} = p_{lj}$ regarding all factor levels j of a factor B (in Section 10.4.3, it was measurement values of the multi-categorical character of interest) in all $i = 1, 2, \ldots, a$ factor levels of a factor A (it was samples in Section 10.4.3), or respectively in all pairs of factor levels i and l; and H_A: $p_{ij} \neq p_{lj}$ for at least one j and one pair of factor levels i and l, respectively. Thus, these hypotheses refer exactly to the interaction of A and B. If we abstain from an exact test and consequently work with the test statistic of Formula (9.3), we would therefore not have to distinguish between model I, model II, and the mixed model. Main effects that are possibly of interest have to be tested separately with the test statistic of Formula (8.11).

10.4.7 Three-way analysis of variance

In three-way analysis of variance, in addition to factor A with the levels $A_1, A_2, \ldots, A_a$ and factor B with the levels $B_1, B_2, \ldots, B_b$, we consider a third factor, C, with the levels C_k, i.e. $C_1, C_2, \ldots, C_c$. Again, the levels of the three factors can be combined in two ways, either according to a cross classification or according to a nested classification. It is of course possible that some combinations of factor levels do not occur; in particular it does not always have to be the case that $n_{ijk} = n$. The outcomes of the random variables $\boldsymbol{y}_{ijkv}$ are then y_{ijkv}, $i = 1, 2, \ldots, a; j = 1, 2, \ldots, b; k = 1, 2, \ldots, c; v = 1, 2, \ldots, n_{ijk}$. Besides the obvious generalization of cross classification and nested classification, there are two mixed classifications in which both cross classification and nested classification are given. This means there exists the case of the cross classification $A \times B \times C$, and the case of the nested classification $A \succ B \succ C$ as well as the two mixed classifications $(A \times B) \succ C$ and $(A \succ B) \times C$. It should be noted that each of these cases can be given as model I, model II, or as a mixed model. Table 10.22 summarizes all variants.

Table 10.22 Classifications in combination with the models of three-way analysis of variance. '×' indicates a cross classification, '≻' a nested classification; bold notation indicates a random factor, otherwise fixed factors are represented.

$A \times B \times C$
$A \times B \times \mathbf{C}$
$A \times \mathbf{B} \times \mathbf{C}$
$\mathbf{A} \times \mathbf{B} \times \mathbf{C}$
$A \succ B \succ C$
$\mathbf{A} \succ B \succ C$
$A \succ \mathbf{B} \succ C$
$A \succ B \succ \mathbf{C}$
$\mathbf{A} \succ \mathbf{B} \succ C$
$\mathbf{A} \succ B \succ \mathbf{C}$
$A \succ \mathbf{B} \succ \mathbf{C}$
$\mathbf{A} \succ \mathbf{B} \succ \mathbf{C}$
$(A \times B) \succ C$
$(A \times \mathbf{B}) \succ C$
$(\mathbf{A} \times \mathbf{B}) \succ C$
$(A \times B) \succ \mathbf{C}$
$(A \times \mathbf{B}) \succ \mathbf{C}$
$(\mathbf{A} \times \mathbf{B}) \succ \mathbf{C}$
$(A \succ B) \times C$
$(\mathbf{A} \succ B) \times C$
$(A \succ \mathbf{B}) \times C$
$(\mathbf{A} \succ \mathbf{B}) \times C$
$(A \succ B) \times \mathbf{C}$
$(\mathbf{A} \succ B) \times \mathbf{C}$
$(A \succ \mathbf{B}) \times \mathbf{C}$
$(\mathbf{A} \succ \mathbf{B}) \times \mathbf{C}$

Doctor Model equations for the different types of classifications and models of three-way analysis of variance are given in Table 10.23.

Table 10.23 Model equations for all classifications in combination with model I, model II, and mixed model in three-way analysis of variance; the factor levels of nested factors are set in the index in brackets.

Structure	Model equation
$A \times B \times C$	$\mathbf{y}_{ijkv} = \mu + a_i + b_j + c_k + (ab)_{ij} + (ac)_{ik} + (bc)_{jk} + (abc)_{ijk} + \mathbf{e}_{ijkv}$
$A \times B \times \mathbf{C}$	$\mathbf{y}_{ijkv} = \mu + a_i + b_j + \mathbf{c}_k + (ab)_{ij} + (\mathbf{ac})_{ik} + (\mathbf{bc})_{jk} + (\mathbf{abc})_{ijk} + \mathbf{e}_{ijkv}$
$A \times \mathbf{B} \times \mathbf{C}$	$\mathbf{y}_{ijkv} = \mu + a_i + \mathbf{b}_j + \mathbf{c}_k + (\mathbf{ab})_{ij} + (\mathbf{ac})_{ik} + (\mathbf{bc})_{jk} + (\mathbf{abc})_{ijk} + \mathbf{e}_{ijkv}$
$\mathbf{A} \times \mathbf{B} \times \mathbf{C}$	$\mathbf{y}_{ijkv} = \mu + \mathbf{a}_i + \mathbf{b}_j + \mathbf{c}_k + (\mathbf{ab})_{ij} + (\mathbf{ac})_{ik} + (\mathbf{bc})_{jk} + (\mathbf{abc})_{ijk} + \mathbf{e}_{ijkv}$
$A \succ B \succ C$	$\mathbf{y}_{ijkv} = \mu + a_i + b_{j(i)} + c_{k(ij)} + \mathbf{e}_{ijkv}$

Table 10.23 (*Continued*)

Structure	Model equation
$\boldsymbol{A} \succ B \succ C$	$\boldsymbol{y}_{ijkv} = \mu + \boldsymbol{a}_i + b_{j(i)} + c_{k(ij)} + \boldsymbol{e}_{ijkv}$
$A \succ \boldsymbol{B} \succ C$	$\boldsymbol{y}_{ijkv} = \mu + a_i + \boldsymbol{b}_{j(i)} + c_{k(ij)} + \boldsymbol{e}_{ijkv}$
$A \succ B \succ \boldsymbol{C}$	$\boldsymbol{y}_{ijkv} = \mu + a_i + b_{j(i)} + \boldsymbol{c}_{k(ij)} + \boldsymbol{e}_{ijkv}$
$\boldsymbol{A} \succ \boldsymbol{B} \succ C$	$\boldsymbol{y}_{ijkv} = \mu + \boldsymbol{a}_i + \boldsymbol{b}_{j(i)} + c_{k(ij)} + \boldsymbol{e}_{ijkv}$
$\boldsymbol{A} \succ B \succ \boldsymbol{C}$	$\boldsymbol{y}_{ijkv} = \mu + \boldsymbol{a}_i + b_{j(i)} + \boldsymbol{c}_{k(ij)} + \boldsymbol{e}_{ijkv}$
$A \succ \boldsymbol{B} \succ \boldsymbol{C}$	$\boldsymbol{y}_{ijkv} = \mu + a_i + \boldsymbol{b}_{j(i)} + \boldsymbol{c}_{k(ij)} + \boldsymbol{e}_{ijkv}$
$\boldsymbol{A} \succ \boldsymbol{B} \succ \boldsymbol{C}$	$\boldsymbol{y}_{ijkv} = \mu + \boldsymbol{a}_i + \boldsymbol{b}_{j(i)} + \boldsymbol{c}_{k(ij)} + \boldsymbol{e}_{ijkv}$
$(A \times B) \succ C$	$\boldsymbol{y}_{ijkv} = \mu + a_i + b_j + (ab)_{ij} + c_{k(ij)} + \boldsymbol{e}_{ijkv}$
$(A \times \boldsymbol{B}) \succ C$	$\boldsymbol{y}_{ijkv} = \mu + a_i + \boldsymbol{b}_j + (\boldsymbol{ab})_{ij} + c_{k(ij)} + \boldsymbol{e}_{ijkv}$
$(\boldsymbol{A} \times \boldsymbol{B}) \succ C$	$\boldsymbol{y}_{ijkv} = \mu + \boldsymbol{a}_i + \boldsymbol{b}_j + (\boldsymbol{ab})_{ij} + c_{k(ij)} + \boldsymbol{e}_{ijkv}$
$(A \times B) \succ \boldsymbol{C}$	$\boldsymbol{y}_{ijkv} = \mu + a_i + b_j + (ab)_{ij} + \boldsymbol{c}_{k(ij)} + \boldsymbol{e}_{ijkv}$
$(A \times \boldsymbol{B}) \succ \boldsymbol{C}$	$\boldsymbol{y}_{ijkv} = \mu + a_i + \boldsymbol{b}_j + (\boldsymbol{ab})_{ij} + \boldsymbol{c}_{k(ij)} + \boldsymbol{e}_{ijkv}$
$(\boldsymbol{A} \times \boldsymbol{B}) \succ \boldsymbol{C}$	$\boldsymbol{y}_{ijkv} = \mu + \boldsymbol{a}_i + \boldsymbol{b}_j + (\boldsymbol{ab})_{ij} + \boldsymbol{c}_{k(ij)} + \boldsymbol{e}_{ijkv}$
$(A \succ B) \times C$	$\boldsymbol{y}_{ijkv} = \mu + a_i + b_{j(i)} + c_k + (ac)_{ik} + (bc)_{j(i)k} + \boldsymbol{e}_{ijkv}$
$(\boldsymbol{A} \succ B) \times C$	$\boldsymbol{y}_{ijkv} = \mu + \boldsymbol{a}_i + b_{j(i)} + c_k + (\boldsymbol{ac})_{ik} + (bc)_{j(i)k} + \boldsymbol{e}_{ijkv}$
$(A \succ \boldsymbol{B}) \times C$	$\boldsymbol{y}_{ijkv} = \mu + a_i + \boldsymbol{b}_{j(i)} + c_k + (ac)_{ik} + (\boldsymbol{bc})_{j(i)k} + \boldsymbol{e}_{ijkv}$
$(\boldsymbol{A} \succ \boldsymbol{B}) \times C$	$\boldsymbol{y}_{ijkv} = \mu + \boldsymbol{a}_i + \boldsymbol{b}_{j(i)} + c_k + (\boldsymbol{ac})_{ik} + (\boldsymbol{bc})_{j(i)k} + \boldsymbol{e}_{ijkv}$
$(A \succ B) \times \boldsymbol{C}$	$\boldsymbol{y}_{ijkv} = \mu + a_i + b_{j(i)} + \boldsymbol{c}_k + (\boldsymbol{ac})_{ik} + (\boldsymbol{bc})_{j(i)k} + \boldsymbol{e}_{ijkv}$
$(\boldsymbol{A} \succ B) \times \boldsymbol{C}$	$\boldsymbol{y}_{ijkv} = \mu + \boldsymbol{a}_i + b_{j(i)} + \boldsymbol{c}_k + (\boldsymbol{ac})_{ik} + (\boldsymbol{bc})_{j(i)k} + \boldsymbol{e}_{ijkv}$
$(A \succ \boldsymbol{B}) \times \boldsymbol{C}$	$\boldsymbol{y}_{ijkv} = \mu + a_i + \boldsymbol{b}_{j(i)} + \boldsymbol{c}_k + (\boldsymbol{ac})_{ik} + (\boldsymbol{bc})_{j(i)k} + \boldsymbol{e}_{ijkv}$
$(\boldsymbol{A} \succ \boldsymbol{B}) \times \boldsymbol{C}$	$\boldsymbol{y}_{ijkv} = \mu + \boldsymbol{a}_i + \boldsymbol{b}_{j(i)} + \boldsymbol{c}_k + (\boldsymbol{ac})_{ik} + (\boldsymbol{bc})_{j(i)k} + \boldsymbol{e}_{ijkv}$

We do not deal with the assumptions for the derivation of hypothesis testing in detail; they are analogous to the previous tracts. We will also not give explicitly the hypotheses themselves or the empirical and theoretical analysis of variance tables, from which the F-tests would be derived directly (but see Rasch, Herrendörfer, Bock, Victor, & Guiard, 2008). Rather, we want to lead the reader over to the software programs, which will calculate the appropriate F-tests, given the design is correctly specified by the researcher as regards cross or nested classification on the one hand and with regards to model I, model II, or the mixed model on the other hand. However, we want to illustrate the application of two variants by using examples.

Regarding planning a study, again the **R**-package `OPDOE` offers all variants (see Rasch, Pilz, Verdooren, & Gebhardt, 2011).

Doctor

Example 10.15 The extent of differences in the effectiveness of psychotherapists will be examined in dependence of their sex, as well as in dependence of the client's sex.

In an extension of the research question in Examples 10.8 and 10.14, this is now about the extent of differences in the effectiveness of psychotherapists with respect to their sex and with respect to the client's sex; in particular regarding the sex relationship of both. We thus distinguish the already considered n_i clients between 'male' and 'female'. The character *sex of the therapist* continues to be the fixed factor A. The psychotherapists represent the levels of the nested factor B (*therapist*), a random factor. And finally, the third factor C represents the *sex of the client*, which is also a fixed factor. Therefore, this is the mixed classification $(A \succ \mathbf{B}) \times C$. The structure of the data – if y_{ijkv} is the vth outcome in the ith level of A, the jth level of B, and the kth level of C ($i = 1, 2; j = 1, 2, \ldots, b; k = 1, 2; v = 1, \ldots, n_{ijk}$) – is shown in Table 10.24 (n has to be replaced by n_{ijk} in the given example); to make the $a \cdot b = 2 \cdot 11 = 22$ psychotherapists distinguishable regarding their identification, we number them $1, 2, \ldots, p, p+1, p+2, \ldots, 2p$ (with $p = b$). We determine the type-I risk with $\alpha = 0.01$.

Table 10.24 Data structure of the $2 \times b \times 2 \times n$ outcomes in Example 10.15.

	A_1 Therapist male				A_2 Therapist female			
	B_{11} Therapist 1	B_{12} Therapist 2	...	B_{1b} Therapist p	B_{21} Therapist $p+1$	B_{22} Therapist $p+2$	...	B_{2b} Therapist $2p$
C_1 Client male	y_{1111}	y_{1211}	...	y_{1b11}	y_{2111}	y_{2211}	...	y_{2b11}
	y_{1112}	y_{1212}	...	y_{1b12}	y_{2112}	y_{2212}	...	y_{2b12}
	$\vdots$	$\vdots$	$\vdots$	$\vdots$	$\vdots$	$\vdots$	$\vdots$	$\vdots$
	y_{111n}	y_{121n}	...	y_{1b1n}	y_{211n}	y_{221n}	...	y_{2b1n}
C_2 Client female	y_{1121}	y_{1221}	...	y_{1b21}	y_{2121}	y_{2221}	...	y_{2b21}
	y_{1122}	y_{1222}	...	y_{1b22}	y_{2122}	y_{2222}	...	y_{2b22}
	$\vdots$	$\vdots$	$\vdots$	$\vdots$	$\vdots$	$\vdots$	$\vdots$	$\vdots$
	y_{112n}	y_{122n}	...	y_{1b2n}	y_{212n}	y_{222n}	...	y_{2b2n}

If there is primary interest in the null hypothesis regarding factor A, n does not matter again as concerns planning the study. Then $b = 17$ once more. However, if there is primary interest in the null hypothesis regarding factor C, then $b = 9$ will be needed, for the same precision requirements.

In **R**, we can see this by applying the package OPDOE in the following way. We type

```
> size_b.three_way_mixed_cxbina.model_4_c(alpha = 0.05, beta = 0.2,
+                                         delta = 1, a = 2, c = 2,
+                                         n = 1, cases = "maximin")
```

i.e. we apply the function `size_b.three_way_mixed_cxbina.model_4_c`, using the precision requirements $\alpha = 0.05$, $\beta = 0.20$, and $\delta = 1\sigma$ (`delta = 1`) as arguments. With `a = 2` and `c = 2` we determine the number of factor levels; thus $a = b = 2$. Furthermore, we indicate, with `n = 1`, that the sample size can be as small as possible, and with `cases = "maximin"` we request the sample size for the worst case.

As a result, we get:

```
[1] 9
```

The analysis is conducted as follows – beforehand we check the presumption of variance homogeneity, analogous to Examples 10.12 and 10.14; the proportion of largest to smallest standard deviation is 2.620/1.835 = 1.428 < 1.5, which is the plausible limit. Therefore, the application of analysis of variance seems to be justified.

In **R**, we type

```
> aov.7 <- aov(effect ~ sex_ther + sex_clie +
+             sex_clie:sex_ther +
+             Error(therapist + therapist:sex_clie))
> summary(aov.7)
```

i.e. we apply the function `aov()` and request the analysis of the character *effectiveness* (`effect`) with regard to the fixed factor *sex of the therapist* (`sex_ther`), the fixed factor *sex of the client* (`sex_clie`), the random (`Error()`) factor `therapist`, the interaction between *sex of the client* and *sex of the therapist* (`sex_clie:sex_ther`), and the interaction between `therapist` and *sex of the client*. We assign the result of the analysis to the object `aov.7`. Finally, we submit this object to the function `summary()`.

As a result, we get:

```
Error: therapist
          Df Sum Sq Mean Sq F value Pr(>F)
sex_ther   1   0.10  0.0983  0.0051 0.9438
Residuals 20 386.29 19.3147

Error: therapist:sex_clie
                  Df  Sum Sq Mean Sq F value Pr(>F)
sex_clie           1   6.883  6.8830  0.6147 0.4422
sex_ther:sex_clie  1   9.620  9.6203  0.8591 0.3650
Residuals         20 223.952 11.1976

Error: Within
                  Df Sum Sq Mean Sq F value Pr(>F)
Residuals        544   3007  5.5276
```

Regarding the main effect of the factor *sex of the therapist* (`sex_ther`), we thus obtain with $F = 0.0051$ the p-value 0.9438. Concerning the main effect of the factor *sex of the client* (`sex_clie`), we get with $F = 0.6147$ the p-value 0.4422, and regarding the interaction between the factors *sex of the therapist* (`sex_ther`), and *sex of the client* (`sex_clie`), we obtain with $F = 0.8591$ the p-value 0.3650.

We calculate manually the F value of interest for the main effect of the factor `therapist`; namely: $F = 19.3147 / 11.1976 = 1.725$, with $df_1 = 20$ and $df_2 = 20$ degrees of freedom. Now, we can determine the critical value $F(20, 20{:}\ 0.99) = 2.938$ by consulting Table B4 in Appendix B. Alternatively, we can calculate the p-value which corresponds to the value $F = 1.725$ using **R**, in the following way:

```
> pf(1.725, df1 = 20, df2 = 20, lower.tail = FALSE)
```

i.e. we apply the function `pf()` and use as the first argument the observed F value of `1.725`. As the second and third argument we use the degrees of freedom, each being `20`. With `lower.tail = FALSE`, we request the corresponding p-value.

As a result, we get:

```
[1] 0.1156951
```

We calculate the F value of interest for the interaction between the factor `therapist` and the factor *sex of the client* (`sex_clie`) as $F = 11.1976 / 5.5276 = 2.026$, with $df_1 = 20$ and $df_2 = 544$ degrees of freedom. We calculate the corresponding p-value using **R** in the following way:

```
> pf(2.026, df1 = 20, df2 = 544, lower.tail = FALSE)
```

i.e. we apply the function `pf()` and use the F value as the first argument and the degrees of freedom df_1 and df_2 as the second and the third argument. With `lower.tail = FALSE`, we request the corresponding p-value.

As a result, we get:

```
[1] 0.005414724
```

In SPSS, we proceed (Analyze – General Linear Model – Univariate. . .) as shown in Example 10.14, and additionally drag and drop the character sex_clie to the field Fixed Factor(s):. We click Model. . ., choose again Type I in Sum of squares:, and come back to the previous window by clicking Continue. A click on Paste brings us to a new window with SPSS-syntax, in which we change the last line to match the one shown in Figure 10.14. Finally, we select All in the pull-down menu Run, and obtain the result shown in Table 10.25.

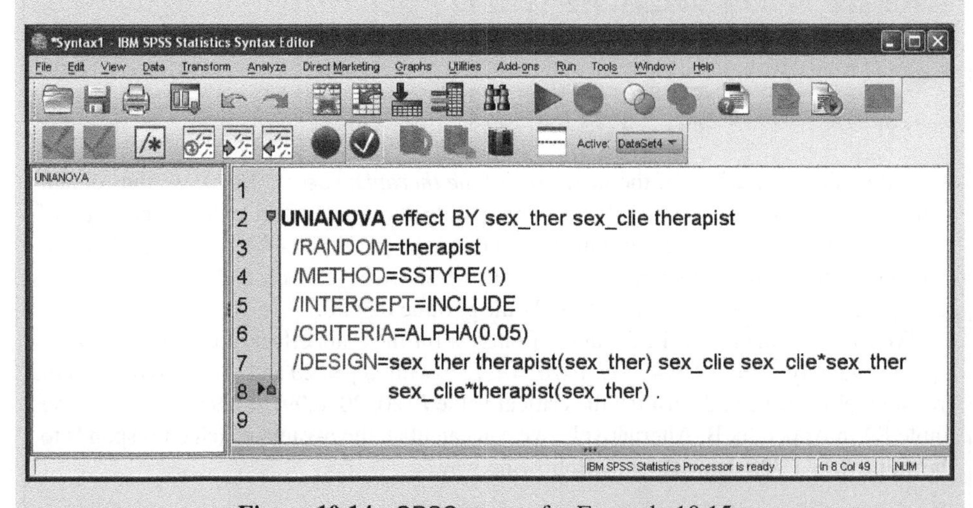

Figure 10.14 SPSS-syntax for Example 10.15.

Table 10.25 SPSS-output of the table of variances for the tree-way analysis of variance with nested classification (mixed model) in Example 10.15.

Tests of Between-Subjects Effects

Dependent Variable:effectiveness

Source		Type I Sum of Squares	df	Mean Square	F	Sig.
Intercept	Hypothesis	14908.014	1	14908.014	757.122	.000
	Error	387.850	19.697	19.690[a]		
sex_ther	Hypothesis	.098	1	.098	.005	.944
	Error	387.850	19.697	19.690[b]		
therapist(sex_ther)	Hypothesis	386.294	20	19.315	1.725	.116
	Error	223.952	20	11.198[c]		
sex_clie	Hypothesis	6.883	1	6.883	.606	.446
	Error	221.133	19.480	11.352[d]		
sex_ther * sex_clie	Hypothesis	9.620	1	9.620	.847	.369
	Error	221.133	19.480	11.352[e]		
sex_clie * therapist (sex_ther)	Hypothesis	223.952	20	11.198	2.026	.005
	Error	3007.019	544	5.528[f]		

a. 1.027 MS(therapist(sex_ther)) - .027 MS(Error)
b. 1.027 MS(therapist(sex_ther)) - .027 MS(Error)
c. MS(sex_clie * therapist(sex_ther))
d. 1.027 MS(sex_clie * therapist(sex_ther)) - .027 MS(Error)
e. 1.027 MS(sex_clie * therapist(sex_ther)) - .027 MS(Error)
f. MS(Error)

There are unimportant differences between **R** and SPSS regarding the results of the two F-tests for examining the main effect of the factor *sex of the client* and the interaction effect of the factors *sex of the therapist* and *sex of the client*.

We notice that only the interaction effect between the client's and the psychotherapist's sex is significant. Therefore, certain psychotherapists are systematically more effective for females using their therapy; certain other psychotherapists are systematically more effective for males. The significant main effect of factor B, *therapist*, which was found in the Examples 10.8 and 10.14 can now be explained by the specific interaction.

One may be further interested in finding out between which psychotherapists and which sex there are differences and also in which direction the differences are. Then we have to create a graph similar to Example 10.13.

Doctor

Example 10.16 The items of a psychological test will be calibrated in accordance with the Rasch model

The Rasch model (see Section 15.2.3.1) implies certain properties of the items of a psychological test. These can be used for empirical data, to verify whether the items of a given test actually correspond to the Rasch model; then, and only then, it is justifiable to count the number of solved items as the test score per person.

The most important property in this context is that there is no interaction between certain groups of persons; for instance with regard to the sex and specific items. Let us therefore consider the *testee's sex* as factor A (with two factor levels), the *testees* themselves as the factor levels of factor B, which is nested to factor A; and the *items* as factor levels of a factor C. Then A and C are fixed, and B is random; in the calibration of a psychological test, it is exactly about the specifically designed items; it is also exactly about the respective levels of the partition into subgroups of testees, but it is not about certain testees, but instead about those of a random sample. This is therefore the same variant as in Example 10.15, the variant $(A \succ \mathbf{B}) \times C$. However, the interesting character, *test-achievement*, is a dichotomous one, having only the two measurement values 'solved' and 'not solved'; its modeling as a normally distributed random variable is therefore implausible. Nevertheless, Kubinger, Rasch, & Yanagida (2009) demonstrated, using simulation studies, that if a three-way analysis of variance is applied on such data, it largely meets the nominal type-I risk (with regard to the interaction of interest between A and C) – under certain realistic conditions with a practically useful power of $1 - \beta$. As a result, not only a new model test for the Rasch model is obtained, but, in contrast to any other model tests for the Rasch model, it is possible to determine the necessary number of testees b per factor level of A to meet any desired precision requirements; i.e. planning a study for the calibration of a psychological test (see Example 15.8 for illustration).

Doctor

Example 10.17 The extent of faking in personality questionnaires will be investigated

Several studies suggest that the extent to which persons fake personality questionnaires in selection situations in the sense of a more 'attractive' view of their personality (see the psychometric quality criterion of unfakeability e.g. in Kubinger, 2009b) depends on three factors among others: the *answer format* (factor A), in particular with the factor levels 'dichotomous response format' and 'analogue scale response format'; the *time pressure* (factor B), with the factor levels 'response time limited' and 'response time not limited'; and the *instruction* (factor C), with the factor levels 'warning is given that faking response behavior can be detected using ingenious computer software' and 'no such warning'. The three factors can be cross classified; they are all fixed and dichotomous. Thus, this is variant $A \times B \times C$. Planning the study will be based on $\alpha = 0.05$, $\beta = 0.05$, and $\delta = \sigma$.

In **R**, we type, using the package OPDOE

```
> size_n.three_way_cross.model_1_axbxc(alpha = 0.05, beta = 0.05,
+                                       delta = 1, a = 2, b = 2,
+                                       c = 2, cases = "maximin")
```

i.e. we apply the function `size_n.three_way_cross.model_1_axbxc()`, using the arguments $\alpha = 0.05$, $\beta = 0.05$, and $\delta = \sigma$ (`delta = 1`). With `a = 2`, `b = 2`, and `c = 2`, we indicate the number of factor levels $a = b = c = 2$, and with `cases = "maximin"`, we request the sample size n for the worst case.

As a result, we get:

```
[1] 27
```

According to this, we have to present the personality questionnaire in an experiment to $n = 27$ subjects in each combination of factor levels; we therefore need a total of 216 subjects who are randomly assigned to the combinations of factor levels.

We use the data from Khorramdel & Kubinger (2006) for this analysis; these are the data from *Example 10.17* (see Chapter 1 for its availability). For the sake of example, we restrict ourselves to the character *internalization* – it measures the extent of 'internal locus of control of reinforcement' (see footnote 1 in Chapter 4).

We again examine the proportion of largest to smallest standard deviation. By analogy e.g. to Example 10.12, we obtain the ratio $1.723/1.221 = 1.411$ both with **R** and SPSS, which is smaller than 1.5 and in accordance with the abovementioned rule of thumb. Thus, application of analysis of variance again appears to be justifiable.

In **R**, we load the package `car`, which we have already used in Example 9.7, by applying the function `library()`. Furthermore, we enable access to the database *Example_10.17* (see Chapter 1) with the help of the function `attach()`. Now, we type

```
> options(contrasts = c("contr.sum", "cont.poly"))
> aov.8 <- aov(inter ~ format * time * instruct)
> Anova (aov.8, type = 3)
```

i.e. we apply the function `options()` and set `contrasts = c("contr.sum", "cont.poly")` to request the same computational procedure as SPSS. Next we use the function `aov()` and request the analysis of the character *internality* (`inter`) with regard to the (fixed) factors *response format* (`format`), *time pressure* (`time`), and *instruction* (`instruct`), including all main effects and interaction effects. We assign the result of the analysis to the object `aov.8`. Next, we submit this object as the first argument to the function `Anova()`. With `type = 3`, we decide on type III.

As a result, we get (shortened output):

```
Anova Table (Type III tests)

Response: inter
Sum Sq  Df   F value Pr(>F)
(Intercept)           800.49   1 381.2850 <2e-16
format                  8.68   1   4.1364 0.0433
time                    0.15   1   0.0694 0.7925
instruct                2.73   1   1.3027 0.2551
format:time             0.53   1   0.2511 0.6169
format:instruct         5.25   1   2.5010 0.1154
time:instruct           3.81   1   1.8124 0.1798
format:time:instruct    0.13   1   0.0619 0.8038
Residuals             417.79 199
```

For calculating the means, we type

```
> model.tables(aov.8, type = "means")
```

i.e. we apply the function `model.tables()`, and submit the object `aov.8` to the function. With `type = "means"`, we request the calculation of the means.

As a result, we get (shortened output):

```
Tables of means
Grand mean

2.10628

format
      dichotomous analogue
            2.319    1.835
```

In SPSS, we reach the window shown in Figure 10.7 using the same sequence of commands (Analyze – General Linear Model – Univariate...) as in Example 10.8. There, we drag and drop the character internalization to the field Dependent Variable:. Next, we drag and drop the characters answer format, time pressure, and instruction to the field Fixed Factor(s):. We click Options... and tick Descriptive statistics in the window shown in Figure 10.8. A click on Continue brings us back to the previous window. After clicking OK, we obtain the results shown in Table 10.26 and 10.27.

Table 10.26 SPSS-output of the table of variances for the three-way analysis of variance (model I) in Example 10.17.

Tests of Between-Subjects Effects

Dependent Variable:internalization

Source	Type III Sum of Squares	df	Mean Square	F	Sig.
Corrected Model	23.872[a]	7	3.410	1.624	.130
Intercept	800.486	1	800.486	381.285	.000
format	8.684	1	8.684	4.136	.043
time	.146	1	.146	.069	.792
instruct	2.735	1	2.735	1.303	.255
format * time	.527	1	.527	.251	.617
format * instruct	5.251	1	5.251	2.501	.115
time * instruct	3.805	1	3.805	1.812	.180
format * time * instruct	.130	1	.130	.062	.804
Error	417.789	199	2.099		
Total	1360.000	207			
Corrected Total	441.662	206			

a. R Squared = .054 (Adjusted R Squared = .021)

Table 10.27 SPSS-output of the means for the factor levels of the significant main effect in Example 10.17 (shortened output).

Descriptive Statistics

Dependent Variable:internalization

answer format	time pressure	instruction	Mean	Std. Deviation	N
dichotomous	Total	Total	2.32	1.524	116
analogue	Total	Total	1.84	1.344	91

All main effects and interaction effects, with the exception of factor A (*answer format*) are not significant. By considering the means we notice that there is a larger test score (in the direction of *internalization*) for the 'dichotomous response format' than for the 'analogue scale response format'. Since a higher internal locus of control of reinforcement is more socially desirable, this result implies that there is a larger extent of fakeability in personality questionnaires given a dichotomous response format.

Regarding the analysis of more than three-way classifications, we refer to Hartung, Elpelt, & Voet (1997).

Summary

In *analysis of variance*, we have to distinguish between *fixed* and *random factors*, i.e. those whose factor levels were fixed, or determined by chance. It is a *model I* if there are only fixed factors, a *model II* if there are only random factors, and a *mixed model* if there are both types of factors. Additionally, two or more factors can be combined in certain ways; namely as a *cross classification* or as a *nested classification*. In terms of the latter, not every factor level of the one factor is combined with all factor levels of another, but every factor level of a nested factor appears in only one factor level of another factor. The basic difference between one-way and multi-way analysis of variance is that it is possible to examine *interaction effects* in addition to the *main effects* with regard to differences between the means. Any type of analysis of variance leads to an *F-test*. The application of a two-way analysis of variance for ordinal-scaled characters is uncommon. A bivariate comparison of two nominal-scaled characters leads to the χ^2-test again.

References

Bechhofer, R. E. (1954). A single sample multiple decision procedure for ranking means of normal populations with known variances. *Annals of Mathematical Statistics*, *25*, 16–39.

Brunner, E. & Munzel, U. (2002). *Nicht-parametrische Datenanalyse* [Non-parametric Data Analysis]. Heidelberg: Springer.

Cattell, R. B. (1987). *Intelligence: Its Structure, Growth, and Action.* New York: Elsevier.

Hartung, J., Elpelt, B., & Voet, B. (1997). *Modellkatalog Varianzanalyse. Buch mit CD-Rom* [Guide to Analysis of Variance. Book with CD-Rom]. Munich: Oldenbourg.

Hsu, J. C. (1996). *Multiple Comparisons: Theory and Methods*. New York: Chapman & Hall.

Khorramdel, L. & Kubinger, K. D. (2006). The effect of speededness on personality questionnaires – an experiment on applicants within a job recruiting procedure. *Psychology Science*, *48*, 378–397.

Kubinger, K. D. (2009b). *Psychologische Diagnostik – Theorie und Praxis psychologischen Diagnostizierens* (2nd edn) [Psychological Assessment – Theory and Practice of Psychological Consulting]. Göttingen: Hogrefe.

Kubinger, K. D., Rasch, D., & Yanagida, T. (2009). On designing data-sampling for Rasch model calibrating an achievement test. *Psychology Science Quarterly*, *51*, 370–384.

Lord, F. M. & Novick, M. R. (1968). *Statistical Theories of Mental Test Scores*. Reading, MA: Addison-Wesley.

Moder, K. (2007). How to keep the type I error rate in ANOVA if variances are heteroscedastic. *Austrian Journal of Statistics*, *36*, 179–188.

Moder, K. (2010). Alternatives to F-test in one way ANOVA in case of heterogeneity of variances. *Psychological Test and Assessment Modeling*, *52*, 343–353.

Rasch, D. (1995). *Einführung in die Mathematische Statistik* [Introduction to Mathematical Statistics]. Heidelberg: Barth.

Rasch, D., Herrendörfer, G., Bock, J., Victor, N., & Guiard, V. (2008). *Verfahrensbibliothek Versuchsplanung und -auswertung. Elektronisches Buch* [Collection of Procedures in Design and Analysis of Experiments. Electronic Book]. Munich: Oldenbourg.

Rasch, D., Pilz, J., Verdooren, R. L., & Gebhardt, A. (2011). *Optimal Experimental Design with R*. Boca Raton: Chapman & Hall/CRC.

Rasch, D., Rusch, T., Šimečková, M., Kubinger, K. D., Moder, K., & Šimeček, P. (2009). Tests of additivity in mixed and fixed effect two-way ANOVA models with single sub-class numbers. *Statistical Papers*, *50*, 905–916.

World Health Organization (1993). *The ICD-10 Classification of Mental and Behavioural Disorders. Diagnostic Criteria for Research*. Geneva: WHO.

Part IV

DESCRIPTIVE AND INFERENTIAL STATISTICS FOR TWO CHARACTERS

After the introduction of essential procedures for planning a study and its statistical analysis with respect to a single character, we will now refer to the simultaneous ascertainment of two characters – taken from the same sample. Generally, in this part a single sample or population is of interest, but two samples are of interest only by exception.

Insofar as we were concerned with inferential statistics up to now (particularly with hypothesis testing), the scientific question was about differences. Furthermore, mostly differences of means of samples and populations, respectively, were of interest. In the following chapter, however, the question is completely different: it is about the statistical relationship of two characters.

11

Regression and correlation

In this chapter we polarize deterministic relationships and dependencies on the one side and stochastic relationships of two characters on the other side – thereby stochastic means dependent on chance. Within the field of psychology only the latter relationships are of interest. We will deal at first with the graphical representation of corresponding observation pairs as a scatter plot. We will focus mainly on the linear regression function, which grasps the relationship of two characters, modeled by two random variables. Then we will consider several correlation coefficients, which quantify the strength of relationship. Depending on the scale type of the characters of interest, different coefficients become relevant. In addition to (point) estimators for the corresponding parameters, this chapter particularly deals with hypothesis testing concerning these parameters.

11.1 Introduction

The term 'relationship' can be explained by means of its colloquial sense. We know from everyday life, for instance, phrases and statements of the following type: 'Children's behavior is related to parents' behavior', 'Learning effort and success in exams are related', 'Birth-rate and calendar month are related', and so on.

Bachelor **Example 11.1** Type and strength of the relationship between the ages of the mother and father of last-born children is to be determined

From everyday life experience, as well as for obvious reasons, ages of both parents are related. Although there are some extreme cases, for example a 40-year-old mother together with a 25-year-old father of a child, or a 65-year-old father with a 30-year-old mother, nevertheless in our society a preponderance of parents with similar ages can be observed.

In contrast to relationships of variables in mathematics, in psychology relationships do not follow specific mathematical functions. Thus exact functions like $y = f(x)$ do not exist for relationships in the field of psychology.

Bachelor In many mathematical functions, every value x from the domain of a function f corresponds to exactly one value y from the co-domain. In this case, the function is then unique. The graph of such a function is a curve in the plane (x, y). In such a function, x is called the *independent variable* and y is called the *dependent variable*. This terminology is common also in the field of psychology, mainly in experimental psychology, and can also be found in relevant statistical software packages.

Thus, if we talk about a relationship, an interdependence between two characters or a dependency of one character on the other is meant. We have to distinguish strictly between the terms of a functional relationship in mathematics, the *stochastic dependency* in statistics, and the colloquial term of relationship in psychology, which means a dependency 'in the mean'. However, we will see that psychology is concerned exactly with the (stochastic) dependencies with which statistics deals. We also call them 'statistical dependencies'. Cases which deal with *functional relationships,* like in mathematics, hardly exist in empirical sciences and are not discussed in this book.

Bachelor **Example 11.2** Some simple mathematical functions

Between the side lengths, l, of a square and its circumference, c, the following functional relationship exists: $c = 4l$. Here, this function is a linear function ('equation of a straight line'). To calculate the area A from the side lengths, the formula $A = l^2$, a quadratic function, has to be used. Of course it is possible to combine both formulas and to calculate A from c by means of the formula: $A = \frac{c^2}{16}$.

Obviously relations between two quantitative characters, between two ordinal-scaled, and two nominal-scaled characters are of interest, but also all cases of mixtures (e.g. a quantitative character on one hand and an ordinal-scaled character on the other hand).

Example 11.3 The dependency of characters from Example 1.1

Aside from *gestational age at birth*, which has been addressed several times before, and which trivially is a ratio-scaled character, mainly the test scores of all subtests are interval scaled. It could be of interest, for instance, to what extent test scores from the first and the second test date are related, or scores from different subtests.

Besides this, data from Example 1.1 includes ordinal-scaled characters like *social status* or *sibling position,* as well as nominal-scaled characters like *marital status of the mother*, *sex of the child, urban/rural,* and *native language of the child.* Of interest could be the relationships of *gestational age at birth* (ratio-scaled) and *sibling position* (ordinal-scaled), of *social status* (ordinal-scaled) and *urban/rural* (nominal-scaled, polychotomous), or of *marital status of the mother* (nominal-scaled, polychotomous) and the *native language of the child* (nominal-scaled, dichotomous).

However, relationships often also result between an actually interesting character and a factor or a noise factor (see Chapter 4, as well as Section 12.1.1 and Section 13.2).

Example 11.4 Does a relationship exist between the native language of a child and its test scores in the intelligence tests battery in Example 1.1?

When it is a question of a difference in the (nominal-scaled, dichotomous) factor *native language of the child* related to, for instance, the test score of subtest *Everyday Knowledge*, then the two-sample Welch test, which has already been discussed, even tests the relationship between the test score (as a quantitative character) and the *native language of the child*. If we furthermore suppose that the (nominal-scaled, dichotomous) character *sex of the child* is a noise factor, then the interest is in the relationship between *sex of the child* or *native language of the child* and the test score.

For the purpose of discovering whether certain relationships of (quantitative) characters exist, we illustrate the phenomenon of a relation by representing the pairs of observations in a rectangular coordinate system. In the case of two characters x and y, this is done by means of representing every research unit v by a point, with the observations x_v and y_v as coordinates. We thus obtain a *diagram*, which represents all pairs of observations as a *scatter plot*.

Bachelor **Example 11.5** In which way are the test scores in the characters *Applied Computing, 1st test date* and *Applied Computing, 2nd test date* related?

We assume that intelligence is a stable character, thus a character which is constant over time and over different situations. In this case, the relationship should be strict. For illustrating the relation graphically, we represent the test scores at the first test date on the x-axis (axis of the abscissa) and the test scores at the second test date on the y-axis (axis of the ordinates). Both characters are quantitative.

In **R**, we first enable access to the data set *Example_1.1* (see Chapter 1) by using the function `attach()`. Then we type

```
> plot(sub3_t1, sub3_t2,
+       xlab = "Applied Computing, 1st test date (T-Scores)",
+       ylab = "Applied Computing, 2nd test date (T-Scores)")
```

i.e. we use both characters, *Applied Computing, 1st test date* (`sub3_t1`) and *Applied Computing, 2nd test date* (`sub3_t2`), as arguments in the function `plot()`, and label the axes with `xlab` and `ylab`.

As a result, we get a chart identical to Figure 11.1.

In SPSS, we use the same command sequence (Graphs – Chart Builder...) as shown in Example 5.2 in order to open the window shown in Figure 5.5. Here we select Scatter/Dot in the panel Choose from: of the Gallery tab; next we drag and drop the symbol Simple Scatter into the Chart preview above. Then we move the character Applied Computing, 1st test date to the field X-Axis? and the character Applied Computing, 2nd test date to the field Y-Axis?. After clicking OK we obtain the chart shown in Figure 11.1

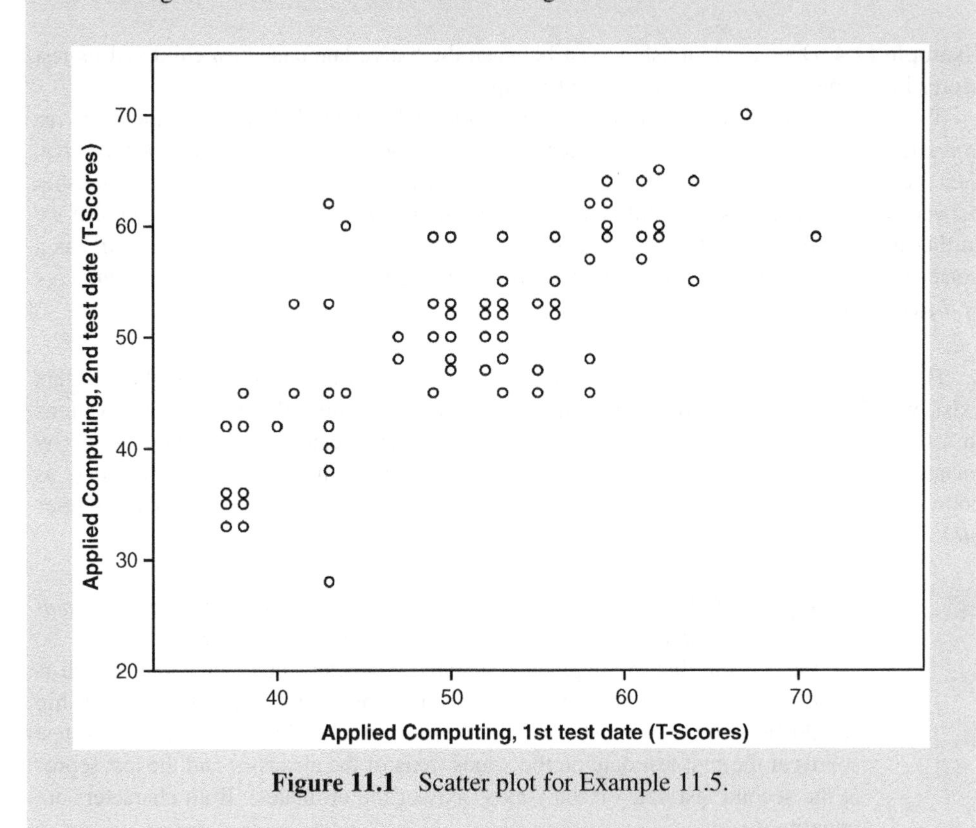

Figure 11.1 Scatter plot for Example 11.5.

We clearly distinguish from the scatter plot the following tendency: 'The bigger the score is at the first test date, the bigger the score is at the second test date', but obviously there is no functional relationship. In that case, all of the points would have to be situated exactly on the graph of any determined function, which is hard to imagine for the given scatter plot. Nevertheless, one comes to the conclusion that this scatter plot can be described, at least roughly, by a straight line (with a positive slope).

Although, in empirical research in psychology, only relationships by trend and no functional relationships exist, in the case of two (quantitative) characters it seems to be meaningful to place a straight line or, if necessary, some other type of curve through the points of a scatter plot in the way that all deviations of the points from the resulting line (or curve) are as small as possible – on the whole. The procedure of adjusting a line or, more generally stated, a curve, to the points of a scatter plot concerns the 'type' of relationship. The question is if the relationship is linear or nonlinear; and in the case that it is linear, the slope and the intercept of the linear function are of interest.

Bachelor Obviously, it is always feasible to fit a mathematical function (e.g. a straight line) in the best way possible to a scatter plot. However, the essential question is to what extent the adjustment succeeds. It is always about how distant the single points of the scatter plot are from the curve that belongs to the function. In the case of a relationship of 100%, all of the points would be situated exactly on it.

The question of the type of relationship is the subject of the so-called *regression analysis*. Initially the appropriate model is selected; that is the determination of the type of function. In psychology this is almost always the linear function. The next step is to concretize the chosen function; that is, to find that one which adjusts to the scatter plot as well as possible. If this concretization were done by appearance in a subjective way, the result would be too inexact. An exact determination can be made by means of a certain numeric procedure, the least squares method (see Section 6.5). From a formal point of view, this means that the *slope* and *intercept* of the so-called *regression line* have to be determined in the best way possible.

Bachelor After having exactly determined the type of relationship between two characters, as concerns future research units the outcome for one of the characters can be predicted from the outcome of the other character. For that purpose, the graph of the mathematical function, which has finally been chosen, has to be positioned in a way that the overall differences (over all pairs of observations) between the true value and that value determined by the other character based on the given relationship, becomes a minimum. That is how the term 'regression' derives from Latin ('move backwards'), in the way that a certain point on the abscissa is projected, via the respective mathematical function, onto the corresponding value on the ordinate.

Doctor Generally, in regression analysis two different cases have to be distinguished: there are two different types of pairs of outcomes. In the case which is usual in psychology, both these values of each research unit were ascertained (mostly simultaneously) by observation, and result as a matter of fact (so-called model II). In contrast, in the other type (model I), the pair of outcomes consists of one value (e.g. the value x), which is chosen (purposely) in advance by the researcher but only the other value (y) is observed as a matter of course. This type is rarely found in psychology. An example of this model is the observation of children's growth between the age of 6 and 12 years. The investigator thus determines the age at which the body size of the children is measured. In this case, the x-values are not measured, but determined *a priori*.

The type of the model influences the calculation of slope and intercept of the regression line. More detailed information concerning model I in combination with regression analysis can be found for example in Rasch, Verdooren, & Gowers (2007).

11.2 Regression model

If we model the characters x and y by two random variables $\boldsymbol{x}$ and $\boldsymbol{y}$, then $(\boldsymbol{x}, \boldsymbol{y})$ is called a *bivariate random variable*. Then the regression model is

$$\boldsymbol{y}_v = f(\boldsymbol{x}_v) + \boldsymbol{e}_v \quad (v = 1, 2, \ldots, n) \tag{11.1}$$

or

$$\boldsymbol{x}_v = g(\boldsymbol{y}_v) + \boldsymbol{e}'_v \quad (v = 1, 2, \ldots, n) \tag{11.2}$$

The function f is called a *regression function* from $\boldsymbol{y}$ onto $\boldsymbol{x}$ and the function g is called a regression function from $\boldsymbol{x}$ onto $\boldsymbol{y}$. In function f, $\boldsymbol{y}$ is the *regressand* and $\boldsymbol{x}$ is the *regressor*, and vice versa in function g. The random variables $\boldsymbol{e}_v$ or $\boldsymbol{e}'_v$ describe error terms. The model particularly assumes that these errors are distributed with mean 0 and the same (mostly unknown) variance in each research unit v. It is also assumed that the error terms are independent from each other, as well as that the errors are independent from the variables $\boldsymbol{x}$ or $\boldsymbol{y}$, respectively.

If functions f and g are linear functions, the following regression models result:

$$\boldsymbol{y}_v = \beta_0 + \beta_1 \boldsymbol{x}_v + \boldsymbol{e}_v \quad (v = 1, 2, \ldots, n) \tag{11.3}$$

or

$$\boldsymbol{x}_v = \beta'_0 + \beta'_1 \boldsymbol{y}_v + \boldsymbol{e}'_v \ (v = 1, 2, \ldots, n) \tag{11.4}$$

respectively. If the random variables $\boldsymbol{x}$ and $\boldsymbol{y}$ are (two-dimensional) normally distributed, the regression function is always linear.

Usually the two functions $\boldsymbol{y} = \beta_0 + \beta_1 \boldsymbol{x}$ and $\boldsymbol{x} = \beta'_0 + \beta'_1 \boldsymbol{y}$ differ from each other. They are identical exclusively in the case that the strength of the relationship is maximal.

We thus have defined the regression model for the population. Now, we have to estimate the adequate parameters β_0 and β_1 or β'_0 and β'_1, respectively, with the help of a random sample taken from the relevant population. With regard to the content, this is of interest mainly because, in practice, sometimes only one of the values (e.g. x-value) is known and we want to 'predict' the corresponding one (y-value) in the best way possible. For both of the random variables $\boldsymbol{x}$ and $\boldsymbol{y}$ we will denote, in addition to the parameters of the regression model, means and variances as follows: μ_x and μ_y, σ_x^2 and σ_y^2.

Starting from the need to place the regression line through the scatter plot in the way that the overall differences between the true value and the value determinable from the other character's value (based on the given relationship) are minimized, the estimates for the unknown parameters in the population, for instance for $y_v = \beta_0 + \beta_1 x_v + \boldsymbol{e}_v$, can be given as follows:

$$b_0 = \hat{\beta}_0 = -b_1 \bar{x} \tag{11.5}$$

$$b_1 = \hat{\beta}_1 = \frac{\frac{1}{n-1} \sum_{v=1}^{n} (x_v - \bar{x})(y_v - \bar{y})}{s_x^2} \tag{11.6}$$

Here s_x^2 is the estimate of σ_x^2. The equation for the estimated regression line is now $\hat{\boldsymbol{y}} = b_1\boldsymbol{x} + b_0$; thus for a concrete research unit v, $\hat{y}_v = b_1 x_v + b_0$.

Doctor

Replacing the outcomes x_v and y_v in Formula (11.5) and (11.6) by the random variables $\boldsymbol{x}_v$ and $\boldsymbol{y}_v$, by which these outcomes are modeled, we obtain the corresponding estimators; these are unbiased.

Master

What was sloppily called overall 'difference' between the true value, y_v, and that value determined by the other character based on the given relationship, $\hat{y}_v$, is, from a formal point of view, a sum of squares. The sum of squares has basically the same function here as the absolute value of differences, but it possesses particular statistical advantages compared to the absolute value. Thus β_0 and β_1 are estimated by $\hat{\beta}_0$ and $\hat{\beta}_1$ such that

$$\sum_{v=1}^{n} (y_v - \hat{y}_v)^2 = \sum_{v=1}^{n} \left[y_v - (\hat{\beta}_1 x_v + \hat{\beta}_0)\right]^2$$

takes a minimum. The solution is found by partial differentiation and setting the first derivative to zero. The resulting values, $b_0 = \hat{\beta}_0$ and $b_1 = \hat{\beta}_1$, thus constitute the estimates.

So it is not the distances in the sense of the shortest distance between a scatter plot's point and the respective regression line's point which are determined (and squared and added up), but the difference in a line segment parallel to the y-axis (see Figure 11.2). The distances in the sense of the shortest distance would be obtained by dropping down the perpendicular of a point to the regression line.

The differences on a section of line parallel to the x-axis, which are generated when x and y as regressor and regressand are interchanged, have also been drawn in Figure 11.2. Thus different quantities are minimized: either the dotted or the

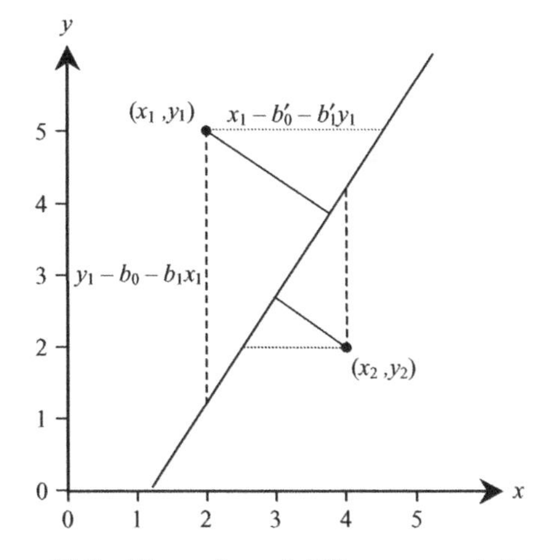

Figure 11.2 Illustration of differences and distances.

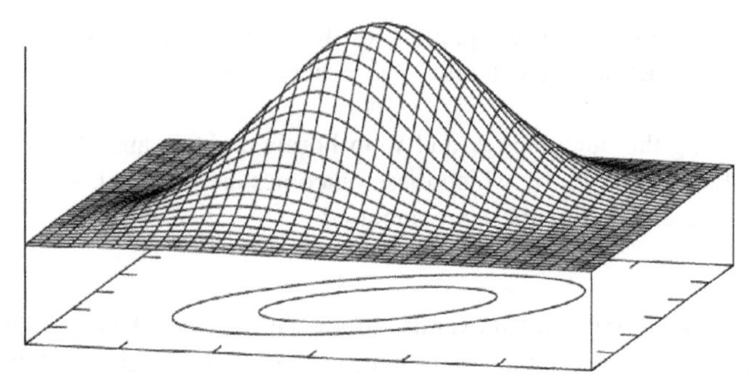

Figure 11.3 The density function of a two-dimensional normal distribution.

dashed sections of line. This leads to two different regression lines except for the case of a perfect relationship, in which all of the points are situated exactly on the regression line.

The numerator of Formula (11.6) is called the (sample) *covariance*. It describes the extent to which the two characters 'vary together'. In Formula (11.6) this is the estimate $s_{xy} = s_{yx} = \hat{\sigma}_{xy}$ of the parameter $\sigma_{xy} = \sigma_{yx}$ in the underlying population. Covariance in the population characterizes the two-dimensional normal distributed variable $(\boldsymbol{x}, \boldsymbol{y})$ in addition to the means and variances. However, the condition of distribution, which is met here, will only be relevant for questions of inferential statistics (see Section 11.4). Anyway, we give here a graphical illustration of such a two-dimensional normal distribution (Figure 11.3).

Master From a formal point of view this results in the option to write Formula (11.6) also as: $b_1 = \frac{s_{xy}}{s_x^2}$ (or $b_1' = \frac{s_{xy}}{s_y^2}$, respectively). For the parameters analogously we may write, for instance, $\beta_1 = \frac{\sigma_{xy}}{\sigma_y^2}$.

The difference between observation y_v at the point of observation x_v and the y-value of the regression line $\hat{y}_v$, thus $(y_v - \hat{y}_v) = y_v - (\hat{\beta}_1 x_v + \hat{\beta}_0)$, is called the *residual*. A careful *residual analysis* is generally part of model fitting; that is, it gives information whether the chosen model fits or several pairs of values are striking.

Bachelor **Example 11.5 – continued**

We consider the relationship of test scores in subtest *Applied Computing* between first and second test date. With regard to the content, it is logical to predict the latter by the former.

In **R**, we first type

```
> lm.1 <- lm(sub3_t2 ~ sub3_t1)
> summary(lm.1)
```

i.e. we apply the function `lm()` and submit the formula `sub3_t2 ~ sub3_t1` as an argument, meaning that *Applied Computing, 2nd test date* (`sub3_t2`) is the regressand

and *Applied Computing, 1st test date* (sub3_t1) the regressor; we assign the result of this regression analysis to the object lm.1. Finally we submit this object to the function summary().

As a result, we get (shortened output):

```
Coefficients:
               Estimate  Std. Error t value  Pr(>|t|)
(Intercept)     7.10534    3.53079   2.012   0.0469
sub3_t1         0.85313    0.06907  12.351   <2e-16
```

Hence the slope b_1 of the regression line is 0.8531, the intercept b_0 is 7.1053; at this point we will not discuss the remaining values.

Next we save the predicted values for each x-value, as well as the residuals and the standardized residuals, by typing

```
> resid <- cbind("PRE_1" = lm.1$fitted.values,
+                "RES_1" = lm.1$residuals,
+                "ZRE_1" = rstandard(lm.1))
> summary(resid)
```

i.e. we extract the needed values from object lm.1, namely with $fitted.values the predicted values and with $residuals the residuals; with the function rstandard() we ascertain the standardized residuals by submitting lm.1 as an argument. We label the results "PRE_1", "RES_1" and "ZRE_1" and combine the single result vectors column by column with the function cbind(). We assign the results to the object resid, which we submit to the function summary() as an argument.

As a result, we get:

```
     PRE_1            RES_1                   ZRE_1

Min.   :38.67   Min.   :-1.579e+01   Min.   :-2.8926572
1st Qu.:43.79   1st Qu.:-3.042e+00   1st Qu.:-0.5577114
Median :51.47   Median : 5.984e-02   Median : 0.0109406
Mean   :50.18   Mean   : 8.790e-17   Mean   :-0.0007525
3rd Qu.:54.88   3rd Qu.: 2.827e+00   3rd Qu.: 0.5169367
Max.   :67.68   Max.   : 1.821e+01   Max.   : 3.3359756
```

Next, we want to plot the standardized residuals for each x-value (*Applied Computing, 1st test date*) in a rectangular coordinate system; hence we type

```
> plot(sub3_t1, resid[, "ZRE_1"],
+      xlab = "Applied Computing, 1st test date (T-Scores)",
+      ylab = "Standardized residuals")
```

i.e. we use the character *Applied Computing, 1st test date* (sub3_t1) and the standardized residuals (resid[, "ZRE_1"]) as arguments in the function plot(); xlab and ylab label the axes.

As a result, we get the chart in Figure 11.4.

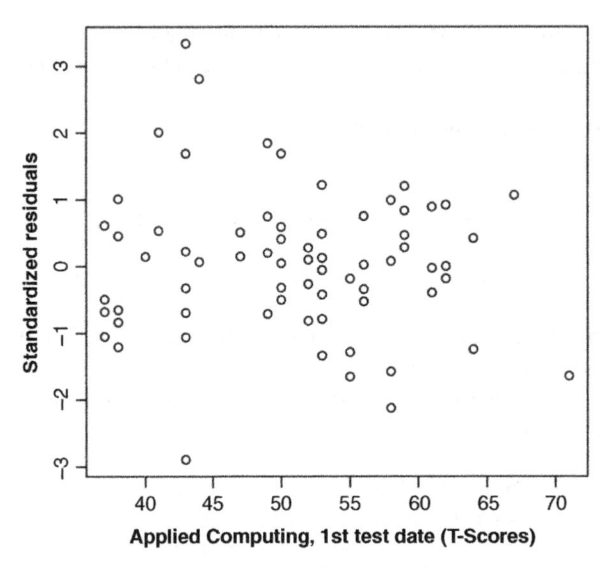

Figure 11.4 R-output of a scatter plot showing the outcomes and residuals in Example 11.5.

It is advisable to check whether $-2 <$ `ZRE_1` < 2 is true at all times; if not, one has to assume observation errors or simply an unfitting model. This rule of thumb arises from the fact that, if the random errors are standard normal distributed, then more than 95% of all values would be within the above-specified range. In our case, 6 out of 100 research units are outside these bounds. You detect this by typing

```
> sum(rstandard(lm.1) < −2 | rstandard(lm.1) > 2)
```

i.e. we select those standardized residuals which are either smaller ('<') than −2 or ('|') bigger ('>') than 2, and count their number with the function `sum()`.

As a result, we get:

```
[1] 6
```

Next, we want to illustrate the scatter plot along with the regression line to answer the actual question of the relation between the test scores in subtest *Applied Computing* on the first and second test date; hence we type

```
> plot(sub3_t1, sub3_t2,
+       xlab = " Applied Computing, 1st test date (T-Scores)",
+       ylab = " Applied Computing, 2nd test date (T-Scores)")
> abline(lm.1)
```

i.e. we use the function `plot()` as before, and label the axes appropriately. Next, we use the regression model in object `lm.1` as an argument in the function `abline()` and thereby amend the chart with a line according to the model parameters. The completed chart conforms to Figure 11.10.

In SPSS, we use the sequence of commands

Analyze
  Regression
    Linear...

from the menu and select the character **Applied Computing, 2nd test date** and move it to the field **Dependent:**. Next we select **Applied Computing, 1st test date** and move this character to the field **Independent(s):** (see Figure 11.5). Next, we click **Save...** and get to the window shown in Figure 11.6, where the required fields are already checked. Hence we request the predicted values for each x-value as well as the residuals and the standardized residuals (these are the residuals divided by the estimated standard deviation). The new variables PRE_1 (predicted values), RES_1, and ZRE_1 (residuals and standardized residuals) are added to the file (see Figure 11.7). A click on **Continue**, followed by **OK** gets us the result shown in Table 11.1.

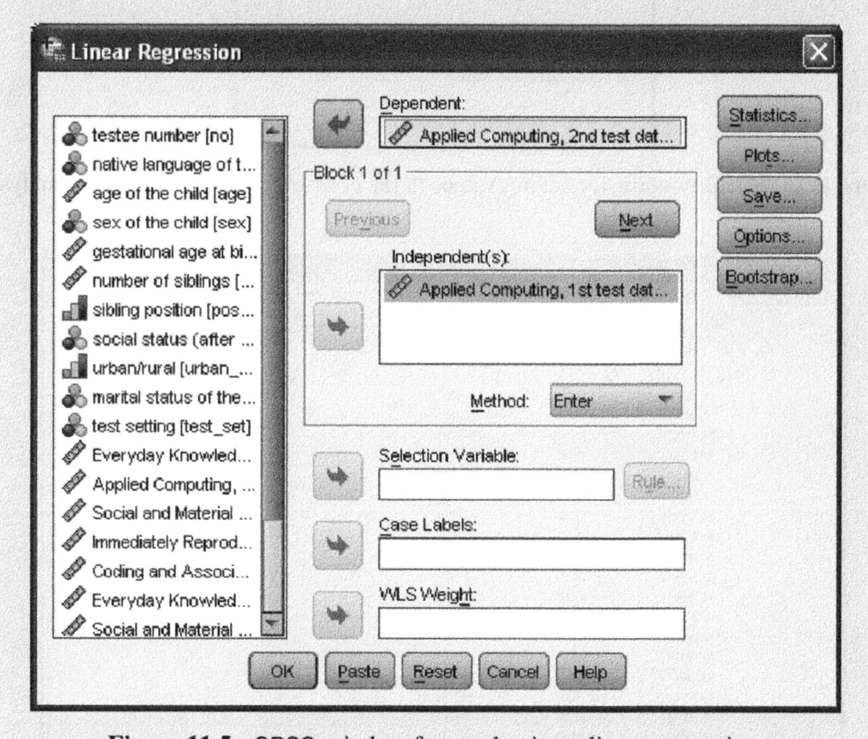

Figure 11.5 SPSS-window for conducting a linear regression.

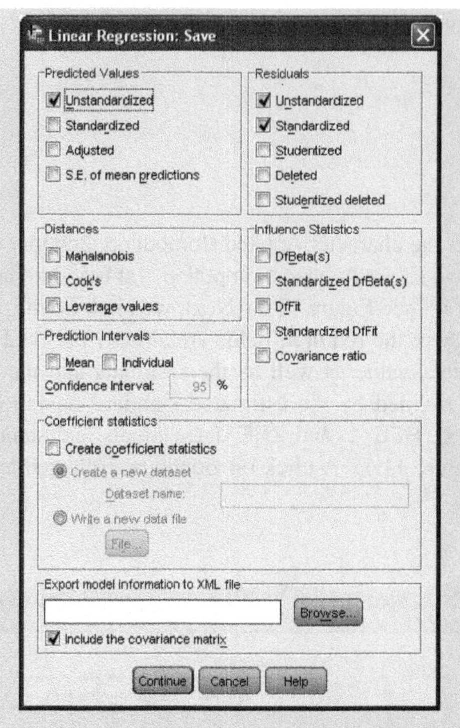

Figure 11.6 SPSS-window for saving values in the course of a linear regression analysis.

*Example 1.1.sav [DataSet1] - IBM SPSS Statistics Data Editor

Visible: 26 of 26 Variables

	PRE_1	RES_1	ZRE_1
1	54.88076	.11924	.02164
2	57.44016	4.55984	.82745
3	54.02763	-7.02763	-1.27526
4	47.20257	.79743	.14470
5	49.76197	-1.76197	-.31973
6	57.44016	2.55984	.46452
7	52.32136	-.32136	-.05832
8	48.90883	1.09117	.19801
9	59.14642	-.14642	-.02657
10	59.99956	5.00044	.90740
11	43.79004	-5.79004	-1.05068
12	47.20257	2.79743	.50763
13	51.46823	1.53177	.27796
14	49.76197	2.23803	.40612
15	59.14642	-2.14642	-.38950
16	43.79004	-3.79004	-.68775
17	56.58703	.41297	.07494
18	47.20257	2.79743	.50763
19	43.79004	1.20996	.21956
20	38.67124	-2.67124	-.48473

Data View | Variable View

IBM SPSS Statistics Processor is ready

Figure 11.7 SPSS-Data View after saving the residuals in Example 11.5 (section).

Table 11.1 SPSS-output of the regression coefficients in Example 11.5 (shortened output).

Coefficients[a]

Model		Unstandardized Coefficients		Standardized Coefficients		
		B	Std. Error	Beta	t	Sig.
1	(Constant)	7.105	3.531		2.012	.047
	Applied Computing, 1st test date (T-Score)	.853	.069	.780	12.351	.000

a. Dependent Variable: Applied Computing, 2nd test date (T-Scores)

In Table 11.1 we find the slope b_1 of the regression line in the last row of column Unstandardized Coefficients in B (hence in the row Applied Computing, 1st test date). The intercept b_0 is called Constant in SPSS and is found one row above. We will not discuss the remaining values in the table at this point. The regressand is referred to as Dependent Variable below the table.

Since we have requested the residuals, SPSS primarily returns the minimum and maximum of the residuals in Table 11.2. The residuals themselves are added, like the predicted values and the standardized residuals, to the data set (cf. again Figure 11.7).

Table 11.2 SPSS-output showing statistics of the residuals.

Residuals Statistics[a]

	Minimum	Maximum	Mean	Std. Deviation	N
Predicted Value	38.67	67.68	50.18	6.841	100
Residual	-15.790	18.210	.000	5.483	100
Std. Predicted Value	-1.682	2.558	.000	1.000	100
Std. Residual	-2.865	3.304	.000	.995	100

a. Dependent Variable: Applied Computing, 2nd test date (T-Scores)

Next, we want to plot the standardized residuals of each x-value (*Applied Computing, 1st test date*) in a rectangular coordinate system. We can draw such a chart by following the command sequence (Graphs – Chart Builder...) as before, but need to select Standardized Residual for the field Y-Axis?. This way we get a chart analogous to Figure 11.4.

It is advisable to check whether $-2 <$ ZRE_1 < 2 is true at all times; if not, you have to assume observation errors or simply an unfitting model. This rule of thumb arises from the fact that, if the random errors were standard normal distributed, then more than 95% of all values would be within the above-specified range. To ascertain the count of values outside these bounds, we would need to create a frequency table as shown in Example 5.2, from which we refrain here. That such cases even exist can be seen in Table 11.2, in the Row Std. Residual in the columns Minimum and Maximum.

Next, we finally want to illustrate the scatter plot along with the regression line to answer the actual question of the relation between the test scores in subtest *Applied Computing* on the first and second test date; hence we select

Analyze
Regression
Curve Estimation...

and get to the window shown in Figure 11.8. Here we keep the default selection of Linear, move the character Applied Computing, 2nd test date to the field Dependent(s): and the character Applied Computing, 1st test date to the field Variable: in the panel Independent. With OK we get the desired chart (Figure 11.9). Double-clicking the chart starts the Chart Editor (see Figure 6.4), which we can use to, e.g., rescale the chart, so that the zero value for both variables is shown. We do that by selecting Edit from the menu bar and clicking Select X Axis. Subsequently a new window pops up (not illustrated), where we change the value in row Minimum in field Custom to 0, Apply this setting and close the window. We repeat this procedure for the sequence of commands Edit – Select Y Axis. Now we can close the Chart Editor and, as a result, we get Figure 11.10.

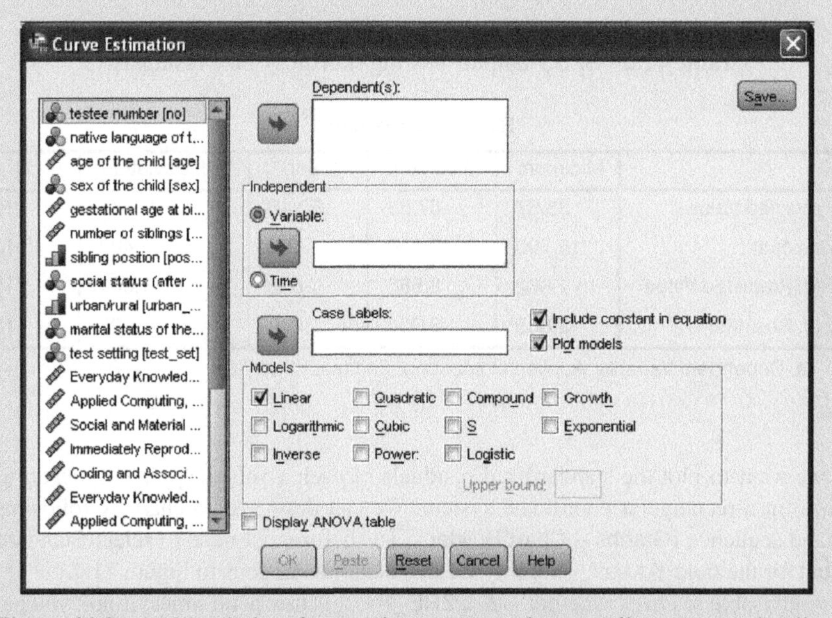

Figure 11.8 SPSS-window for creating a scatter plot as well as a regression line.

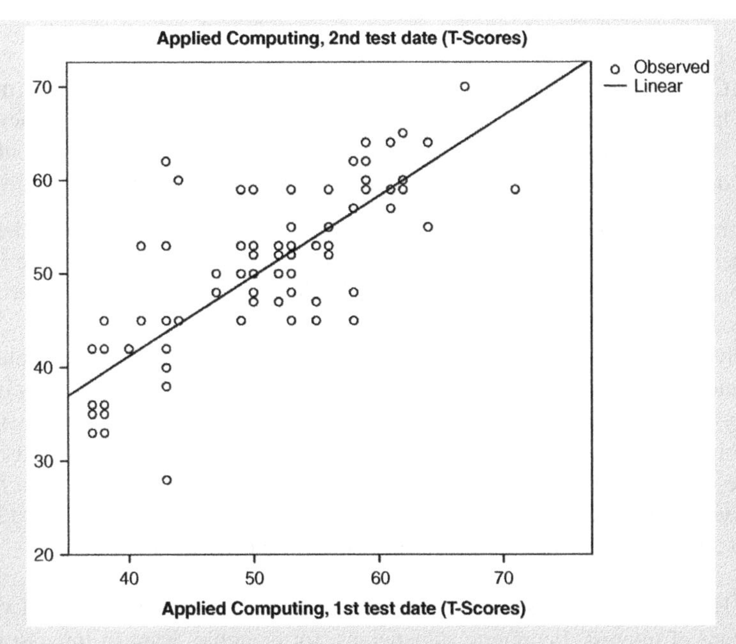

Figure 11.9 SPSS-output of the scatter plot and the regression line in Example 11.5 (scaled by SPSS).

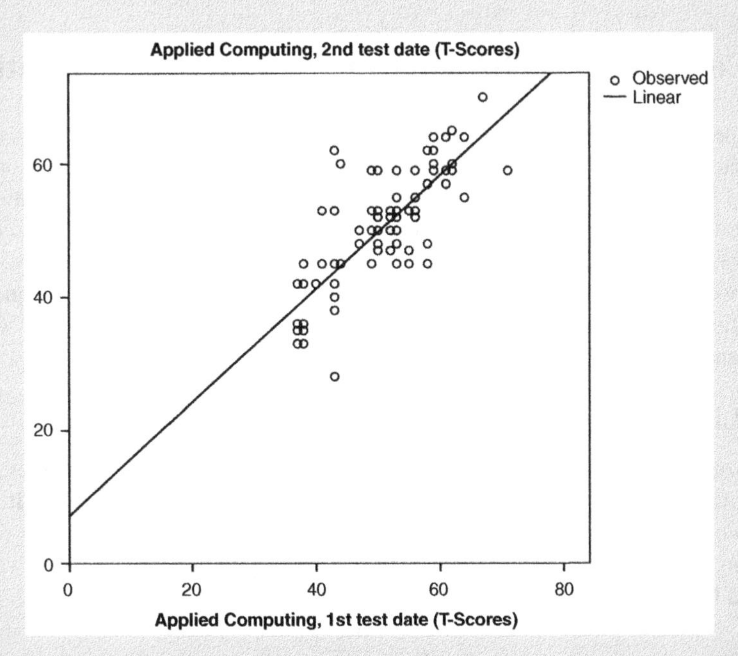

Figure 11.10 SPSS-output of the scatter plot and the regression line in Example 11.5 (rescaled by the user).

Thus the regression line for *Applied Computing, 2nd test date,* to *Applied Computing, 1st test date* is: $\hat{y}_v = 0.853x_v + 7.105$. In Figure 11.4 no particular abnormality is noticeable, at the most that residuals fluctuate somewhat more in small test scores. If these fluctuations were more distinct, we would have to doubt the model condition that all random variables $\boldsymbol{e}_v$ have the same variance.

The equation of the estimated regression line $\hat{y} = b_1x + b_0$ now actually can be used for (future) observations x_{n+t}, $t = 1, 2, \ldots$ to best determine the corresponding value in variable $\boldsymbol{y}$: $\hat{y}_{n+t} = b_1x_{n+t} + b_0$. Thus $\hat{y}_{n+t}$ is the predicted value on the regression line for the points for which no x-values are given.

Generally, it is only legitimate or reasonable to use, for a prediction, x-values which fall in the interval between the biggest and the smallest observed x_v-value. This is not only scientifically logical, but is mainly due to the fact that the linear regression often empirically is valid within a certain well-defined area, but in other areas a nonlinear relationship exists. For instance, in infants, body weight has a relatively strong linear relation with body size. However, it would obviously be meaningless to try to predict the body weight of a 12-year-old by means of a linear regression, which was determined in babyhood.

Master However, often it is exactly predictions outside the observed area of values that are of interest. Economic statisticians, for example, want to forecast the future economic trend from the trend of the past years. The problems talked about above can often be solved by using an adequate growth function.

11.3 Correlation coefficients and measures of association

After the type of statistical relationship between two characters has been defined, the regression model is determined and it is possible to estimate the parameters of the regression function by means of a sample of pairs of outcomes. Of course the strength of the correlation is of as great or greater interest. The measure for the strength of correlation is called the *correlation coefficient* or, more generally stated, a *measure of association*.

So far, we have been concerned exclusively with linear functions and consequently with the correlation between two quantitative characters. We will continue in this way in the following, and discuss other types of associations subsequently.

11.3.1 Linear correlation in quantitative characters

A measure of the strength of the linear correlation between two random variables $\boldsymbol{x}$ and $\boldsymbol{y}$ is *Pearson's correlation coefficient* or the *product-moment correlation coefficient*. It is defined as the ratio of covariance and the product of standard deviations. For didactical reasons, in the following we will first consider the situation in the population and then afterwards come back to data from a sample:

$$\rho = \frac{\sigma_{xy}}{\sigma_x \sigma_y} \qquad (11.7)$$

In contrast to the regression coefficient, the correlation coefficient is a measure which is symmetric regarding $\boldsymbol{x}$ and $\boldsymbol{y}$. It takes values in the interval between -1 and 1. If $\rho > 0$, we say that a positive correlation exists between the two random variables. Then high values of $\boldsymbol{x}$ are related (by trend) with high values of $\boldsymbol{y}$. Depending on the strength of the relationship, thus the absolute value of ρ, the relation is more or less consistent. If $\rho < 0$, we say that a negative correlation exists between the two random variables. Then high values of $\boldsymbol{x}$ are related (by trend) with small values of $\boldsymbol{y}$. If the two random variables are independent from each other, thus 'uncorrelated', the covariance is $\sigma_{xy} = 0$, and therefore also the correlation coefficient $\rho = 0$ and both of the regression coefficients β_1, β_1'. Nevertheless, if, for two random variables, $\rho = 0$, then they are not necessarily independent from one another. But, a linear relation does not exist in any case.

Master

Pearson's correlation coefficient can be deduced intuitively as follows.

First we reflect on the question of for which cases we would consider a correlation between two characters as relevant, for instance relevant to an extent that it seems to be meaningful to predict an unknown value y_i from x_i by means of the regression line. Now, the correlation would surely be considered as relevant and such a prediction as meaningful, if the character x, modeled by $\boldsymbol{x}$, could 'explain' 'why' the observations of the other character y, modeled by $\boldsymbol{y}$, take, exactly, certain realizations y_v, $v = 1, 2, \ldots, n$.

We assume here that we are interested in the correlation between intelligence (intelligence test with test score x) and school achievement (school achievement test with test score y), which is indeed the subject of many debates. School achievement thus is to be modeled by a random variable, which will certainly not lead to the same outcomes in all children, but to outcomes with a certain variability, measured by variance. If intelligence, as the other modeled random variable, is capable of explaining this variability in the way that high school achievement often is related to high intelligence and low school achievement to low intelligence, then the following question arises: to what extent exactly is the intelligence 'responsible' for the variability in school achievement of these children? It is important to note that causality between the two characters is not presumed thereby. That is to say that it is equally possible, with regard to the content, that, inversely, differences in school achievement cause the corresponding differences in the intelligence test. It is also possible that the differences between the children concerning school achievement as well as intelligence test achievement are due to differences between the children in the trait 'achievement motivation'.

Thus the question is about the extent of variance σ_y^2 in school achievement, which is determined (we also say: described) solely by the variance in intelligence. Provided that the regression from $\boldsymbol{y}$ onto $\boldsymbol{x}$ is given, we can determine the value $\hat{\boldsymbol{y}}_v = \hat{\boldsymbol{y}}(x_v) = \boldsymbol{b}_0 + \boldsymbol{b}_1 x_v$ for each research unit v. Calculating the variance $s_{\hat{y}}^2$ of all of these values $\hat{\boldsymbol{y}}_v$, we obtain the estimation for $\sigma_{\hat{y}}^2$ of $\hat{\boldsymbol{y}}$ of that fraction of variance of $\boldsymbol{y}$, σ_y^2, which can be described (or predicted, respectively) or determined for the given population from $\boldsymbol{x}$. It is of interest now to relate this variance $\sigma_{\hat{y}}^2$ to the total variance of $\boldsymbol{y}$ (within the population), thus to calculate the relative proportion of variance explained by $\boldsymbol{x}$: $\frac{\sigma_{\hat{y}}^2}{\sigma_y^2}$.

Since

$$\sigma_{\hat{y}}^2 = E\,[\hat{\boldsymbol{y}} - E\,(\hat{\boldsymbol{y}})]^2 = E\,[\boldsymbol{b}_0 + \boldsymbol{b}_1\boldsymbol{x} - E\,(\boldsymbol{b}_0 + \boldsymbol{b}_1\boldsymbol{x})]^2 = \beta_1^2 \cdot E\,[\boldsymbol{x} - E(\boldsymbol{x})]^2$$

$$= \left(\frac{\sigma_{xy}}{\sigma_x^2}\right)^2 \cdot \sigma_x^2 = \frac{\sigma_{xy}^2}{\sigma_x^2}$$

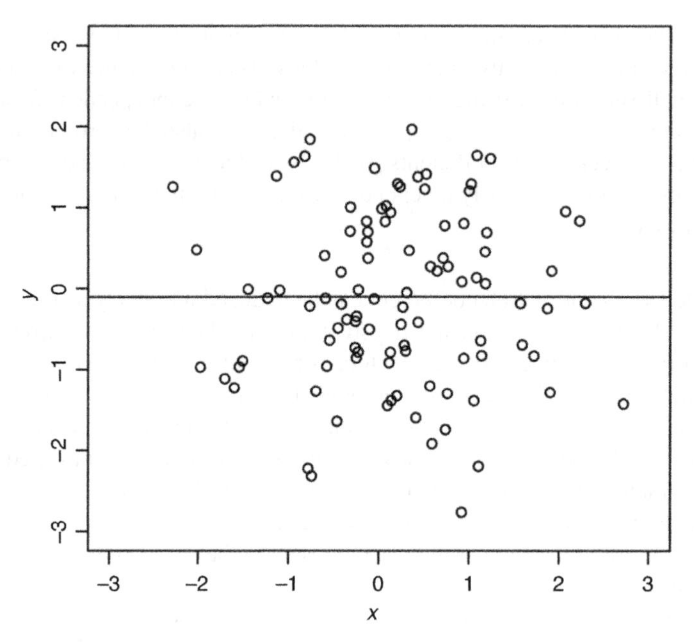

Figure 11.11 The regression line for a scatter plot with Pearson's correlation coefficient of 0.00031.

this relative proportion of the variance of $\boldsymbol{y}$ explained by $\boldsymbol{x}$ equals the square of Pearson's correlation coefficient:

$$\frac{\sigma_{\hat{y}}^2}{\sigma_y^2} = \frac{\sigma_{xy}^2}{\sigma_x^2} \cdot \frac{1}{\sigma_y^2} = \rho^2.$$

Ideally, $\hat{\boldsymbol{y}}$ and $\boldsymbol{y}$ coincide precisely for all v. Then, numerator and denominator are identical in the quotient ρ^2; thus $\rho^2 = 1$. Given the case that $\boldsymbol{y}$ and $\boldsymbol{x}$ are not (linearly) correlated, the regression line from $\boldsymbol{y}$ to $\boldsymbol{x}$ has to proceed in parallel to the x-axis. That is, the same value, μ_v, has to be predicted for every x_v in order to minimize the sum of squares (see the example in Figure 11.11). Therefore $\beta_1 = \beta_1' = 0$, and thus $\rho^2 = 0$ as well. As a consequence, the value of ρ^2 is between 0 and 1. The stronger the relationship is, the closer the value of ρ^2 is to 1.

The square of Pearson's correlation coefficient is called the *coefficient of determination*. As a symbol for this parameter (concerning the population) we use ρ^2. An appropriate point estimator is $\boldsymbol{B} = \boldsymbol{r}^2 = \hat{\boldsymbol{\rho}}^2$, in which $\boldsymbol{r} = \boldsymbol{r}_{xy} = \hat{\boldsymbol{\rho}}$ constitutes a point estimator for the parameter ρ on its part:

$$\boldsymbol{r} = \frac{\frac{1}{n-1}\sum\limits_{v=1}^{n}(\boldsymbol{x}_v - \bar{\boldsymbol{x}})(\boldsymbol{y}_v - \bar{\boldsymbol{y}})}{\boldsymbol{s}_x \boldsymbol{s}_y} \tag{11.8}$$

Master The estimator $\boldsymbol{r}$ for ρ in Formula (11.8) is not unbiased. An unbiased estimator also exists, but as the bias is not too big, we recommend using Formula (11.8), nevertheless.

Bachelor A coefficient of determination of, for instance, $B = 0.81$ (thus $r = 0.9$) means that 81% of the variance in $\boldsymbol{y}$ can be explained by the variance in $\boldsymbol{x}$ (and vice versa). Then, 19% of the variance (the variability) of $\boldsymbol{y}$ depends on other, unknown (influencing) factors. A coefficient of determination of $B = 0.49$ (thus $r = 0.7$) indicates that almost half of the variance of the variables is explained mutually. This can be referred to as a medium-sized correlation. If the correlation coefficient equals 0.3 and therefore the coefficient of determination $B = 0.09$, not even 10% of the variance is being explained mutually. In most cases, this relationship is not relevant in a practical sense.

For Lecturers:

If the two regression lines are drawn into a rectangular coordinate system, they intersect in the area of observations (see Figure 11.12). The angle $\omega = \omega_{xy} - \omega_{yx}$ which is embedded in the two lines is clearly related to the correlation coefficient (this is true for the unknown regression lines of the population and ρ on the one hand and for the estimated regression lines and r on the other hand). If $\rho = 1$ or $\rho = -1$, this angle equals zero: both lines are identical. If $\rho = 0$, the lines are perpendicular to each other and the angle equals 90°. It can be shown that

$$\rho^2 = 1 - \frac{\sin \omega}{\cos \omega_{yx} \sin \omega_{xy}}$$

If $\omega = 90°$, $\sin \omega = 1$ and $\rho^2 = 0$. Contrariwise, if both lines are identical, ω and $\sin \omega$ equal 0 and $\rho^2 = 1$. That is, ρ equals +1 if the common line ascends, and otherwise −1.

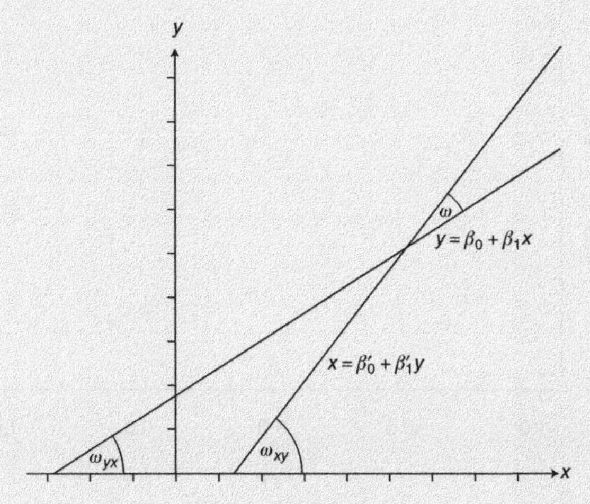

Figure 11.12 All angles of the two regression lines for an arbitrary example.

Generally, we caution against calculating the correlation coefficient uncritically; that is without making sure in advance by means of a graph that the assumed relationship is in fact linear.

Bachelor **Example 11.6** Example for calculation without regard to content

We want to illustrate graphically the following pairs of observations (x, y) in order to determine a possible relationship between $\boldsymbol{x}$ and $\boldsymbol{y}$.

x	0	1	0	−1	0.707107	−0.707107	0.707107	−0.707107
y	1	0	−1	0	0.707107	−0.707107	−0.707107	0.707107

In **R**, we first create two vectors corresponding to the pairs of values by typing

```
> x1 <- c(0, 1, 0, -1, 0.707107, -0.707107, 0.707107, -0.707107)
> y1 <- c(1, 0, -1, 0, 0.707107, -0.707107, -0.707107, 0.707107)
```

i.e. we use the function `c()` to concatenate the values in each row into a vector, which we assign to objects `x1` and `y1`, respectively. Next we type

```
> plot(x1, y1)
```

i.e. we submit `x1` and `y1` to the function `plot()`.

As a result, we get Figure 11.13.

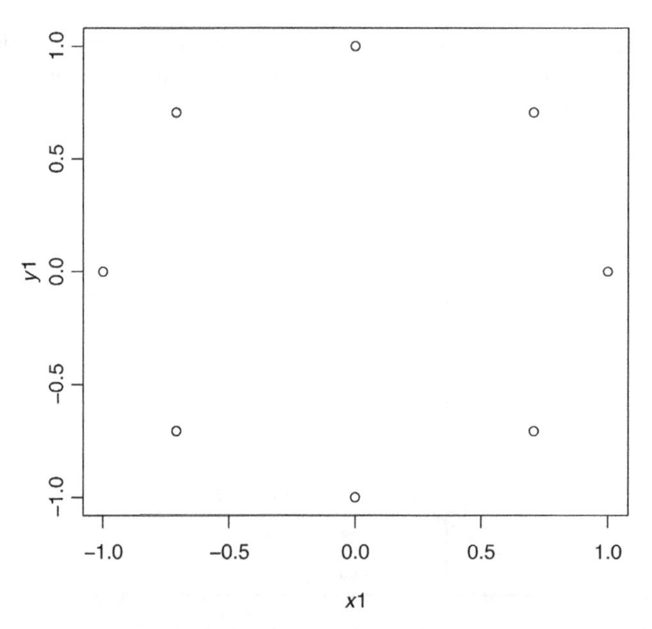

Figure 11.13 Scatter plot for Example 11.6.

Next, we ascertain the extent of the linear relationship by typing

```
> cor(x1, y1, method = "pearson")
```

i.e. we use both variables, `x1` and `y1`, as arguments in the function `cor()`; with `method = "pearson"` we select Pearson's correlation coefficient.

As a result, we get:

```
[1] 0
```

In SPSS, we open a new data sheet (File – New – Data) and start by typing in the characters x and y according to the given table, above. Next we draw a scatter plot with the command sequence (Graphs – Chart Builder...) from Example 11.5 and move x to the field X-Axis? and y to the field Y-Axis?. With OK, we get a chart analogous to Figure 11.13. Now we use the sequence of commands:

Analyze
  Correlate
    Bivariate...

and, in the resulting window (see Figure 11.14), move x and y to the field Variables:, confirm with OK, and get as a result that Pearson's correlation coefficient is 0 (see Table 11.3). Looking at the scatter plot in Figure 11.13, which we create analogously to Example 11.5, obviously no linear relationship exists; however a 100% nonlinear relationship does: all eight points match the circle equation $x^2 + y^2 = 1$ of the unit circle and therefore all y-values are uniquely determined by the x-values. Linear correlation simply is unfit to quantify the existing 100% relationship. The result '1' in the output states the respective correlation coefficient of each variable with itself, which naturally is 1.0.

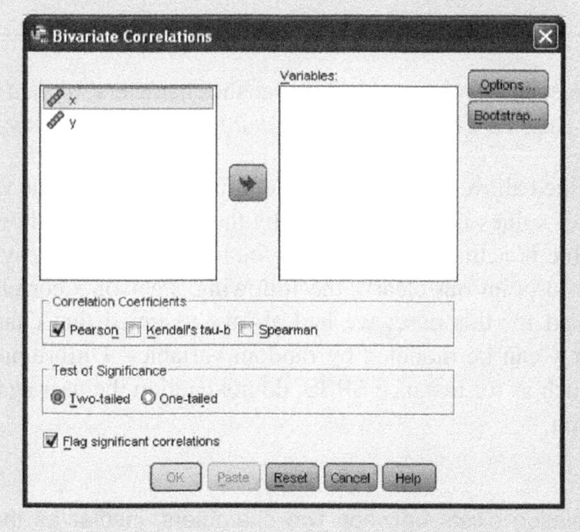

Figure 11.14 SPSS-window for computing correlation coefficients.

Table 11.3 SPSS-output showing Pearson's correlation coefficients for the data given in the table of Example 11.6.

Correlations

		x	y
x	Pearson Correlation	1	.000
	Sig. (2-tailed)		1.000
	N	8	8
y	Pearson Correlation	.000	1
	Sig. (2-tailed)	1.000	
	N	8	8

Bachelor **Example 11.5 – continued**

The question is about a relationship between the test scores in the characters *Applied Computing, 1st test date* and *Applied Computing, 2nd test date*.

In **R**, we type

```
> cor(sub3_t1, sub3_t2, method = "pearson")
```

i.e. we use the characters *Applied Computing, 1st test date* (sub3_t1) and *Applied Computing, 2nd test date* (sub3_t2) as arguments in the function cor().

As a result, we get:

```
[1] 0.7802943
```

In some respects a relationship between the characters *Applied Computing, 1st test date* and *Applied Computing, 2nd test date* exists, but no exact linear correlation.

Doctor As described above, the case of model I, in which one of the values (e.g. x-value) per pair of values is predetermined by the investigator and only the corresponding y-value is actually observed, is found very rarely in psychology. However, we want to point out clearly the following: Pearson's correlation coefficient is not defined for this case; we had always assumed that character x as well as character y can be modeled by random variables. Unfortunately, software programs, such as for instance SPSS, do not caution the user against such mistaken application.

Summary

Often a statistical relation exists between two characters, insofar as the *scatter plot* of a research unit's pairs of observations, which are represented in a rectangular coordinate system, can be described rather precisely by a straight line. The optimal line is determined by means of *regression analysis.* We aim for one of the characters, the *regressand*, to be predicted as precisely as possible by the other one, the *regressor*, with the help of the sought-after line. The looked-for parameters (of the line) are called *regression coefficients.* A measure of the strength of correlation is *Pearson's correlation coefficient*. The square of this coefficient is called the *determination coefficient*; it indicates the percentage of variance of the characters modeled by random variables, which they mutually explain due to their relationship.

11.3.2 Monotone relation in quantitative characters and relation between ordinal-scaled characters

If no linear, but a (strictly) monotone relation between two characters is given, Pearson's correlation coefficient underestimates the actual relationship. A monotone relation means that **y** increases or decreases when **x** increases. Therefore, in this case other correlation coefficients are considered; that is mainly such ones as are applicable for ordinal-scaled characters. The best known of these coefficients is *Spearman's rank correlation coefficient/Spearman's* r_s, which is characterized by the parameter ρ_S for the population and by the statistic $r_S = \hat{\rho}_S$ for the sample. It can be calculated very easily with the aid of statistical software programs.

Master

Spearman's rank correlation coefficient in the sample is derived from Pearson's correlation coefficient by ranking the values per variable **x** and **y** according to the rules in Section 5.3.2. With these ranks Pearson's correlation coefficient is calculated.

The interpretation of the algebraic sign, thus of the direction of the relationship, is done in the following way: a positive Spearman's rank correlation coefficient indicates that, with higher ranks in one character, the rank in the other character tends to increase.

Also for Spearman's rank correlation coefficient, the coefficient of determination $B = r_S^2$ is frequently calculated. The interpretation is more difficult in this case: it describes the percentage of variance explained by the ranks – not by the originally quantitative outcomes.

In analogy to Pearson's correlation coefficient, if $\rho_S = 0$ or $r_S = 0$, this does not mean that no relationship between **x** and **y** exists. But certainly, no monotone relationship does exist.

Bachelor

Example 11.6 – continued

Spearman's rank correlation coefficient results with $r_S = 0$ as well for the data from Example 11.6.

In **R**, we type

```
> cor(x1, y1, method = "spearman")
```

i.e. we use the characters `x1` and `y1` as arguments in the function `cor()`; with `method = "spearman"` we select the Spearman's rank correlation coefficient.

As a result, we get:

```
[1] 0
```

In SPSS, we proceed analogously to computing Pearson's correlation coefficient, but in Figure 11.14 select Spearman instead of the default Pearson.

Spearman's rank correlation coefficient is applied in the following cases:

- Two quantitative characters are given, but the question is not about a linear, but about a monotone relationship.
- At least one of the characters is an ordinal-scaled character; the other character is either quantitative or ordinal scaled as well.

Bachelor **Example 11.7** Does a relationship exist between the test score in subtest *Applied Computing, 1st test date* and *social status* in Example 1.1?

The test scores in subtest *Applied Computing* are interval-scaled, but *social status* is merely ordinal-scaled – given that we exclude all of the children with 'single mother in household', as we did in former examples. Consequently, Pearson's correlation coefficient does not come into consideration, but Spearman's rank correlation coefficient does. The analysis is performed in the same way as in Example 11.6. The result is: $r_S = -0.274$. The relationship is to be described as marginally (negative); the test score in subtest *Applied Computing* and the *social status* are hardly associated.

Besides Spearman's rank correlation coefficient, another correlation coefficient based on ranks is used more and more: *Kendall's* τ. Due to the given symbol, it is difficult to differentiate between the parameter in the population and the statistic in the sample as an appropriate estimator for the parameter, but this should always become clear from the context.

Master By its concept, Kendall's approach differs from Spearman's approach: the latter is based on the differences between the ranks of the research units in the two characters, and the former is based on the number of permutations in the ranking of the research units with respect to one character compared to the ranking of the research units in the other character. The square of Kendall's τ thus cannot be interpreted as the percentage of explained variance.

Other measures to determine the strength of relationship in ordinal-scaled characters exist as well. None of them are established in psychology (but see, for example, Kubinger, 1990).

11.3.3 Relationship between a quantitative or ordinal-scaled character and a dichotomous character

Sometimes it is recommended to use Pearson's correlation coefficient also in the case of one quantitative and one dichotomous character. Analogously, Spearman's rank correlation coefficient or Kendall's τ would be applicable, if one character was ordinal-scaled and the other one a dichotomous character. Here, it is less critical that in the second case numerous ties arise (this fact would be irrelevant due to the obligatory correction for ties which is included in the formula), than the fact that, in both cases, despite an 'ideal' relationship, in many instances the correlation coefficients cannot achieve the value of 1.

Master

Example 11.8 Numeric example without regard to content, for the artificial use of Pearson's and Spearman's rank correlation coefficient or Kendall's τ, respectively, in the case of one dichotomous character

For the data mentioned below, the correlation coefficients do not equal 1, although data was chosen in such a way that the maximum possible relationship is given. In one case the y_v are quantitative outcomes; in the other case they are already into ranks transformed values.

Person number		1	2	3	4	5	6	7	8	9	10
	y_v	1	2	3	4	5	6	7	8	9	10
	x_v	1	1	1	1	1	2	2	2	2	2

In this example Pearson's correlation coefficient has to coincide with Spearman's correlation coefficient. The result is: $r = r_S = 0.870$. Kendall's τ results as $\tau = 0.745$ (calculations are performed as in Example 11.6).

In **R**, we begin by creating two vectors, representing the values of the characters x and y; hence we type

```
> y2 <- 1:10
> x2 <- rep(1:2, each = 5)
```

i.e. we assign the values `1` to `10` to the object `y2` and a sequence of numbers created by the function `rep()` to the object `x2`; namely, the values `1` to `2` are repeated five times, each (`each = 5`). Next, we ascertain the different correlation coefficients by typing

```
> cor(y2, x2, method = "pearson")
> cor(y2, x2, method = "spearman")
> cor(y2, x2, method = "kendall")
```

i.e. we use the characters `y2` and `x2` as arguments in the function `cor()` and set the respective correlation coefficient, in fact Pearson's correlation coefficient, Spearman's rank correlation coefficient, and Kendall's τ, with the argument `method`.

As a result, we get:

```
[1] 0.8703883
[1] 0.8703883
[1] 0.745356
```

In SPSS, we open a new data sheet (File – New – Data) and begin by typing in the values of the two characters, x and y (as y and x). Next, we select Kendall's tau-b in Figure 11.14.

As the reader can easily verify, each of these coefficients equals 1 only if for $x_v = 1$ all of the observations in character y are the same and also for $x_v = 2$ all of the observations in character y are identical.

Master

Hence, different variants of Kendall's τ exist, *inter alia* the statistics τ_b and τ_c. The latter is preferable when the number of (realized) measurement values of the two characters, which are to be correlated, are not the same, because otherwise the strength of the relationship may be underestimated.

Master

Example 11.8 – continued

We now determine the coefficient τ_c as well.

In **R**, we create a new function to determine the coefficient τ_c; hence we type

```
> tau.c <- function(x, y) {
+   x <- table(x, y)
+   x <- matrix(as.numeric(x), dim(x))
+   con.x <- sum(x*mapply(function(r, c) {sum(x[(row(x) > r) &
+               (col(x) > c)])}, r = row(x), c = col(x)))
+ dis.x <- sum(x*mapply(function(r, c) {sum(x[(row(x) > r) &
+               (col(x) < c)])}, r = row(x), c = col(x)))
+   m <- min(dim(x))
+   return((m*2*(con.x-dis.x))/((sum(x)^2)*(m-1)))
+ }
```

i.e. we use the function `function()` and set the two the characters that we would like to correlate as arguments. The sequence of commands inside the braces defines the inner workings of the function and will not be discussed.

Next, we type

```
> tau.c(y2, x2)
```

i.e. we use the previously created characters `y2` and `x2` as arguments in the new function `tau.c()`.

As a result, we get:

```
[1] 1
```

In SPSS, we create a cross tabulation with the command sequence (Analyze – Descriptive Statistics – Crosstabs...) explained in Example 5.13, and, in the window in Figure 5.28, click the button Statistics...; in the resulting window (see Figure 11.15) we select Kendall's tau-c in the panel Ordinal. With Continue and OK we get the result: 1.0.

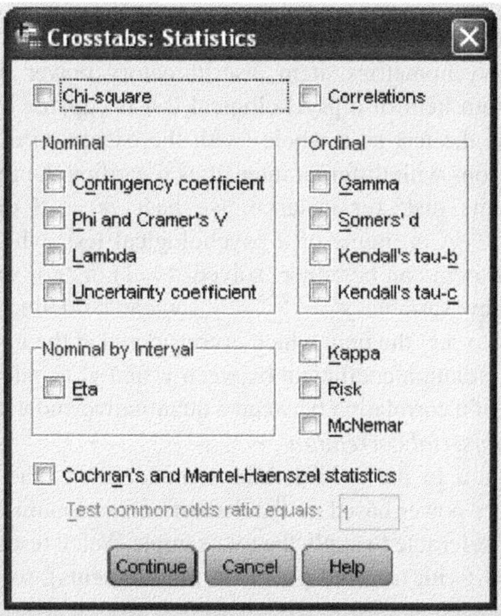

Figure 11.15 SPSS-window for computing Kendall's τ_c.

The reader may calculate other examples in order to find that, even in the case of unequal numbers of observations of the two characters, $\tau_b < \tau_c$ – taking absolute values – does not always result. If the relationship is marginal, sometimes – taking absolute values – $\tau_c < \tau_b$ (see, for instance, data from Example 1.1 and the two characters *urban/rural* and *number of siblings*: $\tau_b = -0.021$, $\tau_c = -0.010$.

Generally, the problem of determining the relationship does not exist: for the dichotomous character as a factor with the factor levels '+' and '–', or '1' and '2', or the like, in the case of a quantitative character ***y***, the two-sample Welch test can be applied, and in the case of an ordinal-scaled character, Wilcoxon's rank-sum test. Then, however, we test a hypothesis immediately. If the null hypothesis has to be rejected, the distributions of the variable ***y*** in the two groups (factor levels) differ (in the case of planning the study, to the extent which was regarded as relevant). In this respect, a relationship between the factor levels in ***x*** and the observations in ***y*** exists. More exactly we could say that, equivalently to the null hypothesis H_0: $\mu_1 = \mu_2 = \mu$ in the two-sample Welch-test, the hypothesis H_0: $\rho = 0$ (with the alternative hypothesis H_A: $\rho \neq 0$ or H_A: $\rho > 0$ or H_A: $\rho < 0$) is being tested. And equivalently to the null hypothesis H_0: $F(z) = G(z)$ in Wilcoxon's rank sum test, we have H_0: 'a relationship between ***x*** and ***y*** does not exist' (with the alternative hypothesis H_A: 'a relationship between ***x*** and ***y*** exists', or H_A: 'a relationship between ***x*** and ***y*** exists in such a way that higher values in ***x*** are associated with higher values in ***y***', or H_A: 'a relationship between ***x*** and ***y*** exists in such a way that higher values in ***x*** are associated with lower values in ***y***'). Herewith not only the inferential statistics subject of hypothesis testing concerning the relationship between two characters – for the given scale types of the characters – is completed, but also the subject planning and sequential testing (see Chapter 9).

Master

Example 11.9 Item discriminatory power

Within psychometrics, item discriminatory power describes the extent to which a certain item of a psychological test is capable of discriminating in the same way as the test as a whole (with the exception of the item in question) between persons with different intensities regarding the measured ability.

We assume that, for instance, we have $m = 8$ dichotomous characters y_l, $l = 1, 2, \ldots, m$, items of a psychological test, which were observed in n testees. Each item can be either 'solved' ($= 1$) or 'not solved' ($= 0$). Then, for example, a new variable $\boldsymbol{y}_l^* = \sum_{j=1, j \neq l}^{8} \boldsymbol{y}_l$ can be defined, that is the sum of all solved items except the item which is considered at the moment, l. Subsequently, Pearson's correlation coefficient between $\boldsymbol{y}_l$ and $\boldsymbol{y}_l^*$ is calculated. Sometimes this special case of a correlation between a quantitative and a dichotomous variable is called *point-biserial correlation.*

With regard to the considerations made above, the interpretation of item discriminatory power based on the theoretical maximum value of 1 is not useful. It would be preferable to apply the two-sample Welch test and, if the study has not been planned (with targeted precision requirements), to interpret the estimated effect size.

11.3.4 Relationship between a quantitative character and a multi-categorical character

A method for quantifying the relationship between an ordinal-scaled and a multi-categorical (non-dichotomous) character does not exist in statistics. In this case, we have to downgrade the ordinal-scaled variable to the next-lower scale type; thus to exploit only the information of a nominal scale (equal or unequal). This results in using that association measure which is available for two multi-categorical characters (see below in Section 11.3.5).

However, if the question is about a quantitative and a multi-categorical character, *Fisher's correlation ratio/eta-squared* is an appropriate association measure – again, because of the given symbol, it is difficult to differentiate between the parameter in the population and the statistic in the sample as an appropriate estimator for the parameter. This coefficient can be easily calculated with the help of statistical software as well, and it also often cannot reach the value of 1 even in the case of an 'ideal' correlation.

Master
Doctor

With reference to Table 10.3, Fisher's correlation ratio is defined by

$$\eta^2 = \frac{SS_A}{SS_t} \tag{11.9}$$

thus by the relation of the sum of squares of the level means $\bar{y}_i$ concerning the overall mean $\bar{y}$ and the sum of squares of the observations y_{iv} concerning the overall mean $\bar{y}$; thereby, levels mean the different categories of the nominal-scaled character (see Section 10.4.1.1). The possible values of η^2 are $0 \leq \eta^2 \leq 1$. $\eta^2 = 0$ if $\bar{y}_i = \bar{y}$ for all i; then SS_A or the variance of the level means, respectively, equals zero. $\eta^2 = 1$ results only if, for all v, $y_{iv} = \bar{y}_i$ happens to occur; thus if the character y does not possess any variance per factor level i (compare the analogy to Section 11.3.3).

Master

Example 11.8 – continued

Though the structure of data in this example is just a special case of all possible applications of Fisher's correlation ratio, it is nevertheless appropriate to demonstrate that this coefficient cannot reach a value of 1, despite an 'ideal' relationship.

In **R**, we need to compute Fisher's correlation ratio step by step; we start with typing

```
> summary(aov(y2 ~ x2))
```

i.e. we conduct, using the function `aov()`, a simple analysis of variance (cf. e.g. Example 10.1) and request the summarized results with the function `summary()`.

This yields (shortened output):

```
            Df Sum Sq Mean Sq F value    Pr(>F)
x2           1   62.5    62.5      25 0.001053
Residuals    8   20.0     2.5
```

From this table of variances, we extract that SS_A and SS_{res} are `62.5` and `20.0`, respectively; now, we can compute Fisher's correlation ratio. Hence we type:

```
> sqrt(62.5/(62.5 + 20))
```

i.e. we use the function `sqrt()` to calculate the square root in Formula 11.9.

As a result, we get:

```
[1] 0.8703883
```

In SPSS, we simply select Eta in Figure 11.15, and Continue, followed by OK, gets us the same result as the Spearman's rank correlation coefficient; that is $\eta = 0.870$.

Doctor

Fisher's correlation ratio is associated with several other problems as well. η^2 reaches the maximum value of 1 if only one single observation is given per category in the nominal-scaled character (factor level). Then, numerator and denominator in Formula (11.9) are equal.

Doctor

η^2 is also related in a certain way to the *intra-class correlation coefficient*, which is applied in genetics as the *heritability coefficient*; that is the proportion of the phenotypic variance which is covered by the genotypic variance.

The intra-class correlation coefficient is assignable in all models of analysis of variance (see Chapter 10), in which at least one random factor exists. The simplest case is the case of model II of analysis of variance. We will limit our consideration to this case at first. With the components of variance σ_a^2 and σ^2, which have been introduced in 10.4.1.3, the intra-class correlation coefficient is defined by

$$\rho_I = \frac{\sigma_a^2}{\sigma_a^2 + \sigma^2}$$

The estimate of ρ_I is obtained from the estimates for the components of variance by

$$r_I = \hat{\rho}_I = \frac{s_a^2}{s_a^2 + s^2}$$

In the case of equal cell frequencies n, we can also write

$$r_I = \frac{s_a^2}{s_a^2 + s^2} = \frac{\dfrac{MS_A - MS_{\text{res}}}{n}}{\dfrac{MS_A - MS_{\text{res}}}{n} + MS_{\text{res}}}.$$

The term 'correlation' is to be understood here in the way that, with the intra-class correlation coefficient, the correlation of the random variables $\boldsymbol{y}_{iv}$ is measured within the same factor level, or in other words within the same class. It can be defined analogously for two- and three-way cross classification and in all nested classifications. In psychology, it plays an important role mainly in this last sort, which are called 'hierarchical linear models' in this context (see Section 13.3).

Obviously, a strong relation exists between the one-way analysis of variance (see Section 10.4.1.1) and Fisher's correlation ratio, thus the association measure for a quantitative character and a multi-categorical character. Hence, one realizes (as already in Section 11.3.3) that the questions about 'differences' and the questions about 'relationships' do not aim for something basically different. If (significant) differences in the means exist between a levels of a fixed factor A with regard to character y, a relationship between this character and the nominal-scaled factor exists in the respect that the (mean) observations of y are associated with certain categories of A. Contrariwise, if a relationship of the nominal-scaled factor A and the quantitative character y exists, this means that, in certain categories of A, the (mean) observations in y differ.

That is, Fisher's correlation ratio or the determination coefficient, η^2, respectively, is generally capable of quantifying the significant difference of means, concerning the interesting factor, revealed by analysis of variance (or two-sample Welch test). The determination coefficient η^2 thus describes the resulting effect; that is, it estimates the effect size. An effect size defined in this way is, though, not comparable directly with the (relative) effect size as it was much more illustratively defined for the *t*-test by the relevant difference for content δ (see, for instance, in Section 8.3 the relative effect size $E = (\mu_1 - \mu_0)/\sigma$), but can nevertheless be interpreted as an absolute measure, as well. The effect size E or its estimate $\hat{E}$ as defined up to now expresses the (mean) differences in units of the character's standard deviation. In contrast, the determination coefficient η^2, as (estimated) effect size, expresses the percentage of sum of squares (sum of squared differences), of all observations with respect to the overall mean, which can be explained by the sum of squares of the all factor level means with respect to the overall mean. While in the first case the effect size can become (much) bigger than 1, η^2 can maximally reach the value of 1, and even this occurs only in the extremely infrequent case described above. In this respect η^2 as an (estimated) effect size runs the risk of being compared in interpretation with an unrealistic optimal value.

Bachelor **Example 10.1 – continued**

The question was, if children differ in their test scores in *Everyday knowledge, 1st test date,* dependent on their *sibling position.*

In Section 10.4.1.1, we have found for this example that significant differences do not exist. Therefore, it was unnecessary to estimate an effect size. We want to supplement this here, and show how to calculate it with the help of software programs in case of a significant result of the one-way analysis of variance.

In **R**, we compute η^2 analogously to Example 11.8 by typing

```
> 253.9/(253.9 + 9651)
```

i.e. we use the values of SS_A and SS_{res} from Table 10.4 and put them in Formula 11.9.

As a result, we get:

```
[1] 0.02563378
```

In SPSS, we cannot proceed as in the first place (via One-Way ANOVA...) to get Fisher's correlation ratio, but instead use the command sequence (Analyze – General Linear Model – Univariate...) described in Example 10.8 to open the window in Figure 10.7 – children with a sibling position higher than the fourth are already excluded from analysis. Here, after moving the character *Everyday Knowledge, 1st test date* and the factor *sibling position* to Fixed Factor(s):, we click Options... and the window in Figure 10.8 pops up. Finally, we select Estimates of effect size; clicking Continue and OK gets us the, in comparison to Table 10.4, extended Table 11.4. From the column Partial Eta Squared we get $\eta^2 = 0.026$.

Table 11.4 SPSS-output showing the table of variances of the analysis of variances (model I) in Example 10.1 (shortened output).

Tests of Between-Subjects Effects

Dependent Variable:Everyday Knowledge, 1st test date (T-Scores)

Source	Type III Sum of Squares	df	Mean Square	F	Sig.	Partial Eta Squared
Corrected Model	253.949[a]	3	84.650	.772	.513	.026
Intercept	170041.946	1	170041.946	1550.475	.000	.946
pos_sibling	253.949	3	84.650	.772	.513	.026
Error	9651.040	88	109.671			
Total	263375.000	92				
Corrected Total	9904.989	91				

a. R Squared = .026 (Adjusted R Squared = -.008)

Obviously, the effect is negligible. Only 2.6% of the sum of squares of the test scores in subtest *Everyday Knowledge* can be explained by the sum of squares of the mean test score per sibling position.

Fisher's correlation ratio can be calculated as an estimate for the effect size for all other types of analysis of variance with cross classification, too.

Master Doctor

Example 11.7 – continued

If we want (because of the significant result) to determine the result-based type-II risk, we have to calculate η^2 first.

In **R**, we do that by typing

```
> 386.392/(386.392 + 3247.474)
```

i.e. we use the values of SS_A and SS_{res} from e.g. Table 10.7, and put them in Formula 11.9. As a result, we get:

```
[1] 0.1063308
```

In SPSS, we have to repeat the procedure, but this time we need to click Options... in Figure 10.7. This opens the window in Figure 10.8, where we select Estimates of effect size. After Continue and OK we get the result from the column Partial Eta Squared.

The estimated effect size is $\eta^2 = 0.106$.

Master Doctor

As opposed to the calculation according to Formula (11.9), in SPSS a 'partial' η^2 is calculated as an estimated effect size in analysis of variance; in the case of one-way analysis of variance, this effect size coincides with η^2. The underlying idea is to estimate, in the case of multi-way analyses of variance, the separated effect for each case, when the corresponding null hypothesis is rejected. For instance, for the two-way analysis of variance of Table 10.10, the following three measures of partial coefficients result:

$$\eta_A^2 = \frac{SS_A}{SS_A + SS_{\text{res}}}, \quad \eta_B^2 = \frac{SS_B}{SS_B + SS_{\text{res}}}, \quad \text{and} \quad \eta_{AB}^2 = \frac{SS_{AB}}{SS_{AB} + SS_{\text{res}}}$$

So, for all partial coefficients we have $\eta_{\text{part}}^2 \geq \eta^2$. Thus, the not very illustrative interpretation of partial η^2 is: the corresponding percentage of variance (to be formally correct: sum of squares) of all observations of the character of interest – corrected by the other factors' contribution, the interaction effects included – which is explained by the variance (to be formally correct: sum of squares) of the factor level means.

The ostensible advantage of estimation of the specific effect by the effect size of partial η^2 is accompanied by the problem that, unlike for η^2, the sum of all estimated effect sizes does not equal 1.

Referring to analysis of variance, the question of hypothesis testing concerning Fisher's correlation ratio has also been solved, and herewith also the question of planning a study.

11.3.5 Correlation between two nominal-scaled characters

In the case of relationships between two nominal-scaled characters, we have to distinguish generally between the case of both characters being dichotomous on one hand and the case of at least one character being multi-categorical on the other hand.

In any case, the observation pairs (x_v, y_v) can be represented very clearly in a *contingency table*, as we already did in previous chapters (see mainly Section 9.2.3); more exactly speaking, we talk about a two-dimensional contingency table, because two characters are given.

In the more general case, the character which was modeled by the random variable $\boldsymbol{x}$ has r different categories, and the character modeled by $\boldsymbol{y}$ possesses c categories. When arranged in a matrix (table), $r \times c$ cells result, for example r rows times c columns. The number of observed combinations of categories is entered in these cells. If n research units are given, the frequency per combination of categories is described by n_{ij}, $i = 1, 2, \ldots, r$ and $j = 1, 2, \ldots, c$. The special case of $r = c = 2$ is of course included here. In this case, the table is often referred to as a *fourfold table* or *2 × 2 table*. In contrast, the general contingency table then is called an *$r \times c$ table*.

At first, we consider the general case with $r > 2$ and/or $c > 2$.

If the number of categories r and number of categories c is not very big, any tendencies of a relationship can be detected already from the contingency table. In doing so, it is important to keep in mind that, because of the nominal-scaled categories, no direction of 'arrangement of order' can be ascertained; all of the rows and all of the columns are generally interchangeable with each other arbitrarily.

Bachelor **Example 9.9 – continued**

The question was about the difference in *marital status of the mother* between children with German or Turkish as their native language. Using the χ^2-test, we determined a significant difference. For this purpose we interpreted the character *native language of the child* as a factor with the two factor levels 'German' and 'Turkish'. If a difference in the distribution on the categories of marital status exists between children with German and children with Turkish as their native language, then a relationship between the two characters exists in so far as certain categories of marital status occur more frequently and other categories less frequently in children whose native language is German then in children whose native language is Turkish.

Table 9.5 shows mainly that children with Turkish as their native language have considerably more married mothers than the children with German as their native language. Contrariwise, the former have only a fraction of mothers with the marital status 'divorced' compared to the latter.

The common measure for the general case of a contingency table is the so-called *contingency coefficient*, more precisely Pearson's contingency coefficient. Consequently, it is an unsigned measure and it is based on the statistic of the χ^2-test from Formula (9.3):

$$C = \sqrt{\frac{\chi^2}{\chi^2 + n}}$$

As it facilitates comprehension, we will use C for the statistic in the sample, and ζ for the parameter in the population, which is estimated by C; although this notation is not common practice. As this coefficient also cannot achieve the value of 1 in many cases, dependent on $r \cdot c$, it is preferable to calculate (in the sample) the corrected contingency coefficient:

$$C_{\text{corr}} = \sqrt{\frac{t}{t-1}}\,C \tag{11.10}$$

in which t is the smaller of the two values r and c.

Bachelor **Example 9.9 – continued**

We want to quantify the relationship determined by means of the χ^2-test between *marital status of the mother* and *native language of the child*.

In **R**, we apply the package vcd, which we load after its installation (see Chapter 1) using the function library(). Next, we type

```
> assoc.1 <- assocstats(table(marital_mother, native_language))
> summary(assoc.1)
```

i.e. we create, using the function table(), a contingency table of the two characters *marital status of the mother* (marital_mother) and *native language of the child* (native_language), and submit it as an argument to the function assocstats();

the result of this analysis is assigned to the object `assoc.1`. Finally, we request the summarized results with the function `summary()`.

As a result, we get:

```
Number of cases in table: 100
Number of factors: 2
Test for independence of all factors:
        Chisq = 16.463, df = 3, p-value = 0.0009112
        Chi-squared approximation may be incorrect
                    X^2 df   P(> X^2)
Likelihood Ratio 17.338  3 0.00060216
Pearson          16.463  3 0.00091122

Phi-Coefficient   : 0.406
Contingency Coeff.: 0.376
Cramer's V        : 0.406
```

To compute C_{corr} we type

```
> sqrt(2/(2-1))*assoc.1$contingency
```

i.e. we conduct the computation according to Formula (11.10), and use the contingency coefficient `$contingency` in the object `assoc.1`.

As a result, we get:

```
[1] 0.53171
```

In SPSS, we produce a cross tabulation (see Example 5.13), where we click Statistics... and get to Figure 11.15. Next, we select Contingency coefficient. As a result, we get Table 11.5: the contingency coefficient C is 0.376. Since SPSS states neither the maximum of C nor C_{corr} directly, we compute C_{corr} manually:

$$C_{\text{corr}} = \sqrt{\frac{t}{t-1}}C = \sqrt{2} \cdot 0.376 = 0.532$$

The p-value 0.001 in Table 11.5 in Approx. Sig. corresponds to that of the χ^2-test earlier in Example 9.9

Table 11.5 SPSS-output of the contingency table in Example 9.9

Symmetric Measures

		Value	Approx. Sig.
Nominal by Nominal	Contingency Coefficient	.376	.001
N of Valid Cases		100	

Compared to the maximum possible relationship, the observed relationship between *native language of the child* and *marital status of the mother* is 53%; that is, a relationship of medium strength.

Other measures of association exist as well for this case, but they hardly play any role in psychology (but see Kubinger, 1990). Nevertheless, some of them are output 'automatically' by SPSS.

Doctor If, in the case of contingency tables, the question about the relationship between two characters focuses on regression analysis, we generally have to differentiate between two models, as in analysis of variance. One of them is found only incidentally in psychology, but then mostly is simply not recognized at all by the researcher as an entirely different statistical problem. The common case refers to research units being sampled by chance and for which the values in both characters have been ascertained accordingly (model II). In the other case (model I), one value per pair of outcomes (e.g. the value x) is determined or in other words selected in a systematic way, and only the corresponding y value is ascertained accordingly. An example for model I is the survey of Example 1.1 concerning the children with German or Turkish as their native language: the researcher determined that 50 children from the first population and 50 children from the second population would be investigated. In this case, the values of x are not ascertained, but determined *a priori*. Therefore, the cases of application in Chapter 9 differ from those in this chapter. There, the different rows (or columns, respectively) representing several populations have been determined with regard to their marginal totals prior to data acquisition; only the respective other marginal totals were random. In the case examined here we have only one sample (of research units), and both marginal totals, of rows as well as of columns, are random. However, this difference is of interest only theoretically: practically, the calculation procedures for the association measures described here and for the χ^2-test are the same.

For Lecturers:

In contingency tables, a model III also exists; this is given in a famous example by Ronald A. Fisher (for more detailed information see e.g. Salsburg, 2002). It is about an English lady who pretends to be able to find out in which order tea and milk had been put in a cup. Fisher designed an experiment, in which eight different cups of tea were given to the lady for tasting, of which four were filled in one way and four cups in the other way. This was also told to the lady. Therewith the marginal sums of, for instance, of the rows of the contingency table, are fixed. Given that the lady is intelligent, the marginal sums of the columns are also fixed, because the lady was told how many cups had been filled using each method of preparation.

Table 11.6 General 2 × 2 table.

		Character 2		Marginal totals of rows
		B	$\bar{B}$	
Character 1	A	a	b	$a+b$
	$\bar{A}$	c	d	$c+d$
Marginal totals of columns		$a+c$	$b+d$	$n=a+b+c+d$

In the case of a 2 × 2 table, there is also an association measure, which is related to the χ^2-test from Formula (9.3). This is the *ϕ-coefficient*:

$$\sqrt{\frac{\chi^2}{n}}$$

Again because of the given symbol, it is difficult to differentiate between the parameter in the population and the statistic in the sample as an appropriate estimator for the parameter. As this coefficient also cannot achieve the value of 1 in many cases, even if an 'ideal' relationship is given, it is preferable to determine the corrected ϕ-coefficient (within the sample):

$$\phi_{corr} = \frac{\phi}{\phi_{\max}} \tag{11.11}$$

It has to be noted that $\phi_{\max}$ cannot be calculated easily. If o_{11} to o_{22} in Table 9.2 are renamed as a to d (see the general 2 × 2 table in Table 11.6), $\phi_{\max}$ results as follows:

$$\phi_{\max} = \sqrt{\frac{(a+b)(b+d)}{(c+d)(a+c)}} \tag{11.12}$$

with $(a+b) \leq (c+d)$, $(b+d) \geq (a+c)$, and $(c+d) \geq (b+d)$.

Master A ϕ-coefficient exists in a general form as well; that is the relation between the contingency coefficient and ϕ-coefficient is as following:

$$C = \sqrt{\frac{\chi^2}{\chi^2 + n}} = \sqrt{\frac{\phi^2}{\phi^2 + 1}}$$

The reason why there are two association measures at the same time is, on the one hand, that, for $r \times c$ tables, the ϕ-coefficient can also become bigger than 1: namely maximally $\sqrt{t-1}$; for $r = c = 3$, this is 1.414, for instance. On the other hand, though Pearson's contingency coefficient is indeed the oldest association measure based on the χ^2-test, the ϕ-coefficient is the immediate special case of Pearson's correlation coefficient for the case of two dichotomous characters.

Using the frequencies a to d, from Formula (11.8) as well as from Formula (9.3) it can easily be derived:

$$\phi = \frac{ad - bc}{\sqrt{(a+b)(c+d)(a+c)(b+d)}} \tag{11.13}$$

Master

Example 11.10 Relationship of two items of a psychological test

It can be demonstrated easily that the ϕ-coefficient cannot achieve the value of 1 even in the case of an 'ideal' relationship. At first we give, in Table 11.7 (upper, left-hand 2×2 table), a numeric Example 1 for data from 100 testees who solved (+) or did not solve (−) two items of a psychological test. According to (11.12), $\phi = 0.40$. This value of an association measure, of Pearson's correlation coefficient, would have to be interpreted as rather small, because only 16% of the variance is mutually determined. However, we see that the difficulty of the two items differs. While item x was solved by only 20 out of the 100 testees, item y was solved by 50 out of 100 testees. Even if an ideal relationship existed in the empirically given marginal totals of rows and columns, the ϕ-coefficient would not equal 1. Instead of using Formula (11.11) – in which it would be necessary to interchange rows or columns, or to skip over the 2×2 table, respectively, because of the side conditions $(a + b) \leq (c + d)$, $(b + d) \geq (a + c)$, and $(c + d) \geq (b + d)$ – we use the following calculation procedure according to Kubinger (1995) to determine ϕ_{max}.

Table 11.7 2×2 tables for Example 11.10.

Numeric Example 1

		Item x +	Item x −	
Item y	+	18	32	50
	−	2	48	50
		20	80	

$\phi = 0.40$

Ideal relationship for given marginal totals

		Item x +	Item x −	
Item y	+	20	30	50
	−	0	50	50
		20	80	

$\phi_{max} = 0.50$

Example 2
Ideal relationship

		Item x +	Item x −	
Item y	+	20	0	20
	−	0	80	80
		20	80	

$\phi_{max} = 1.00$

Example 3
Ideal relationship

		Item x +	Item x −	
Item y	+	30	20	50
	−	0	50	50
		30	70	

$\phi_{max} = 0.65$

In the intersection of row and column of each biggest marginal total of items $\boldsymbol{x}$ and $\boldsymbol{y}$, the smaller marginal total has to be selected as a' or d', respectively. In order to get the observed smaller marginal total correct, b' or c' (depending on which of the marginal totals was the smaller one), respectively, has to be set equal to 0. Thereby also all other values of the contingency table are determined. The calculation according to Formula (11.12) is conducted with these values instead of the originally given a, b, c, and d.

For Example 1 the following results are derived with this procedure: the smaller of the two bigger marginal totals (in the rows this is 50; in the columns this is 80) is 50; this is the value for d'. For this reason c' has to set equal to 0, so that the marginal total remains 50. Then $b' = 30$ and $a' = 20$ necessarily (see the upper, right-hand 2×2 table, ideal relationship for given marginal totals, in Table 11.7). If these values for a, b, c, and d are inserted into Formula (11.12), the result is $\phi_{max} = 0.50$. With regard to this, the ϕ-coefficient with $\phi = 0.40$ is not that small. When we calculate $\phi_{corr} = \frac{0.40}{0.50} = 0.80$, exactly this fact becomes evident. It is easy to realize that only when the marginal totals of both items are identical can the ϕ-coefficient actually become 1 (see the lower, left-hand 2×2 table in Table 11.7). Example 3 (see the lower, right-hand 2×2 table in Table 11.7) illustrates additionally that the ϕ-coefficient can become bigger the more similar the marginal totals of both items are: instead of $\phi_{max} = 0.50$ in Example 1, the maximum of the ϕ-coefficient is now $\phi_{max} = 0.65$.

Master

Interestingly, although it is a fact that Pearson's contingency coefficient (in quadratic contingency tables) is necessarily always smaller than 1 as a consequence of non-equivalence between the two marginal totals of x and y, this is not reflected in literature. The reader can easily verify this analogously to Example 11.10.

As the ϕ-coefficient represents a special case of Pearson's correlation coefficient for the case of two dichotomous characters, $\phi^2 = B$ has to conform to the coefficient of determination. Nevertheless, ϕ^2 should not be used for interpretation, precisely because it often underestimates the strength of association. It is preferable to interpret ϕ_{corr} as the relative strength of relationship, which is given compared to the theoretically maximum possible relationship. Squaring ϕ_{corr}, a pseudo-coefficient of determination is obtained: compared to the maximum possible percentage of explained variance, ${\phi^2}_{corr} \cdot 100\%$ was observed.

Bachelor

Example 9.7 – continued

The question is, if the frequency of only-children is identical in families with children with German as their native language and in families with children with Turkish as their native language. Again, the question for differences is generally equivalent to the question for a relation between *native language of the child* and the character 'only child' versus 'child with siblings' which arises from the character *number of siblings*.

We refer once more to Table 9.4. We already know that a significant difference exists according to the results of the χ^2-test. Thus, a relationship exists. We want to quantify its strength by means of the ϕ-coefficient.

In **R**, we use the package `vcd` (cf. Example 9.9 above); since a vector of the character values 'only child' and 'child with siblings' has already been created (cf. Example 9.7), we type:

```
> assoc.2 <- assocstats(table(native_language, only_child))
> summary(assoc.2)
```

i.e. we use the contingency table of *native language of the child* (native_language) and *number of siblings* (only_child), which we create with the function table(), as an argument in the function assocstats(); the result is assigned to the object assoc.2. Finally, we request the summarized results with the function summary().

As a result, we get:

```
Number of cases in table: 100
Number of factors: 2
Test for independence of all factors:
        Chisq = 8.575, df = 1, p-value = 0.003407
                    X^2 df  P(> X^2)
Likelihood Ratio 9.1852  1 0.0024399
Pearson          8.5755  1 0.0034072

Phi-Coefficient   : 0.293
Contingency Coeff.: 0.281
Cramer's V        : 0.293
```

Since **R** neither states $\phi_{\max}$ nor ϕ_{corr}, we need to compute both manually. From Table 9.3, we see that the conditions $(a+b) \leq (c+d)$, $(b+d) \geq (a+c)$, and $(c+d) \geq (b+d)$ are not satisfied without interchanging rows and columns. After several failed trials we find the solution: the contingency table needs to be flipped 90° to the right; thus $a = 3$, $b = 14$, $c = 47$, and $d = 36$. Now the rules apply: $(a+b) = 17 \leq (c+d) = 83$; $(b+d) = 50 \geq (a+c) = 50$; and $(c+d) = 83 \geq (b+d) = 50$. Hence $\phi_{\max} = \sqrt{\frac{(a+b)(b+d)}{(c+d)(a+c)}} = \sqrt{\frac{17 \cdot 50}{83 \cdot 50}} = 0.453$ and therefore, $\phi_{\mathrm{corr}} = \frac{0.293}{0.453} = 0.647$.

In SPSS, for the variable only_child already created in Example 9.7, we use the same sequence of commands (Analyze – Descriptive Statistics – Crosstabs...) and press Statistics..., and select Phi and Cramer's V (as well as Chi-square for checking purposes) – we cannot disable the output of the hardly ever used coefficient Cramer's V. We click Continue and conclude the entries with OK. As a result, we get the already known contingency table, as well as Table 11.8 including the ϕ-coefficient and the result of the χ^2-test.

Table 11.8 SPSS-output of the ϕ-coefficient in Example 9.7 (shortened output).

Symmetric Measures

		Value	Approx. Sig.
Nominal by Nominal	Phi	.293	.003
N of Valid Cases		100	

Since SPSS neither states $\phi_{\max}$ nor ϕ_{corr}, we need to compute both manually. From Table 9.3, we see that the conditions $(a+b) \leq (c+d)$, $(b+d) \geq (a+c)$, and $(c+d) \geq (b+d)$

are not satisfied without interchanging rows and columns. After several failed trials we find the solution: the contingency table needs to be flipped 90° to the right; thus $a = 3$, $b = 14$, $c = 47$, and $d = 36$. Now the rules apply: $(a + b) = 17 \leq (c + d) = 83$; $(b + d) = 50 \geq (a + c) = 50$; and $(c + d) = 83 \geq (b + d) = 50$. Hence $\phi_{\max} = \sqrt{\frac{(a+b)(b+d)}{(c+d)(a+c)}} = \sqrt{\frac{17 \cdot 50}{83 \cdot 50}} = 0.453$ and therefore, $\phi_{\text{corr}} = \frac{0.293}{0.453} = 0.647$.

The strength of the relationship, compared to the maximum possible relationship (for the given marginal totals), comes up to 64.7%. The calculated strength of relationship is considerable: children with Turkish as their native language are only-children much less frequently than children with German as their native language. Squaring ϕ_{corr}, we obtain a pseudo-coefficient of determination of $0.647^2 = 0.419$. Compared to the maximum possible percentage of variance, we observed 41.9%.

Referring to the χ^2-test, the topic of hypothesis testing concerning relationships of two characters – given the corresponding scaling of the characters – is completed as well: the test statistic in Formula (9.3) is valid for any two multi-categorical characters. Then, the null hypothesis is, expanding or supplementary to Section 9.2.3 (with H_0: $p_{1j} = p_{2j}$ for all j) in the case of an $r \times c$ contingency table, H_0: $p_{hj} = p_{gj}$ for all j and all h and g (or, equivalently, H_0: $p_{hj} = p_{hl}$ for all h and all j and l) or now H_0: $\zeta = 0$, with the alternative hypothesis H_A: $p_{hj} \neq p_{gj}$, for at least one j and $h \neq g$ (or, equivalently, H_A: $p_{hj} \neq p_{hl}$ for at least one h and $j \neq l$) or now H_A: $\zeta > 0$. For the ϕ-coefficient the same is true analogously; we just cannot differentiate between statistics and parameters in this case.

Contrariwise, with the reference of χ^2-tests to measures of association, there is the possibility to define a corresponding effect size and to estimate it for the case that the research was not planned. Thus, if any χ^2-distributed statistic is given, because of $\phi = \sqrt{\frac{\chi^2}{n}}$, in addition to Fisher's correlation ratio η^2 is a further association- or correlation-based effect size that can be estimated: $\phi^2 = \chi^2 / n$. However, interpretation is difficult, because, as described above, ϕ^2 often cannot achieve the value of 1 even in cases of optimal explanation of variance due to the given distributions of marginal totals. By the way, the (estimated) effect size ϕ^2 is suited generally as well for all standard normal-distributed statistics. As known from mathematical statistics, the squared standard normal-distributed z^2 is χ^2-distributed with $df = 1$ degree of freedom. From that point of view the following effect size can be estimated as well: $\phi^2 = z^2 / n$.

A particular question regarding the association of two nominal-scaled characters concerns the concordance of two raters with respect to a certain character. In psychology, mainly in psychological assessment, this is very frequently of interest, for instance with regard to the objectivity of psychological testing or that is to say with regard to test administrator (examiner) effects (see Section 2.3). Somewhat misleadingly, one then talks about *inter-rater reliability*. It can be quantified by means of the *kappa coefficient*. This coefficient corrects the percentage of observed concordances over all categories by that percentage of accordance which would be expected to occur in random ratings. Additionally, the resulting percentage is related to the maximum possible percentage which cannot be explained by chance. Using the notation for the absolute frequencies in a quadratic contingency table (with $r = c$) such as, for example, in Section 9.2.3, and $n = \sum_i^c \sum_j^c o_{ij}$, this results in:

$$\kappa = \frac{\sum_{i=1}^{c} \frac{o_{ii}}{n} - \sum_{i=1}^{c} \left(\frac{\sum_{j=1}^{c} o_{ij}}{n} \right) \left(\frac{\sum_{j=1}^{c} o_{ji}}{n} \right)}{1 - \sum_{i=1}^{c} \left(\frac{\sum_{j=1}^{c} o_{ij}}{n} \right) \left(\frac{\sum_{j=1}^{c} o_{ji}}{n} \right)} \tag{11.14}$$

In pertinent program packages even a test statistic for testing the null hypothesis H_0: $\kappa = 0$ (against the alternative hypothesis H_A: $\kappa > 0$) is calculated; this test statistic is asymptotically normally distributed. However, in applying the kappa coefficient one aims only for a description of how close it comes to the ideal value of $\kappa = 1$, thus a complete concordance.

Bachelor **Example 11.11** How strong is the concordance between two raters in an assessment center?

Within an assessment center, 25 participants in a seminar are rated by two assessors with respect to their competence in a presentation exercise, according to three criteria: 'convinces by content' (1); 'seems to be respectable, but does not convince by content' (2); and 'seems to be incompetent' (3). The exact results are the following (ratings which are listed one upon the other stem from the same participant. See Chapter 1 for the availability of the data):

Assessor 1	1	2	2	1	1	2	2	1	2	1	1	3	1	1	1	1	1	2	1	3	2	1	2	1	1
Assessor 2	1	2	3	1	2	2	1	1	1	1	1	2	1	1	1	2	1	1	1	3	2	1	1	1	1

In **R**, we type

```
> assess1 <- c(1, 2, 2, 1, 1, 2, 2, 1, 2, 1, 1, 3, 1, 1, 1, 1,
+              1, 2, 1, 3, 2, 1, 2, 1, 1)
> assess2 <- c(1, 2, 3, 1, 2, 2, 1, 1, 1, 1, 1, 2, 1, 1, 1, 2,
+              1, 1, 1, 3, 2, 1, 1, 1, 1)
```

i.e. we define the single values of each assessor as a separate vector by using the function `c()`, and assign them to the objects `assess1` and `assess2`, respectively. Next, we use the appropriate function in the package `vcd`, namely

```
> Kappa(table(assess1, assess2))
```

i.e. we use `assess1` and `assess2` as arguments in the function `table()` to create a contingency table, which we submit as an argument to the function `Kappa()`.

As a result, we get (shortened output):

```
Unweighted 0.3710692
```

In SPSS, we open a new data sheet (File – New – Data) and begin by typing in the two characters assess1 and assess2 according to the values above. Following the command sequence (Analyze – Descriptive Statistics – Crosstabs...), we create a cross tabulation and use the character assess1 in the field Row(s): and the character assess2 in the field Column(s):. By clicking Statistics... we open the window in Figure 8.11 (cf. Example 8.14), where we select Kappa, Continue, and confirm with OK. The result is $\kappa = 0.371$.

Compared to the maximum possible strength of concordance, the two assessors conform to each other to 37.1%.

11.3.6 Nonlinear relationship in quantitative characters

In the case that the regression function f in (11.1) is plausibly not linear, a *polynomial regression function* – namely *quadratic regression function* – or *logistic regression function* comes into consideration.

Doctor The quadratic regression function is a polynomial of degree two. The corresponding regression model is

$$\boldsymbol{y}_v = \beta_0 + \beta_1 \boldsymbol{x}_v + \beta_2 \boldsymbol{x}_v^2 + \boldsymbol{e}_v \tag{11.15}$$

While the quadratic regression function is appropriate (better than the linear one) when the slope of the function is steeper in smaller values of $\boldsymbol{x}$ than in medium-sized values, and flatter for high values of $\boldsymbol{x}$, the logistic regression function is appropriate when an inflection point exists in the course of the relation between $\boldsymbol{x}$ and $\boldsymbol{y}$. The logistic regression results in the regression model

$$\boldsymbol{y}_v = \frac{\beta_0}{1 + \beta_1 e^{\beta_2 \boldsymbol{x}_v}} + \boldsymbol{e}_v \quad \text{with } \beta_0, \beta_1, \beta_2 \neq 0 \tag{11.16}$$

Here β_0 indicates the asymptotic value of y_v, thus at the point where $x_v \to \infty$; β_1 indicates the shift along the abscissa, and β_2 the extent of curvature.

Doctor **Example 11.12** Numeric example for quadratic and logistic regression functions without regard to content

We use the analysis of Example 11.5. This is about the relationship between the character *Applied Computing, 1st test date* and *Applied Computing, 2nd test date*. We would like to represent the scatter plot including a linear, a quadratic, and a logistic regression function.

In **R**, we type

```
> lm.lin <- lm(sub3_t2 ~ sub3_t1)
> lm.quad <- lm(sub3_t2 ~ 1 + sub3_t1 + I(sub3_t1^2))
> nls.log <- nls(sub3_t2 ~ SSlogis(sub3_t1, Asym, xmid, scal))
```

i.e. we use the function `lm()` to create regression models for the linear, the quadratic, and the logistic regression function, respectively. *Applied Computing, 1st test date* (`sub3_t1`) is the regressor; *Applied Computing, 2nd test date* (`sub3_t2`) is the regressand. We specify the linear regression function with `sub3_t2 ~ sub3_t1`, and the quadratic with `sub3_t2 ~ 1 + sub3_t1 + I(sub3_t1^2)` – where the statement in `I()` represents the quadratic term. In order to create the logistic regression function, we use the function `nls()` in **R** with `sub3_t2 ~ SSlogis(sub3_t1, Asym, xmid, scal)` – where the statement in `SSlogis()` represents the logistic term. The resulting regression models from the function `lm()` and `nls()` are assigned to the objects `lm.lin`, `lm.quad`, and `nls.log`. Next we illustrate the scatter plot; therefore, we type

```
> plot(sub3_t1, sub3_t2,
+      xlab = "Applied Computing, 1st test date (T-Scores)",
+      ylab = "Applied Computing, 2nd test date (T-Scores)")
```

i.e. we use `sub3_t1` and `sub3_t2` as arguments in the function `plot()` and label the axis with `xlab` and `ylab`. We add the regression functions to the chart, by typing

```
> abline(lm.lin)
> lines(seq(35, 75, 0.1), predict(lm.quad,
+        newdata = data.frame(sub3_t1 = seq(35, 75, 0.1))), lty = 5)
> lines(seq(35,75, 0.1), predict(nls.log,
+        newdata = data.frame(sub3_t1 = seq(35, 75, 0.1))), lty = 4)
> legend("bottomright", lty = c(1, 5, 4),
+        c("Linear", "Quadratic", "Logistic"))
```

i.e. we submit the linear regression function in the object `lm.lin` as an argument to the function `abline()`. The function `lines()` draws a line using the coordinates from the first argument, a number series from `35` to `75` in steps of `0.1` produced by the function `seq()`, and the second argument, the corresponding predicted values (`newdata = data.frame`) of the quadratic regression function in the object `lm.quad` using the function `predict()`; we proceed analogously for the logistic function. In doing so, we predict the values of the character *Applied Computing, 2nd test date* from the possible values `35` to `75` of the character *Applied Computing, 1st test date* (`sub3_t1`); with `lty = 5` and `lty = 4` we select the line types. Lastly, we create a `legend()`, set its position with the first argument, in this case the lower right corner of the chart (`"bottomright"`), and amend it with the labels `"Linear"`, `"Quadratic"`, and `"Logistic"` using the function `c()`.

As a result, we get Figure 11.16.

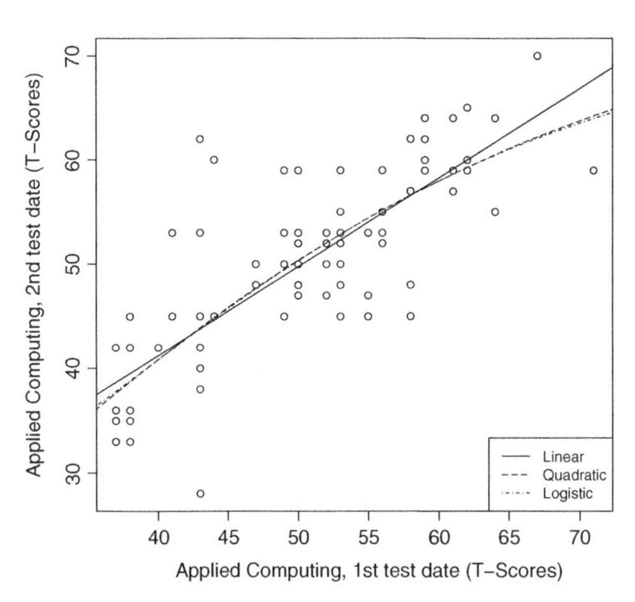

Figure 11.16 **R**-output showing the linear, quadratic, and logistic regression function in Example 11.12.

Next, we plot the corresponding regression coefficients, by typing

```
> coef(lm.lin)
> coef(lm.quad)
> coef(nls.log)
```

i.e. we use the respective regression objects in the function `coef()`.

As a result, we get:

```
(Intercept)      sub3_t1
  7.1053387    0.8531325

  (Intercept)        sub3_t1  I(sub3_t1^2)
-15.946484522    1.795332804  -0.009384195

    Asym      xmid      scal
72.73897 35.58403 17.81893
```

To obtain the corresponding regression coefficients for the logistic regression function according to formula (11.16), we rearrange the resulting parameters `xmid` and `scal`. We type

```
> exp(coef(nls.log)[2]/(coef(nls.log)[3]))
> -1/coef(nls.log)[3]
```

As a result, we get:

```
    xmid
 7.36677
    scal
-0.0561201
```

Hence, we obtain $\beta_0 = 72.739$; $\beta_1 = 7.367$; $\beta_2 = -0.056$.

In SPSS, we use the command sequence (Analyze – Regression – Curve Estimation...), analogously to Example 11.5, to open the window in Figure 11.8, and once again select Linear and, additionally, Quadratic. After clicking OK we get the result in Table 11.9 as well as a chart analogous to Figure 11.16, just without the logistic regression function. We obtain the coefficient of determination as a measure of the relationship size from Table 11.9 in column R Square.

Table 11.9 SPSS-output of the regression coefficients in Example 11.12.

Model Summary and Parameter Estimates

Dependent Variable:Applied Computing, 2nd test date (T-Scores)

	Model Summary					Parameter Estimates		
Equation	R Square	F	df1	df2	Sig.	Constant	b1	b2
Linear	.609	152.549	1	98	.000	7.105	.853	
Quadratic	.615	77.472	2	97	.000	-15.946	1.795	-.009

The independent variable is Applied Computing, 1st test date (T-Scores).

In order to obtain the logistic regression function in SPSS, we use the command sequence Analyze – Regression – Nonlinear... (not shown here), where we type beta0/(1+beta1*exp(beta2 * u3_t1)) in the panel Model Expression: and drop the character *Applied Computing, 2nd test date* in the panel Dependent:. The parameter estimation process requires pertinent starting values; namely the highest observation value of the regressand for β_0; that is 70 in this case; for β_1 and β_2 we choose the arbitrary 0.9 and −0.1. To enter the starting values, we click Parameters.... As a consequence a small window (not shown) appears, where we define the parameters one after another by entering the Name: (beta0, beta1, and beta2) and the Starting Value: (70, 0.9, and -0.1), each of which we confirm with Add, and click on Continue after we have defined the last parameter. After clicking on OK, we obtain the results: $\beta_0 = 72.739$; $\beta_1 = 7.367$; $\beta_2 = -0.056$; the coefficient of determination B is 0.613.

The estimated regression functions are for the linear, the quadratic, and the logistic function:

$$\hat{y}_v = 7.1053 + 0.8531 x_v,\ \hat{y}_v = -15.9465 + 1.7953 x_v - 0.0094 x_v^2,$$
$$\text{and } \hat{y}_v = \frac{72.74}{1 + 7.37 \cdot e^{-0.056}}$$

We recognize that the linear regression describes the data hardly any worse than do the two selected *nonlinear regressions;* see Chapter 14 for more details.

11.4 Hypothesis testing and planning the study concerning correlation coefficients

We have described the methods of regression and correlation mainly for the purpose of descriptive statistics, so far. However, when dealing with the regression model (and Pearson's correlation coefficient), the difference between the result in the sample and the true relationship in the corresponding population already has been denoted. Naturally, as for the case of just one character, in inferential statistics the subjects of (point) estimation, confidence intervals, and hypothesis testing, and (mainly with respect to the last of these), planning of the study and sequential testing, become relevant.

The point estimator for Pearson's correlation coefficient has already been discussed above (see Formula (11.8)), and actually also the one for regression coefficients by explicitly indicating the estimates in Formulas (11.5) and (11.6). For the association measures discussed here, it was addressed at least implicitly that the statistics from the sample yield point estimates for the underlying parameters. Similarly, the calculated correlation coefficients for ordinal-scaled characters are to be regarded as estimates. As for confidence intervals, most software packages output standard errors, by means of which the user can generally calculate the confidence intervals for β_0 and β_1. However, in psychological research the confidence interval for the parameters of a regression line is hardly of interest. The same applies to hypothesis testing with regard to the parameters of the regression line and therewith also for the corresponding planning of the study. In contrast, these subjects are important for Pearson's correlation coefficient in psychological research practice. They are discussed in detail in the following. As we remember, we have to assume that the two characters of interest, which are modeled by random variables, are normally distributed.

Doctor In Table 11.1 (or in the corresponding summary of results in the calculation with **R**, respectively) of Example 11.5, in addition to the regression coefficients for the example (of which the content is not of interest here), after analysis conducted by means of SPSS, three supplementary columns can be found: the standard error, the value of a *t*-distributed statistic (whose formula we will not reproduce here), and the *p*-value. This last refers implicitly to the null hypotheses H_0: $\beta_0 = 0$ and H_0: $\beta_1 = 0$, respectively. If the type-I risk has been determined by $\alpha = 0.05$ previously, then both null hypotheses have to be rejected. Using **R**, the same information is obtained. For calculation of the confidence intervals for β_0 and β_1, the indicated standard error can be used. For β_1 it corresponds to the term with the square root in the following formula for the lower limit $\boldsymbol{L}$ and the upper limit $\boldsymbol{U}$:

$$\boldsymbol{L} = \boldsymbol{b}_1 - t\left(n-2, 1-\frac{\alpha}{2}\right)\sqrt{\frac{1}{n-2}\left(\frac{\boldsymbol{s}_y^2}{\boldsymbol{s}_x^2} - \boldsymbol{b}_1^2\right)};$$

$$\boldsymbol{U} = \boldsymbol{b}_1 + t\left(n-2, 1-\frac{\alpha}{2}\right)\sqrt{\frac{1}{n-2}\left(\frac{\boldsymbol{s}_y^2}{\boldsymbol{s}_x^2} - \boldsymbol{b}_1^2\right)} \qquad (11.17)$$

The determination of a confidence interval for Pearson's correlation coefficient is easily possible with **R**; we do not explain the planning of the study with regard to the confidence interval here, because in psychological research the main focus is rather on hypothesis testing.

Master Doctor

SPSS delivers a confidence interval for Pearson's correlation coefficient via bootstrapping (see Section 14.4). An approximate confidence interval with the lower limit $\boldsymbol{L}$ and the upper limit $\boldsymbol{U}$ can be determined as follows:

$$\boldsymbol{L} = \frac{1}{2}\ln\frac{1+\boldsymbol{r}}{1-\boldsymbol{r}} - \frac{z\left(1-\frac{\alpha}{2}\right)}{\sqrt{n-3}}; \quad \boldsymbol{U} = \frac{1}{2}\ln\frac{1+\boldsymbol{r}}{1-\boldsymbol{r}} + \frac{z\left(1-\frac{\alpha}{2}\right)}{\sqrt{n-3}} \tag{11.18}$$

Master Doctor

Example 11.13 How strong is the relationship between *Everyday Knowledge* and *Applied Computing* (at the first test date) in Example 1.1?

We are not content to calculate a point estimate, but want to calculate a confidence interval for ρ with a confidence coefficient of $1-\alpha = 0.95$.

In **R**, we type

```
> cor.test(sub1_t1, sub3_t1, alternative = "two.sided",
+          method = "pearson", conf.level = 0.95)
```

i.e. we use the two characters *Everyday Knowledge, 1st test date* (`sub1_t1`) and *Applied Computing, 1st test date* (`sub3_t1`) as arguments in the function `cor.test()`; with `alternative = "two.sided"` we select the two sided alternative hypothesis H_A: $\rho \neq 0$; with `method = "pearson"` we request the product-moment correlation coefficient; and `conf.level = 0.95` sets the confidence coefficient.

As a result, we get:

```
        Pearson's product-moment correlation

data:  sub1_t1 and sub3_t1
t = 4.7371, df = 98, p-value = 7.33e-06
alternative hypothesis: true correlation is not equal to 0
95 percent confidence interval:
 0.2570199 0.5789775
sample estimates:
     cor
0.431647
```

For SPSS, we create a complementary syntax. We begin with the command sequence (Analyze – Correlate – Bivariate...) described in Example 11.6. The result is $r = \hat{\rho} = 0.432$. Next, we open a new data sheet (File – New – Data) and, in the Variable View, define the characters r and n; then, we change to Data View and type the number 0.432 in the first row in column r, and the number 100 in column n. Now we create the syntax; hence we select

File

New

Syntax

and type the program code shown in Figure 11.17 into the resulting window. Subsequently we click Run and select All to start the computation. As a result, we get Figure 11.18; the confidence interval of the unknown Pearson's correlation coefficient ρ has the bounds 0.26 and 0.58.

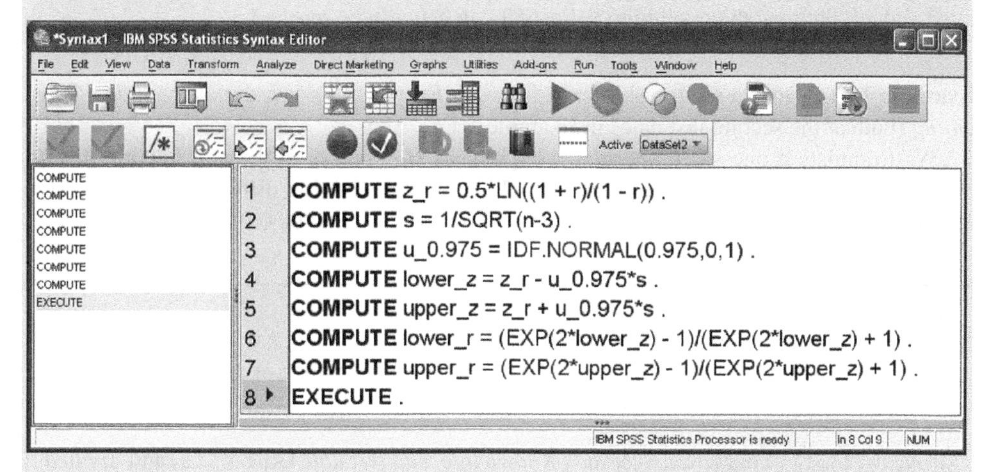

Figure 11.17 SPSS-syntax for computing the 95% confidence interval of ρ.

*Untitled1 [DataSet0] - IBM SPSS Statistics Data Editor

	r	n	z_r	s	u_0.975	lower_z	upper_z	lower_r	upper_r
1	.432	100	.46	.10	1.96	.26	.66	.26	.58

Figure 11.18 SPSS-Data View after running the syntax in Figure 11.17 in Example 11.13.

In the case of another confidence coefficient α, for example $\alpha = 0.01$, you have to replace the number 0.975 four times by the number 0.995 in the Syntax.

Traditionally, only one null hypothesis is in question; that is H_0: $\rho = 0$ (with the alternative hypothesis H_A: $\rho \neq 0$ or H_A: $\rho > 0$ or H_A: $\rho < 0$, respectively). For this case an exact test exists, and the test statistic

$$t = \frac{r\sqrt{n-2}}{\sqrt{1-r^2}} \tag{11.19}$$

is t-distributed with $df = n - 2$ degrees of freedom.

We reject the null hypothesis in favor of the two-sided alternative hypothesis H_A: $\rho \neq 0$, if $|t| > t(n-2, 1-\frac{\alpha}{2})$, otherwise the null hypothesis is accepted. In the case of a one-sided alternative hypothesis (given that the sign indicates the right direction), $t(n-2, 1-\frac{\alpha}{2})$ has to be replaced by $t(n-2, 1-\alpha)$. If the null hypothesis is to be rejected, we say that a 'significant correlation' exists. As an estimated effect size, $B = r^2$ comes into consideration analogously to η^2 and ϕ^2, but now the maximum value of 1 can actually be reached.

Example 11.14 Does a relationship exist between *Everyday Knowledge* and *Applied Computing* (both at the second test date) in Example 1.1?

We formulate a one-sided alternative hypothesis H_A: $\rho > 0$ to the null hypothesis H_0: $\rho = 0$, because it can be deduced from intelligence theories that different components of intelligence are positively correlated. The type-I risk is chosen to be 0.05.

In **R**, we type

```
> cor.test(sub1_t2, sub3_t2, alternative = "greater",
+          method = "pearson")
```

i.e. we use the two characters *Everyday Knowledge, 2nd test date* (`sub1_t2`) and *Applied Computing, 2nd test date* (`sub3_t2`) as arguments in the function `cor.test()`; with `alternative = "greater"` we select the appropriate one-sided alternative hypothesis, and with `method = "pearson"` we request Pearson's correlation coefficient.

As a result, we get

```
        Pearson's product-moment correlation

data:  sub1_t2 and sub3_t2
t = 5.1332, df = 98, p-value = 7.222e-07
alternative hypothesis: true correlation is greater than 0
95 percent confidence interval:
 0.3191661 1.0000000
sample estimates:
      cor
0.4603284
```

In SPSS, we proceed analogously to Example 11.6 and use the same command sequence (Analyze – Correlate – Bivariate...). However, this time we select One-tailed in the panel Test of Significance – knowing that in SPSS we can't ascertain Pearson's correlation coefficient without testing the null hypothesis H_0: $\rho = 0$ at the same time. Hence, we uncheck Flag significant correlations. Selecting this option just abets the not-to-be-recommended process of selecting the type-I error rate in hindsight, since SPSS states different type-I errors according to the p-value (in the row Sig. (1-tailed)).

As a result, we get $r = \hat{\rho} = 0.460$; the p-value is stated as .000.

As $\alpha > p$, the result is significant. The null hypothesis H_0: $\rho = 0$ has to be rejected. A significant correlation between *Everyday Knowledge* and *Applied Computing* (at the second test date) exists; that is, in the population also a correlation of $\rho > 0$ is to be anticipated. As this is quite an 'empty' statement (the true correlation indeed is bigger than zero, but in the end still so small that it is not of practical relevance) we add: the best estimate for Pearson's correlation coefficient in the population is $\hat{\rho} = 0.460$. We can also say that the estimated effect equals $B = 0.2116$; thus that somewhat more that one-fifth of the variance is explained mutually by the relationship.

For Lecturers:

In the case that the p-value (e.g. in SPSS) is output as .000, and thus does not contain a decimal place with a number other than 0, the following wording is often advised: '$p < 0.001$'. Behind this is the well-meant intention to communicate explicitly to the reader that we do not have $p = 0$. Certainly, an output of the type '.000' means that it is a truncated number, which differs from zero only in a later decimal place. This is totally clear for the mathematically competent reader. The wording '$p < 0.001$' is, however, really crucial, because it gives the impression that the researcher has chosen a type-I risk of $\alpha = 0.001$. This problem would not exist at all if it were written: $p < \alpha = 0.05$ (e.g. for a previously determined type-I risk of $\alpha = 0.05$).

In publication in the field of psychology it can be found that mostly merely the null hypothesis H_0: $\rho = 0$ is tested. However, this is almost always unsatisfying and lacking practically all information. It is quite possible then that the true correlation coefficient, the one in the population, equals only $\rho = 0.1$, for instance, or even less, which will probably not lead to any particular scientific consequences.

Bachelor Mainly in combination with the significance of a correlation coefficient, in statistically unprofessional psychological literature the following argument is often found: 'the correlation indeed is not very high at 0.3, but significant'; here the given significance is stated in an effort to compensate for the lack in strength of the correlation coefficient. This is completely meaningless, because a significant correlation indicates just that it does not equal 0 (in the population) – a correlation coefficient is relevant only when the coefficient of determination actually shows a strength of relationship which is relevant with regards to content.

The question of whether the relationship of (at least) two characters reaches a certain size $\rho_0 \neq 0$ (thus H_0: $\rho = \rho_0$, with for instance (at least) $\rho_0 = 0.7$, to explain (almost) 50% of the variance) is much more interesting than testing the null hypothesis H_0: $\rho = 0$. However, for the case of $\rho_0 \neq 0$, there exists only a statistic for which the distribution is known merely with regard to its asymptotic behavior. In order to use this statistic legitimately, a sample size of $n \geq 50$ is required; and even in this case the specified type-I and type-II risks hold just approximately. We will limit our considerations here to the use of pertinent software packages (the corresponding formula can be found e.g. in Figure 11.19).

The planning of the study is carried out in analogy to the previous chapters. After the null hypothesis, thus ρ_0, has been specified (and, where indicated, $\rho_0 = 0$ as well), type-I risk α and type-II risk β as well as δ, the (minimal) relevant difference between ρ_0 and ρ, have to be determined.

Example 11.15 Determination of inter-rater reliability

In psychometrics, the precision of measurement of a test is often ascertained by administrating the test twice to the same testees within quite a short period of time. According to the German standards for 'Proficiency Assessment Procedures' (DIN, Deutsches Institut für Normung e.V., 2002), Pearson's correlation coefficient – here called reliability – would have to be at least equal to 0.70, in order to be applicable for vocational aptitude assessment. The question is about the required sample size for testing the null hypothesis that Pearson's correlation coefficient between character x and character y equals $\rho_0 = 0.8$; that is H_0: $\rho \geq 0.8$; H_A: $\rho < 0.8$. The type-I risk is determined by $\alpha = 0.05$; the type-II risk by $\beta = 0.2$. A deviation of $\delta = -0.1$ is relevant; that is, if ρ is below the interval [0.7; 0.8], the type-II risk should not be bigger than 0.2.

In **R**, we use the package `OPDOE` and type

```
> size_n.regII.test_rho_2(side = "one", alpha = 0.05, beta = 0.2,
+                         rho = 0.8, delta = 0.1)
```

i.e. we apply the function `size_n.regII.test_rho_2` using `side = "one"` as first argument to specify the alternative hypothesis as one-sided, and `alpha = 0.05` as second and `beta = 0.2` as third argument to set the type-I error and type-II error, respectively; finally, with `delta = 0.1` we state the value that marks the deviation limit for the given $-\delta$.

As a result, we get

```
[1] 119
```

Thus, we have to recruit a sample of the size $n = 119$.

Master
Doctor

Example 11.15 – continued

We assume that in our sample of the size $n = 119$ the Pearson's correlation coefficient of $r = 0.69$ results. As described above, the null hypothesis H_0: $\rho \geq 0.8$ is to be tested.

In **R**, we define a new `function()`; hence, we type

```
> cor.p0 <- function(r, r0, n) {
+   z <- 0.5*log((1+r)/(1-r))
+   zeta <- 0.5*log((1+r0)/(1-r0))
+   u <- (z-zeta)*sqrt(n-3)
+   p <- pnorm(u, lower.tail = FALSE)
+   return(list("u" = u, "pval" = p))
+ }
```

i.e. we set `r`, for the observed, and `r0`, for the hypothetical, correlation coefficient, as arguments in the function `function()`; `n` states the sample size. The sequence of commands inside the braces defines the inner workings of the function and will not be discussed. The newly created function is assigned to the object `cor.p0`. Next, we type

```
> cor.p0(r = 0.69, r0 = 0.8, n = 119)
```

i.e. we use the function `cor.p0()` and apply the arguments `0.69` for r, `0.8` for ρ, and `119` for n.

As a result, we get:

```
$u
[1] -2.699653

$pval
[1] 0.9965294
```

Hence, a probability of 0.9965 corresponds to the underlying standard normal-distributed statistic $z = -2.6997$.

In SPSS, testing the null hypothesis H_0: $\rho \geq \rho_0$ is not possible; Kubinger, Rasch, & Šimečkova (2007) published an applicable syntax on that account, which we will display here. To use it, we proceed analogously to Example 11.13 and open a new data sheet (File – New – Data) to define, via Variable View, the characters r, rho, and n; we change to Data View and type, in the first row in column r, the number 0.69; in column rho the number 0.8; and in column n the number 119. Next, we type the program in Figure 11.19 into the window Syntax Editor. Subsequently we click Run and select All and thus start the computation. The result can be obtained from the amended Data View (see Figure 11.20) in the columns u (here the symbol "u" used instead of "z" as usual) and p.

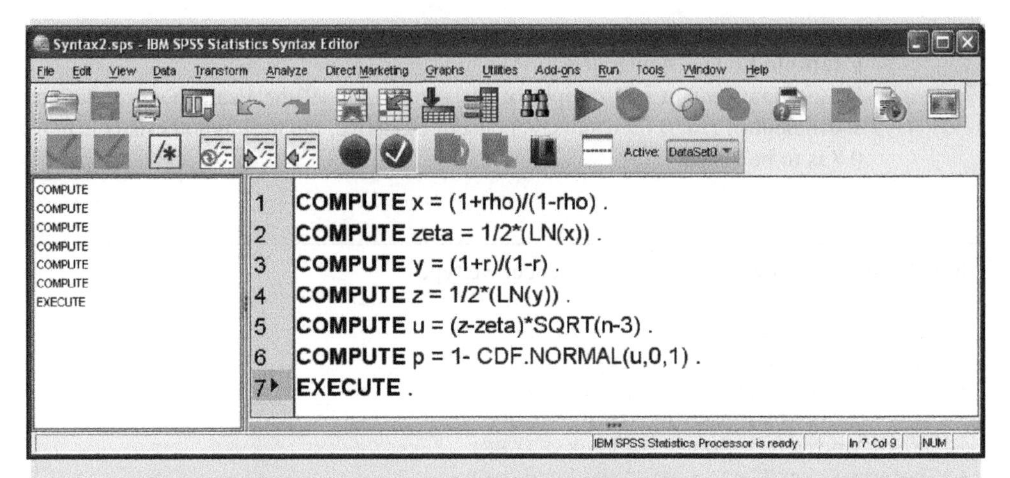

Figure 11.19 SPSS-syntax for testing the null hypothesis H_0: $\rho \geq \rho_0$.

*Untitled2 [DataSet2] - IBM SPSS Statistics Data Editor

File Edit View Data Transform Analyze Direct Marketing Graphs Utilities Add-ons Window Help

Visible: 9 of 9 Variables

	r	rho	n	x	zeta	y	z	u	p
1	.69	.80	119	9.00	1.10	5.45	.85	-2.70	1.00
2									
3									

Data View | Variable View

IBM SPSS Statistics Processor is ready

Figure 11.20 SPSS-Data View after running the syntax in Figure 11.19 in Example 11.15.

The statistic shows with $z = -2.70$, thus with a negative sign, that the empirical result is in the direction of the one-sided alternative hypothesis. In this case the calculated probability value of 0.9965 has to be subtracted from 1 in order to determine the p-value. Thus, the p-value equals $1 - 0.9965 = 0.0035$. The null hypothesis has to be rejected.

The reader may recalculate that the observed Pearson's correlation coefficient of 0.69 would not have led to the rejection of the null hypothesis in question for $n = 20$, for instance, with a p-value of 0.1507.

Sequential testing of hypotheses concerning Pearson's correlation coefficient is still not available for routine use in pertinent software packages.

It has hardly been investigated what are the consequences of a violation of the two-dimensional normal distribution; in particular as concerns type-I risk. Apart from this, there are hardly any routinely applied procedures to test this requirement. For Spearman's rank correlation coefficient, simply the same statistic (11.14) as for Pearson's correlation coefficient is used, but different suggestions concerning sample size are given; however, a sample size of $n > 30$, which is realistic within the field of psychology, should be sufficient (see e.g. Kubinger, 1990). For *Kendall's* τ a special statistic exists, which is not reported here (see e.g. Kubinger, 1990).

Summary

For monotone nonlinear relationships and for relationships with ordinal-scaled characters, the *Spearman correlation coefficient* and *Kendall's* τ are appropriate. For the determination of the strength of a relationship between a quantitative and a multi-categorical character, *Fisher's correlation ratio/eta-squared* exists. For two dichotomous characters, the *ϕ-coefficient* is available; for two nominal-scaled characters with at least one multi-categorical character there is the *contingency coefficient*. For all of these last three coefficients, it always has to be taken into account that they often cannot achieve the value of 1 even in cases of an 'ideal' relationship. For all coefficients, mainly for *Pearson's correlation coefficient*, hypothesis testing is possible, which in praxis mostly concerns whether a relationship between the two characters of interests exists at all. However, hypothesis testing with regard to the question of whether a relationship of a certain strength exists is much more meaningful.

11.5 Correlation analysis in two samples

Occasionally it is also of interest whether two correlation coefficients, ρ_1 and ρ_2, which have been calculated in the same characters x and y, but in two different populations 1 and 2, differ. Then, the null hypothesis is H_0: $\rho_1 = \rho_2$, and the two-sided alternative hypothesis is H_A: $\rho_1 \neq \rho_2$ (the one-sided alternative hypothesis is either H_A: $\rho_1 < \rho_2$ or H_A: $\rho_1 > \rho_2$).

Master Doctor

Again, the distribution of the statistic in question is known merely asymptotically, so that in this case, as well, mostly sample sizes of at least $n = 50$ are required. For this reason we do not address the subject of planning of a study here.

Master Doctor

Example 11.16 Does Pearson's correlation coefficient (inter-rater reliability) between the first and the second test date of subtest *Everyday Knowledge* differ between children with German and children with Turkish as their native language in Example 1.1?

As the children with Turkish native language have been tested in the given study partly in German and partly in Turkish at the first test date (see the character *test setting* in Table 1.1) – and inversely at the second test date – we have to suspect that Pearson's correlation coefficient between the two characters *Everyday Knowledge, 1st test date* and *Everyday Knowledge, 2nd test date* does not constitute a suitable measure for retest reliability for children with Turkish as their native language. The test scores could differ considerably between the two test dates in several children, depending on whether they (already) mastered the German language or if they (still) know Turkish at all, respectively. In contrast, in children with German native language differences between the first and second test date should be merely due to random effects.

We assume that the effect of changing the language has to be considered as relevant only if it causes a shifting of Pearson's correlation coefficient of at least $\delta = 0.2$; the latter as concerns a reduction of the strength of retest reliability. We determine the type-I risk by $\alpha = 0.05$.

We calculate the two correlation coefficients in question as in Example 11.6. For this we have to select the relevant data: the data from children with German as native language on the one hand and the data from children with Turkish as native language on the other hand. This selection can be processed as in Example 5.4, which there concerned the character *test setting*, but here concerns the character *native language of the child*. The results are $r_G = 0.944$ for the children with German as native language and $r_T = 0.614$ for the children with Turkish as native language.

In **R**, we define a new function using the function `function()`; hence, we type

```
> cor.diff <- function(r1, r2, n1, n2) {
+   z_r1 <- 0.5*log((1+r1)/(1-r1))
+   z_r2 <- 0.5*log((1+r2)/(1-r2))
+   u <- (z_r1-z_r2)/sqrt(1/(n1-3)+1/(n2-3))
+   p <- 2*min(pnorm(u), pnorm(u, lower.tail = FALSE))
+ return(list("u" = u, "pval" = p))
+ }
```

i.e. we set `r1` and `r2` for the two correlation coefficients as well as `n1` and `n2` for the two sample sizes as arguments in the function `function()`. The sequence of commands inside the braces defines the inner workings of the function and will not be discussed. We assign the new function to the object `cor.diff`. Next, we type

```
> cor.diff(r1 = 0.944, r2 = 0.614, n1 = 50, n2 = 50)
```

i.e. we use the arguments `0.944` and `0.614` for the two observed product-moment correlation coefficients, and `50` (twice) for the two sample sizes in the function `cor.diff()`.

As a result, we get:

```
$u
[1] 5.1301

$pval
[1] 2.895877e-07
```

A probability of almost zero corresponds to the underlying standard normal-distributed statistic $z = 5.1301$.

In SPSS, this hypothesis test is not available. Hence, we have to use the Syntax provided here in Figure 11.21. For this purpose, we create a new data sheet (File – New – Data) and in the Variable View define the characters r1, r2, n1, and n2. We change to the Data View and type the value 0.944 in the column r1 and the value 0.614 in the column r2; we type 50 in the column n1 as well as in the column n2. Subsequently, we click Run and select All to start the computation. The result can be obtained from the amended Data View (see Figure 11.22): we get the value 0.00 in column p.

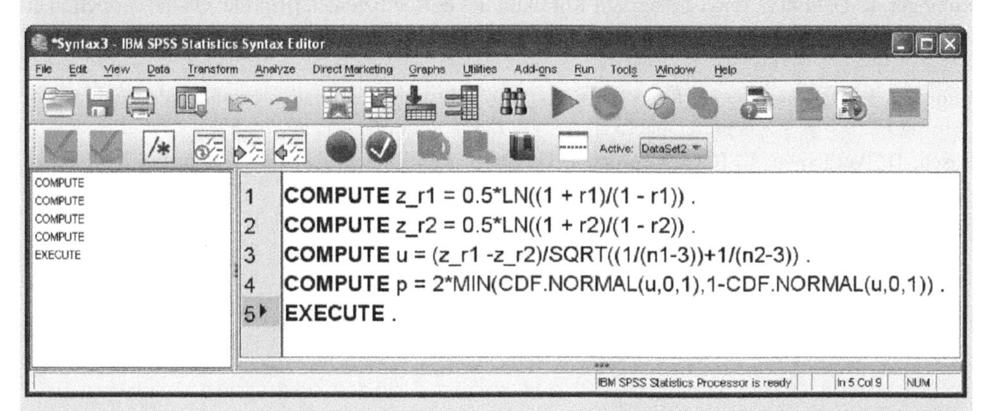

Figure 11.21 SPSS-syntax for computing a statistic for comparing two Pearson's correlation coefficients.

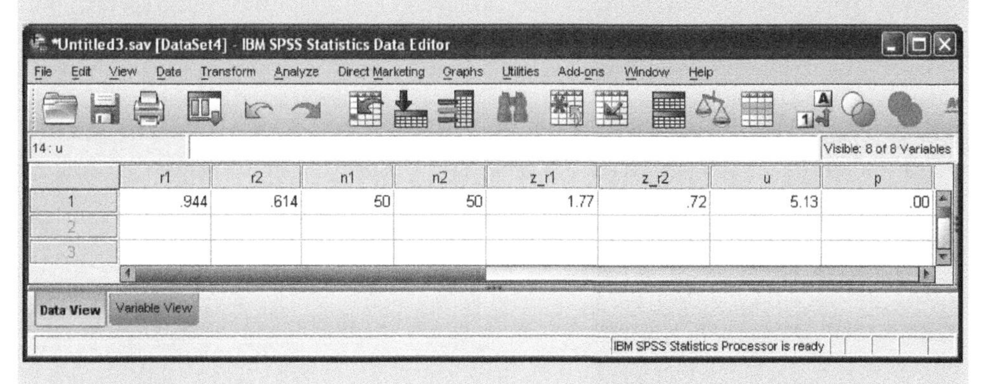

Figure 11.22 SPSS-Data View after running the syntax in Figure 11.21 for Example 11.16.

With the resulting p-value of 0.00, the null hypothesis has to be rejected. The retest reliability for subtest *Everyday Knowledge* is significantly greater in children with German as their native language than in children with Turkish as their native language.

It is also possible to compare correlation coefficients ρ_i $(i = 1, 2, \ldots, a)$ between the characters x and y in more than two samples. As this very rarely plays a role in psychology, we do not discuss this case here.

References

DIN, Deutsches Institut für Normung e.V. (2002). *DIN 33430: Anforderungen an Verfahren und deren Einsatz bei berufsbezogenen Eignungsbeurteilungen.* [Requirements for Proficiency Assessment Procedures and Their Implementation]. Berlin: Beuth.

Kubinger, K. D. (1990). Übersicht und Interpretation der verschiedenen Assoziationsmaße [Review of measures of association]. *Psychologische Beiträge, 32*, 290–346.

Kubinger, K. D. (1995). Entgegnung: Zur Korrektur des ϕ-Koeffizienten [Riposte: On the correction of the ϕ-coefficient]. *Newsletter der Fachgruppe Methoden in der deutschen Gesellschaft für Psychologie, 3*, 3–4.

Kubinger, K. D., Rasch, D., & Šimečkova, M. (2007). Testing a correlation coefficient's significance: using H_0: $0 < \rho \leq \lambda$ is preferable to H_0: $\rho = 0$. *Psychology Science, 49*, 74–87.

Rasch, D., Verdooren, L. R., & Gowers, J. I. (2007). *Fundamentals in the Design and Analysis of Experiments and Surveys* (2nd edn). Munich: Oldenbourg.

Salsburg, D. (2002). *The Lady Tasting Tea*. New York: Owl Books.

Part V

INFERENTIAL STATISTICS FOR MORE THAN TWO CHARACTERS

The comparison of two means with the two-sample t-test is simply a special case of a one-way analysis of variance for the case of only two observed samples (or actually: factor levels), instead of at least three. Having said that, it has been shown that the approach in inferential statistics cannot be generalized, that simply, from the case of two samples to the case of at least three samples. This applies analogously if we want to cross from the case of two simultaneously observed characters over to the case of at least three characters. The generalization is not trivial. However, the procedures for two characters can easily be regarded as special cases of at least three characters. Nevertheless, it would be more appropriate from a didactic point of view to introduce the case of exactly two samples as well as the case of exactly two characters separately.

In this section, we will discuss all generalizations of the methods explained up to now for more than two characters. Firstly, for one sample from only one population again; then for at least two samples each from a (different) population.

To start with, we note that the topics of planning the research study and sequential testing are generally of secondary interest, as both are easiest to implement in the case of many characters by focusing on one character as the most important one. (Exceptions to this, in other words the planning of a research study which takes more than two characters into consideration, can be found e.g. in Rasch, Herrendörfer, Bock, Victor, & Guiard, 2008.)

12

One sample from one population

In this chapter we will discuss procedures for the combined analysis of more than two characters, which were all ascertained from the research units of one single sample. Besides this, special approaches to regression and correlation analysis will be presented, mainly for quantitative characters. As concerns hypothesis testing, we will primarily discuss procedures for the comparison of means, or 'homological' methods for ordinal-scaled characters, respectively.

12.1 Association between three or more characters

We first consider the case of exclusively quantitative characters. One single sample from a certain population is of interest. However, we now not only consider the two characters x and y modeled by random variables, but also the character z. If we have more than three characters, we choose the notations y_q, $q = 1, 2, \ldots, m$ for the particular characters.

Bachelor **Example 12.1** How are the characters *Everyday Knowledge, 1st test date*, *Applied Computing, 1st test date*, and *Social and Material Sequencing, 1st test date*, related in Example 1.1?

First, we illustrate the (three-dimensional) scatter plot.

In **R**, we use the package `scatterplot3d`, which we load after its installation (see Chapter 1) by using the function `library()`; additionally, we use the function `attach()` in order to make the database *Example_1.1* (see Chapter 1) available. Now, we type

```
> scatterplot3d(sub1_t1, y = sub3_t1, z = sub4_t1, xlim = c(20, 80),
+               ylim = c(20, 80), zlim = c(20, 80),
```

Statistics in Psychology Using R and SPSS, First Edition. Dieter Rasch, Klaus D. Kubinger and Takuya Yanagida.

```
+             xlab =
+                       "Everyday Knowledge, 1st test date (T-Scores)",
+             ylab =
+                       "Applied Computing, 1st test date (T-Scores)",
+   zlab =
+      "Social and Material Sequencing, 1st test date (T-Scores)")
```

i.e. we apply the function `scatterplot3d()` to the three characters *Everyday Knowledge, 1st test date*, *Applied Computing, 1st test date*, and *Social and Material Reasoning, 1st test date*; with `xlim`, `ylim`, and `zlim`, we define the respective axes. As a result, we obtain Figure 12.1a.

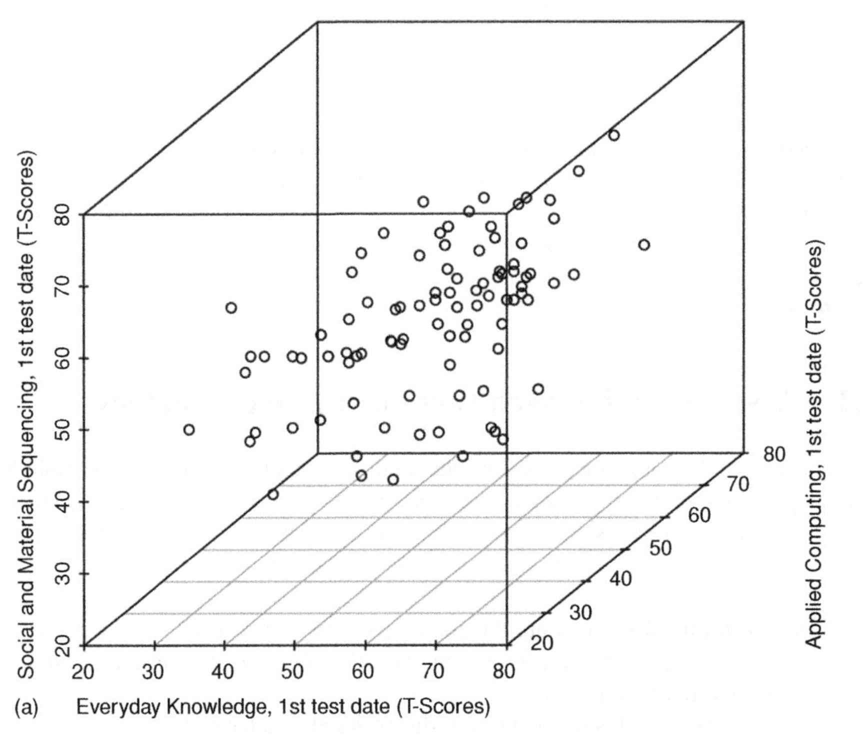

Figure 12.1a **R**-output showing the three-dimensional scatter plot in Example 12.1.

In SPSS, following the steps described in Example 5.2 (Graphs – Chart Builder...), we open the window in Figure 5.5. There, we click Scatter/Dot and drag and drop the symbol for Simple 3-D-Scatter (first row, third symbol) into the Chart preview. Now, we drag and drop Everyday Knowledge, 1st test date into the field X-Axis?, then Applied Computing, 1st test date into the field Y-Axis? and Social and Material Sequencing, 1st test date into the field Z-Axis?. After clicking OK, we obtain Figure 12.1b.

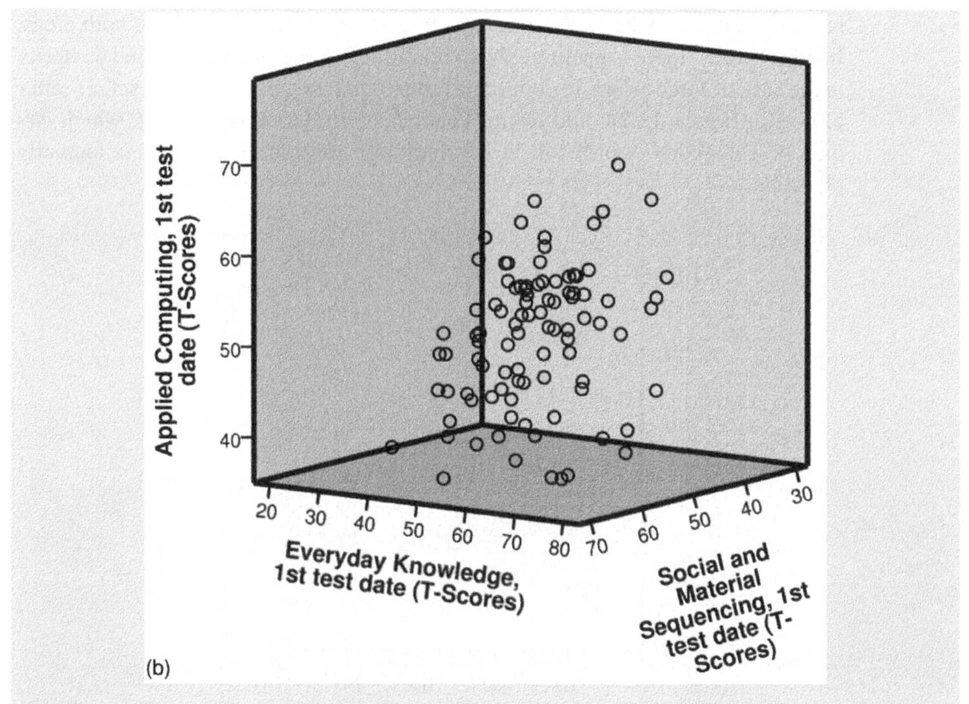

Figure 12.1b SPSS-output showing the three-dimensional scatter plot in Example 12.1.

One realizes – at least readers with sufficient spatial imagination do – that the three-dimensional scatter plot takes a form similar to an ellipsoid. Pearson's correlation coefficients can be calculated in a pair-wise manner as in Example 11.6.

12.1.1 Partial correlation coefficient

The correlation between two characters of interest x and y is often 'disturbed', 'overlain', or 'obscured' by a third character, z.

Bachelor **Example 12.2** How strong is the association between oral fluency and dexterity?

Let us imagine that a researcher tests $n > 30$ three-year-old children with pertinent psychological tests. Developmental psychology leads us to expect that three-year-old children will achieve a lower proficiency level concerning both oral fluency and dexterity, compared to children of school-going age. Based on other scientific findings in developmental psychology, we also expect that children will fundamentally differ such that some children are more gifted with respect to linguistic abilities and others are more practically gifted, so that some children perform much better in one area than in the other. Furthermore, there are also probably children who perform relatively well or relatively poorly in both areas. In other words, there is presumably no relation between oral fluency and dexterity in the case of three-year-old children. If illustrated on a fictitious data set, a scatter plot as in Figure 12.2 could result. Pearson's correlation coefficient, which can then be calculated, will result in a value close to zero. This result is factually plausible.

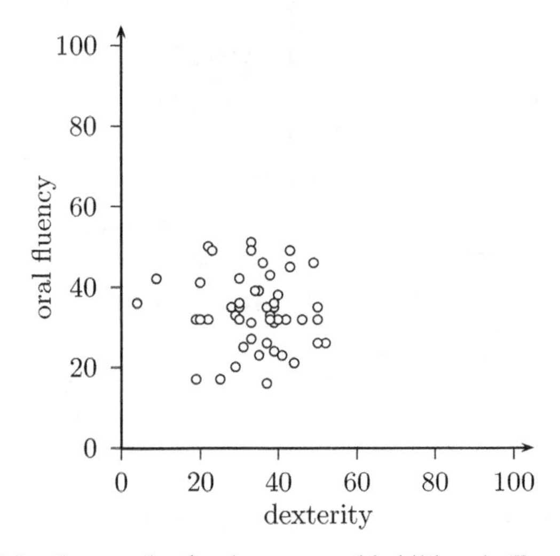

Figure 12.2 Scatter plot for three-year-old children in Example 12.2.

If the researcher were to make more effort, he would also investigate the question of the association between oral fluency and dexterity in six-year-old children at the same time – three-year-old children are, on their own, not typical in order to answer the question *per se*. We would indeed also expect the two characters to be unrelated in this case; thus a similar scatter plot should result (see Figure 12.3). However, if the researcher analyzed the data as a whole and thus represented both of the scatter plots in one single chart (see Figure 12.4), a clear trend would result; namely that the greater the oral fluency is, the greater the dexterity is: Pearson's correlation coefficient would be considerably greater than zero (actually the estimate is $r = 0.674$).

If the researcher also assessed four- and five-year-old children, the results would be even more problematic. Again, there is no correlation if an age-specific analysis is done. However, if all partial samples are included over the four levels of age, a seemingly even more clear relationship between oral fluency and dexterity in the graph results (see Figure 12.5), and Pearson's correlation coefficient is $r = 0.592$. According to this example, if one were to create a scatter plot each time before modeling a linear regression in order to gain an impression about the nature of the relationship, this would indeed be successful in the case of three- and six-year-old children exclusively, but it would fail if all four age groups were observed together. In this example the 'true' relationship between oral fluency and dexterity is being overlain by a third character, the age.

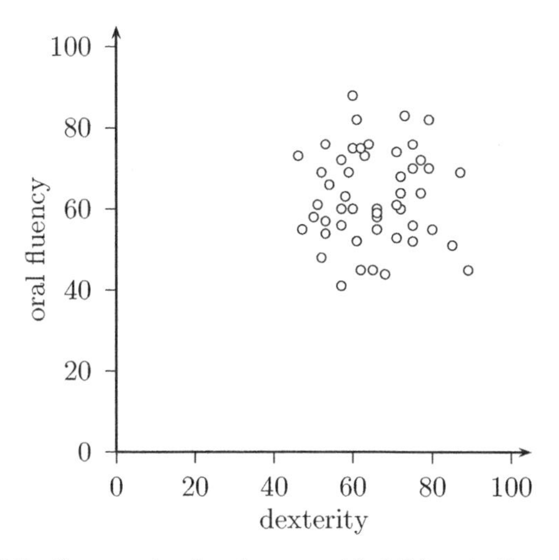

Figure 12.3 Scatter plot for six-year-old children in Example 12.2.

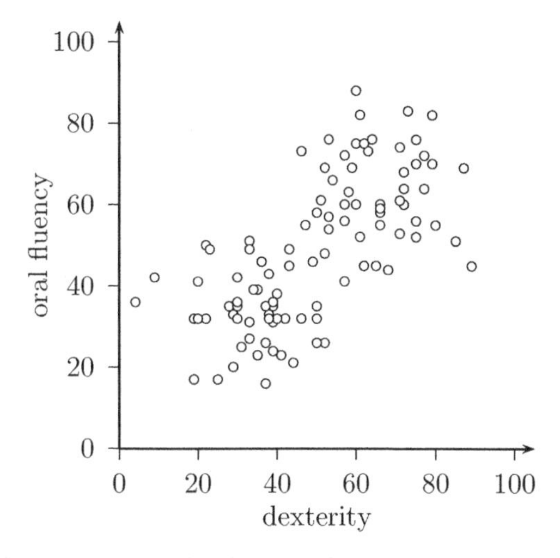

Figure 12.4 Scatter plots for three- and six-year-olds in Example 12.2.

We call this a noise factor (see Section 4.5). Even when no (linear) correlation exists between two characters, a correlation can appear to be present, as both of the characters depend on a third one. In this case, one refers to a 'spurious correlation'. Apart from the fact that a direct, 'true' relationship between the characters does not exist, but instead an artificial one, a pseudo-relationship, though mathematically correct, misleads us to an incorrect interpretation. Under no circumstances should one conclude that there is a causal relationship between the two characters if Pearson's correlation coefficient is high. However, sometimes an artificial result is easy to recognize: if, for instance, the birth rate at the Austrian Lake Neusiedl over the calendar months is correlated with the number of storks which nest in the area, the resulting correlation coefficient is also considerably larger than zero. Of course, there is no causal relationship between these two characters, birth rate, on one hand, and number of storks, on

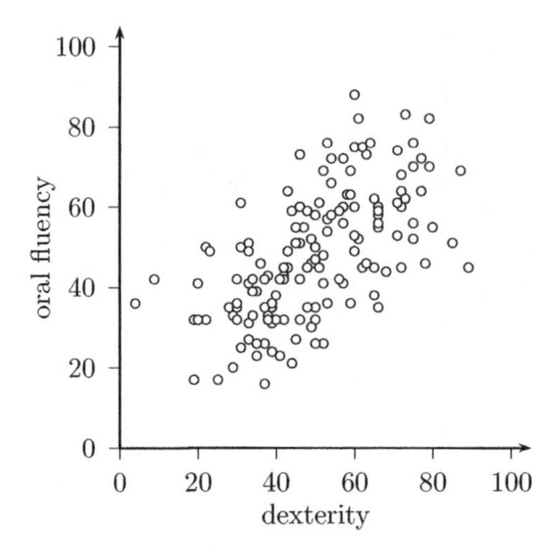

Figure 12.5 Combined scatter plots for three-, four-, five-, and six-year-old children in Example 12.2.

the other hand; a relatively high Pearson's correlation coefficient is obtained artificially, as both characters depend on a third one, the calendar month (birth rates still show highs in the calendar months of March to May/June in Central Europe). On the other hand, there are also cases where the mutual dependency of the two characters of interest on a third one causes a correlation of almost zero, even though the two characters themselves correlate relatively highly.

Master

Example 12.3 Relationship between the 'dose' and the effect of a certain psychotherapeutic treatment

The following deals with criminal offenders, who are selected at random to undergo psychotherapy. We consider the character x, the 'dose' of the psychotherapy conducted – operationalized by the number of therapy units actually conducted per offender within one year. Additionally, we assess the character y; this is the difference for a certain test score from a questionnaire for assessing *Acceptance of social values and norms*, which is assessed at the beginning and then again at the end of the therapy. The effect of the psychotherapeutic treatment would be great, if the value of the difference were (positive and) high. Given linearity of the relationship, Pearson's correlation coefficient between the two modeled variables $\boldsymbol{x}$ and $\boldsymbol{y}$ of $r_{xy} = 0.1$ might result. This result initially argues against the usefulness of the therapy; the result indicates, in particular, that a longer treatment does not lead to a greater effect. However, this conclusion is relativized if one takes into account that the character *Compliance*, modeled by the random variable $\boldsymbol{z}$, can also be assessed by using a particular questionnaire and correlates with the number of therapy sessions with $r_{xz} = 0.9$. We suppose that *Compliance* also correlates with *Acceptance of social values and norms* with $r_{yz} = -0.2$. While this last correlation coefficient again hardly indicates any relationship, it is interesting to see that the more therapy sessions take place, the higher the compliance is. Therefore, perhaps the low correlation between 'dose' and effect stems from the fact that several offenders do not show any compliance with respect to therapy?

The influence of such a noise factor z can be eliminated statistically. This is done by calculating the *partial correlation coefficient* (*of first order*) with the help of the pair-wise determined Pearson's correlation coefficients r_{xy}, r_{xz}, and r_{yz} in the following way:

$$r_{xy.z} = \frac{r_{xy} - r_{xz} r_{yz}}{\sqrt{\left(1 - r_{xz}^2\right)\left(1 - r_{yz}^2\right)}} \tag{12.1}$$

This coefficient can be interpreted as the extent of the association between the characters x and y, corrected for the noise factor. It thus expresses the strength of the linear correlation between the two characters on condition that all observations are somehow technically 'standardized' to a certain value for a third character, z. This coefficient can again be interpreted as an estimation of the correlation coefficient ρ_{xy} in the population which is actually in question.

Master

Formula (12.1) is obtained by initially trying to predict the two variables $\boldsymbol{x}$ and $\boldsymbol{y}$ from $\boldsymbol{z}$: $\hat{x}_v = b_1 z_v + b_0$ and $\hat{y}_v = b_1^* z_v + b_0^*$. Subsequently, the differences $x_v - \hat{x}_v$ and $y_v - \hat{y}_v$ are calculated and correlated. In other words, the correlation or the coefficient of determination, respectively, is calculated. It is thus computed, to what extent the variability of the observations in $\boldsymbol{x}$ (corrected by the variability which can be predicted accordingly by $\boldsymbol{z}$) can be explained by the variability of the observations in $\boldsymbol{y}$ (also corrected by the variability which can be predicted accordingly by $\boldsymbol{z}$). Algebraic transformations finally lead to Formula (12.1).

The outline of the mathematical derivation of a partial correlation coefficient implies that the application of Formula (12.1) for Spearman's rank correlation coefficient or even Kendall's τ in the case of at least one ordinal-scaled character is not appropriate: the calculation of differences does not make sense in this case.

When we have $k = 4$ characters, of which 2 are to be regarded as noise factors, it is possible to calculate the partial correlation coefficient of second order with the help of the corresponding partial correlation coefficients of first order in the following way. (Here we use the notations '1' to '3' instead of x, y, and z, and for the fourth character the notation '4')

$$r_{12.34} = r_{12.43} = \frac{r_{12.4} - r_{13.4} r_{23.4}}{\sqrt{\left(1 - r_{13.4}^2\right)\left(1 - r_{23.4}^2\right)}} \tag{12.2}$$

We do not present partial correlation coefficients of higher order here.

Master

Example 12.3 – continued

Inserting the calculated correlation coefficients in Formula (12.1), we obtain

$$r_{xy.z} = \frac{0.1 - (-0.2) \cdot 0.9}{\sqrt{(1 - 0.04)(1 - 0.81)}} = 0.66$$

That is, if we only conducted the therapy with offenders with the same *Compliance* – ideally probably with high *Compliance* – a relationship of medium strength ($B = 0.66^2 = 0.43$) between the 'dose' and effect of the applied psychotherapy can be expected: the higher the number of sessions, the greater (by trend) is the difference of the test scores in *Acceptance of social values and norms*.

Bachelor **Example 12.4** We are interested in the relationship between the subtests *Immediately Reproducing – numerical, 1st test date* and *Coding and Associating, 1st test date*

Pearson's correlation coefficient is obtained analogously to Example 11.6; the estimate is $r = 0.525$. The coefficient is thus not particularly high; however, it is considerable, as the underlying test concept aims for all subtests to not correlate with each other or, if they do, then only slightly. We now have to take into consideration that the extent of this relationship could be influenced by *gestational age at birth*. Psychological research indicates that premature infants show deficits in certain areas of cognitive development, which cannot be compensated for until the age of 10 or 11 years. Perhaps *gestational age at birth* thus overlies the 'true' relationship between the two subtests.

In **R**, we use the package `ggm`, which we load after its installation (see Chapter 1) by using the function `library()`. Subsequently, we type

```
> parcor(var(cbind(sub5_t1, sub7_t1, age_birth)))
```

i.e. we use the function `cbind()` in order to merge the characters *Immediately Reproducing – numerical, 1st test date* (`sub5_t1`), *Coding and Associating, 1st test date* (`sub7_t1`), and *gestational age at birth* (`age_birth`) into a matrix, which we in turn use as the argument in the function `var()`; this calculates the variance–covariance matrix (see Section 13.2), which we in turn submit into the function `parcor()`.

As a result, we get:

```
             sub5_t1    sub7_t1  age_birth
sub5_t1    1.0000000 0.51274555 0.15968154
sub7_t1    0.5127456 1.00000000 0.03522305
age_birth  0.1596815 0.03522305 1.00000000
```

The required value is to be found at the intersection of (`sub5_t1`) and (`sub7_t1`) in the output matrix: the partial correlation coefficient is 0.51.

In SPSS, we select

Analyze

Correlate

Partial...

and, first of all, in the window produced (Figure 12.6), drag and drop the characters Immediately Reproducing, 1st test date and Coding and Associating, 1st test date into the field Variables:; gestational age at birth goes into the field Controlling for:. After clicking OK, we obtain the result $r_{xy.z} = 0.513$.

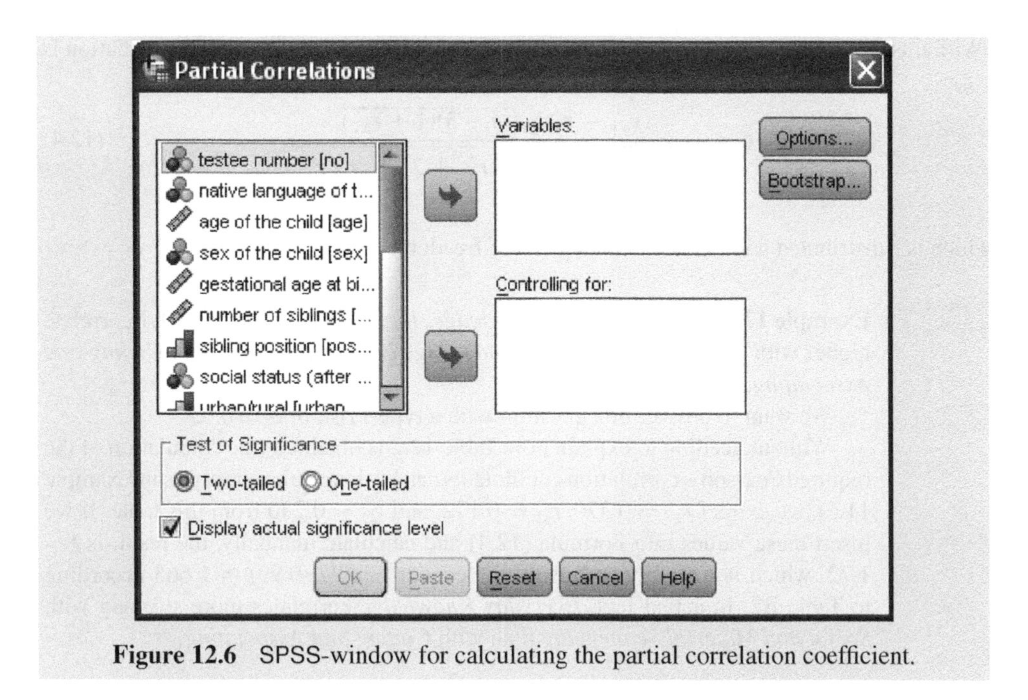

Figure 12.6 SPSS-window for calculating the partial correlation coefficient.

This means that, even if the influence of *gestational age at birth* is eliminated, a certain linear relationship between the two subtests exists.

If one only wants to test the null hypothesis H_0: $\rho_{xy.z} = 0$ for the partial correlation coefficient (e.g. against the two-sided alternative hypothesis H_A: $\rho_{xy.z} \neq 0$), then an exact test is possible in analogy to the statistic in Formula (11.19) (it is robust with respect to deviations from the fundamental requirement of a three-dimensional normal distribution):

$$t = \frac{r_{xy.z}\sqrt{n-3}}{\sqrt{1 - r_{xy.z}^2}} \tag{12.3}$$

is t-distributed with $df = n - 3$ degrees of freedom. However, in this case the question of whether the relationship between two characters (eliminating a third one) reaches a certain size $\rho_0 \neq 0$ is of more interest; thus H_0: $\rho_{xy.z} = \rho_0$. As in Section 11.4 there is also a test for this – the indicated syntax for SPSS from Example 11.15 can be used, whereby the term SQRT(n – 3) has to be replaced by SQRT(n – 4).

12.1.2 Comparison of the association of one character with each of two other characters

Sometimes it is also of interest whether the strength of the association between a character y and a character x is greater than that between y and the character z. The null hypothesis is then H_0: $\rho_{xy} = \rho_{yz}$, and the alternative hypothesis H_A: $\rho_{xy} > \rho_{yz}$. As can be shown

(Williams, 1959), the test statistic for each respective two-dimensional normal distribution is

$$\boldsymbol{t} = \frac{(\boldsymbol{r}_{xy} - \boldsymbol{r}_{yz})\sqrt{(n-3)(1+\boldsymbol{r}_{xz})}}{\sqrt{2(1 - \boldsymbol{r}_{xy}^2 - \boldsymbol{r}_{yz}^2 - \boldsymbol{r}_{xz}^2 + 2 \cdot \boldsymbol{r}_{xy}\boldsymbol{r}_{yz}\boldsymbol{r}_{xz})}} \tag{12.4}$$

which is t-distributed with $df = n - 3$ degrees of freedom.

Bachelor **Example 12.5** Does *Everyday Knowledge, 1st test date* in Example 1.1 correlate higher with *Social and Material Sequencing, 1st test date* than with *Coding and Associating, 1st test date?*

We want to answer this question with a type-I risk of $\alpha = 0.05$.

Without needing to explain how Table 12.2 is obtained (the calculation of the required Pearson's correlation coefficients can be done quite simply as in Example 11.6), we extract $r_{xy} = 0.370$, $r_{yz} = 0.172$, and $r_{xz} = 0.230$ from this table. If we insert these values into Formula (12.4) and calculate manually, the result is $t = 1.72$, which is a significant result on account of $t(97, 0.99) \approx 1.665$ according to Table B2. In actual fact, *Everyday Knowledge* correlates more strongly with *Social and Material Sequencing* than with *Coding and Associating*.

12.1.3 Multiple linear regression

In the case of more than two characters, a regression analysis can also be of interest in the sense that one character may be predicted by at least two others. This is called *multiple linear regression.*

Doctor In multiple linear regression we can again differentiate between model I and model II (see Section 11.1). In the case of model II, which is mostly relevant in psychology, it is a matter of tuples of outcomes, all of which are observed (simultaneously and) randomly in the respective research unit and not selected by the investigator in any targeted manner.

The linear regression function in the general case is:

$$\boldsymbol{y}_v = \beta_0 + \beta_1 \boldsymbol{x}_{1v} + \beta_2 \boldsymbol{x}_{2v} + \cdots + \beta_m \boldsymbol{x}_{mv} + \boldsymbol{e}_v \; (v = 1, 2, \ldots, n) \tag{12.5}$$

The $\boldsymbol{e}_v$ are defined again as normally distributed errors in analogy to the case of a simple regression, for which all other assumptions from Section 11.2 should also be fulfilled. The following topics generally exist for this: estimation of the regression coefficients $\beta_0, \beta_1, \ldots, \beta_m$, prediction of y by means of $\hat{y}(x_1, \ldots, x_m) = \hat{y}$ for given values $x_1, \ldots, x_m$, confidence intervals for each of the regression coefficients, hypothesis testing concerning the regression coefficients, and the determination of the strength of the association – the last of these again primarily by means of some coefficient of determination.

In particular, the estimation of regression coefficients as well as the determination of the *multiple (linear) correlation coefficient* in the sample is easy to do with relevant software programs.

Master

The easiest way of deducing the multiple correlation coefficient is with the coefficient of determination, by determining the proportion of variance of the regressand which is explained by the regressors by means of the multiple (linear) regression. The multiple correlation coefficient R is thus defined in turn as the square root of the coefficient of determination. In this way, no arithmetic sign results because a (multi-dimensional) 'direction' for the dependency of the regressand on the regressors is not possible. Obviously, this multiple correlation coefficient is not a symmetric measure for the relationship between several characters, but always depends on which character is defined as the regressand.

Master

Example 12.6 Can the test score in *Everyday Knowledge, 1st test date* in Example 1.1 be sufficiently explained by the test scores from the other four subtests?

Subtests of the like of *Everyday Knowledge* are often regarded as typical and universal representatives for measuring general cognitive abilities ('intelligence'); if this were true, the test score in this subtest at the very least would have to be explained well by all other test scores collectively.

In **R**, we type

```
> summary(lm(sub1_t1 ~ sub3_t1 + sub4_t1 + sub5_t1 + sub7_t1))
```

i.e. we apply the function `lm()` to our regression formula by having *Everyday Knowledge, 1st test date* (`sub1_t1`) predicted by all the other subtests (`sub3_t1`, `sub4_t1`, `sub5_t1`, `sub7_t1`); we have the results summarized by the function `summary()`.

This yields (shortened output):

```
Call:
lm(formula = sub1_t1 ~ sub3_t1 + sub4_t1 + sub5_t1 + sub7_t1)

Coefficients:
             Estimate Std. Error
(Intercept)   9.80978    6.93208
sub3_t1       0.41036    0.11402
sub4_t1       0.25919    0.09794
sub5_t1       0.32344    0.09755
sub7_t1      -0.14256    0.09121

Multiple R-squared: 0.3507
```

In SPSS, we proceed analogously to Example 11.5 (Analyze – Regression – Linear...) in order to open the window in Figure 11.5, where we drag and drop the character Everyday Knowledge, 1st test date into the field Dependent:. Now, we drag and drop all in all four characters into the field Independent(s):; namely Applied Computing, 1st test date, Social and Material Sequencing, 1st test date, Immediately Reproducing – numerical, 1st test date, and Coding and Associating, 1st test date. By clicking OK, we obtain as a result the estimated regression coefficients displayed in Table 12.1; we also find the multiple linear correlation coefficient $R = 0.592$ in the results output.

Table 12.1 SPSS-output of the regression coefficients in Example 12.6 (shortened output).

Coefficients [a]

Model		Unstandardized Coefficients	
		B	Std. Error
1	(Constant)	9.810	6.932
	Applied Computing, 1st test date (T-Scores)	.410	.114
	Social and Material Sequencing, 1st test date (T-Scores)	.259	.098
	Immediately Reproducing - numerical, 1st test date (T-Scores)	.323	.098
	Coding and Associating, 1st test date (T-Scores)	-.143	.091

a. Dependent Variable: Everyday Knowledge, 1st test date (T-Scores)

We thus obtain the estimation $\hat{y}_v = \hat{x}_{1v}$ for the test score of person v in the subtest *Everyday Knowledge, 1st test date*: $\hat{x}_{1v} = 0.41 \cdot x_{2v} + 0.259 \cdot x_{3v} + 0.323 \cdot x_{4v} - 0.143 \cdot x_{5v} + 9.810$. With a coefficient of determination of $0.592^2 = 0.351$, just slightly more than one-third of the variance is explained; this appears to be far too little if *Everyday Knowledge* is to be considered typical for other cognitive abilities.

As in the case of a linear regression with only a single regressor, it is of course forbidden to apply the model of multiple linear regression when at least one of the characters is ordinal-scaled, because differences in the respective measurement values are not empirically founded.

12.1.4 Intercorrelations

In order to clearly illustrate all pair-wise correlations (in psychology, these collectively are frequently termed *intercorrelations*), it is possible to systematically arrange them in a matrix.

Bachelor **Example 12.7** Intercorrelations of the test scores from the five subtests in Example 1.1

We have assessed the test scores from each of the five subtests *Everyday Knowledge, Applied Computing, Social and Material Sequencing, Immediately Reproducing – numerical*, and *Coding and Associating* at two different test dates. We can now correlate these $m = 10$ (interval-scaled) characters in a pair-wise manner with each other – this results in $\binom{10}{2} = 45$ Pearson's correlation coefficients according to Section 7.2.1.

In **R**, we type

```
> cor(cbind(sub1_t1, sub3_t1, sub4_t1, sub5_t1, sub7_t1, sub1_t2,
+           sub3_t2, sub4_t2, sub5_t2, sub7_t2))
```

i.e. we create a matrix by applying the function `cbind()` to all five subtests at both test dates. We use this matrix as an argument in the function `cor()`.

The results corresponds to the ones in Table 12.2

In SPSS, using the sequence of commands (Analyze – Correlate – Bivariate...) described in Example 11.6, we open the window in Figure 11.14 in order to simply drag and drop all 10 characters into the field Variables:. In SPSS there is a preset default determination of the significance of the correlation coefficient, which should definitely be deactivated by removing the check mark at the option Flag significant correlations. This would otherwise lead to the labeling of coefficients significantly different from zero with one to three asterisks in the output. As already discussed in Section 8.2.3, this leads to a misinterpreted type-I risk; additionally, in a practical context, it would hardly be interesting to know if a correlation coefficient deviates from zero, but rather if it deviates from a certain previously defined value ρ_0 (see Section 11.4). Nevertheless, by doing this we cannot prevent the SPSS-output from stating the p-value with regards to the null hypothesis H_0: $\rho = 0$ at Sig. (2-tailed). By clicking OK, we obtain the results; i.e. a table whose content corresponds to Table 12.2.

A matrix with 10 rows and 10 columns thus results. In fact, the upper or lower triangular matrix would have been sufficient for illustration because the matrix is naturally symmetrical to the main diagonal. The value is always 1 in the main diagonal as the correlation of a character with itself always equals 1.00. All correlation coefficients are positive; that is, children with larger test scores in one subtest (at one test date), also tend to have larger test scores in other subtests or at a different test date, respectively. The strongest relationship exists between the two test scores in subtest *Coding and Associating* ($r = 0.908$), and the weakest one between *Everyday Knowledge* and *Coding and Associating* (1st test date; $r = 0.172$). In this concrete case, we thus obtain two different types of correlations: on the one hand, those between the several subtests concerning the similarity of the abilities assessed, and, on the other hand, the stability of the measured abilities over the two dates in time (the so-called *retest reliability*).

An intercorrelation matrix can of course also be created for Spearman's rank correlation coefficients and also for the coefficients of Kendall's τ, and, in actual fact, for all measures of association.

We now recognize that relationships between more than two characters in the case of exclusively quantitative characters cannot be described by a single statistic. The same applies if all of the characters are at least ordinal-scaled. However, if we are dealing with several characters, of which all except one are nominal-scaled, then it is of course possible to examine more complex relationships by means of the corresponding interaction effects in a multiple analysis of variance. This is even more so the case when we deal with more than two solely nominal-scaled characters (see Section 12.1.6.).

Table 12.2 Matrix of Pearson's correlation coefficients for all test scores in Example 12.7.

	Everyday Knowledge, 1st test date	*Applied Computing, 1st test date*	*Social and Material Sequencing, 1st test date*	*Immediately Reproducing – numerical, 1st test date*	*Coding and Associating, 1st test date*	*Everyday Knowledge, 2nd test date*	*Applied Computing, 2nd test date*	*Social and Material Sequencing, 2nd test date*	*Immediately Reproducing – numerical, 2nd test date*	*Coding and Associating, 2nd test date*
Everyday Knowledge, 1st test date	1	0.432	0.370	0.447	0.172	0.768	0.362	0.423	0.336	0.261
Applied Computing, 1st test date	0.432	1	0.176	0.337	0.289	0.280	0.780	0.272	0.268	0.295
Social and Material Sequencing, 1st test date	0.370	0.176	1	0.339	0.230	0.440	0.230	0.920	0.405	0.268
Immediately Reproducing – numerical, 1st test date	0.447	0.337	0.339	1	0.525	0.303	0.268	0.418	0.853	0.560
Coding and Associating, 1st test date	0.172	0.289	0.230	0.525	1	0.313	0.339	0.291	0.590	0.908
Everyday Knowledge, 2nd test date	0.768	0.280	0.440	0.303	0.313	1	0.460	0.480	0.319	0.394
Applied Computing, 2nd test date	0.362	0.780	0.230	0.268	0.339	0.460	1	0.321	0.315	0.377
Social and Material Sequencing, 2nd test date	0.423	0.272	0.920	0.418	0.291	0.480	0.321	1	0.501	0.327
Immediately Reproducing – numerical, 2nd test date	0.336	0.268	0.405	0.853	0.590	0.319	0.315	0.501	1	0.642
Coding and Associating, 2nd test date	0.261	0.295	0.268	0.560	0.908	0.394	0.377	0.327	0.642	1

12.1.5 Canonical correlation coefficient

Sometimes we are interested in the relationship of m_1 (interval-scaled) characters, on the one hand, and, m_2 other (interval-scaled) characters on the other hand. When all of these $m_1 + m_2 = m$ characters in n research units from one sample have been observed, we can try to find the largest achievable linear relationship between the two groups of characters. The solution for this problem is the *canonical correlation coefficient*.

Doctor

If one of the groups of characters consists of the characters $x_1, x_2, \ldots, x_{m_1}$, and the other of the characters $z_1, z_2, \ldots, z_{m_2}$, a linear combination is formed from each of these groups; thus a kind of 'meta-character' with the weights $c_1, c_2, \ldots, c_{m_1}$ or $d_1, d_2, \ldots, d_{m_2}$, respectively: $x^* = c_1x_1 + c_2x_2 + \cdots + c_{m_1}x_{m_1}$ and $z^* = d_1z_1 + d_2z_2 + \cdots + d_{m_2}z_{m_2}$. For this, all weights are to be determined in the way that the Pearson's correlation coefficient between x^* and z^* reaches a maximum value. The result is the canonical correlation coefficient.

This is applied very rarely in psychology. Therefore, we will not do the calculation for an example here. The calculation itself can easily be carried out using pertinent software programs.

Doctor

Example 12.8 How great is the association between the test scores in the test battery *T1* and the test scores in the test battery *T2?*

Assume that, in a research study, the test scores of five-year-old children in the m_1 subtests of a (developmental) test battery *T1* were assessed in the year x, and in the year + 1 the test scores of the same, now six-year-old children, were assessed in the m_2 subtests of an (intelligence) test battery *T2*. Both test batteries are age specific; in other words, we suppose that the test battery *T1* can not be applied for six-year-old children and contrariwise the test battery *T2* cannot be applied for five-year-olds. Nevertheless, it is important for counseling centers to know how well the results in the test battery *T1* at preschool age can predict the (intelligence) test scores in the test battery *T2* at school age. Thus, the canonical correlation coefficient between the two test batteries is of interest, or even more, the exact regression coefficients. If the canonical correlation coefficient is almost 1, then the test battery *T1* is suitable for predictions. It is then of interest which regression coefficients (in each one of the two linear combinations) contribute the most to the correlation.

12.1.6 Log-linear models

If all of the characters of interest are nominal-scaled, then the data can be summarized in a multi-dimensional contingency table; that is, the contingency table discussed in Section 11.3.5 becomes a 'contingency cube' in the case of three characters, and so on. In the general case, we refer to it as a k-dimensional contingency table.

Master

The method which is discussed in the following is also sometimes used for ordinal-scaled characters. In this case, it is important to bear in mind that if this method is used, the additional information due to having ranked categories (measurement values) is lost.

Master
Doctor

With the help of the so-called *log-linear models*, it is now possible to test various relationships between any number of nominal-scaled characters as to whether they exist or do not exist. The only limitation is that the frequencies per combination of all categories have to be sufficient. Already in the case of three dichotomous characters, for instance, $2^3 = 8$ combinations are possible. If we want to analyze only 6 items of a questionnaire with 4 answer categories, the number of combinations already equals $4^6 = 4096$.

Therefore, we will restrict ourselves here to only three characters – in accordance with the case of a maximum of three factors in the analysis of variance discussed in Chapter 10. The three characters are denoted as A, B, and C – in analogy to the discussion there. A shall possesses a categories (levels) A_i; B the b categories (levels) B_j; and C the c categories (levels) C_k. The n research units yield outcomes, for which we count the numbers of 'hits' per combination of categories $A_i \times B_j \times C_k$. So, n_{ijk}, for instance, means that the combination $A_i \times B_j \times C_k$ has been realized n_{ijk} times within the n research units.

Master

Example 12.9 The relationship between *native language of the child*, *social status*, and the character *only child vs. child with siblings* in Example 1.1 is of interest

We determine the character *only child vs. child with siblings* as demonstrated in Example 9.7. Subsequently, we count the frequencies per combination of categories.

In **R**, we obtain the three-dimensional contingency table by typing

```
> table(native_language, social_status, only_child)
```

i.e. we apply the function `table()` to the three characters `native_language`, `social_status`, and *only child vs. child with siblings* (`only_child`). As a result, we get:

```
, , only_child = 1

                social_status
native_language upper classes middle classes lower middle class
        German              3              2                  3
       Turkish              1              0                  1
                social_status
```

```
native_language upper lower class lower classes
         German                 2             0
        Turkish                 0             0
               social_status
native_language single mother in household
         German                          4
        Turkish                          1

, , only_child = 2

               social_status
native_language upper classes middle classes lower middle class
         German             7             11                  7
        Turkish             1             17                 14
               social_status
native_language upper lower class lower classes
         German                 5             2
        Turkish                 9             3
               social_status
native_language single mother in household
         German                          4
        Turkish                          3
```

In SPSS, we obtain the three-dimensional contingency table by once again following the steps (Analyze – Descriptive Statistics – Crosstabs...) described in Example 5.13; thus opening the window in Figure 5.28. There, we drag and drop native language of the child into the field Row(s):, social status into the field Column(s):, and only_child into the field Layer 1 of 1. By clicking OK, we obtain Table 12.3, among other results.

Table 12.3 SPSS-output displaying the three-dimensional contingency table in Example 12.9.

native language of the child * social status (after Kleining & Moore according to occupation of father/ alternatively of the single mother) *only_child Crosstabulation

Count

			social status (after Kleining & Moore according to occupation of father/ alternatively of the single mother)						
only_chy_child			upper classes	middle classes	lower middle class	upper lower class	lower classes	single mother in household	Total
Yes	native language of the child	German	3	2	3	2		4	14
		Turkish	1	0	1	0		1	3
	Total		4	2	4	2		5	17
No	native language of the child	German	7	11	7	5	2	4	36
		Turkish	1	17	14	9	3	3	47
	Total		8	28	21	14	5	7	83

Doctor

Again, the frequencies n_{ijk}, that is the data observed in the realized sample, can be modeled as being caused by certain parameters in the corresponding population. In other words, these observed frequencies are to be defined as random variables $\boldsymbol{n}_{ijk}$. They are induced by the expected values η_{ijk} or the probabilities p_{ijk}, respectively, so that the n_{ijk} also constitute the estimates $\hat{\eta}_{ijk}$ of the estimator $\hat{\boldsymbol{\eta}}_{ijk} = N\hat{\boldsymbol{p}}_{ijk}$. The random variables $\boldsymbol{n}_{ijk}$ can be explained in various ways under different null or alternative hypotheses, and thus should be modeled specifically. Again one aims for a linear model approach as this is the simplest one. However, this cannot be carried out in a trivial manner as the parameters of the random variables are now probabilities or rely directly on probabilities, respectively. Therefore, in statistics the following approach is chosen:

$$p_{ijk} = e^{\sum_l \lambda_l}$$

or, as this can be written more easily, $\ln p_{ijk} = \sum_l \lambda_l$. If we now switch from parameter p_{ijk} to parameter η_{ijk}, then $\ln \eta_{ijk} = \ln np_{ijk} = \ln n + \ln p_{ijk} = \ln n + \sum_l \lambda_l$ results.

Master Doctor

In the given case, the following log-linear model can now be formulated:

$$\begin{aligned}\ln(\eta_{ijk}) + \boldsymbol{e}_{ijk} = \mu + \alpha_i + \beta_j + \gamma_k + (\alpha\beta)_{ij} + (\alpha\gamma)_{ik} + (\beta\gamma)_{jk} \\ + (\alpha\beta\gamma)_{ijk} + \boldsymbol{e}_{ijk} \qquad (12.6)\end{aligned}$$

whereby the symbols on the right have the same meaning as in the three-way analysis of variance in Section 10.4.7.

The model in Formula (12.6) is called the *saturated model* as it contains all possible interaction effects.

Statistical tests can verify whether the saturated model explains the data better than, for instance, a model which is reduced to certain interaction effects. Or one (immediately) compares two different special cases of Formula (12.6) and tests whether the model with fewer parameters explains the data not much worse than the model with more parameters. That is to say, these types of tests examine diverse null hypotheses that certain interaction effects equal zero or do not exist, respectively. If this is the case, the respective null hypothesis can be retained. Then, (with respect to the given type-I risk) a relationship between the characters concerned does not exist. In the reverse case, a (significant) relationship exists.

Doctor

It is only possible to compare hierarchically subordinate models with superordinate models. For instance, the model $\mu + \alpha_i + \beta_j + \gamma_k + (\beta\gamma)_{jk}$ can be compared to the model $\mu + \alpha_i + \beta_j + \gamma_k + (\alpha\gamma)_{ik} + (\beta\gamma)_{jk}$. In this way, the interaction $(\alpha\gamma)_{jk}$ is tested against zero. In contrast, the two models $\mu + \alpha_i + \beta_j + \gamma_k + (\beta\gamma)_{jk}$ and $\mu + \alpha_i + \beta_j + \gamma_k + (\alpha\beta)_{ij} + (\alpha\gamma)_{ik}$ cannot be compared because, in doing so, no definable hypothesis is tested.

Master
Doctor

This principal of hypothesis testing is that of a goodness of fit test (see Section 14.2.1). These tests are generally used for testing the given, actual observed frequencies (of combinations of categories) and the hypothetically expected frequencies with regard to congruence. In Chapters 8 to 10 different variants of χ^2-tests have already been discussed; these are in actual fact (also) tests of goodness of fit in this respect. Asymptotically χ^2-distributed test statistics also always result with regard to the log-linear models – in which the number of degrees of freedom equals the number of the independent summands of the relevant test statistic minus the number of estimated parameters, minus 1.

In the given case of three characters A, B, and C, the following models can be tested (successively) with respect to their goodness of fit or can be compared to the saturated model, respectively.

Null hypothesis 1, the following model is true:
$\ln(\eta_{ijk}^{(1)}) = \mu + \alpha_i + \beta_j + \gamma_k + (\alpha\beta)_{ij} + (\alpha\gamma)_{ik} + (\beta\gamma)_{jk}$; thus H_0^1 : $(\alpha\beta\gamma)_{ijk} = 0$, for all i, j, and k

Null hypothesis 2a, the following model is true:
$\ln(\eta_{ijk}^{(2a)}) = \mu + \alpha_i + \beta_j + \gamma_k + (\alpha\gamma)_{ik} + (\beta\gamma)_{jk}$; thus $H_0^{2a} : (\alpha\beta)_{ij} = 0$ and $(\alpha\beta\gamma)_{ijk} = 0$, for all i, j, and k

Null hypothesis 2b, the following model is true:
$\ln(\eta_{ijk}^{(2b)}) = \mu + \alpha_i + \beta_j + \gamma_k + (\alpha\beta)_{ij} + (\beta\gamma)_{jk}$; thus $H_0^{2b} : (\alpha\gamma)_{ik} = 0$ and $(\alpha\beta\gamma)_{ijk} = 0$, for all i, j, and k

Null hypothesis 2c, the following model is true:
$\ln(\eta_{ijk}^{(2c)}) = \mu + \alpha_i + \beta_j + \gamma_k + (\alpha\beta)_{ij} + (\alpha\gamma)_{jk}$; thus $H_0^{2c} : (\beta\gamma)_{jk} = 0$ and $(\alpha\beta\gamma)_{ijk} = 0$, for all i, j, and k

Null hypothesis 3a, the following model is true:
$\ln(\eta_{ijk}^{(3a)}) = \mu + \alpha_i + \beta_j + \gamma_k + (\beta\gamma)_{jk}$; thus $H_0^{3a} : (\alpha\beta)_{ij} = (\alpha\gamma)_{ik} = 0$ and $(\alpha\beta\gamma)_{ijk} = 0$, for all i, j, and k

Null hypothesis 3b, the following model is true:
$\ln(\eta_{ijk}^{(3b)}) = \mu + \alpha_i + \beta_j + \gamma_k + (\alpha\beta)_{ij}$; thus $H_0^{3b} : (\alpha\gamma)_{ik} = (\beta\gamma)_{jk} = 0$ and $(\alpha\beta\gamma)_{ijk} = 0$, for all i, j, and k

Null hypothesis 3c, the following model is true:
$\ln(\eta_{ijk}^{(3c)}) = \mu + \alpha_i + \beta_j + \gamma_k + (\alpha\gamma)_{ik}$; thus $H_0^{3c} : (\alpha\beta)_{ij} = (\beta\gamma)_{jk} = 0$ and $(\alpha\beta\gamma)_{ijk} = 0$, for all i, j, and k

Null hypothesis 4, the following model is true:
$\ln(\eta_{ijk}^{(4)}) = \mu + \alpha_i + \beta_j + \gamma_k$; thus $H_0^4 : (\alpha\beta)_{ij} = (\alpha\gamma)_{ik} = (\beta\gamma)_{jk} = 0$ and $(\alpha\beta\gamma)_{ijk} = 0$, for all i, j, and k.

Doctor

If the η_{ijk} are now estimated as $\hat{\eta}_{ijk}^{(g)}$ for each model (g), the corresponding null hypothesis can be tested by means of the following test statistic; thus the goodness of fit of the model (g) to the actual frequencies in the three-dimensional contingency table (that is, to the saturated model):

$$2I(H_0^{(g)}) = 2\sum_i \sum_j \sum_k n_{ijk} \ln\left(\frac{n_{ijk}}{\hat{\eta}_{ijk}^{(g)}}\right) \tag{12.7}$$

This test statistic is asymptotically χ^2-distributed – the number of degrees of freedom will not be discussed here.[1]

Differences between the test statistics of the different models can be statistically compared: if model (h) is a model subordinated to model (g), then $2I(H_0^{(g-h)}) = 2I(H_0^{(g)}) - 2I(H_0^{(h)})$is also asymptotically χ^2-distributed, with a number of degrees of freedom which equals the difference of the degrees of freedom of the two test statistics.

Example 12.9 – continued

We want to examine whether a three-factor interaction effect exists between the characters *native language of the child*, *social status*, and the character *only child vs. child with siblings* in Example 1.1. The null hypothesis is H_0: $(\alpha\beta\gamma)_{ijk} = 0$, for all i, j, and k; the alternative hypothesis is H_1: $(\alpha\beta\gamma)_{ijk} \neq 0$ for at least two triples $ijk \neq lmp$. We take a type-I risk of $\alpha = 0.05$.

In **R**, we use the package `MASS`, which we load after its installation (see Chapter 1) by applying the function `library()`. Then, we type

```
> loglm(~(native_language + social_status + only_child)^2,
+         data = table(native_language, social_status, only_child))
```

i.e. we apply the function `loglm()`, specifying that all two-way (`^2`) interactions between the three characters *native language of the child* (`native_language`), `social_status`, and *only child vs. child with siblings* (`only_child`) are to be modeled, however excluding the three-way interaction. Additionally, we request the output of a three-dimensional frequency table of all three characters with `table(native_language, social_status, only_child)`, which we assign to the object `data`.

As a result, we get:

```
Call:
loglm(formula = ~(native_language + social_status + only_child)^2,
    data = table(native_language, social_status, only_child))

Statistics:
                      X^2 df   P(> X^2)
Likelihood Ratio 4.524669  5 0.4765896
Pearson               NaN  5       NaN
```

[1]The denotation 'I' refers to the so-called Fisher's information within the frame of parameter estimators (see Section 15.2.3.1).

i.e. the interesting χ^2-(goodness of fit)-test under `Pearson` is not being calculated because the expected values of at least one combination of categories are too small. Nevertheless, the value stated under `Likelihood Ratio`, which is also asymptotically χ^2-distributed, usually does not differ from the regular χ^2-test value (for more details on the likelihood-ratio test, see Section 14.2.2).

In SPSS, we select the sequence of commands

Analyze
    Loglinear
        Model Selection...

in order to open the window in Figure 12.7, where the characters *native language of the child, social status*, and *only_child* (from Example 9.7) have already been dragged and dropped into the field Factor(s):. Afterwards, one has to use the button Define Range... in order to define which groups are to be analyzed. In this case, we type a 1 into the field Minimum: (into a small window which opens after the respective character has been marked) for the characters native_language and only_child, and a 2 into the field Maximum:; the character social_status ranges from 1 to 6. Now, we click Paste in order to open the SPSS Syntax-Editor. In Figure 12.8, we already deleted everything except the first two lines in this window and complemented the syntax respectively – we only need the first Design command for our example; in order to show the reader how other null hypotheses might be tested hierarchically, the other Design commands are listed as well. By clicking Run and All, we obtain very extensive output results. The relevant result can be found in Table 12.4: the test value of the χ^2-(goodness of fit)-test under Pearson.

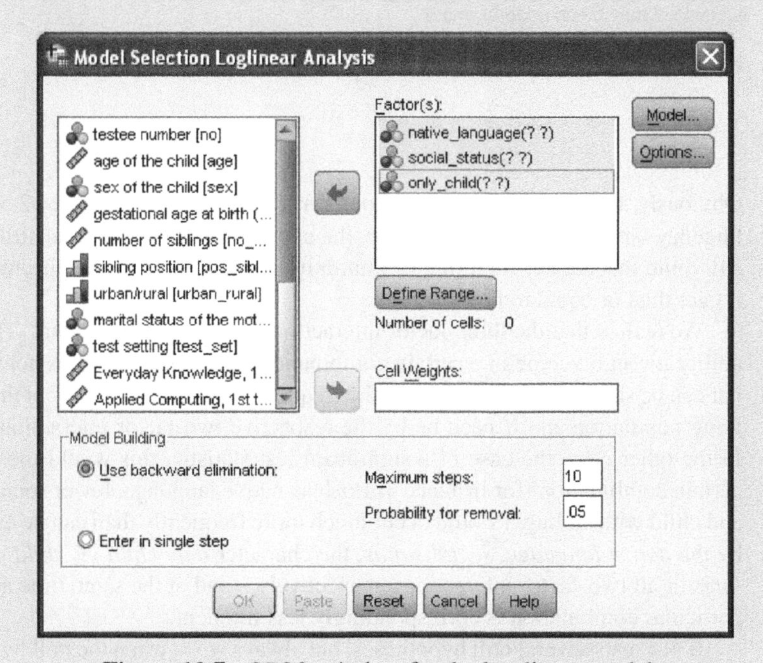

Figure 12.7 SPSS-window for the log-linear models.

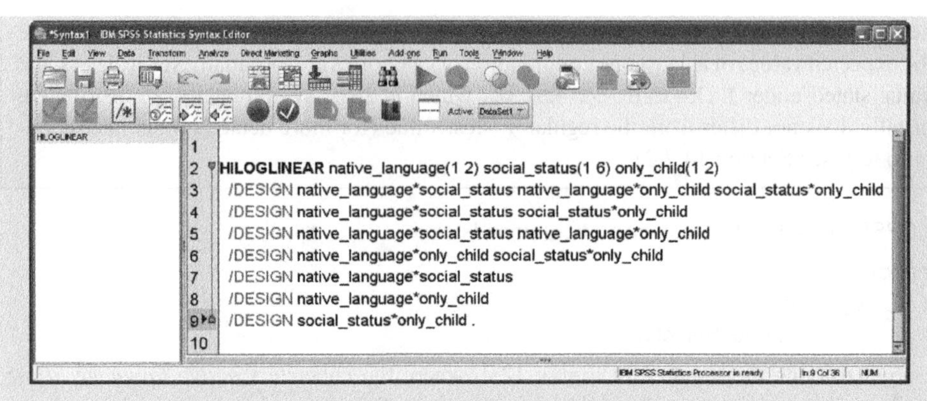

Figure 12.8 SPSS-syntax for Example 12.9.

Table 12.4 SPSS-output of the χ^2-(goodness of fit)-test in Example 12.9.

Goodness-of-Fit Tests

				Adjusted	
	Chi-Square	df	Sig.	df[a]	Sig.
Likelihood Ratio	4.525	5	.477	3	.210
Pearson	4.907	5	.427	3	.179

a. One degree of freedom is subtracted for each cell with an expected value of zero. The unadjusted df is an upper bound on the true df, while the adjusted df may be an underestimate.

Obviously, the number of observations for a three-dimensional $2 \times 2 \times 6$ contingency table is very low. Therefore, the approximation to the χ^2-distribution is still quite inaccurate. As a rule of thumb it is advisable to have expected values bigger than or equal to five.

We realize that the three-factor interaction effect is not significant. The probability for an outcome in a certain combination of categories is thus not specific, but can be sufficiently explained by the frequencies of each category in the underlying population and if need be by the respective two-factor interaction effects. In the other case, the case of a significant test statistic, this would mean that a certain combination (for instance Turkish as native language, lower social status, and child with siblings) would occur much more frequently than can be explained by the *native language*, *social status*, the character *only child vs. child with siblings* or all two-factor interactions, respectively – and at the same time any other particular combination is correspondingly less frequent.

If one tests several null hypotheses, but always stops when the null hypothesis is rejected for the first time in the hierarchical sequence of such tests, then the study-wise type-I risk holds the nominal value of α.

Summary

If more than two quantitative characters are ascertained in research units of a single sample, the *partial correlation coefficient* can determine the strength of the relationship between the two of them, eliminating the influence of one or more other characters (on these two characters) – the third or all other characters are potential noise factors. *Multiple linear regression* refers to the case that one character should be predicted from at least two other characters; *canonical correlation coefficient* refers to the case that a group of at least two characters should be correlated with another group of at least two other characters. *Log-linear models* test the relationships between at least three nominal-scaled characters.

12.2 Hypothesis testing concerning a vector of means μ

Although rather rare in psychology, one can also test hypotheses with regard to the underlying mean in the multivariate case; thus for the m characters $y_1, y_2, \ldots, y_m$, this can be done in complete analogy to Section 8.2.3. A presupposition is that the quantitative characters y_m can be modeled by an m-variate normally distributed random variable, in which we arrange the m (univariate, normally distributed) random variables $\boldsymbol{y}_q$ in a vector notation; it is easier to write the transposed vector $\vec{\boldsymbol{y}}^T = (\boldsymbol{y}_1, \boldsymbol{y}_2, \ldots, \boldsymbol{y}_m)$. Likewise, we can define a (transposed) mean vector: $\vec{\mu}^T = (\mu_1, \mu_2, \ldots, \mu_m)$. The null hypothesis is H_0: $\vec{\mu} = \vec{\mu}_0$, where $\vec{\mu}_0^T = (\mu_{01}, \mu_{02}, \ldots, \mu_{0m})$. The alternative hypothesis is H_A: $\vec{\mu} \neq \vec{\mu}_0$; that is to say, $\mu_{1l} \neq \mu_{0l}$ is true for any l.

Master Doctor

A test formally requires the examination of the so-called *covariance matrix/variance–covariance matrix/variance matrix* as well as the *determinant* of this matrix and the *trace* in this context. The covariance matrix for m characters appears as follows:

$$\Sigma = \begin{pmatrix} \sigma_1^2 & \sigma_{12} & \ldots & \sigma_{1m} \\ \sigma_{12} & \sigma_2^2 & \ldots & \sigma_{2m} \\ \vdots & \vdots & \ldots & \vdots \\ \sigma_{1m} & \sigma_{2m} & \ldots & \sigma_m^2 \end{pmatrix} \tag{12.8}$$

That is to say, the variances per random variable in the population are arranged in the main diagonal and all pair-wise covariances, which are the numerators of Pearson's correlation coefficients (see Section 11.2), are off the main diagonal. Analogously, in the case of $n > m$ observations per character, we would obtain a matrix $S = \hat{\Sigma}$, in which we insert the corresponding estimates instead of the parameters (or the matrix $\boldsymbol{S} = \hat{\boldsymbol{\Sigma}}$ for the estimators, respectively).

The estimation of the parameters and the determination of confidence intervals will not be discussed here. Our discussion will be limited to hypothesis testing.

It can be shown that Hotelling's statistic

$$\boldsymbol{T}^2 = n\left(\hat{\boldsymbol{\mu}} - \vec{\mu}\right)^T \boldsymbol{S}^{-1}(\hat{\boldsymbol{\mu}} - \vec{\mu}) \tag{12.9}$$

is distributed in accordance with *Hotelling's distribution* – this particular distribution or

$$\frac{n-m}{m(n-1)}\boldsymbol{T}^2$$

respectively, can be transformed into an F-distribution with $df_1 = m$ and $df_2 = n - m$ degrees of freedom. In Formula (12.9), $\boldsymbol{S}^{-1}$ is the inverse matrix of $\boldsymbol{S}$, and $\hat{\vec{\mu}}$ is the estimator of the parameter vector $\vec{\mu}$.

The test can easily be conducted using pertinent computer programs. The assumption of the m-dimensional normal distribution of the outcomes is, however, difficult to test.

Master Doctor

Example 12.10 Is the acquired sample of children with German as their native language in Example 1.1 representative with respect to the test scores in the five subtests?

The subtests were scaled in such a way that the mean equals $\mu_q = 50$ in the population (at the first test date). The question is now whether the acquired sample is in accordance with the population in so far as that the obtained means do not differ significantly from the test score 50. Naturally, we could apply the procedure for each subtest individually as in Example 8.11. However, in this case, we would run $m = 5$-times the type-I risk. With the test statistic (12.9) on the other hand, we test the null hypothesis H_0: $\mu^T = \mu_0^T = (50, 50, 50, 50, 50)$ against the alternative hypothesis H_A: $\mu^T \neq (50, 50, 50, 50, 50)$ with the chosen type-I risk study-wise; we choose $\alpha = 0.05$.

In **R**, we type

```
> sub1_t1.ce <- sub1_t1-50
> sub3_t1.ce <- sub3_t1-50
> sub4_t1.ce <- sub4_t1-50
> sub5_t1.ce <- sub5_t1-50
> sub7_t1.ce <- sub7_t1-50
> sub.ce <- cbind(sub1_t1.ce, sub3_t1.ce, sub4_t1.ce, sub5_t1.ce,
+                 sub7_t1.ce)[native_language == "German", ]
```

i.e. we subtract the value `50` from each of the observed values of each subtest (`sub1_t1` through `sub7_t1`) and assign the results to one new object each. By using the function `cbind()`, we merge the objects to a matrix column by column; with `[native_language == "German",]`, we select the children with German as a native language and assign the matrix to the object `sub.ce`. In order to conduct the analysis, we type

```
> summary(manova(sub.ce ~ 1), test = "Hotelling-Lawley",
+         intercept = TRUE)
```

i.e. we use the function `manova()`, to request that the matrix be analyzed in the specified way; with `test = "Hotelling-Lawley"`, we select the desired test value, and with `intercept = TRUE`, we request hypothesis testing. Finally, we have the analysis summarized by applying the function `summary()`.

As a result, we get (shortened output):

```
            Df Hotelling-Lawley approx F num Df den Df   Pr(>F)
(Intercept)  1           0.5365   4.8285      5     45 0.001282
Residuals   49
```

In SPSS, we conduct a transformation of the observed values of all subtests from the first test date by selecting

File
  New
    Syntax

and typing in the command lines from Figure 12.9 in the resulting window. Now, we click Run and select All in order to conduct the calculation.

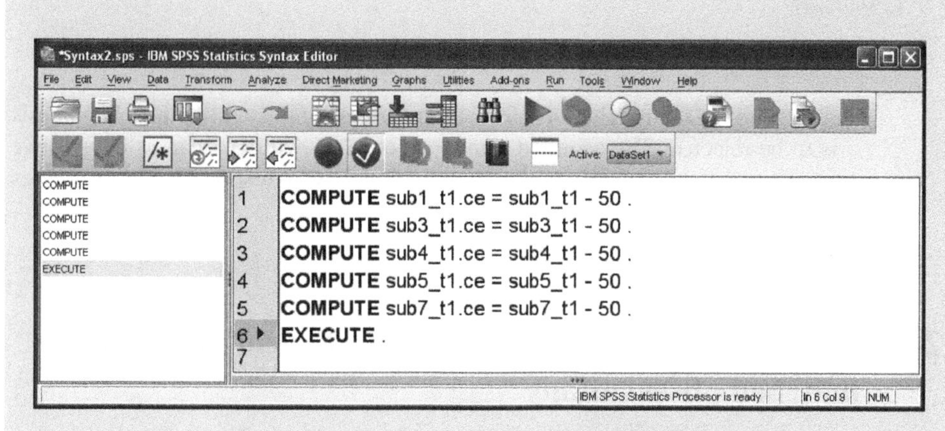

Figure 12.9 SPSS-syntax for Example 12.10.

Next, following the sequence of commands (Data – Select Cases...) shown in Example 5.4, we select the children with German as a native language by clicking the button If condition is satisfied in the window in Figure 5.14 and clicking the button If... in order to type native_language = 1 into the resulting window Select Cases: If seen in Figure 5.15. We confirm this by clicking Continue and OK. Now, we conduct the analysis by selecting

Analyze
  General Linear Model
    Multivariate...

In the resulting window (not shown here), we drag and drop the characters sub1_t1.ce through sub7_t1.ce into the field Dependent Variables:. Now, we select Options... and set a check mark at Estimates of effect size. After clicking Continue and OK, we obtain the results in Table 12.5, of which we are only interested in the line concerning Hotelling's Trace.

Table 12.5 SPSS-output for Hotelling's T^2 in Example 12.10.

Tests of Between-Subjects Effects

Dependent Variable:intelligence first-born

Source		Type III Sum of Squares	df	Mean Square	F	Sig.
Intercept	Hypothesis	5041155.169	1	5041155.169	27979.629	.000
	Error	28467.229	158	180.172[a]		
Index1	Hypothesis	547.913	2	273.956	1.355	.259
	Error	63890.633	316	202.186[b]		
urban_rural	Hypothesis	5024.602	1	5024.602	27.888	.000
	Error	28467.229	158	180.172[a]		
id(urban_rural)	Hypothesis	28467.229	158	180.172	.891	.792
	Error	63890.633	316	202.186[b]		
Index1 * urban_rural	Hypothesis	805.454	2	402.727	1.992	.138
	Error	63890.633	316	202.186[b]		

a. MS(id(urban_rural))
b. MS(Error)

According to the p-value of 0.001, the result is significant; the null hypothesis is to be rejected. The acquired sample of children with German as their native language is to be considered as not representative with respect to their test scores in the five subtests.

12.3 Comparisons of means and 'homological' methods for matched observations

As has already been discussed several times in previous chapters, the following deals with the case of more than one character; however each of the given observations is clearly matched to another one across the same research unit. This problem has already been solved for exactly two such characters; however not for the case of at least three characters. As we are interested in both quantitative and ordinal-scaled variables – in the latter case (only) with regard to differences in the position of the underlying distributions – it is not only about the comparisons of means, but also about 'homological' methods for ordinal-scaled characters.

12.3.1 Hypothesis testing concerning means

If a sample has two characters, which practically mean the same thing in terms of content, but which have been ascertained under different conditions for the respective research units, then in actual fact we again obtain one single character by forming the differences of the outcomes per person v, like $d_v = x_v - y_v$ as already shown in Sections 8.2.3 and 9.1. The methods designed for one single character (in only one, or two or more samples) can be applied for these differences. When at least three characters are given, the approach of forming differences is, however, no longer applicable. Instead we have to take approaches such as that in Section 8.2.4 (see Example 8.6), where the paired sample t-test has been described as an alternative method of analysis.

Thus, the case of interest is that a tuple of outcomes, definitely matched to one other, from the m modeled random variables $\boldsymbol{y}_1, \boldsymbol{y}_2, \ldots, \boldsymbol{y}_m$ is given. We assume a normal distribution for each random variable. It is important to note that the research units do not necessarily have to deal with the same person; the only thing of importance is the definite association of the outputs per given research unit.

Bachelor Master

Example 12.11 The influence of sibling order on intelligence will be investigated

The results of several research studies with impressively large sample sizes have repeatedly shown that the higher the sibling order of the tested person, the lower that person's test scores are in relevant intelligence tests. Nevertheless, there are several methodological objections, which argue against the binding impact of the statements made in these research studies. It can be criticized, amongst other things, that independent samples have been used in all research studies, in other words all children from a family have never been examined. Only when it can be significantly confirmed on average that, for instance, first born are more intelligent than last born, using matched tuples of outcomes, thus across the children from one family in each case, do the results gain validity. In the case of unrelated tuples of outcomes, it can generally be criticized that the noise factor *circumstances which account for which sibling from a family becomes part of the sample* overlies the association of interest; namely that between birth order and intelligence.

If we suppose that all three children (aged between 16 and 30 years) from a family of three children are tested with an intelligence test, the tuples of outcomes are matched because it can generally be expected that the test scores of the siblings are correlated based on the factors of genetic constitution and environmental conditions. Therefore, the test scores are dependent on one another. Thus, a one-way analysis of variance according to Section 10.4 does not come into consideration.

An *analysis of variance for matched samples* provides the solution to this problem. It analyses the data structure given in Table 12.6. It is evident that this is a special case of the one provided

Table 12.6 Data structure of a cross classification of factor A, reserach unit, with a levels, and of factor B, with b factor levels, in a single cell allocation ($n_{ij} = n = 1$).

	Levels of B			
Levels of A	B_1	B_2	...	B_b
A_1	y_{11}	y_{12}	...	y_{1b}
A_2	y_{21}	y_{22}	...	y_{2b}
$\vdots$	$\vdots$	$\vdots$	$\vdots$	$\vdots$
A_a	y_{a1}	y_{a2}	...	y_{ab}

in Table 10.9: two factors, namely the factor A with a factor levels of the different research units and the actual factor of interest B, with b factor levels designated to several (treatment) conditions, are cross classified, with the special case of $n = 1$ per combination of factor levels. Factor A is a random factor because we assume that the a research units have been randomly taken from a defined population; as a rule factor B is a fixed factor as we are actually interested in the selected factor levels ($b = m$ characters). It is thus a special case of the mixed model from Section 10.4.4.3. As $n = 1$, the interaction effect cannot be tested in a common way.

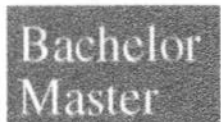

Example 12.11 – continued

We want to investigate the given research question using the data set *Example 12.11* (see Chapter 1 for its availability). The null hypothesis is H_0: 'There are no significant differences between the different sibling positions with regard to the character *intelligence*'. We choose a type-I risk of $\alpha = 0.05$.

In **R**, we first have to rearrange the data from the database Example 12.11. In order to do this, we type

```
> Example_12.11.1 <- reshape(Example_12.11,
+                             varying = list(c("int1", "int2",
+                                              "int3"),
+                                            c("r1", "r2", "r3")),
+                             direction = "long")
```

i.e. we use the function `reshape()`, setting the database `Example_12.11` as the first argument. With the second argument, `varying`, by using the function `c()` we set the characters which we associate to each other as vectors of the their respective observed values; that is `int1`, `int2`, and `int3` on the one hand, and `r1`, `r2`, and `r3` on the other hand. These two vectors are submitted as a list by applying the function `list()`. With `direction = "long"`, we select the desired rearrangement. We assign the rearranged database to the object `Example_12.11.1`. Next we set the character `time` to a factor by typing

```
> Example_12.11.1$id <- as.factor(Example_12.11.1$id)
> Example_12.11.1$time <- as.factor(Example_12.11.1$time)
```

i.e. we apply the function `as.factor()` to the objects `Example_12.11.1$id` and `Example_12.11.1$time`, respectively, and assign the factor to `$id` and `$time` respectively in the data set `Example_12.11.1`.

For the following analysis, we use the package `ez`, which we load after installation (see Chapter 1) by applying the function `library()`. Afterward, we type

```
> ezANOVA(Example_12.11.1, dv = .(int1), wid = .(id),
+         within = .(time))
```

i.e. we apply the function `ezANOVA()` to the database in the object `Example_12.11.1`; with the second argument, `dv = .(int1)`, we submit the information on which of the character groups is to be analyzed (`int` or `u`), and with the third argument `wid = .(id)`, the information that the research units are identifiable by the label id. Finally, with the fourth argument, within = .(time), we indicate the factor of the observed values associated to each other.

As a result we get (shortened output):

```
Note: model has only an intercept; equivalent type-III tests
      substituted.
$ANOVA
  Effect DFn DFd        F        p p<.05         ges
2   time   2 318 1.346574 0.261609       0.005549277

$'Mauchly's Test for Sphericity'
  Effect         W         p p<.05
2   time 0.9951725 0.6822916

$'Sphericity Corrections'
  Effect       GGe     p[GG] p[GG]<.05      HFe    p[HF] p[HF]<.05
2   time 0.9951957 0.2615938           1.007781 0.261609
```

In this, the value in column ges refers to a so-called Generalized Eta-Squared measure of effect size.

In SPSS, we first have to rearrange the data in the database Example 12.11. In order to do this, we select the same sequence of commands as in the continuation of Example 10.1 (Data – Restructure...) and, in the resulting window (Figure 10.3), confirm the preset setting Restructure selected variables into cases by clicking Next. In the next window (not shown here), we click on More than one, leave the preset 2 at How Many? and confirm this by clicking Next. In the following window (not shown here), we drag and drop the characters intelligence first-born, intelligence second-born, and intelligence third-born into the field Target Variable: trans1 under Variables to be Transposed; we have to proceed analogously for trans2 – which we select via Target Variable: – and for the characters rating intelligence first-born, rating intelligence second-born, and rating intelligence third-born. By clicking Next, we get to another window, where we simply confirm all of the above with Next; in the next window, we once again confirm by clicking Next and, in the final window, click on Finish. Eventually, we arrive at a small window with a warning, to which we respond by clicking OK. Finally, we obtain a new data sheet with which we can continue the analysis.

Now, we proceed analogously (Analyze – General Linear Model – Univariate...) to, for example, Example 10.8 in order to mark Dependent Variable: trans1, displayed as intelligence first-born [trans1], in the window in Figure 10.7. We drag and drop the character Index1 into the field Fixed Factor(s): and the character id into the field Random Factor(s):. If we also want the estimated magnitude of effect η^2, we have to additionally click Options... in order to set a check mark at Estimates of effect size in the window in Figure 10.8. By clicking OK, we obtain the results in Table 12.7. The resulting p-value concerning the character Index1, i.e. the *sibling position*, amounts to 0.262.

Table 12.7 SPSS-output of the analysis of variance for matched outcomes in Example 12.11.

Tests of Between-Subjects Effects

Dependent Variable:intelligence first-born

Source		Type III Sum of Squares	df	Mean Square	F	Sig.	Partial Eta Squared
Intercept	Hypothesis	5041155.169	1	5041155.169	23932.513	.000	.993
	Error	33491.831	159	210.640[a]			
Index1	Hypothesis	547.913	2	273.956	1.347	.262	.008
	Error	64696.088	318	203.447[b]			
id	Hypothesis	33491.831	159	210.640	1.035	.394	.341
	Error	64696.088	318	203.447[b]			
Index1 * id	Hypothesis	64696.088	318	203.447			1.000
	Error	.000	0	[c]			

a. MS(id)
b. MS(Index1 * id)
c. MS(Error)

In SPSS, it is also possible to select

Analyze

General Linear Model

Repeated Measures...

in order to conduct the same analysis; however, we have to use the original database *Example 12.11* for this. In the window Repeated Measures Define Factor(s) in Figure 12.10, we type the name of the character for which we have repeated observations into the field Within-Subject Factor Name:; in our case, for example, pos_sibling. We type 3 into the field Number of Levels: and subsequently click Add (the results of this are already displayed in Figure 12.10). By clicking Define, we get to the next window (Figure 12.11), where we drag and drop the three levels of the factor *sibling position*, namely intelligence first-born, intelligence second-born, and intelligence third-born, into the field Within-Subjects Variables (which has already been done in Figure 12.11). If we also want the estimated magnitude of effect η^2, we additionally have to click Options... in order to set a check mark at Estimates of effect size in the resulting window (quite similar to the one in Figure 10.8).

Now, we continue by clicking OK and obtain, among other things, a composition of results on the three levels of the factor *sibling position* (see Table 12.8). The p-value of the respective F-test is again 0.262.

Finally, in SPSS, it is possible to conduct the same analysis by following the steps described in Example 10.1 (Analyze – Scale – Reliability Analysis...); in the window in Figure 10.4, we drag and drop the characters intelligence first-born, intelligence second-born, and intelligence third-born into the field Items: and click Statistics.... In the following window (not shown here), we mark the option F-Test in ANOVA Table, return to the previous window by clicking Continue and confirm by clicking OK. The output then also contains the p-value 0.262.

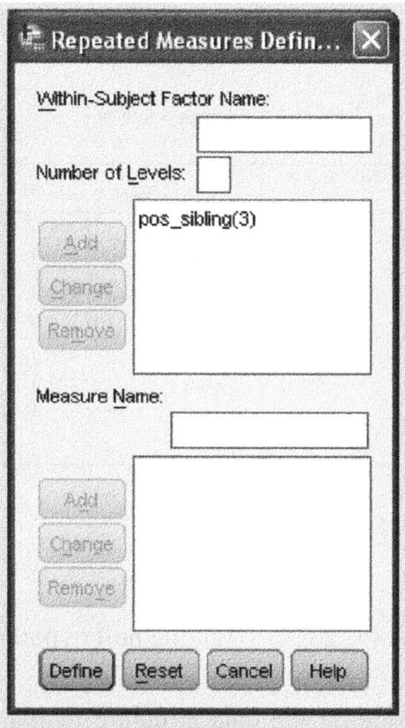

Repeated Measures Defin...
Within-Subject Factor Name:
Number of Levels:
pos_sibling(3)
Add
Change
Remove
Measure Name:
Add
Change
Remove
Define Reset Cancel Help

Figure 12.10 SPSS-window for computing an analysis of variance for matched outcomes.

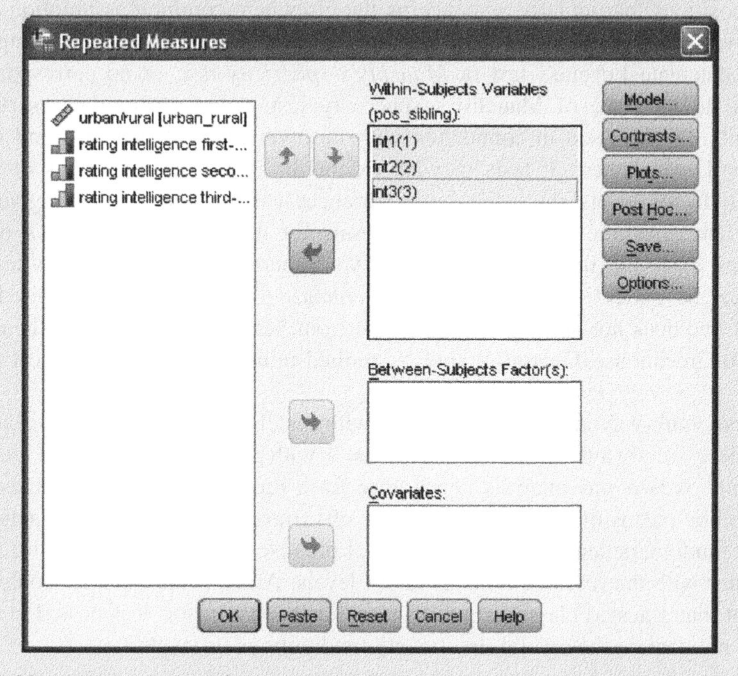

Figure 12.11 SPSS-window for re-structuring the data in order to apply an analysis of variance for matched outcomes.

Table 12.8 SPSS-output of the analysis of variance for matched outcomes in Example 12.11.

Tests of Within-Subjects Effects

Measure:MEASURE_1

Source		Type III Sum of Squares	df	Mean Square	F	Sig.	Partial Eta Squared
pos_sibling	Sphericity Assumed	547.912	2	273.956	1.347	.262	.008
	Greenhouse-Geisser	547.912	1.990	275.279	1.347	.262	.008
	Huynh-Feldt	547.912	2.000	273.956	1.347	.262	.008
	Lower-bound	547.912	1.000	547.912	1.347	.248	.008
Error(pos_sibling)	Sphericity Assumed	64696.087	318	203.447			
	Greenhouse-Geisser	64696.087	316.472	204.429			
	Huynh-Feldt	64696.087	318.000	203.447			
	Lower-bound	64696.087	159.000	406.894			

Due to the p-value of $0.262 > 0.05$, the null hypothesis can be accepted.

We have previously only considered the data structure of matched samples and discussed which of the models of analysis of variance, as explained above, is relevant for this case – additionally we have shown possible analyses in an example. We have not yet referred to the presuppositions, particularly as concerns the elsewhere-common assumption of homogeneity of variances. Depending on which method of analysis is chosen, computer programs will calculate Levene's test or *Mauchly's sphericity test*, or no corresponding test is provided. In the case of Mauchly's sphericity test, there are similar reservations to the ones initially discussed in connection with the two-sample Welch test and illustrated in detail for Levene's test. It tests to what extent the pair-wise differences of outcomes of each two factor levels show the same variances (we do not explicitly give the null hypothesis here). All the same, several proposals for the correction of the F-distributed test statistics exist for the case that Mauchly's sphericity test has a significant result (for instance the correction according to *Greenhouse–Geisser*; see e.g. Table 12.8). We suggest – if one does not use our rule of thumb from Section 10.4.1.1 – that the correction according to Greenhouse–Geisser simply be applied immediately, but no pretest should be conducted.

Of course, mult-way analyses of variance with matched outcomes are possible as well. Basically these are always special cases for $n = 1$ with an additional random factor of the research units. A two-way analysis of variance for a tuple of matched outcomes can then be calculated by means of a three-way analysis of variance, in which the two actual factors are fixed or random, depending on the content of the research question. The third factor is a random factor with the research units as factor levels. At the same time, it should be taken into account that a nested classification is given for the last factor: it is nested in the factor which does not contain the several (treatment) conditions as its levels.

Doctor **Example 12.12** The influence of the birth order on intelligence will be investigated, whereby the interaction with the character city/rural is of particular interest.

Expanding on Example 12.11, we want to investigate the given research question using the data set *Example 12.11* (see Chapter 1 for its availability). We have one fixed factor *birth order* with $a = 3$ levels, the random factor *family* with b levels, and the factor *residence* with the $c = 2$ levels 'city' and 'rural'. The null hypothesis is identical to the one in Example 12.11 with regard to the factor *birth order*; with regard to the interaction effect the null hypothesis is H_0: 'There are no combinations of "city" or "rural", respectively, and birth order, which show particularly high or particularly low mean test scores in the character intelligence.' We decide on a type-I risk of $\alpha = 0.05$.

In **R**, we keep on using the rearranged database from Example 12.11 as well as the package `ez`, which we already loaded in Example 12.11. We type

```
> ezANOVA(Example_12.11.1, dv = .(int1), wid = .(id),
+         within = .(time), between = .(residence))
```

i.e. we use the function `ezANOVA()`, adding the argument `between = .(residence)` to the command from Example 12.11, which causes the additional factor `residence` to be included into the analysis.

As a result, we get (shortened output):

```
$ANOVA
          Effect DFn DFd         F            p p<.05         ges
2      residence   1 158 27.887756 4.194872e-07     * 0.051596580
3           time   2 316  1.354974 2.594484e-01       0.005897507
4 residence:time   2 316  1.991869 1.381496e-01       0.008645615

$'Mauchly's Test for Sphericity'
          Effect         W         p p<.05
3           time 0.9963584 0.7509703
4 residence:time 0.9963584 0.7509703

$'Sphericity Corrections'
          Effect       GGe     p[GG] p[GG]<.05      HFe     p[HF]
3           time 0.9963716 0.2594412          1.009075 0.2594484
4 residence:time 0.9963716 0.1383308          1.009075 0.1381496
```

In SPSS, analogously to Example 12.11, we can proceed in two ways. We start by describing the second approach, because the data does not have to be rearranged for it. Following the sequence of commands (Analyze – General Linear Model – Repeated Measures...) in Example 12.11, we open the window in Figure 12.10 and Figure 12.11, additionally dragging and dropping the character residence into the field Between-Subject Factor(s): in the latter window. If we also want the estimated magnitude of effect η^2, we have to additionally click Options... in order to set a check mark at Estimates of effect size in the following window (quite similar to the one in Figure 10.8). By clicking Continue and OK, we obtain the results in Table 12.9 and Table 12.10. The p-values under Sig. are the ones to be interpreted.

Table 12.9 SPSS-output of the two-way analysis of variance (Tests of Within-Subjects Effects) for observed values associated with each other in Example 12.12.

Tests of Within-Subjects Effects

Measure:MEASURE_1

Source		Type III Sum of Squares	df	Mean Square	F	Sig.	Partial Eta Squared
pos_sibling	Sphericity Assumed	547.912	2	273.956	1.355	.259	.009
	Greenhouse-Geisser	547.912	1.993	274.954	1.355	.259	.009
	Huynh-Feldt	547.912	2.000	273.956	1.355	.259	.009
	Lower-bound	547.912	1.000	547.912	1.355	.246	.009
pos_sibling * residence	Sphericity Assumed	805.454	2	402.727	1.992	.138	.012
	Greenhouse-Geisser	805.454	1.993	404.194	1.992	.138	.012
	Huynh-Feldt	805.454	2.000	402.727	1.992	.138	.012
	Lower-bound	805.454	1.000	805.454	1.992	.160	.012
Error(pos_sibling)	Sphericity Assumed	63890.633	316	202.186			
	Greenhouse-Geisser	63890.633	314.853	202.922			
	Huynh-Feldt	63890.633	316.000	202.186			
	Lower-bound	63890.633	158.000	404.371			

Table 12.10 SPSS-output of the two-way analysis of variance (Tests of Between-Subjects Effects) for observed values associated with each other in Example 12.12.

Tests of Between-Subjects Effects

Measure:MEASURE_1
Transformed Variable:Average

Source	Type III Sum of Squares	df	Mean Square	F	Sig.	Partial Eta Squared
Intercept	5041155.169	1	5041155.169	27979.629	.000	.994
residence	5024.602	1	5024.602	27.888	.000	.150
Error	28467.229	158	180.172			

Neither the main effect with respect to the factor *birth order* nor the interaction effect result in significance. We were not interested in the main effect with respect to the factor *residence*; that, however, is significant.

If we were to be further interested in the main effect concerning *residence*, we could now once again conduct a calculation in order to determine the means per respective factor level. For this, we would have to click Options... in Figure 12.11 in order to drag and drop the character residence into the field Display Means for: in the resulting window (not shown here); by clicking Continue and OK, we obtain the required means. With the first approach in Example 12.10, we get to the same results. However, one needs to rearrange the data, as was shown above. The random factor *family* is now nested in the factor *residence*; the b levels are respectively divided into b_1 and b_2 levels. We proceed with the sequence of orders (Analyze – General Linear Model – Univariate...) as in, for example,

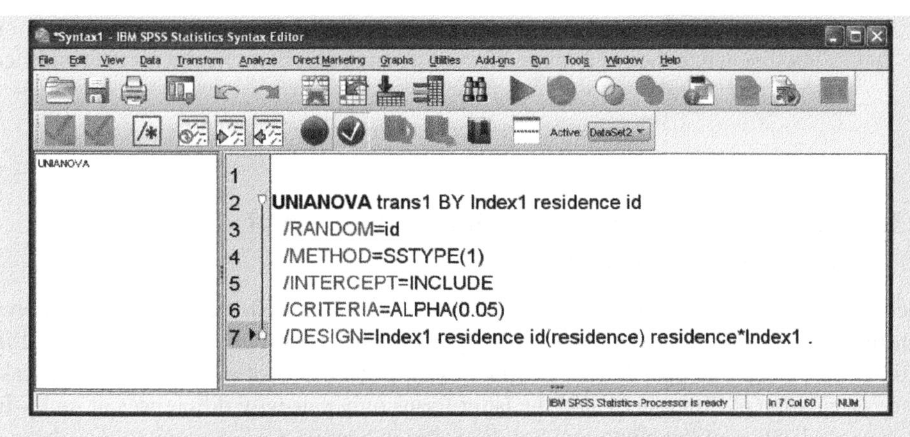

Figure 12.12 SPSS-syntax for Example 12.12.

Table 12.11 SPSS-output of the two-way analysis of variance for observed values associated with each other in Example 12.12.

Tests of Between-Subjects Effects

Dependent Variable:intelligence first-born

Source		Type I Sum of Squares	df	Mean Square	F	Sig.
Intercept	Hypothesis	5041155.169	1	5041155.169	27979.629	.000
	Error	28467.229	158	180.172[a]		
Index1	Hypothesis	547.913	2	273.956	1.355	.259
	Error	63890.633	316	202.186[b]		
residence	Hypothesis	5024.602	1	5024.602	27.888	.000
	Error	28467.229	158	180.172[a]		
id(residence)	Hypothesis	28467.229	158	180.172	.891	.792
	Error	63890.633	316	202.186[b]		
Index1 * residence	Hypothesis	805.454	2	402.727	1.992	.138
	Error	63890.633	316	202.186[b]		

a. MS(id(residence))
b. MS(Error)

Example 10.8, and mark Dependent Variable: trans1, displayed as intelligence first-born [trans1], in the window in Figure 10.7. Now, we drag and drop the characters Index1 as well as residence into the field Fixed Factor(s):, and the character id into the field Random Factor(s):. However, in order to specify that it is nested, we have to change the syntax. To do this, we click – still in the window in Figure 10.7 – the button Paste, and obtain the window in Figure 12.12. There, one has to replace the last line as follows: /DESIGN = Index1 residence id(residence) residence*Index1. This has already been done in Figure 12.12. Now, we click Run and select All, which causes the computation to be conducted. The results can be found in Table 12.11.

It has to be concluded that the birth order does not have a significant effect on intelligence, either as a main effect or as an interaction effect with the residence ('city' *vs.* 'rural').

12.3.2 Hypothesis testing concerning the position of ordinal-scaled characters

When the observed characters $y_1, y_2, \ldots, y_m$ are ordinal scaled, neither the procedures from the preceding section nor Wilcoxon's signed-ranks test, which has been discussed in Section 8.5.5, come into question. The former assume characters of at least an interval scale for the calculation of differences, and the latter forms only pair-wise differences of ranks. In fact, a generalization of the binominal test discussed in Section 8.5.3 is required. *Friedman's test* is an example of such a generalization. In applying such a test, the idea is to provide the tuples of outcomes per research unit v with ranks $r_{v1}, r_{v2}, \ldots, r_{vm}$; if ties appear, we proceed as in Section 5.3.2. If all given ranks are summed per character q, all of these (R_q) should be (approximately) equal when the null hypothesis is true. As the same number of observations a is given for every character q, it is actually not necessary to determine the means of the ranks $\bar{R}_q$; however these can be regarded as estimations $\hat{P}$ for the unknown mean ranks in the population. Thus, the null hypothesis is: H_0: $P_q = P_l = P$ for all q and l; the alternative hypothesis is H_A: $P_q \neq P_l$ for at least one $q \neq l$.

The calculation of the Friedman's test is very simple using pertinent computer programs.

Master It can be shown that the test statistic

$$\chi^2 = \frac{12}{am(m+1)} \sum_{q=1}^{m} \left[\sum_{v=1}^{a} \left(r_{vq} - \frac{m+1}{2} \right)^2 \right] \tag{12.10}$$

is asymptotically χ^2-distributed with $df = m - 1$ degrees of freedom.

Bachelor **Example 12.13** The influence of the birth order on the rating of intelligence (by a parent)

We investigate this research question using the data set *Example 12.11*. We have already tested the differences between the birth order with regard to objective test scores in an intelligence test in Example 12.11 and 12.12, respectively. We are now interested in the parent's subjective ratings in the same sample; thus the characters *rating intelligence first-born*, *rating intelligence second-born*, and *rating intelligence third-born*. The null hypothesis is: 'There are no differences in the rating of intelligence with regard to birth order.' According to the explanations in Chapter 5, ratings are generally ordinal-scaled data; we thus apply the Friedman's test. We choose a type-I risk of $\alpha = 0.05$.

In **R**, we use the rearranged database from Example 12.11 and type

```
> friedman.test(r1 ~ time | id, data = Example_12.11.1)
```

i.e. we use the function `friedman.test()`, submitting to it the instruction that the character `r1` is to be analyzed with regards to the factor `time`; with `| id`, we state that the character `time` is hierarchically subordinate to the factor of the observation units (`id`). Finally, we state the database to be used with `data = Example_12.11.1`.

As a result, we get:

```
        Friedman rank sum test

data:  r1 and time and id
Friedman chi-squared = 2.9759, df = 2, p-value = 0.2258
```

For determining the middle ranks, we use the original database *Example 12.11* again. We type

```
> with(apply(apply(cbind(r1, r2, r3), 1, rank), 1, mean),
+       data = Example_12.11)
```

i.e. we use the function `apply()`; the first argument being the new matrix which we create with `cbind(r1, r2, r3)` from the characters *rating intelligence first-born* (`r1`), *rating intelligence second-born* (`r2`), and *rating intelligence third-born* (`r3`); the second argument, `1`, being the instruction that the ranking (`rank`) is to be conducted line by line. We submit the result to the function `apply()` as the first argument; the second argument, `1`, stipulating that the mean (`mean`) is to be computed line by line. By applying the function `with()`, we state the database to be used; namely `Example_12.11`.

As a result, we get:

```
      r1       r2       r3
2.103125 1.981250 1.915625
```

In SPSS, we use the original database *Example 12.11* and select

Analyze
    Nonparametric Tests
        Related Samples...

in order to select Fields in the upper bar in the resulting window (not shown here), so that we arrive at a window very similar to the one in Figure 8.12. We mark the characters rating intelligence first-born, rating intelligence second-born, and rating intelligence third-born, and drag and drop them into the field Test Fields:. Now, we select Settings in the upper bar and thus get to the window in Figure 8.13. There, we click the button Customize tests and set a check mark at Friedman's 2-way ANOVA by ranks (k samples). After clicking Run, we obtain the table Hypothesis Test Summary in the output window (not shown here), which we simply double-click; following this, a window opens, from which we extract the relevant result: the test value according to Formula (12.11) amounts to $\chi^2 = 2.976$, with $df = 2$ degrees of freedom; the respective p-value being 0.226. Additionally, we can see the middle ranks in this window: the middle rank for the first-born is 2.10, the one for the second-born 1.98 and the one for the third-born is 1.92.

With a p-value of 0.226 the result is not significant. In other words, there is no difference in the parent's subjective rating of their children's intelligence with regard to their birth order.

Mult-way rank variance analyses for matched outcomes do not exist. This should therefore be taken into consideration during the planning of a research study.

Summary

If at least three matched outcomes per research unit are given, then hypothesis testing is fundamentally different to the case that the respective outcomes stem from different research units. Nevertheless, certain variations of the analysis of variance apply. *Friedman's test* exists for ordinal-scaled characters.

References

Rasch, D., Herrendörfer, G., Bock, J., Victor, N., & Guiard, V. (2008). *Verfahrensbibliothek Versuchsplanung und -auswertung. Elektronisches Buch* [Collection of Procedures in Design and Analysis of Experiments. Electronic Book]. Munich: Oldenbourg.

Williams, E. J. (1959). The comparison of regression variables. *Journal of the Royal Statistical Society B*, *21*, 396–399.

13

Samples from more than one population

This chapter is about scenarios where at least two characters are observed on research units of samples from at least two populations. Predominantly, the discussed methods can only be used for quantitative characters. In particular, the generalization of the analysis of variance with more than one character will be discussed. Thereby even noise factors are taken into account. Finally, it will be shown how characters (in combination) can be selected, from a relatively large pool of characters, in order to best discriminate between two or more groups.

Basically, this chapter is about the generalizations of the research questions from Chapter 12; that is, for the case of more than one sample or population, respectively. Further, it is about the generalization of research questions from Chapter 10; that is, comparing more than one or two samples with respect to at least two characters instead of only a single character. Regression and correlation analysis will also be dealt with, since this is obligatory whenever at least two characters are under consideration. First of all, the *general linear model* will be introduced, as all of the respective methods can be incorporated there.

13.1 General linear model

So far, we considered, as one of the most complex cases of the general linear model, for instance the three-way analysis of variance (see Table 10.23):

$$\boldsymbol{y}_{ijkv} = \mu + a_i + b_j + c_k + (ab)_{ij} + (ac)_{ik} + (bc)_{jk} + (abc)_{ijk} + \boldsymbol{e}_{ijkv} \qquad (13.1)$$

Statistics in Psychology Using R and SPSS, First Edition. Dieter Rasch, Klaus D. Kubinger and Takuya Yanagida.

If we now label the vector of all main and interaction effects θ (a, b, (ab), etc.) as $\vec{\theta}$, then the general linear model[1] can be written quite simply as ($\vec{y}$ and $\vec{e}$ are also vectors):

$$\vec{y} = \vec{\theta} + \vec{e} \tag{13.2}$$

In the simplest case of one character in just a single random sample, the model can be written as: $\boldsymbol{y}_v = \mu + \boldsymbol{e}_v$; $v = 1, 2, \ldots, n$.

The model in (13.2) can also be written as follows (with X as a matrix and β as a vector):

$$\vec{y} = X\vec{\beta} + \vec{e} \tag{13.3}$$

For example, the simple analysis of variance (model I) can be written as: $\boldsymbol{y}_{iv} = \mu + a_i + \boldsymbol{e}_{iv}$; $i = 1, 2, \ldots, a$; $v = 1, 2, \ldots, n_i$ (see Formula (10.1)).

Doctor For example, for $a = 3$ and $n_i = n = 2$, $\vec{y}$ is the vector of all $\boldsymbol{y}_{iv}$. In lexicographical order, this is written (transposed) as: $\vec{y}^T = (\boldsymbol{y}_{11}, \boldsymbol{y}_{12}, \boldsymbol{y}_{21}, \boldsymbol{y}_{22}, \boldsymbol{y}_{31}, \boldsymbol{y}_{32})$. Analogously, $\vec{e}$ is the vector of all $\boldsymbol{e}_{iv}$, and $\vec{\beta}$ the vector of the four parameters μ, a_1, a_2, a_3. X has the form

$$X = \begin{pmatrix} 1 & 1 & 0 & 0 \\ 1 & 1 & 0 & 0 \\ 1 & 0 & 1 & 0 \\ 1 & 0 & 1 & 0 \\ 1 & 0 & 0 & 1 \\ 1 & 0 & 0 & 1 \end{pmatrix}$$

Formula (13.3) then is

$$\begin{pmatrix} \boldsymbol{y}_{11} \\ \boldsymbol{y}_{12} \\ \boldsymbol{y}_{21} \\ \boldsymbol{y}_{22} \\ \boldsymbol{y}_{31} \\ \boldsymbol{y}_{32} \end{pmatrix} = \begin{pmatrix} 1 & 1 & 0 & 0 \\ 1 & 1 & 0 & 0 \\ 1 & 0 & 1 & 0 \\ 1 & 0 & 1 & 0 \\ 1 & 0 & 0 & 1 \\ 1 & 0 & 0 & 1 \end{pmatrix} \cdot \begin{pmatrix} \mu \\ a_1 \\ a_2 \\ a_3 \end{pmatrix} + \begin{pmatrix} \boldsymbol{e}_{11} \\ \boldsymbol{e}_{12} \\ \boldsymbol{e}_{21} \\ \boldsymbol{e}_{22} \\ \boldsymbol{e}_{31} \\ \boldsymbol{e}_{32} \end{pmatrix}$$

In analysis of variance, X is a matrix consisting of zeros and ones. In simple linear regression, the model equation is $\boldsymbol{y}_v = \beta_0 + \beta_1 \boldsymbol{x}_v + \boldsymbol{e}_v$; $v = 1, \ldots, n$; it follows $\vec{\beta} = \begin{pmatrix} \beta_0 \\ \beta_1 \end{pmatrix}$, and the matrix X has n rows and two columns, the vth line being $(1\ x_v)$. In the regression analysis, the matrix X thus contains the outcomes of the regressor.

Despite the fact that – if one permits random factors and regressors – this covers a large number of models, the general notation does offer some advantages. Most notably, estimation by the method of least squares or derivation of the distribution of test statistics can be handled

[1] The general linear model was defined mainly as model I of the analysis of variances, as well as of the regression analysis, so that both approaches could be handled consistently.

within the general model, but for specific cases then can be easily determined for special models. The analysis of covariance, factor analysis, and discriminant analysis can also be described using the general linear model.

13.2 Analysis of covariance

The analysis of covariance, a special case of the general linear model, is related to regression analysis as well as to analysis of variance. The analysis of covariance is about the mean differences between the (at least two) levels of a particular factor A with respect to the character y; these factor levels can again be seen as different populations. It is also about the assumption/fear that the differences between the factor levels $A_1, A_2, \ldots, A_i, \ldots, A_a$ with respect to the (means of) the character y (i.e. the relationship of $\boldsymbol{y}$ and $E(\boldsymbol{y}) = \mu$, respectively, and A) are 'disturbed', 'superimposed', or 'hidden' by another character, z. In this respect, the analysis of covariance differs from the partial correlation coefficient in Section 12.1.1 simply by replacing the quantitative character x by the qualitative factor A. In both cases, the characters y and z are quantitative. The factor can be fixed or random. The character or to say noise factor z, modeled as $\boldsymbol{z}$, is called a *covariate*. This factor can be fixed (in that case we write z instead of $\boldsymbol{z}$) or random. In psychological applications it is, however, usually random.

Master Doctor

The characters y and z are modeled by the normally distributed random variables $\boldsymbol{y}$ and $\boldsymbol{z}$. For the purpose of hypothesis testing, a two-dimensional normal distribution of these variables must be assumed. The data structure is shown in Table 13.1. The difference from Table 10.1 is the additional column in which the outcomes z_{iv} are listed.

Table 13.1 Data structure of a (one-way) analysis of covariance

level A_1		level A_2		...	level A_a	
y_{11}	z_{11}	y_{21}	z_{21}	...	y_{a1}	z_{a1}
y_{12}	z_{12}	y_{22}	z_{22}	...	y_{a2}	z_{a2}
$\vdots$	$\vdots$	$\vdots$	$\vdots$	$\vdots$	$\vdots$	$\vdots$
y_{1n_1}	z_{1n_1}	y_{2n_2}	z_{2n_2}	...	y_{an_a}	z_{an_a}

We only consider the case of a fixed factor A. The model equation is derived from formula (10.1); however, in the expression $\boldsymbol{y}_{iv} = \mu + a_i + \boldsymbol{e}_{iv}$, the a_i is co-determined by the ith of the a regression functions from $\boldsymbol{z}$ onto $\boldsymbol{y}$: $a_i = a_i^* - (\beta_{0i} + \beta_{1i}\,\boldsymbol{z}_{iv}) = \beta_{0i}^* + \beta_{1i}\,\boldsymbol{z}_{iv}$. Assuming $\beta_{i1} = \beta_1$ it follows:

$$\boldsymbol{y}_{iv} = \mu + \beta_{0i}^* + \beta_1\,\boldsymbol{z}_{iv} + \boldsymbol{e}_{iv}\ (i = 1, \ldots, a; v = 1, \ldots, n_i) \tag{13.4}$$

With respect, for instance, to the analysis of variances, the following assumptions are made:

- The $\boldsymbol{e}_{iv}$ are normally distributed $N(0, \sigma^2)$; therefore they all have the same expectation 0 and the same variance σ^2.

- The $\boldsymbol{e}_{iv}$ are independent (from each other).
- The z_{iv} and the $\boldsymbol{e}_{iv}$ are independent (from each other).
- $a \geq 2$, and, for all i, $n_i \geq 2$.
- $\sum_{i=1}^{a} a_i = 0$.

Because $\beta_{i1} = \beta_1$, it is particularly presupposed that the regression lines are parallel. Therefore, they only differ in terms of the intercept β_{0i}. It follows that the a levels of the factor A only have influence on the intercept. This assumption is certainly fulfilled if the variance–covariance matrix

$$\Sigma_{(i)} = \begin{pmatrix} \sigma_{y(i)}^2 & \sigma_{yz(i)} \\ \sigma_{yz(i)} & \sigma_{z(i)}^2 \end{pmatrix} \tag{13.5}$$

is the same in all a populations, which means that they are independent from i (if so, $\beta_{i1} = \frac{\sigma_{yz(i)}}{\sigma_{z(i)}^2} = \beta_1 = \frac{\sigma_{yz}}{\sigma_z^2}$). This then is referred to as the homogeneity of the variance–covariance matrix.

Master
Doctor

Example 13.1 The influence of Latin courses on the development of the ability of reasoning is to be examined

Teachers often argue that Latin courses in high school are important not only because they are the basis for other languages, but also because they are an excellent exercise in reasoning. There might be a study in which students in the 11th level of education are tested with a pertinent psychological test for measuring their *reasoning* ability. The factor in question, *Latin courses*, then would have two levels: 'from the 7th level of education with Latin courses' and 'without Latin courses'; thus the factor is fixed. The character of interest is the test score in *reasoning*.

Bear in mind (see Section 3.1), that this is a so-called *ex-post-facto* design. That is, the assignment to the two levels is not due to randomization. It follows that, in the end, any effects detected may have existed from the very beginning: students who choose Latin courses thus might differ fundamentally, and right from the onset, from those who do not. Intelligence is one of the potential noise factors. For this we might have measured the character *intelligence*, too; again with an adequate psychological test. The idea is to 'clean up' the differences between the levels of the factor *Latin courses* in the *reasoning* test scores from the possible contribution of the factor *intelligence*. If the factor *Latin courses* were a quantitative character like the two characters *reasoning* and *intelligence*, we would compute a partial correlation coefficient. This then would adjust the correlation between *Latin courses* and *reasoning* with respect to the noise factor *intelligence*.

A comparison between the means of students with and without *Latin courses* as concerns the character *reasoning* would be unfair, if all those with *Latin courses* were more intelligent than those without Latin courses. To analyze the latter, we assume that if there is any (positive) correlation between the character *reasoning* (y) as the regressand and the character *intelligence* (z) as the regressor, then the slope of the regression line in both groups will be equal. Thus, the effect of *Latin courses* could be estimated by the distance between the two parallel regression lines for any value of the regressor variable, for example for an *intelligence* value of 100. This estimation would then be adjusted with respect to the eventually given difference in the means of *intelligence* in both groups. If we take the (from the content point of view unrealistic) value 0 instead of 100, the estimation is just the difference between the intercepts.

We now use the data set *Example 13.1* (see Chapter 1 for its availability) to answer the given research question. The null hypothesis is: eliminating the effects of z, the means of $\mathbf{y}$ do not differ with respect to the factor levels A_i. Initially, however, it is of importance whether the slopes of the regression lines are the same for both factor levels or not.

In **R**, we first enable access to the database *Example_13.1* (see Chapter 1) by using the function `attach()`. Then we type

```
> coef(lm(rsng ~ intelligence, subset = latin == "without Latin"))
> coef(lm(rsng ~ intelligence, subset = latin == "with Latin"))
```

i.e. we use the function `lm()`, each time specifying that the character reasoning (`rsng`) is to be analyzed with regards to the character `intelligence` as the first argument; in the second arguments, we use `subset`, once to select the children '`without Latin`' and once to select the children '`with Latin`'. We apply the function `coef()` to the results.

This yields:

```
(Intercept) intelligence
  32.0325187    0.6361691

(Intercept) intelligence
  16.1740102    0.9057169
```

In SPSS, we first estimate the regression coefficients for each of the two levels of the factor *Latin course* separately. In order to do this, we split the data by the factor *Latin courses* by using the same sequence of commands (Data – Split File...) as shown in Example 5.11 and, in the following window in Figure 5.23, click the button Compare groups, after which we drag and drop Latin course into the field Groups Based on:. After clicking OK, the following analysis will be conducted separately for children 'without Latin' and 'with Latin'. Following the steps described in Example 11.5 (Analyze – Regression – Linear...), we arrive at Figure 11.5, where we mark the character reasoning in order to drag and drop it into the field Dependent:. We drag and drop the character intelligence into the field Independent(s):. After clicking OK, we obtain the results in Table 13.2, among other things.

Table 13.2 SPSS-output of the regression coefficients in Example 13.1 (shortened output)

Coefficients [a]

Latin courses	Model		Unstandardized Coefficients B	Std. Error
without Latin	1	(Constant)	32.033	14.184
		intelligence	.636	.143
with Latin	1	(Constant)	16.174	10.220
		intelligence	.906	.103

a. Dependent Variable: reasoning

The estimated regression function from *reasoning* onto *intelligence* for students without *Latin courses* is: $\hat{y}_{\text{without Latin}} = 32.033 + 0.636z$; for students with *Latin courses*, $\hat{y}_{\text{with Latin}} = 16.174 + 0.906z$. Although there is a test which could test the hypothesis of equal slopes, we forgo such a test and illustrate the situation only graphically, with a scatter plot.

In **R**, we thus type

```
> plot(intelligence, rsng, col = as.numeric(latin),
+       xlab = "intelligence", ylab = "reasoning")
```

i.e. we apply the function `plot()` and use `intelligence` and *reasoning* (`rsng`) as arguments; we select the color of each data point with `col = as.numeric(latin)` according to the character value of the factor `latin`, wherein we first have to convert them into numeric values by using the function `as.numeric()`. Finally, we label the axes accordingly with `xlab` and `ylab`.

As a result, we get a chart that exactly corresponds to the one in Figure 13.1.

In SPSS, we deactivate the partitioning of the data we needed earlier by using the same sequence of commands (Data – Split File...) as before, and clicking the button Analyze all cases, do not create groups. After clicking OK, the partitioning will be cancelled. Following the steps described in Example 5.2, we open the Chart Builder (see Figure 5.5), where we click the diagram type Scatter/Dot in the gallery tab in order to select a Grouped Scatter (the second chart from the left in the first row; when the cursor hovers over this panel, Grouped Scatter appears); we click on it and drag and drop it into the field Chart preview. Now, we drag and drop the character reasoning into the field Y-Axis?, the character intelligence into the field X-Axis?, and the character Latin courses into the field Set color. After clicking OK, we obtain Figure 13.1 (which can be edited with the context menu Edit Content – In Separate Window).

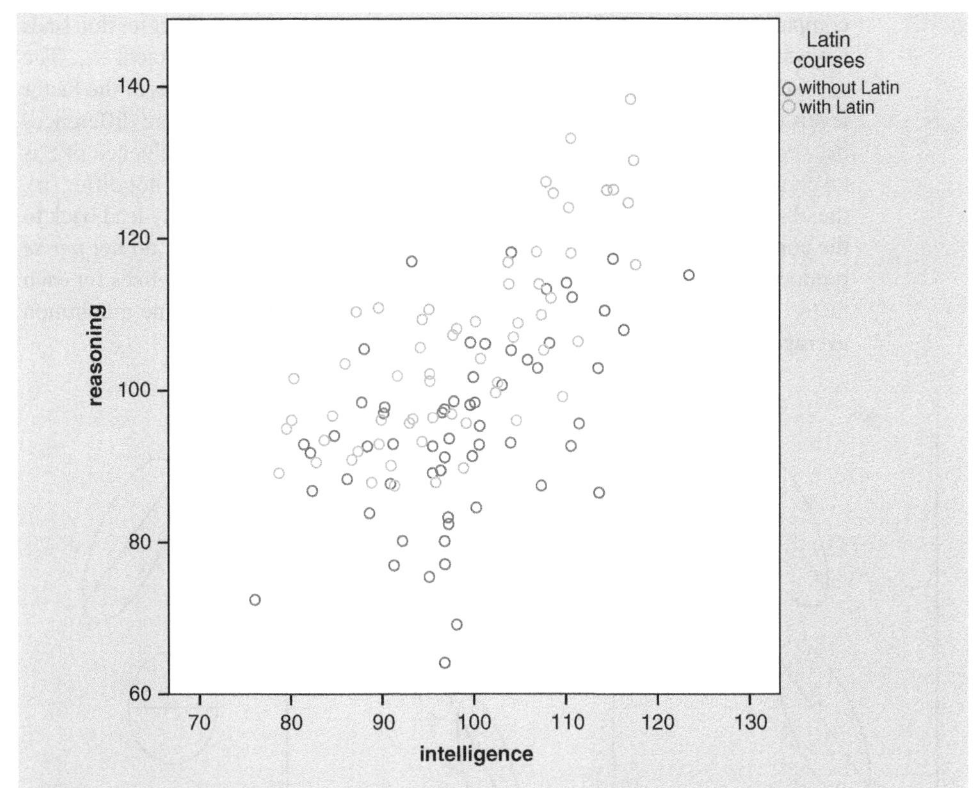

Figure 13.1 SPSS-output showing the scatter plot of the characters *reasoning* and *intelligence* in Example 13.1.

Although the two slopes, in numerical terms, clearly differ from each other, the chart leaves the impression that the regression lines adapted to the two scatter plots run almost parallel. Clearly, however, they run shifted by an additive constant (intercept) on the ordinate. The dots of the students with *Latin courses* are on average higher in the level of *reasoning* than those of students without *Latin courses*.

Master Doctor

Obviously, answering the given research question requires more than descriptive statistics. The null hypothesis is: H_0: $\mu_{i.z} = \mu_{l.z}$ for all $i \neq l$; $i, l = 1, 2, \ldots, a$, and $\mu_{i.z}$ the mean value of $\boldsymbol{y}_i$ with the effects of $\boldsymbol{z}$ eliminated. Basically, this is about a null hypothesis just like that in the one-way analysis of variances. But this time, a regression analysis has to be performed in advance. This is why we do not explicitly give the sum of squares here. Testing this hypothesis is quite simple with the assistance of pertinent computer programs. However, the user has to bear in mind that the presumption of homogeneity of the variance–covariance matrices has to be fulfilled.

The appropriateness of this presumption can be illustrated by looking at the schematic representation in Figure 13.2. Remember the null hypothesis claims that there is no difference between the factor levels of A with respect to the character y – taking into account some dependence of $\boldsymbol{y}$ and A on $\boldsymbol{z}$. It is tested by comparing two slopes. The first concerns the average slopes of all regression lines between $\boldsymbol{y}$ and $\boldsymbol{z}$ for each factor level; that is the regression coefficient $\hat{\beta}_{1i}$. The second concerns the slope resulting from the regression of the means of the factor levels ($\bar{\boldsymbol{y}}_i$ on $\bar{\boldsymbol{z}}_i$). If these two kinds of slopes differ (a), then there are differences between the means of the factor levels of A_i – even though the influence of $\boldsymbol{z}$ is taken into account and eliminated, respectively. If these slopes do not differ (b), the observed differences in the means of the factor levels A_i can only lead back to the common dependence of $\boldsymbol{y}$ and A on $\boldsymbol{z}$; thus, the factor levels A_i do not *per se* produce the differences in the variable $\boldsymbol{y}$. If, however, the regression lines for each factor level are not parallel (c), it is obviously useless to determine a common average slope for them.

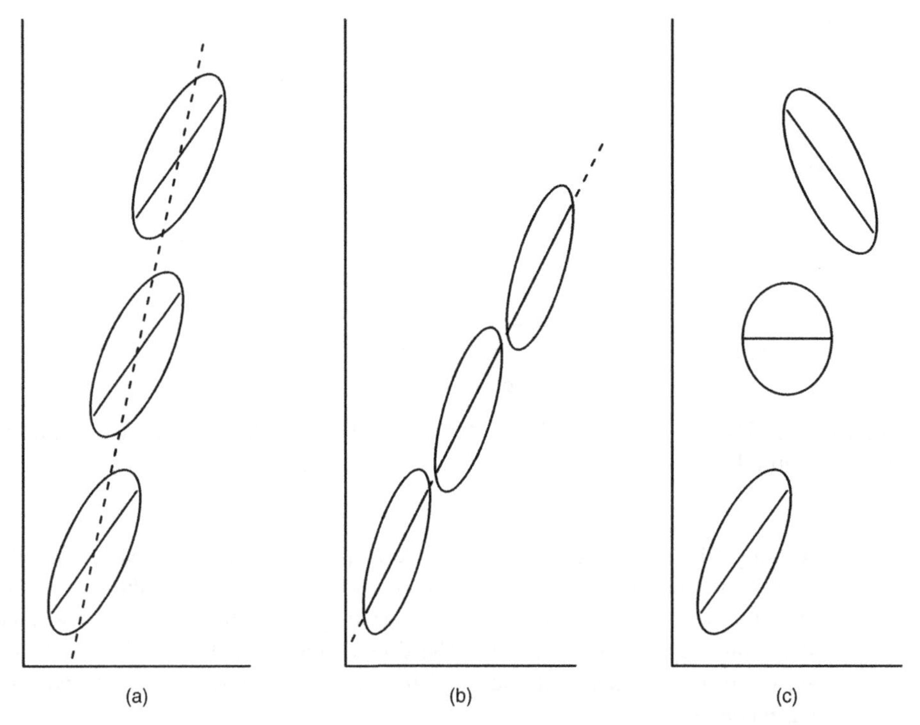

Figure 13.2 Schematic representation of a two-dimensional scatter plot between the interesting character y (ordinate) and the noise factor z (abscissa) for each of the three factor levels A_1 to A_3. It includes the regression line per factor level (solid line) and the regression line for the means of the factor levels (dashed line).

As with the (two-sample) t-test, this again brings along the problem that a pre-test is necessary – apart from the hardly testable presumption of a two-dimensional normal distribution. Despite the methodological inconsequence, combined with the risk of a false decision with the pre-test, such a test is highly recommended here (just think about the risk of an entirely artificial result given a situation like the one in Figure 13.2c). The method of choice is the *Box-M test*. However, this test does not (directly) test the equality (homogeneity) of the slopes of the regression lines, but (in an indirect way via) the homogeneity of the variance–covariance matrices; that is the null hypothesis H_0: $\Sigma_{(i)} = \Sigma$ for every i (H_A: $\Sigma_{(i)} \neq \Sigma_{(l)}$ for any $i \neq l$).

Note, however, that the slopes can be equal even if the variance–covariance matrices differ. We do not go into detail here as concerns the exact test statistic and its distribution; the calculation with pertinent computer programs is easy. The problem is that it does not come at all close to maintaining the type-I error if the presupposition of a two- (or in general: multi-) dimensional normal distribution is violated. A significant result therefore always brings along a higher risk of falsely rejecting the presupposition of homogeneous variance–covariance matrices than stated by the nominal type-I risk α.

Example 13.1 – continued

This is about testing the differences of the means of the character *reasoning* with respect to the factor *Latin courses*. Thereby the effects of the character *intelligence* will be eliminated. The null hypothesis, H_0: $\mu_{i.z} = \mu_{l.z}$ for all $i \neq l$; $i, l = 1, 2, \ldots, a$, shall be tested with a type-I error of $\alpha = 0.05$. Also the null hypothesis H_0: $\Sigma_{(i)} = \Sigma$ for all i is to be tested at this level of significance.

In **R**, we create a function in order to conduct the Box-M test by typing

```
> boxM.test <- function(X, group) {
+   levels <- levels(as.factor(group))
+   p <- length(levels)
+   n <- table(group)
+   m <- ncol(X)
+   N <- sum(n)
+   gamma <- 1 - (2*m^2+3*m-1)/(6*(m+1)*(p-1)) * (sum(1/(n-1)) -
+                 (1/(N-p)))
+   s.k <- lapply(levels, function(i) cov(subset(X,
+                 subset = group == i)))
+   s. <- 1/(N-p) * matrix(colSums(matrix(unlist(lapply(1:p,
+                 function(k) +s.k[[k]]*(n[k]-1))),nrow=p,
+                 byrow=T)),ncol=m)
+   M <- gamma*sum((n-1)*unlist(lapply(s.k,
+                 function(x) log(det(solve(x)%*%s.,
+                 "f")))))
+   correction = ((m-1)*(m+2))/(6*(p-1))*(sum(1/(n-1)^2)
+                 -(1/(N-p)^2))
+   df1 <- (m*(m+1)*(p-1))/2
+   df2 <- trunc((df1+2)/abs(correction-(1-gamma)^2))
+   F <- M/gamma*(1-(1-gamma)-(df1/df2))/df1
+   probf <- pf(F, df1, df2, lower=FALSE)
+   cat("     df1      df2        Box M            F        ProbF\n",
+       formatC(df1, width = 7), formatC(df2, width = 7),
+       formatC(M/gamma, format = "f", width=12, digits=5,
+               flag = " "),
+       formatC(F, format = "f", width=11, digits=5, flag = " "),
+       formatC(probf, format = "f", width = 11, digits=5,
+               flag=" "), "\n")
+    res <- list(Box.M = M/gamma, F = F, df1 = df1, df2 = df2,
+               probf = probf)
+    return(invisible(res))
+ }
```

i.e. we use the function `function()` and set the matrix that is to be analyzed as the first argument with `X` (which contains the observed values column by column), and add as a second argument the factor of interest, with `group`. The sequence of commands inside the curly brackets determines the application flow of the function and will not be discussed. We assign this function to the object `boxM.test`. Now, we type

```
> boxM.test(cbind(rsng, intelligence), group = latin)
```

i.e. we use the function `cbind()` in order to define a matrix combining the character *reasoning* (`rsng`) and the character `intelligence` and submit it as the first argument to the function `boxM.test()`; with `group = latin`, we demand that the matrix is to be analyzed with regards to the character *Latin course*.

As a result, we get:

```
df1      df2          Box M           F        ProbF
  3 2.506e+06       6.08152     1.98995      0.11309
```

For the implementation of the analysis of covariance, we use the package `car`, which we already installed in Example 10.18; we now type

```
> lm.cov <- lm(rsng ~ intelligence + latin)
> Anova(lm.cov, type = 3)
```

i.e. we again use the function `lm()`, submitting to it the request that the character *reasoning* (`rsng`) is to be analyzed with regards to the character `intelligence` and the factor *Latin courses* (`latin`); we assign the results of the analysis to the object `lm.cov`. In the next step, we apply the function `Anova()` to the object `lm.cov`, selecting the type III method of determining the squared sums of variances with `type = 3`.

As a result, we get (shortened output):

```
Anova Table (Type III tests)

Response: rsng
              Sum Sq  Df F value    Pr(>F)
(Intercept)    375.0   1  4.0095   0.04756
intelligence  7551.9   1 80.7383 5.295e-15
latin         3472.3   1 37.1234 1.458e-08
Residuals    10943.6 117
```

In SPSS, we basically use the same sequence of commands (Analyze – General Linear Model – Univariate...) as in Example 10.8 in order to open the window Univariate in Figure 10.7. Although it is, at this point, possible to set a check mark at Homogeneity tests in the respective window via Options..., this is not the order for conducting the Box-M test. Instead, we have to select

Analyze
    General Linear Model
        Multivariate...

in order to open the window in Figure 13.3. There, we drag and drop both characters: the one of interest, *reasoning*, as well as the noise factor, *intelligence*, into the field Dependent Variables:; we drag and drop the factor *Latin courses* into the field Fixed Factor(s):. Now, we choose Options... (without figure here) and set a check mark at Homogeneity tests. After choosing Continue, we return to the previous window, and after clicking OK, obtain various results, of which we are only interested in the ones in Table 13.3.

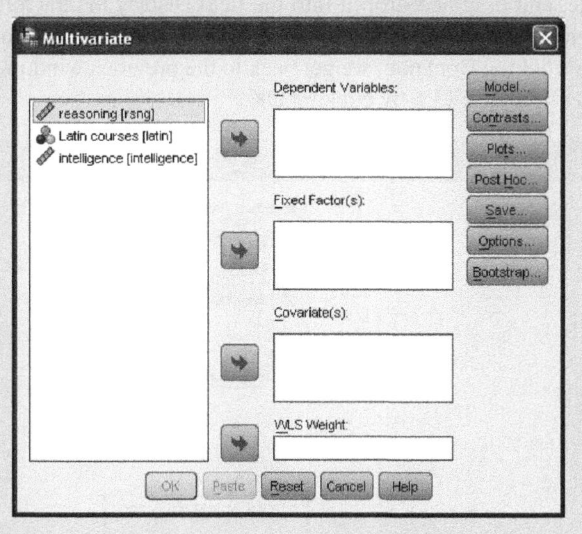

Figure 13.3 SPSS-window for conducting the Box-M test.

Table 13.3 SPSS-output of the Box-M test for checking the homogeneity of the variance–covariance matrix in Example 13.1

Box's Test of Equality of Covariance Matrices[a]

Box's M	6.082
F	1.990
df1	3
df2	2506320.000
Sig.	.113

Tests the null hypothesis that the observed covariance matrices of the dependent variables are equal across groups.

a. Design: Intercept + latin

Because the Box-M test does not yield a significant result, we can continue the calculations – if this was not so, our research question would not be testable with the conventional methods. Analogously to Example 10.8, we proceed to open the window Univariate in Figure 10.7 and conduct the analysis of covariance. In Figure 13.4, we have already dragged and dropped the character reasoning into the field Dependent Variable:, the character Latin courses into the field Fixed Factor(s):, and the character intelligence into the field Covariate(s):. In case of a significant result, we also want to obtain the means for every factor level as adjusted by the influence of the covariate according to the regression analysis; in order to do this, we click the button Options... in the window in Figure 13.4, mark latin in the panel Factor(s) and Factor interactions:, and drag and drop it into the field Display Means for:; if we also want the estimated magnitude of effect η^2, we have to set a check mark at Estimates of effect size in this window. By clicking Continue, we get back to the previous window, where we obtain the results in Tables 13.4 and 13.5 by clicking OK.

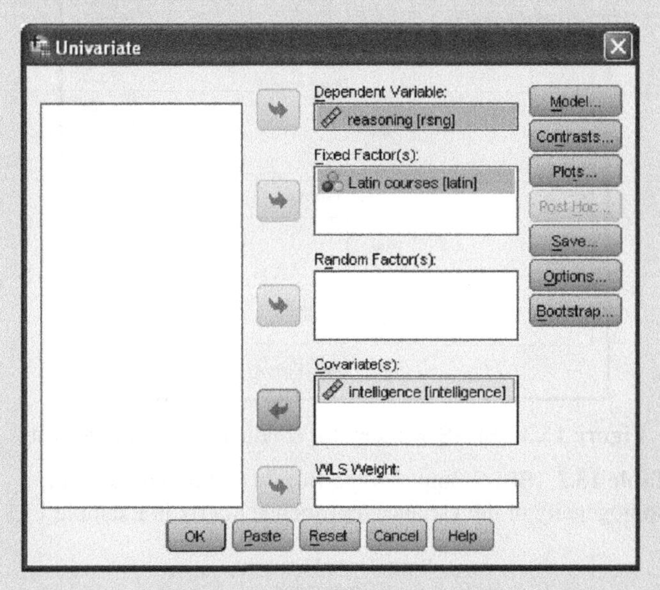

Figure 13.4 SPSS-window for conducting the analysis of covariance.

Table 13.4 SPSS-output of the table of variances in Example 13.1

Tests of Between-Subjects Effects

Dependent Variable:reasoning

Source	Type III Sum of Squares	df	Mean Square	F	Sig.	Partial Eta Squared
Corrected Model	10945.060[a]	2	5472.530	58.508	.000	.500
Intercept	649.738	1	649.738	6.946	.010	.056
intelligence	7551.858	1	7551.858	80.738	.000	.408
latin	3472.341	1	3472.341	37.123	.000	.241
Error	10943.599	117	93.535			
Total	1227076.319	120				
Corrected Total	21888.659	119				

a. R Squared = .500 (Adjusted R Squared = .491)

Table 13.5 SPSS-output of the means in Example 13.1

Latin courses

Dependent Variable:reasoning

Latin courses	Mean	Std. Error	95% Confidence Interval	
			Lower Bound	Upper Bound
without Latin	94.837[a]	1.249	92.364	97.309
with Latin	105.595[a]	1.249	103.123	108.068

a. Covariates appearing in the model are evaluated at the following values: intelligence = 98.74.

As we can see, both the effect of *intelligence* on the character *reasoning* and the (adjusted for the influence of intelligence) effect of *Latin courses* on *reasoning* are significant. More precisely, we can see that *reasoning* is significantly higher in the group with *Latin courses*.

Master Doctor

Of course, the analysis of covariance and the model of the Formula (13.4) can be generalized. Even nonlinear regression approaches are possible but are rarely used in psychology. In contrast, expanding to more than one factor and more than one noise factor is of higher interest. So too are nested classifications of factors possible, as well as taking into account covariates in multivariate analysis of variances, as dealt with in the following section. In those analyses, more than one character y is of simultaneous interest. Though the practical calculation using computer programs is completely the same, we must warn not to use too many factors, characters, and/or covariates: apart from the need of extremely large sample sizes in such cases, the presumption of a multidimensional normal distribution is not only not testable, but simply unrealistic. Likewise unrealistic is the expectation that the variance–covariance matrices are homogeneous.

Doctor

As concerns the aspect of variance analysis and the aspect of regression analysis in Formula (13.4), there are different combinations of model I and model II; that is, different combinations of fixed and random factors and characters, respectively:

- The case of a fixed factor A, and where the measurement values of the character z are fixed as well, is very rare in psychology. Besides, it is better suited to Part IV, with only a single character modeled as a random variable.
- The case of a fixed factor A, and where the character z can be modeled as random variable $\boldsymbol{z}$, we have already dealt with in detail above.
- The case of a random factor $\boldsymbol{A}$, and where the measurement values of the character z are fixed, is also very rare in psychology.
- The case of a random factor $\boldsymbol{A}$, and where the character z can be modeled as random variable $\boldsymbol{z}$, we did not in fact explicate by a formula; to proceed accordingly is, however, easy when using a computer program.

If a second or third factor is of interest, there are also cross and nested classifications, and there are also models I, II, and mixed models, analogous to Chapter 10. For further information see Rasch, Herrendörfer, Bock, Victor, & Guiard (2008). Using computer programs, the application again is easy. One just has to bear in mind that in the case of a multivariate analysis of variance where at least one covariate is considered, the usage of random factors is not possible in SPSS.

13.3 Multivariate analysis of variance

M*ultivariate*[2] *analysis of variance* is the generalization of one-way and multi-way analysis of variance on at least two characters of interest, y_1 and y_2 – in general terms $y_1, y_2, \ldots, y_m$. We will model these characters analogously as (multidimensional) normally distributed random variables $\boldsymbol{y}_1, \boldsymbol{y}_2, \ldots, \boldsymbol{y}_m$. Analogous to Chapter 10, the effect of one (fixed or random) factor or more than one (fixed or random) factors can be analyzed simultaneously. When at least two factors are given, both models can be used; that is models with fixed factors and models with random factors, and also mixed models. The distinction between cross classification and nested classification applies, too. However, similarly to the case of just a single character using SPSS, one can (guided by the menu) only analyze factors in cross classification, but not in nested classifications. With respect to Section 13.2, the multivariate analysis of variance can be generalized to an analysis of covariance, but we will not discuss that in detail.

Master Doctor

Example 13.2 Do the test scores on the subtests of the intelligence test battery (at test date 1) depend on the *social status* (Example 1.1)?

In contrast to similar questions (e.g. for instance Example 10.7), we now want to analyze all of the five subtests simultaneously. That is, we do not just analyze the effect of the factor *social status* on the test scores of a single subtest and on a single variable, respectively.

Master Doctor

As soon as more than one character is of interest, it is not advisable to simply perform the test numerous times for each one singly. Too much information could remain unused as concerns the relationships between the characters in question. In Section 10.3, which was about the multiple comparison of means, we already saw that making numerous comparisons of data within the same research question is problematic, at least with respect to the overall type-I risk. Here we have an analogous problem. A further problem is that if statistical tests are used for different characters, which are all ascertained from the same research units, the tests could reveal significant results, which are then, as a pool, misinterpreted as concerns the contents. That is because, if the characters are related closely enough, then some 'duplications' of the significances occur which entail many of the same consequences/conclusions without it being directly obvious from the content point of view: in fact, the second, third, and so on significant result might not be a new result with respect to the first one.

[2] Often the word 'variate' is used within statistics instead of 'random variable'.

Master Doctor

Example 13.2 – continued

If we apply the one-way analysis of variance as we did in Chapter 10 for all the five subtests one after the other, five independently established significant differences in the means might occur between the children of certain (or all) measurement values of the *social status* – for example, a significant and relevant difference regarding a disadvantage of the 'lower classes'. A possible educational policy consequence could then be that children of those 'lower classes' are to be trained early in all abilities measured by the subtests *Everyday Knowledge, Applied Computing, Social and Material Sequencing, Immediate Reproducing – numerical*, and *Coding and Associating*. Such an enhancement policy does not only mean an enormous national economic effort, but above all also a high energetic-motivational burden for the children themselves and most likely for their social environment, too.

Actually we can imagine that, in particular as concerns children from the 'lower classes', the test scores of the five subtests in question correlate very highly (see the 'general intelligence theory' in which essentially just a single intelligence dimension, the 'general intelligence', is responsible for intelligent performances). If that were the case, the indicated enhancement program, with respect to all those subtests (abilities) resulting in significance, means an unnecessarily high effort: although some general enhancement would suffice, a fivefold enhancement of one and the same ability would apply.

Master Doctor

Hence, when testing for differences in means, the relationships between the characters in question have to be taken into account in order to avoid mutually duplicated results.

On the other hand, differences in the means with respect to each character could be too small to reveal a significant effect. Hence, repeatedly applied statistical tests for different characters, all ascertained from the same research units, would not reveal a single significant result. Whether the different characters correlate with each other or not, the cumulative difference in means, though marginal if looked at individually, could in sum reach a relevant amount; expressed metaphorically, 'Constant dripping wears away stone.' Thus, by carrying out an analysis repeatedly instead of jointly, errors of interpretation from the content point of view could occur in the other direction, too: relevant effects could remain undiscovered.

Master Doctor

Example 13.2 – continued

If the given abilities are in fact largely independent from each other, and the means do not differ with respect to the categories of *social status* in a relevant and significant way, then one may argue for each subtest as follows: according to experience, such a minimal difference does not bring along problems in the (subsequent) school career. If, however, those minimal differences accumulate, massive problems regarding school and development are to be faced.

Master Doctor

To illustrate this, we look at the illustration in Figure 13.5. This shows the bivariate normally distributed random variables $\boldsymbol{y}_1$ and $\boldsymbol{y}_2$. With respect to the factor of interest A with $a = 3$ factor levels or with respect to $a = 3$ samples of the corresponding populations, the result is a bivariate normal distribution as shown

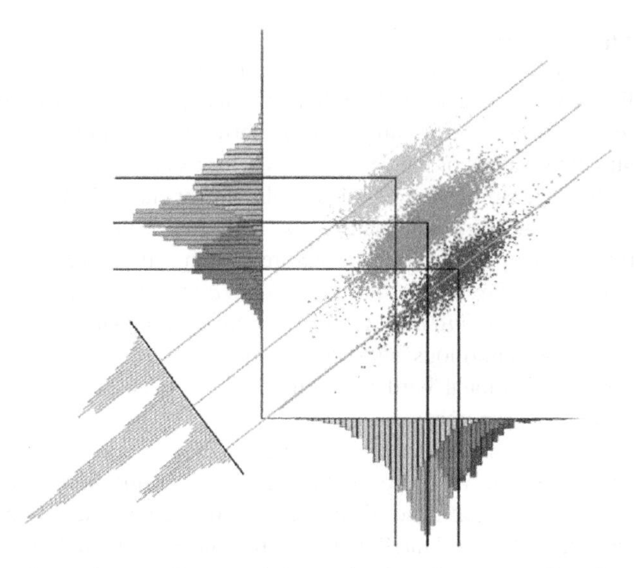

Figure 13.5 Horizontal as well as upright projection from two directions of a bivariate normal distribution including the projection on an optimally discriminating plane.

in Figure 11.3. Illustrated in a horizontal projection, the result is the corresponding density appearing like contour lines on a graphical map: the points along the line (here in an elliptical shape) have the same density. This density is greater/higher the closer the points (that is, the pairs of outcomes (y_{1v}, y_{2v})) lie to the pair of means $(\bar{y}_1, \bar{y}_2)$ or, that is to say, to the bivariate mean point. In the horizontal projection of Figure 13.5, the density of any bivariate normal distribution is, however, only to be identified indirectly, through the concentration of individual points (around the mean point). This is because the vertical projection lines are missing. In this chart both upright projections are included; that is the density distribution for y_1 and y_2 – for a multidimensional normally distributed random variable the density for each variable is a univariate normal distribution. Analogous to the (one-way) analysis of variance and analogous to the analysis of covariance, for testing a hypothesis the following assumptions need to be met. If at all, only the means of the samples (populations) differ, but not their variances and covariances: $N(\mu_{1i}, \mu_{2i}; \sigma^2_{y_{1i}}, \sigma^2_{y_{2i}}; \rho_{y_{1i}y_{2i}}) = N(\mu_{1i}, \mu_{2i}; \sigma^2_{y_1}, \sigma^2_{y_2}; \rho_{y_1y_2})$ for every i. Just for purposes of vivid illustration, Figure 13.5 gives a chart with slightly different variances. Finally, Figure 13.5 shows also a sheer plan.

Figure 13.5 shows that the univariate analysis using a one-way analysis of variance with respect to the character y_1 would hardly reveal a significant difference, and with respect to the character y_2 significance could only be reached when an appropriately high sample size is given. However, by applying the analysis of variance on the data that are projected according to the line (in space) which goes at a right angle to the three regression lines, then clear differences between the means of the three populations (samples) are given.

Establishing such a line statistically in the space is the goal of *(linear) discriminant analysis*. More clearly expressed, discriminant analysis tries to find the line in space (or the projected line), g, within the multidimensional space, that discriminates best between the populations of interest. This analysis will be discussed in detail in the next section.

The multivariate analysis of variance on the other hand only investigates if there are any differences in the localization of the investigated population.

In the case of a one-way multivariate analysis of variance, we use a vector of random variables $\vec{\boldsymbol{y}}$ with the vector of means $\vec{\mu}$. For an easier notation we use the transposed vector $\vec{\boldsymbol{y}}^T = (\boldsymbol{y}_1, \boldsymbol{y}_2, \ldots, \boldsymbol{y}_m)$ and $\vec{\mu}^T = (\mu_1, \mu_2, \ldots, \mu_m)$. The variance–covariance matrix is a generalization of Formula (13.5) as follows:

$$\Sigma = \begin{pmatrix} \sigma^2_{y_1} & \sigma_{y_1 y_2} & \cdots & \sigma_{y_1 y_m} \\ \sigma_{y_2 y_1} & \sigma^2_{y_2} & \cdots & \sigma_{y_2 y_m} \\ \vdots & \vdots & \cdots & \vdots \\ \sigma_{y_m y_1} & \sigma_{y_m y_2} & \cdots & \sigma^2_{y_m} \end{pmatrix} \tag{13.6}$$

The structure of the data is analogous to that in Table 13.1. The only difference is that, instead of two characters, m characters are observed.

As already indicated, for hypothesis testing not only the non-testable presumption of a multidimensional or, in other words, multivariate normal distribution has to be met, but also the homogeneity of the variance–covariance matrix. That is, the matrices $\Sigma_{(i)}$ for each factor level i have to be the same: $\Sigma_{(i)} = \Sigma$. As concerns the vectors of means $\vec{\mu}_{(i)}$ of each factor level, the null hypothesis thus is H_0: $\vec{\mu}_{(i)} = \vec{\mu}_{(l)} = \vec{\mu}$, for every i and l, $i \neq l$. The alternative hypothesis is H_A: $\vec{\mu}_{(i)} \neq \vec{\mu}_{(l)}$ for at least one $i \neq l$. Therefore, the means of at least one character differ with respect to at least two factor levels (samples).

We give, just exceptionally (because calculation by hand is practically impossible), the formulas of the test statistics. Fundamentally, one can imagine that, analogous to the univariate analysis of variance, this is about the comparison of certain 'deviations' (or their squares). Again, particularly the deviations of the sample means and the deviations within each sample are of interest. Computer programs then calculate the respective test statistics, which are approximately distributed in a known way. This is true, for example, for *Wilks' Λ* and for Hotelling's T^2; the latter test statistic has already been mentioned in other Chapters.

Doctor

First of all, the difference vector $\vec{\boldsymbol{d}}$ has to be calculated for $\vec{\boldsymbol{y}}$ and $\vec{\mu}$, and $\vec{\boldsymbol{d}}_{(i)}$ for $\vec{\boldsymbol{y}}$ and $\vec{\mu}_{(i)}$, respectively. Squaring the elements of those vectors, the elements of them equal the squared differences as needed for the variances (sum of squares). However, the covariances are of interest as well. This is why $\vec{\boldsymbol{d}}^T \cdot \vec{\boldsymbol{d}}$ and $\vec{\boldsymbol{d}}^T_{(i)} \cdot \vec{\boldsymbol{d}}_{(i)}$, respectively, were calculated, as a consequence of which there are two types of matrices. As both these matrices are defined for the vector of random variables, both exist for every research unit v. We of course can sum up all these research unit matrices – with every type taken separately. As a result there are two matrices $\boldsymbol{T}$ and $\boldsymbol{R}$; the former for 'total', the latter for 'residual'. They are the matrices of the sums of squares; apart from the missing division by their respective degrees of freedom, these matrices are variance–covariance matrices. Analogous to the univariate case, it can be shown that $\boldsymbol{T} = \boldsymbol{A} + \boldsymbol{R}$, with $\boldsymbol{A}$ being the corresponding matrix of the squared deviations (sums of squares) of the vector of sample means and the vector $\vec{\mu}$. By determining the determinants for $\boldsymbol{T}$ and $\boldsymbol{R}$, $|\boldsymbol{T}|$ and $|\boldsymbol{R}|$ (that is, scalars), some test statistics can be derived. Most notable is Wilks' test statistic Λ; without specifying the degrees of freedom here, it is an approximately F- and χ^2-distributed test statistic, respectively:

$$\Lambda = \frac{|\boldsymbol{R}|}{|\boldsymbol{T}|} \tag{13.7}$$

As can be shown, Λ is in a functional relationship with η^2, that is the (estimated) effect size of the analysis of variance: $\Lambda = 1 - \hat{\eta}^2$. Λ thus describes the percentage of variance (and covariance) of all outcomes that is not caused by the variance (and covariance) of the outcomes as concerns the different samples.

Hotelling's test statistic $\boldsymbol{T}^2$ also is, in the general case, a possibility. We already suggested this test statistic for the special case of only a single character (and more than two samples) in Section 10.4.1.1, and for the special case of at least two characters (observed in only a single sample). Most of the time both test statistics reveal almost exactly the same p-values. They are also relatively robust against deviations from the multivariate normal distribution of characters. A practical presupposition for both test statistics is that, for each sample, $n_i > m + 1$ is valid.

Hotelling's test statistic $\boldsymbol{T}^2$ is attractive for the case of just two samples, because of the fact that planning a study can be done with it. For this, a vector $\vec{\delta}$ has to be defined with the minimum difference for each character that is of relevance; that is, $\vec{\delta}^T = (\delta_1, \delta_2, \ldots, \delta_m)$.

Planning a study according to a multivariate analysis of variance happens either with regard to a certain character that is in some way 'most important', or the researcher calculates the necessary sample size for each character on its own – given certain precision requirements – and then chooses the largest one. However, neither type-I nor type-II risk will hold with regard to the research as a whole (i.e. study-wise risk).

Doctor

Though strongly recommended, but not derived in detail here using formulas, a relatively new test for the case of just two samples is the *principal component test*, going back to Läuter (1996) and Läuter, Glimm, & Kropf (1998). It is based on a certain transformation of the vector $\vec{y}$ for each sample; that is $\vec{y}_{(i)}$ into scalars. Under the null hypothesis, these scalars are distributed in such a way that they can be handled like normally distributed values. For these just the t- and the Welch test, respectively, are applied. This approach entails the advantage that – analogous to the univariate case concerning the Welch test – non-homogeneous variance–covariance matrices can also be covered. This in turn means that the (for many reasons) problematic testing of homogeneity of the variance–covariance matrix can be omitted. Apart from that, this test has the advantage, over the test statistics of Wilks and Hotelling, that one-sided testing is also possible. Furthermore, in the case of a relatively small sample size with respect to the number of characters m, the statistical power is higher. Also, it can still be used even if the number of characters m is larger than the sample size n.

Doctor

Example 13.3 Numeric example for the principal component test without relation to any content

The following data were given:

Sample 1	y_{11v}	7	1	2	9	2	4	4	3	4
	y_{21v}	3	5	5	7	8	2	6	7	6
Sample 2	y_{12v}	7	6	5	4	3	4	5	5	—
	y_{22v}	7	7	5	6	5	5	6	7	—

The null hypothesis, which is to be tested, is

$$H_0 : \begin{pmatrix} \mu_{11} \\ \mu_{12} \end{pmatrix} = \begin{pmatrix} \mu_{21} \\ \mu_{22} \end{pmatrix}$$

That is, the means of the two characters, y_1 and y_2, are equal in both populations. The alternative hypothesis is two-tailed,

$$H_A : \begin{pmatrix} \mu_{11} \\ \mu_{12} \end{pmatrix} \neq \begin{pmatrix} \mu_{21} \\ \mu_{22} \end{pmatrix}$$

The type-I risk is chosen to be $\alpha = 0.05$.

In **R**, we create a new function by typing

```
> pc.t.test <- function(Y, group) {
+   Y <- t(scale(Y, center = TRUE, scale = TRUE))
+   d <- svd(Y)$u[, 1]
+   z <- crossprod(Y, d)
+   result <- t.test(z[group == 1], z[group == 0], var.equal = TRUE)
+   return(list(statistic = result$statistic,
+               p.value = result$p.value))
+ }
```

i.e. we apply the function `function()`, using the matrix `Y` (see below) and a vector determining which observed values belong to which sample (`group`) as arguments. The sequence of commands in the curly brackets determines the application flow of the function and will not be discussed. Now, we create the matrix from the above table, thus typing

```
> y11 <- c(7, 1, 2, 9, 2, 4, 4, 3, 4)
> y21 <- c(3, 5, 5, 7, 8, 2, 6, 7, 6)
> y12 <- c(7, 6, 5, 4, 3, 4, 5, 5)
> y22 <- c(7, 7, 5, 6, 5, 5, 6, 7)
> pc.dat <- cbind(c(y11, y12), c(y21, y22))
> group <- c(rep(0, length = length(y11)),
+            rep(1, length = length(y12)))
```

i.e. we use the function `c()` to define the separate observed values as vectors and assign them to one object each; using the function `cbind()`, we create a matrix in which the lines stand for the persons and the columns for the different characters. We assign the resulting matrix to the object pc.dat. Finally, we use the function rep() in order to create a vector with the values 0 and accordingly 1, including the respective number (length). We assign all this to the object group. Now, we conduct the analysis by typing

```
> pc.t.test(pc.dat, group = group)
```

i.e. we apply the function pc.t.test() and set the matrix pc.dat as the first and group as the second argument.

As a result, we get:

```
$statistic
       t
-1.118597

$p.value
[1] 0.2809066
```

Because of the p-value of 0.2809 the result is not significant; hence the null hypothesis is to be accepted.

Master Doctor

When using multivariate analysis of variance and the test statistics of both Wilks and Hotelling, the homogeneity of the variance–covariance matrices is an explicit presupposition that has to be met. Again, the Box-M test comes into question. Indeed, the generalized Levene test (see Section 9.2.2) offered by SPSS is much more robust against any violation of the multivariate normal distribution, but does not test the essential assumption of homogeneity of the covariances.

Master Doctor

Example 13.2 – continued This example was about whether the test scores (at the first test date) in the subtests of the intelligence test battery in Example 1.1 depend on the *social status* of those tested. We attempt to answer this question in the first instance just for the children with German as a native language, because we suppose the character *native language of the child* to be a noise factor. The null hypothesis H_0: $\vec{\mu}_{(i)} = \vec{\mu}_{(l)} = \vec{\mu}$, for all categories of *social status* i and l, $i \neq l$ (H_A: $\vec{\mu}_{(i)} \neq \vec{\mu}_{(l)}$) is to be tested with a, say, type-I risk of $\alpha = 0.05$. We test the null hypothesis H_0: $\Sigma_{(i)} = \Sigma$ for every i at this level of significance, too.

In **R**, we use the database Example_1.1.g (see Example 8.3), which contains exclusively children with German as a native language. After having enabled access to the database by applying the function attach() (see Chapter 1), we type

```
> y <- cbind(sub1_t1, sub3_t1, sub4_t1, sub5_t1, sub7_t1)
> boxM.test(y, group = social_status)
```

i.e. we apply the function cbind() to all subtests of the intelligence test battery (from the first test date) and assign the resulting matrix to the object y, which we use as an argument in the function boxM.test(); with group = social_status, we determine the factor *social status*.

As a result, we get the error prompt:

```
Error in solve.default(x) :
  system is computationally singular
```

Thus, we have to take a look at the frequency distribution of the factor *social status*; in order to do this, we type

```
> table(social_status)
```

i.e. we set the character `social_status` as an argument in the function `table()`.

As a result, we get:

```
social_status
          upper classes                middle classes
                     10                            13
      lower middle class            upper lower class
                     10                             7
          lower classes single mother in household
                      2                             8
```

Now, it becomes obvious that only two children stem from the 'lower classes'; such a sparsely occupied factor level is apparently not analyzable. Thus, we exclude this category of *social status* from further analysis. In order to do this, we type

```
> soc.ex <- factor(as.character(social_status),
+                  exclude = "lower classes")
```

i.e. we use the function `factor()`, setting the character `social_status` as the first argument, wherein we transform this character into a character string by means of the function `as.character()`, thus explicitly defining the character values as non-quantitative; with the second argument `exclude`, we exclude the category 'lower classes'. We assign the new character to the object `soc.ex`.

Now, we once more conduct the Box-M test by typing

```
> boxM.test(y, group = soc.ex)
```

i.e. we use the matrix `y` as the first argument in the function `boxM.test()`, and with the second argument `group = soc.ex`, we specify that the former character be analyzed with regards to `soc.ex`.

The results are:

```
   df1    df2       Box M          F      ProbF
    60   2660    79.05417    0.94129    0.60465
```

In SPSS, following the steps described in Example 5.4 (Data – Select Cases...), we select the children with German as a native language by choosing the option If condition is satisfied in the window in Figure 5.14 and clicking the button If... in order to enter native_language = 1 in the resulting window (see Figure 5.15). We confirm this with Continue and OK. Using the sequence of commands (Analyze – General Linear Model – Multivariate...) analogous to Example 13.1, we open the window in Figure 13.3. There we drag and drop the characters Everyday Knowledge, 1st test date; Applied Computing, 1st test date; Social and Material Sequencing, 1st test date; Immediately Reproducing – numerical, 1st test date; and Coding and Associating, 1st test date into the field Dependent Variables:. We drag and drop the character social_status into the field Fixed Factor(s): and, again analogous to Example 13.1, click Options... and set a check mark at Homogeneity tests. By clicking Continue and OK, we obtain the results. The Box-M test yields a p-value of 0.605.

Because of the non-significant result of the Box-M test (the p-value of 0.605 is higher than $\alpha = 0.05$), application of multivariate analysis of variance is warranted.

In **R**, we type

```
> manova.1 <- manova(y ~ social_status)
> summary(manova.1, test = "Wilks")
> summary(manova.1, test = "Hotelling-Lawley")
```

i.e. we use the function `manova()`, submitting to it the request that the matrix `y` is to be analyzed with regards to the factor `social_status`. We assign the results of this analysis to the object `manova.1`. Finally, we submit the object `manova.1` to the function `summary()`; with `test = "Wilks"` and `test = "Hotelling-Lawley"`, we select Wilks' Λ and, respectively, Hotelling's T^2.

As a result, we get (shortened output):

```
              Df  Wilks approx F num Df den Df    Pr(>F)
social_status  5 0.3391   2.0288     25 150.10 0.005017
Residuals     44

              Df Hotelling-Lawley approx F num Df den Df    Pr(>F)
social_status  5           1.4283   2.1938     25    192 0.001583
Residuals     44
```

In SPSS, we have already conducted the respective calculation before anyway; if we also wanted to determine the estimated magnitude of effect η^2, we would, at the end, retroactively (again) have to click Options... in the window in Figure 13.3 in order to set a check mark at Estimates of effect size. We take Wilks' Λ as well as Hotelling's T^2 from Table 13.6.

Table 13.6 SPSS-output of the multivariate analysis of variance in Example 13.3 (shortened output)

Multivariate Tests[c]

Effect		Value	F	Hypothesis df	Error df	Sig.	Partial Eta Squared
social_status	Wilks' Lambda	.339	2.029	25.000	150.095	.005	.194
	Hotelling's Trace	1.428	2.194	25.000	192.000	.002	.222

c. Design: Intercept + social_status

The null hypothesis has to be rejected; the *social status* significantly influences the test scores in the given intelligence test battery.

Once the Box-M test is significant, the researcher only has the option of identifying the responsible characters (and/or samples) by means of descriptive statistics. These one(s) are to be excluded from further analysis. Calculating the correlation matrix (see Section 12.1.4) for each sample is highly recommended. In this calculation, one looks for pairs of characters that clearly differ in their Pearson's correlation coefficients in at least two samples. At least one of the two characters then has to be excluded. Bear in mind that in the case of a repeated use of the Box-M test with an ultimately non-significant result, no statement about the type-II risk can be made at all.

For the exact determination of between which levels of a (fixed) factor there are differences in means, one can use the same *post hoc* tests as in the case of a significant result for a multivariate analysis of variance, as mentioned in Section 10.4.1.2. However, bear in mind that the type-II risk can only be determined as a comparison-wise one, but not as a study-wise one. Furthermore, in the multivariate case even this can again only be done by multiple processing; that is for each character. Thus, strictly speaking, the type-I risk is incalculable. At any rate, an interpretation must not be made with respect to the nominal type-I risk as applied to each pair of samples and each character as well.

Doctor

One could also use Hotelling's test statistic T^2 as described in Section 10.4.1.1 as a multivariate *post hoc* test, now according to its original intention; that is with respect to more than a single character. To do so, the researcher has to compare all pairs of factor levels. Although he/she would simultaneously compare all characters, one would have, unlike with the Newman–Keuls procedure, a comparison-wise type-I risk – for this the study-wise type-I risk becomes quite large for more than four factor levels.

Example 13.2 – continued

We also want to establish, for each character, to exactly what extent the means differ with respect to the factor levels of the factor *social status*.

In **R**, we have already enabled access to the database *Example_1.1.g* by using the function `attach()`; we load the package `agricolae` using the function `library()`, which we already installed in Example 10.7. We type

```
> aov.u1 <- aov(sub1_t1 ~ social_status)
> aov.u3 <- aov(sub3_t1 ~ social_status)
> aov.u4 <- aov(sub4_t1 ~ social_status)
> aov.u5 <- aov(sub5_t1 ~ social_status)
> aov.u7 <- aov(sub7_t1 ~ social_status)

> SNK.test(aov.u1, trt = "social_status")
> SNK.test(aov.u3, trt = "social_status")
> SNK.test(aov.u4, trt = "social_status")
> SNK.test(aov.u5, trt = "social_status")
> SNK.test(aov.u7, trt = "social_status")
```

i.e. we conduct an analysis of variance for each character (all subtests from the first test date) by using the function `aov()`, assigning the results to one object each, which we in turn use as arguments in the function `SNK.test()`; with `trt = "social_status"`, we select the factor of interest.

As a result, we obtain groupings for the two traits *Immediately Reproducing – numerical, 1st test date* and *Coding and Associating, 1st test date* (shortened output):

```
Study:

Student Newman Keuls Test
for sub5_t1

Different value for each comparison
Means with the same letter are not significantly different.

Groups, Treatments and means
a         upper classes                             58
ab        middle classes                            52.07692
ab        single mother in household                49.75
ab        lower middle class                        46.9
 b        lower classes                             43.5
 b        upper lower class                         42

Study:

Student Newman Keuls Test
for sub7_t1

Different value for each comparison
Means with the same letter are not significantly different.
Groups, Treatments and means
a         upper classes                             63.6
ab        middle classes                            57.30769
ab        lower middle class                        52.4
ab        single mother in household                51.25
 b        upper lower class                         49.14286
  c       lower classes                             34.5
```

Since groupings are only dispensed for two of the characters, this means that the other characters show no significant differences in their means.

In SPSS, we click the button Post Hoc... in the window in Figure 13.3, drag and drop social_status into the field Post Hoc Tests for: in the resulting window (not shown here), and mark S-N-K. By clicking Continue and OK, we get to the results; however, groupings are only displayed for the characters Immediately Reproducing – numerical, 1st test date and Coding and Associating, 1st test date (see Table 13.7 and Table 13.8).

Table 13.7 SPSS-output of the Newman–Keuls procedure in Example 13.2 (shortened output)

Immediately Reproducing - numerical, 1st test date (T-Scores)

Student-Newman-Keuls[a,b]

social status (after Kleining & Moore according to occupation of father/alternatively of the single mother)	N	Subset	
		1	2
upper lower class	7	42.00	
lower classes	2	43.50	
lower middle class	10	46.90	46.90
single mother in household	8	49.75	49.75
middle classes	13	52.08	52.08
upper classes	10		58.00
Sig.		.240	.112

Means for groups in homogeneous subsets are displayed.
Based on observed means.
The error term is Mean Square(Error) = 66.405.

a. Uses Harmonic Mean Sample Size = 5.743.
b. Alpha = .05.

Table 13.8 SPSS-output of the Newman–Keuls procedure in Example 13.2 (shortened output)

Coding and Associating, 1st test date (T-Scores)

Student-Newman-Keuls[a,b]

social status (after Kleining & Moore according to occupation of father/alternatively of the single mother)	N	Subset		
		1	2	3
lower classes	2	34.50		
upper lower class	7		49.14	
single mother in household	8		51.25	51.25
lower middle class	10		52.40	52.40
middle classes	13		57.31	57.31
upper classes	10			63.60
Sig.		1.000	.381	.084

Means for groups in homogeneous subsets are displayed.
Based on observed means.
The error term is Mean Square(Error) = 73.419.

a. Uses Harmonic Mean Sample Size = 5.743.
b. Alpha = .05.

In contrast to Example 10.7, there are no differences in the results of **R** and SPSS.

The Newman–Keuls procedure thus did not make a grouping with respect to the characters *Everyday Knowledge, Applied Computing*, and *Social and Material Sequencing* (all at the first test date). However, it made a grouping with respect to the characters *Immediate Reproducing – numerical* and *Coding and Associating* (at the first test date). The mean test score in the subtest *Immediate Reproducing – numerical* is thus (study-wise but per character with a type-I risk of 5%) lower in the 'upper lower class' and in the 'lower classes' than it is in the 'upper classes'. The same result occurs concerning the mean of the test scores of the subtest *Coding and Associating*, but here the 'lower classes' additionally have a lower mean than the 'upper lower class'.

Doctor

The principal component test described above was generalized to the case of more than two samples (populations). However, this generalization indirectly implies the presumption of homogeneity of the variance–covariance matrix; actually there is the even stricter presumption that all m characters can be substantially absorbed in some common character (a so-called 'factor' according to the factor analysis; see Section 15.1.2). Therefore, we do not deal with this rarely used method here.

Master
Doctor

So far, we have just dealt with the one-way multivariate analysis of variance (sometimes extended with the element of an analysis of covariance). Of course, multivariate analysis of variance can also be designed as a multi-way one. In doing so, there are cross classifications as well as nested classifications. The procedure of analyzing the data is exactly the same as used in Chapter 10.

Doctor

Recently, so-called *multilevel models* (also called *hierarchical* or *nested linear models* (HLM)) began to play an important role in psychology. At second glance, however, we see that such models are just complex designs of a multi-way multivariate analysis of variance, mostly with nested classifications.

Doctor

Example 13.4 Testing the standards of education

As part of some nationally established standards of education, the level of performance should be ascertained for students in the eighth level of education using appropriate tests. For example, the performance in English and Mathematics. From the population of all pupils in the federal territory, which was stratified into schools with sizes $N_1, N_2, \ldots, N_s$ (see Section 7.2.1), a schools (factor A) are drawn. Within each school there are then b_i eighth-level classes (factor B), which again are either randomly drawn or acquired by means of cluster sampling (again see Section 7.2.1). From these classes n_{ij} students were tested using the tests y and x. Once more, these students might be selected either randomly or using the cluster sampling method. Finally, the sex of each student is also of interest (factor C). As a matter of fact, this results in a three-way classification analogous to Table 10.22, with the factor A, *school*, the factor B that is nested in A, *class*, and the cross classified factor C, *sex*. The resulting test scores thus are: y_{ijkv} and x_{ijkv}; $i = 1, 2, \ldots, a; j = 1, 2, \ldots, b_i; k = 1, 2 = c; v = 1, 2, \ldots, n_{ijk}$. The difference from Table 10.22 – apart from the fact that more than one character is investigated – is unequal sample sizes n_{ijk} and, additionally, that this is either about a classification $(\mathbf{A} \succ \mathbf{B}) \times C$ or $(\mathbf{A} \succ B) \times C$: the factor A is certainly random; the factor C is certainly fixed. But the factor B is either random or fixed, depending on whether a random sample of the eighth-level classes of each school was drawn or all eighth-level classes of each school were taken into account.

Doctor

Multilevel linear models in the classical sense have different extensions that entail special feasibilities of application. Apart from the fact that, realistically, more than one character is observed for each research unit (see Example 13.4), there is also the option to use regression models for several characters within every combination of factor levels; that is one or more characters are regressors onto any criterion character as the regressand. One must then examine to what extent the regression coefficients differ with respect to the various factor levels or combinations of factor levels. Often the effect of numerous characters is then of interest, which can be modeled as in the (linear) structural equation models (see Section 15.2.2). In the context of using hierarchical linear models, it is often of interest to what degree the outcomes within a certain factor level or a certain combination of factor levels come close to each other; in other words, whether they have a smaller variance than the investigated characters in total; that is, over all the factor levels. For doing so, the intra-class correlation coefficient should be used, which was presented in Section 11.3.4.

Concerning more complex HLM approaches, see specialist literature (e.g. Gelman & Hill, 2006).

13.4 Discriminant analysis

The (linear) discriminant analysis is based on the (one-way) multivariate analysis of variance. Given significant differences between the samples and factor levels, respectively, with respect to the characters' means, it attempts to put all the characters in an optimal relation with a linear function so that the sample differences become as distinct as possible.

Master Doctor

As already mentioned above, discriminant analysis, in addition to the multivariate analysis of variance, is looking for the line in space (projected line), that discriminates optimally between the investigated populations (see Figure 13.5). The projection of all the points in the multidimensional space onto this line reveals a linear combination of all coordinates for each point and can be represented in the so called *(linear) discriminant function*:

$$D_v = d_1 y_{v1} + \cdots + d_m y_{vm} \tag{13.8}$$

$y_{v1}, \ldots, y_{vm}$ are the m outcomes for the m characters of research unit v. In the case of Figure 13.5, present are three samples (populations), but only $m = 2$ characters. If there are $a > 2$ samples (and $m \geq 3$ characters) then $a - 1$ different discriminant functions result, which are projection lines orthogonal to one another. For this, the result of a discriminant analysis can easily and clearly be interpreted if there are only two samples.

The results of a discriminant analysis are the weights d_q, $q = 1, 2, \ldots, m$ of the discriminant function(s).

There are two reasons for using the discriminant function beyond the multivariate analysis of variance:

1. Assigning future research units, with unknown affiliation to the populations in question, as accurately as possible to one of them – using their outcomes in the m characters.

Master Doctor

Example 13.5 Personnel recruitment using psychological tests

Because of the test scores in m psychological tests, a job applicant for a specific profession should be – in terms of a prediction – either assigned to the population of qualified candidates or to the population of unqualified candidates.

Master Doctor

2. Determining the relative contribution (weight) to the discrimination of all the characters.

Master Doctor

Example 13.6 Reasons for the dissolution of life partnerships

Given that several characters, which relate to socio-economic, psycho-social, and emotional aspects of a life partnership could be ascertained quantitatively, then their (explanatory) contributions to the dissolution of a (long-term) life partnership are of interest. For this, partners that come from existing as well as from (recently) dissolved partnerships would have to be investigated/interviewed. A possible result could be that the existence of children is of great importance for maintaining the partnership, but 'alienation' (measured using appropriate psychological tests) and 'work-related diverging development' is of great importance for dissolving. Also, for example, family income and the grade of 'sexual fulfillment in the partnership' (again measured using appropriate psychological tests) could carry weight in certain directions.

Master Doctor

In the second case, the analysis comes to an end with the establishment of the discriminant function, whereby the weights can be tested for their significance. That is, testing the null hypothesis for each character of whether it is relevant or irrelevant for the discrimination of the samples. Basically, this problem can then be solved analogously to the multiple linear regression (see Section 12.1.3), but with a qualitative character being the regressand. In the first case most of the time a *classification analysis* follows. The gathered information using this method goes beyond the information that there are significant differences at all between the investigated samples (factor levels); it also goes beyond the knowledge of the optimal discriminant function. Using the classification analysis, the researcher can get an impression of how accurate the (retrospective) predictions based on the discriminative function are – in terms of scored 'hits'.

The original method of classification analysis uses the discriminant function to retrospectively assign each research unit v to one of the factor levels (populations). That is, D_v is calculated on the basis of the unit's outcomes, and an assignment is made to the population which has the highest density function at the point D_v. This is called the *resubstitution method*. It is not a conclusive method, because high hit scores will be achieved trivially; essentially, the predictive quality of the discriminant function is tested on the basis of just those research units that were themselves exclusively the basis for the actual discriminant function. It is more appropriate to proceed similarly to a cross-validation (see Section 14.2.3 for more details) as follows: only a part of the given data is included in the discriminant analysis; the remaining part is used for testing the obtained discriminant function's predictive fitness. The first part of the data is then called the *training set*; the second part is called the *testing set*.

Classification analysis hence compares the predicted group- or population assignment to the actual realized group membership with a contingency table. This leads to a percentage of 'hits', the explanatory power of which, however, should not be over interpreted. Given $a = 2$ and an equal number of research units per sample (factor level), even guessing (for example by tossing a coin) would very likely produce a hit rate of about 50%. Then, only a hit rate of at least, say, 70% would be important from the content point of view.

In terms of content, because of institutional interests, the two types of possible hits are often of different importance and usefulness. In Example 13.5, there is a hit if the m psychological tests either correctly assign a candidate to the population of 'eligible' ones or correctly assign a candidate to the population of 'not eligible' ones. No hit, but a failure is produced, if the candidate is erroneously assessed as 'eligible' or if the candidate is erroneously assessed as 'not eligible' – failures are actually completely analogous to the two types of errors in hypothesis testing. In the terminology of medicine and clinical psychology, we talk about *specificity*, which concerns the probability of a negative diagnosis given an actual negative state; and we talk about *sensitivity*, which concerns the probability of a positive diagnosis given an actual positive state (see Table 13.9). The classification analysis as a part of discriminant analysis, however, does not differentiate in this regard.

Table 13.9 Sensitivity and specificity of diagnoses

		Test diagnosis	
		positive	*negative*
Actual state	positive	Sensitivity	
	negative		Specificity

Master

Example 13.5 – continued

In particular, this is not about recruitment for a specific job, but about the selection of candidates for a special schooling; for example the training required to be an animal keeper. Assume that we have data from a total of 120 students of three term cohorts who were all taken into schooling without regard to their test scores in the psychological test battery used. Of these, 55 were 'unsuccessful', quitting within the first year of schooling; the others can be called, by definition, 'successful'. The test battery might consist of $m = 8$ subtests. For discriminant analysis the first 80 students from the data set *Example 13.4* (see Chapter 1 for its availability) should be used as the training set; the other students as the testing set. For hypothesis testing we decide on a type-I risk of $\alpha = 0.05$; the null hypothesis asserts that the two groups, the 'successful' ones on the one hand and the 'unsuccessful' ones on the other hand, do not, according to the discriminant function, differ with regard to the characters' means.

In **R**, we use the package MASS, which we already installed in Example 12.9, as well as the package car, which we already installed in Example 9.6. First, we split the research units into a training and a testing set by typing

```
> train.dat <- Example_13.5[1:80, ]
> test.dat <- Example_13.5[81:120, ]
```

i.e. we select the first 80 and the last 40 data lines from the database Example_13.5 with [1:80,] and [81:120,] respectively, and assign them to the objects train.dat and, respectively, test.dat. In order to conduct the Box-M test, we use our function from Example 13.1. Hence, we type

```
> with(boxM.test(cbind(sub1, sub2, sub3, sub4, sub5, sub6, sub7,
+                          sub8),
+        aptitude), data = train.dat)
```

i.e. we use the function cbind() to define the given subtests (sub1 through sub8) as a matrix, which we submit to the function boxM.test() as the first argument; the second argument determines the character to be analyzed as 'unsuccessful' vs. 'successful' (aptitude). By using the function with(), we state the database to be used, namely train.dat.

As a result, we get:

```
     df1       df2          Box M           F        ProbF
      36 1.882e+04       42.37530     1.04645      0.39279
```

The Box-M test yields a non-significant result; thus, the application of the discriminant analysis is legitimate.

Now, we conduct the discriminant analysis; in order to do this, we type

```
> train.lda <- lda(aptitude ~ sub1 + sub2 + sub3 + sub4 +
+                    sub5 + sub6 + sub7 + sub8, data = train.dat)
> print(train.lda)
```

i.e. we use the function lda(), submitting to it the instruction that the character aptitude is to be analyzed with regards to the given subtests (sub1 through sub8); with data = train.dat, we limit the analysis to the training set. We assign the results of this analysis to the object train.lda, which we submit to the function print() as a next step.

As a result, we get:

```
Call:
lda(aptitude ~ sub1 + sub2 + sub3 + sub4 + sub5 + sub6 + sub7 +
    sub8, data = train.dat)

Prior probabilities of groups:
unsuccessful   successful
        0.55         0.45
```

```
Group means:
                  sub1      sub2     sub3     sub4     sub5     sub6
unsuccessful  98.93182  99.43182 100.2727 100.1364 99.54545  94.2500
successful   109.58333 102.47222 100.9444 100.1944 96.30556 108.1111
                 sub7     sub8
unsuccessful 100.7500 99.88636
successful   105.6667 99.22222

Coefficients of linear discriminants:
              LD1
sub1  0.067451948
sub2  0.002632161
sub3  0.017513645
sub4  0.007593686
sub5 -0.001907858
sub6  0.086368202
sub7  0.040055103
sub8  0.010686129
```

Now, we type

```
> Anova(lm(cbind(sub1, sub2, sub3, sub4, sub5, sub6, sub7, sub8) ~
+       aptitude, data = train.dat), test.statistic = "Wilks")
```

i.e. we apply the function `lm()`, requesting that the characters of all eight subtests – defined as a matrix with the help of the function `cbind()` – are to be analyzed with regards to the character `aptitude`; with `test.statistic = "Wilks"`, we select the respective test value. We submit the results of this analysis to the function `Anova()`.

As a result, we get (shortened output):

```
Type II MANOVA Tests: Wilks test statistic
         Df test stat approx F num Df den Df    Pr(>F)
aptitude  1   0.46432   10.239      8     71 2.156e-09
```

We now create a contingency table in which the group membership that was predicted by the discriminant function is opposed to the actual group membership; in order to do this, we type

```
> train.cl <- predict(train.lda, newdata = train.dat)
> table(train.cl$class, train.dat$aptitude)
```

i.e. we apply the function `predict()` and set the object `train.lda` as the first argument; with `newdata = train.dat`, we classify the training sample by means of the obtained discriminant function. We assign the results of this analysis to the object `train.cl`. By applying the function `table()`, we create a two-dimensional frequency table containing the predicted group membership, which we extract from the object `train.cl` with `$class`, and the actual group membership according to the character `aptitude`.

This yields:

```
              unsuccessful successful
unsuccessful            39         10
successful               5         26
```

wherein the lines refer to the predicted group membership and the columns to the actual ones.

Next, we repeat the process, although now using the testing set; hence, we type

```
> test.cl <- predict(train.lda, newdata = test.dat)
> table(test.cl$class, test.dat$aptitude)
```

i.e. in the `predict()` function we merely replace the argument `newdata = train.dat` by `newdata = test.dat` and, also, in the `table()` function, replace `train.cl$class` by `test.cl$class`, and `train.dat$aptitude` by `test.dat$aptitude`.

As a result, we get:

```
              unsuccessful successful
unsuccessful            10          9
successful               1         20
```

In SPSS, we first create a new character, sample, by typing the value 1 into the first 80 lines and the value 2 into the last 40 lines of an additional column in the Data view. In order to confirm the given precondition of the variance–covariance matrix analogously to the multivariate analysis of variance with the Box-M test, we now have to conduct the discriminant analysis. This is accomplished with the steps

Analyze
    Classify
        Discriminant...

which open the window in Figure 13.6. We mark the character aptitude and drag and drop it into the field Grouping Variable:, then click the button Define Range... in order to type 0 into the field Minimum and 1 into the field Maximum in the resulting window (not shown here). With Continue, we get back to the previous window, where we drag and drop the characters subtest 1 through subtest 8 into the field Independents:. Finally, we drag and drop the character sample into the field Selection Variable: and click Value... in order to type 1 in the resulting window (without figure here). With Continue, we get back to the window shown in Figure 13.6. Now, we click Statistics... and set a check mark at Box's M? in the resulting window (without figure here). In case of the discriminant analysis yielding a significant result, one could also have the means per group, the sample of the 'unsuccessful' and the sample of the 'successful' students and per character, dispensed by setting a check mark at Means; or, one could conduct a one-way analysis of variance per character by setting a check mark at Univariate ANOVAs. However, we set a check mark at Unstandardized in order to not only get the automatically dispensed standardized discriminant function coefficients but also the unstandardized ones. With Continue, we get back to Figure 13.6. If we want to save the group membership as predicted by the discriminant function as a new character within the data, we also have to click the button Save... In the resulting window (not shown here), we set a check mark at Predicted group membership. With Continue, we get back again, now clicking Classify... in order to open the window in Figure 13.7. There, we click Summary table. With Continue and OK, we obtain the results displayed in Tables 13.10 through 13.14.

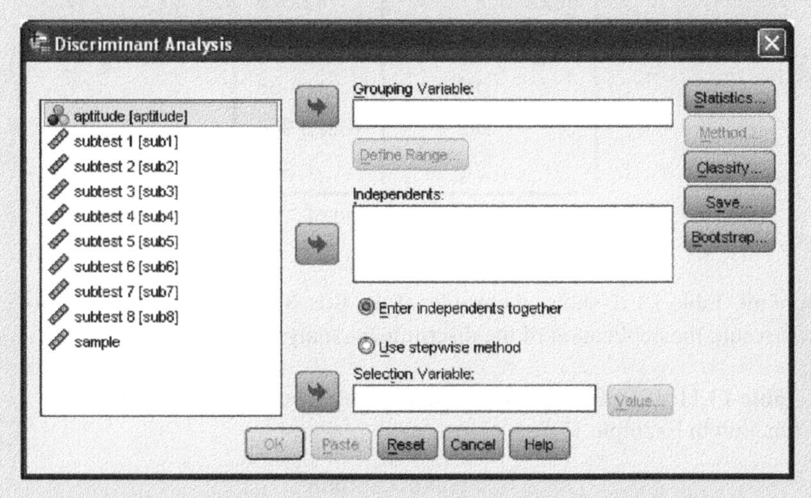

Figure 13.6 SPSS-window for computing the discriminant analysis.

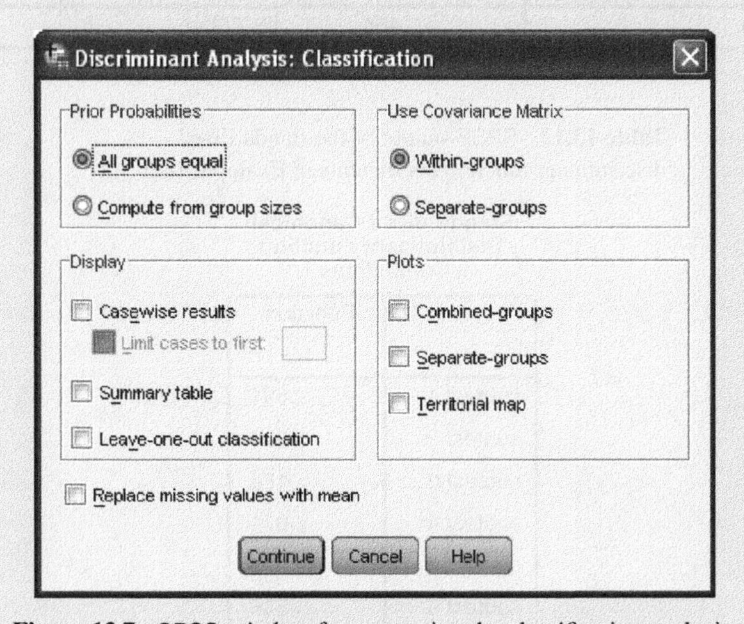

Figure 13.7 SPSS-window for computing the classification analysis.

Table 13.10 SPSS-output of the Box-M test for checking the homogeneity of the variance–covariance matrix in Example 13.5

Test Results

Box's M		42.375
F	Approx.	1.046
	df1	36
	df2	18823.908
	Sig.	.393

Tests null hypothesis of equal population covariance matrices.

First of all, Table 13.10 shows the results of the Box-M test. Because it does not yield a significant result, the application of the discriminant analysis is legitimate.

Table 13.11 SPSS-output of the significance check of the discriminant function in Example 13.5

Wilks' Lambda

Test of Function(s)	Wilks' Lambda	Chi-square	df	Sig.
1	.464	56.771	8	.000

Table 13.12 SPSS-output of the standardized discriminant function coefficients in Example 13.5

Standardized Canonical Discriminant Function Coefficients

	Function
	1
subtest 1	.691
subtest 2	.028
subtest 3	.168
subtest 4	.071
subtest 5	-.019
subtest 6	.846
subtest 7	.345
subtest 8	.118

Table 13.13 SPSS-output of the unstandardized discriminant function coefficients in Example 13.5

Canonical Discriminant Function Coefficients

	Function
	1
subtest 1	.067
subtest 2	.003
subtest 3	.018
subtest 4	.008
subtest 5	-.002
subtest 6	.086
subtest 7	.040
subtest 8	.011
(Constant)	-23.464

Unstandardized coefficients

Table 13.14 SPSS-output of the classification analysis for the training sample (Cases Selected) and for the testing sample (Cases Not Selected) in Example 13.5 (shortened output)

Classification Results[a,b]

				Predicted Group Membership		
			aptitude	unsuccessful	successful	Total
Cases Selected	Original	Count	unsuccessful	39	5	44
			successful	8	28	36
Cases Not Selected	Original	Count	unsuccessful	9	2	11
			successful	8	21	29

a. 83.8% of selected original grouped cases correctly classified.
b. 75.0% of unselected original grouped cases correctly classified.

Most important, first of all we realize that Wilks' test statistic Λ for the multivariate analysis of variance is significant. SPSS further shows (in Table 13.12) the $m = 8$ standardized weights d_q of the discriminant function. The absolute largest contribution for the discrimination of the two groups in question is made by the subtests 6 and 1 (in that order); in particular, the subtests 5, 2, and 4 hardly contribute to discrimination. In Table 13.13, SPSS also indicates those weights, including a constant, by which the outcomes per person have to be weighted and summed up in order to determine the retrospectively predicted group membership. In the case of a positive value, the student has to be assigned to the group which has the larger measurement value; in our case that is 1, namely the group of 'successful' students. In the case of a negative value, the student has to be assigned to the other group. Using **R**, we certainly obtain the same (unstandardized) weights, but not the aforementioned constant. Thus, it is not as easy with **R**, as it is with the results of SPSS, to predict the group membership of future research units.

The results of the classification analysis using **R** and SPSS do not entirely coincide. We focus on the results of SPSS in the following. Table 13.14 shows the comparison of the actual membership of one of the two groups, and the group assignment that was (retrospectively) predicted by the discriminant function – separately for the training set as well as for the testing set. As concerns the training set, 39 plus 29 students were assigned correctly by the discriminant function; 5 plus 8 persons were mismatched; i.e., the hit rate is 83.8%. As concerns the testing set, 9 plus 21 students were assigned correctly and 2 plus 8 were mismatched by the discriminant function obtained independently from another sample; the hit rate here is 75%. This result is not noticeably worse than the hit rate observed in the training set applying the resubstitution method. As a matter of fact, the resulting discriminant function is, for that reason, safeguarded against a merely random fit of the data.

Doctor

Regarding the maximum error of the predicted assignment to one of the groups, even planning a study is possible; i.e., given some probabilities of erroneous assignments, the necessary sample size can be calculated. We will not, however, discuss this here in detail (see Rasch, Herrendörfer, Bock, Victor, & Guiard, 2008).

Master Doctor

Regarding the question of how many of all given characters have to be taken into the discriminant function, in practice the approach of *stepwise* discriminant analysis is often used. In this, the discriminant analysis is repeated either using the *backward* method, namely by successive removal of the character which contributes the least to discrimination; or – considerably more illustrative – using the *forward* method, where successive characters are included for the discriminant function; that is, as long as the character in question still makes a (significant) additional contribution to the discrimination. In this way the researcher obtains a hierarchy of characters with respect to their discriminatory contribution. In the first step of the forward method, the character is selected that has the maximum F-value in the one-way analysis of variance. In the second step, it may occur that not the character with the second largest F-value will be chosen, because this one may correlate highly with the first one. If the (forward) stepwise method breaks down, because no significant additional contribution to discrimination can be obtained for any remaining character, the latter are, from the content point of view, identified as irrelevant. When future research units need to be assigned to the populations in question, these characters can be ignored.

Doctor

The described stepwise approach is not necessarily optimal. It would be better to try all possible variants of combinations of characters to identify the best combination of (as few as possible) characters with regard to the discrimination between the groups. The method of choice would hence be *full enumeration*.

For m characters, this means that we have to determine the respective target function (e.g. Wilks' Λ) for all single-element subsets of characters (i.e. a single character at any one time) as well as for all two-element subsets up to all (m–1)-element subsets of characters, and finally for the total pool of characters. We demonstrate full enumeration for the case of $m = 4$ characters 1, 2, 3, and 4. We thus obtain: [{1},{2},{3},{4}]; [{1,2}]; [{1,3}]; [{1,4}]; [{2,3}]; [{2,4}]; [{3,4}]; [{1,2,3}]; [{1,2,4}]; [{1,3,4}]; [{2,3,4}]; [{1,2,3,4}]. Hence, there are 15 possible partitions. The number of possible combinations of characters grows exponentially with n; in terms of m characters, it amounts to $E_m = 2^m - 1$. For example, we have $E_5 = 31$, $E_6 = 63$, and $E_7 = 127$. Full enumeration for a large number of characters cannot be handled with the computing power of conventional computers at the moment. However, it should be possible with up to $m = 50$ characters.

Master

Example 13.5 – continued

Aiming for maximum efficiency, we want to shorten the given test battery of eight subtests as far as empirically justifiable. For this, we use the data of all 120 students and choose the forward method for a stepwise discriminant analysis. The criterion for a further character to include into the discriminant function will be Wilks' test statistic Λ in each step.

In **R**, we use the package SDDA, which we load by applying the function `library()` after installing it (see Chapter 1). Now, we type

```
> with(sdda(cbind(sub1, sub2, sub3, sub4, sub5, sub6, sub7, sub8),
+       aptitude), data = Example_13.5)
```

i.e. we use the function `cbind()` to define the observed values of all subtests (`sub1` through `sub8`) as a matrix and submit it to the function `sdda()` as the first argument; the second argument determines the character that is to be analyzed with 'unsuccessful' vs. 'successful' (`aptitude`). By applying the function `with()`, we specify the desired database; namely `Example_13.5`.

In contrast to what was described above, this package selects the one character that minimizes the error of predicted group memberships.

As a result, we get:

```
SDDA using dlda.
n = 2 samples and p = 8 variables.
Group levels are: successful unsuccessful
1 variables are chosen in total.
Variables are sub6
```

In SPSS, we now choose the option Use stepwise method instead of Enter independents together in the window in Figure 13.6. We click Method... and thus get to the window in Figure 13.8, where we keep the original settings and, after clicking Continue and OK, obtain the results displayed in Tables 13.15 through 13.17.

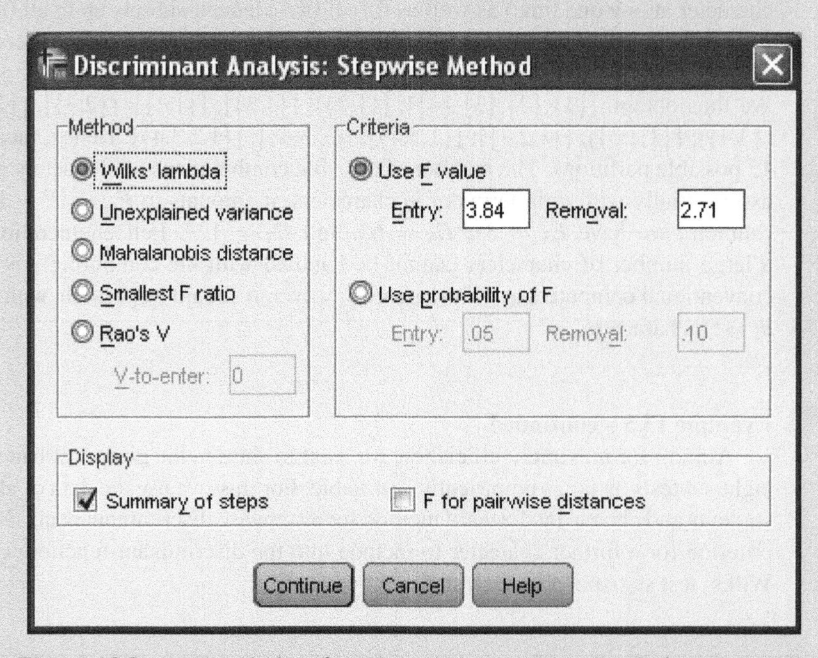

Figure 13.8 SPSS-window for computing a stepwise discriminant analysis.

Table 13.15 SPSS-output of the stepwise-selected characters of the discriminant analysis in Example 13.5 (shortened output)

Variables Entered/Removed[a,b,c,d]

Step	Entered
1	subtest 6
2	subtest 1
3	subtest 7

At each step, the variable that minimizes the overall Wilks' Lambda is entered.

a. Maximum number of steps is 16.
b. Minimum partial F to enter is 3.84.
c. Maximum partial F to remove is 2.71.
d. F level, tolerance, or VIN insufficient for further computation.

Table 13.16 SPSS-output of Wilks' test value Λ of the stepwise discriminant analysis in Example 13.5

Wilks' Lambda

Step	Number of Variables	Lambda	df1	df2	df3	Exact F			
						Statistic	df1	df2	Sig.
1	1	.706	1	1	118	49.238	1	118.000	.000
2	2	.550	2	1	118	47.839	2	117.000	.000
3	3	.530	3	1	118	34.306	3	116.000	.000

Table 13.17 SPSS-output of the classification analysis at only three stepwise-obtained characters in Example 13.5 (shortened output)

Classification Results[a]

		aptitude	Predicted Group Membership		Total
			unsuccessful	successful	
Original	Count	unsuccessful	45	10	55
		successful	15	50	65

a. 79.2% of original grouped cases correctly classified.

The results of **R** and SPSS differ because of different selection criteria. Whereas only one subtest is included in the analysis with the given package in **R**, there are three subtests with the traditional procedure in SPSS; namely those given in Table 13.15. It can be seen from Table 13.16 that the subtests 6, 1 and 7 are successively included in the analysis until Wilks' test statistic Λ finally results with 0.53; this means all in all a significant discrimination. The estimated effect size is $\hat{\eta}^2 = 0.6856$. Table 13.17 finally shows that, instead of the hit rate of 83.8% in terms of using the whole test battery, still 79.2% hits can be achieved using only three subtests.

Master Doctor

It has already been pointed out in Section 13.2 that the presumption of a multivariate normal distribution is not testable (satisfactorily). Basically, as concerns discriminant analysis, optimization of the discriminant function as a method of (only) descriptive statistics is actually unaffected by this presumption – at least, if the observations in all investigated populations have the same distribution. That is to say, classification analysis would also in any case be interpretable given a serious violation of the multivariate normal distribution; the significance test possibly not.

Conversely, it could be concluded from Figure 13.5 that equal covariances (parallel regression lines, tested using the Box-M test) are essential: if the regression lines per population happened to not be parallel, then the multidimensional mean point would not be meaningful enough. Hence, once the Box-M test is significant, we have the following options. Either the researcher uses descriptive statistics methods in order to identify those characters (and/or samples) responsible for this significance and therefore excludes these from the further analysis (see above in Section 13.3). Or he/she uses alternative statistical tests.

For this the logistic regression in its generalized form especially comes into consideration – it has already been described in Section 11.3.6. Unlike in Formula (11.16), we now have the special case of the regressand being a nominal-scaled character. Since, as mentioned above, research questions for discriminant analysis are, anyway, easiest to analyze and to interpret for only two groups (populations) that have to be discriminated, we limit ourselves to the case of $a = 2$ in the following. Thus, we can postulate the estimated probability for one of the two group memberships as a logistic function as follows:

$$\hat{p}_v = \frac{1}{1+x} = \frac{1}{1+e^{-(d_0+d_1 y_{v1}+\cdots+d_m y_{vm})}} \tag{13.9}$$

The analysis, as well as the asymptotically χ^2-distributed test statistic, result analogously to the (linear) discriminant analysis. A partition of the data into a training and a testing set is likewise possible.

What is so special about the logistic regression as a discriminant analytic method is the fact that the regressors can also be dichotomous characters.

Example 13.6 – continued

We now want to answer the given research question with the data set *Example 13.6* (see Chapter 1 for its availability) using the logistic regression. The $m = 3$ characters *alienation*, *work-related diverging development*, and *sexual fulfillment in the partnership* are all quantitative characters. The total sample size of women from existing partnerships and from (recently) dissolved partnerships is $n_1 + n_2 = 50$. The null hypothesis was: 'The three investigated characters do not differ between the two groups of existing and dissolved partnerships.' We decide on a type-I risk of $\alpha = 0.05$.

In **R**, we enable access to the database *Example_13.6* by using the function `attach()` and type

```
> glm.1 <- glm(partnership ~ 1, family = "binomial")
> glm.step <- step(glm.1, scope = ~ p1 + p2 + p3,
+                  direction = "forward")
> summary(glm.step)
```

i.e. we apply the function `glm()` and, as the first argument, specify that the character 'dissolved' vs. 'existing' (`partnership`) is to be analyzed stepwise (`~ 1`) and, because of `family = "binomial"`, by means of the logistic regression analogous to Formula (13.9). We assign the results of this analysis to the object `glm.1`, which we submit to the function `step()` as the first argument in the next step. With the second argument, we specify that the character `partnership` is to be analyzed stepwise and in accordance with the ascending procedure (`direction = "forward"`) with regards to the characters *alienation* (`p1`), *work-related diverging development* (`p2`) and *sexual fulfillment in the partnership* (p3). We assign the results to the object `glm.step`, after which we apply the function `summary()` to it in order summarize the results.

This yields (shortened output):

```
Start:  AIC=63.09
partnership ~ 1

       Df Deviance    AIC
+ p1    1   44.922 48.922
<none>      61.086 63.086
+ p2    1   60.971 64.971
+ p3    1   61.004 65.004

Step:  AIC=48.92
partnership ~ p1

       Df Deviance    AIC
+ p2    1   40.051 46.051
<none>      44.922 48.922
+ p3    1   44.871 50.871

Step:  AIC=46.05
partnership ~ p1 + p2

       Df Deviance    AIC
+           40.051 46.051
+ p3    1   39.949 47.949

Coefficients:
             Estimate Std. Error z value Pr(>|z|)
(Intercept) -32.54884   10.47850  -3.106  0.00189
p1            0.19585    0.05975   3.278  0.00105
p2            0.10594    0.05222   2.029  0.04250
```

Now, we create a contingency table which opposes the group membership as predicted by the logistic function to the actual group membership; hence, we type

```
> glm.fit <- fitted(glm.step)
> table(round(glm.fit), partnership)
```

i.e. we apply the function `fitted()` to the object `glm.step`; thus causing the predicted probabilities to be dispensed. We assign these to the object `glm.fit`. Using the function `table()`, we oppose the actual values of the character `partnership` to the predicted ones, wherein the function `round()` rounds up the values in `glm.fit` to whole numbers.

As a result, we get:

```
  partnership
   dissolved existing
0         32        6
1          3        9
```

In SPSS, we select the sequence of commands

Analyze
 Regression
  Binary Logistic...

and proceed to apply the steps analogously to the ones already taken in Figure 13.9. There, we also change to Forward: LR after clicking Method:. After clicking the button Save..., we get to the window in Figure 13.10, where we set a check mark at Group membership. With Continue and OK, we obtain the results displayed in the Tables 13.18 and 13.19.

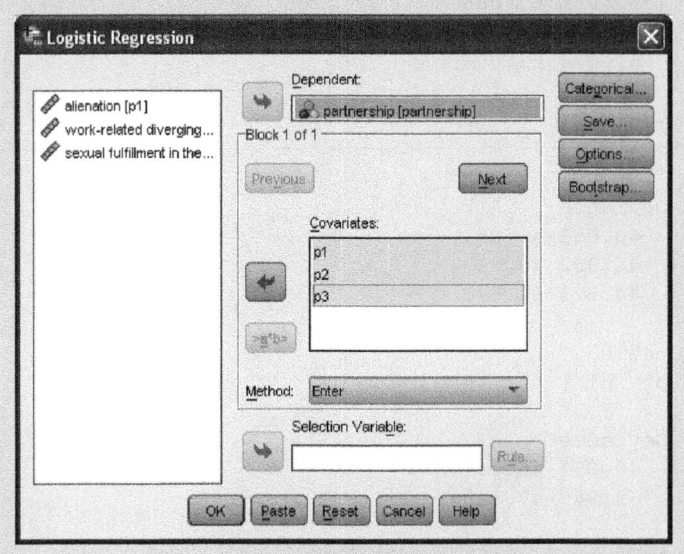

Figure 13.9 SPSS-window for conducting the logistic regression.

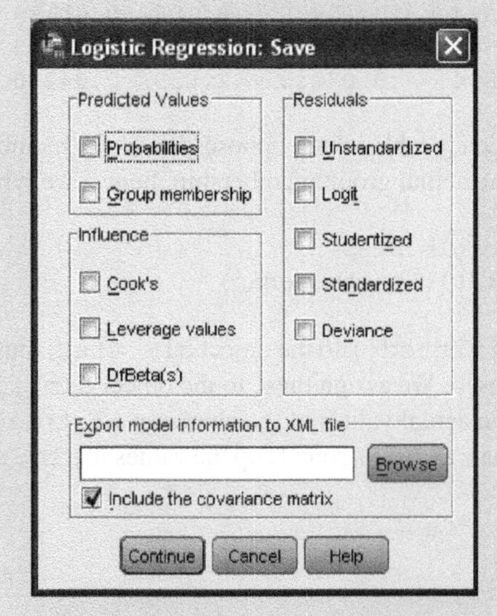

Figure 13.10 SPSS-window for determining the group membership as computed by the logistic regression.

Table 13.18 SPSS-output of the regression coefficients for the stepwise-selected characters in the logistic regression in Example 13.6

Variables in the Equation

		B	S.E.	Wald	df	Sig.	Exp(B)
Step 1[a]	p1	.142	.045	9.855	1	.002	1.153
	Constant	-16.265	5.028	10.466	1	.001	.000
Step 2[b]	p1	.196	.060	10.745	1	.001	1.216
	p2	.106	.052	4.115	1	.042	1.112
	Constant	-32.549	10.478	9.649	1	.002	.000

a. Variable(s) entered on step 1: p1.
b. Variable(s) entered on step 2: p2.

Table 13.19 SPSS-output of the classification analysis at two stepwise-obtained characters in Example 13.6

Classification Table[a]

Observed			Predicted		
			partnership		Percentage Correct
			dissolved	existing	
Step 1	partnership	dissolved	31	4	88.6
		existing	8	7	46.7
	Overall Percentage				76.0
Step 2	partnership	dissolved	32	3	91.4
		existing	6	9	60.0
	Overall Percentage				82.0

a. The cut value is .500

Unlike SPSS, where the best discriminating character is selected stepwise by an approximately χ^2-distributed test statistic (see more details in Section 14.2.2), **R** selects characters using the smallest coefficient *AIC* (see more details in Section 14.2.2). Since the two criteria can be converted into each other, the results are congruent.

It can be seen from Table 13.18 that firstly *alienation* and then *work-related diverging development* were selected; not *sexual fulfillment in the partnership*. From Table 13.18 it follows that the discrimination is significant. Thereby, *alienation* makes the greatest contribution. It certainly can be recognized from the equal sign of the regression coefficients of both selected characters that the effect goes in the same direction, but it cannot be seen in which direction. Finally, in Table 13.19 we find the result of the classification analysis according to the resubstitution method, which shows that 82.0% would be assigned correctly.

For the case of the *m* characters of interest not all being quantitative, but at least one of them being (multi-categorical) nominal or ordinal scaled, the logistic regression also fails, except by introducing so-called *dummy variables* for the categories of the qualitative characters, whereby, however, we lose the information of the categories' ranking, given an ordinal-scaled character. As a dummy variable, we define a separate dichotomous variable for each category; namely with the realizations 'true' and 'not true'.

Master Doctor

Example 13.7 Recoding a *c*-categorical character into *c* dummy variables

In Example 1.1, we can recode the multi-categorical character *marital status of the mother*, with the $c = 4$ character values 'never married' (1), 'married' (2), 'divorced' (3), 'widowed' (4), into four (dichotomous) dummy variables as follows: *never married*, with the measurement values 0 and 1 for 'no, not never married' and 'yes, never married'; *married*, with the measurement values 0 and 1 for 'no, not married' and 'yes, married'; *divorced*, with the measurement values 0 and 1 for 'no, not divorced' and 'yes, divorced'; and *widowed*, with the measurement values 0 and 1 for 'no, not widowed' and 'yes, widowed'.

Master Doctor

If there is (also) interest in multi-categorical nominal-scaled or ordinal-scaled characters, Kubinger (1983) suggested a universally applicable *non-parametric discriminant analysis*; there, the information of the categories' ranking in the case of an ordinal-scaled character is kept. However, this method is not included in relevant computer programs, and thus will not be discussed here (but see Section 14.2.3).

Summary

Any variants of one-way and multi-way analysis of variance can be applied as *multivariate analysis of variance* if there are at least two characters. For this, the essential presumption is homogeneity of the *variance–covariance matrix*, which is best tested by the *Box-M test*. If some noise factors are suspected and if there are also observations of them for each research unit, they can be taken into account using an *analysis of covariance*; their influence on one or several characters and on one or several factors, respectively, can be eliminated.

The *discriminant analysis* is a method to determine the contribution of each investigated character that is made by them (collectively) in order to discriminate between two or more groups of research units and populations, respectively. Thereby, besides hypothesis testing on the basis of multivariate analysis of variance, often a *classification analysis* is performed, which establishes how many research units can be predicted correctly with regard to their group affiliation, by those weights obtained with the *discriminant function.* Most often used in discriminant analysis is a *stepwise* approach; for example, the characters are successively included for the discriminant function until the discrimination cannot be improved any further. For the case of nominal-scaled characters, the *logistic regression* should be used for the same problem.

References

Gelman, A., & Hill, J. (2006). *Data Analysis Using Regression and Multilevel/Hierarchical Models*. Cambridge: University Press.

Kubinger, K. D. (1983). Some elaborations towards a standard procedure of distribution-free discriminant analyses. *Biometrical Journal*, *25*, 765–774.

Läuter, J. (1996). Exact t and F tests for analyzing studies with multiple endpoints. *Biometrics*, *52*, 964–970.

Läuter, J., Glimm, E., & Kropf, S. (1998). Multivariate tests based on left-spherically distributed linear scores. *Annals of Statistics*, *26*, 1972–1988.

Rasch, D., Herrendörfer, G., Bock, J., Victor, N., & Guiard, V. (2008). *Verfahrensbibliothek Versuchsplanung und -auswertung. Elektronisches Buch* [Collection of Procedures in Design and Analysis of Experiments. Electronic Book]. Munich: Oldenbourg.

Part VI

MODEL GENERATION AND THEORY-GENERATING PROCEDURES

Up to this point, basically, we have been discussing methods of statistics (for the purposes of planning as well as for analyzing) that serve not only to answer a particular research question (see Chapter 3), but are, in their application, based on a (specific psychological) theory. Often enough, this 'theory' may be not much more than an everyday heuristic.[1] It would be heuristic, for example, to assume a relationship between learning efforts and learning success. But, psychology as a science is often aimed at (newly) constructing an empirically based theory for certain observable phenomena. For this purpose, special statistical methods are suited.

Because of this, we will look at the general basics of such theory-generating approaches in this Part. Then, we will show specific theory-generating and model-based methods, particularly distinguishing between methods of descriptive and inferential statistics.

[1] Heuristic, from the Greek (ευρίσκω, to find), is a method of cognition which is based on speculations and assumptions, not on fixated algorithms.

14

Model generation

In this chapter, we will first introduce the generalized linear model and explain the principle of models with latent variables. Then, we will look at different methods of determining the quality and excellence of a model. And last, we will show how statistical problems can be solved by simulation studies when a solution is not possible by analytical means.

14.1 Theoretical basics of model generation

We have already generated quite a few statistical models. This was in connection with modeling the interesting character as a (multivariate) normally distributed random variable. For many statistical methods, however, the presumption of normal distribution has proved to be not that essential; either they prove to be quite robust against violation of this presumption, or testing it seemed unpractical or even impossible. This statistical model generation had, however, with regards to content little or no consequences for psychology as a science.

This is different when generating statistical models in connection with the methods of regression analysis or analysis of variance. For regression analysis with only a single regressor, as well as for multiple regression analysis, we mostly model linear relationships between two characters or between one character and a group of other characters; for the canonical correlation coefficient, between two groups of characters. Such a determination of a model is indeed influential to the finding which is gained through the empirical results, and thus with regards to contents; that is to psychological theories. The same is true for the (linear) discriminant analysis and, because the effects are linearly modeled, also for the logistic regression in Section 13.4. All in all, these methods offer four ways of judging the quality and excellence of a defined model. Either the linear model is not questioned at all; this is, for example, true for the discriminant analysis if no classification analysis is applied on a testing set (see Section 13.4); then, one merely tries to determine the (significant) contributions of each character, so that they explain the group memberships in a linearly optimal manner. Or

Statistics in Psychology Using R and SPSS, First Edition. Dieter Rasch, Klaus D. Kubinger and Takuya Yanagida.

as a second way, one estimates at least a determinant (the effect size in terms of the coefficient of determination) to asses the generalized validity of the model-based results; but, then very often results have been used for binding conclusions which explain the data rather poorly – apart from the fact that a linear relationship is assumed unchecked. Then, there is a third way; that is to compare a model, e.g. linear modeling, with an alternative model in order to recognize a better or the best fit of the model to the data; we only demonstrated this once, in Section 11.3.6 in Example 11.12, by comparing the linear regression model with the quadratic and the logistic one. And finally, there is basically a fourth way, which is to actually test the model in itself; this will be discussed in Section 15.2.2.1. This is concerned with the (absolute) validity of the model and not only with the (relative) validity in relation to other, competing models (and also not only in relation to how well the model is able to describe the data).

With analysis of variance, we are basically dealing with the third technique, though once again focused more on the definition of a statistical model. Again, this is about linear models. This time, though, it is about opposing two hypotheses and thus about comparing two 'models'. Depending on whether the hypothesis is accepted or rejected, this contributes to the generation of psychological theories in one or the other direction. Here, interaction effects are essential: most of the time, their existence is allowed but their non-existence is hypothesized (modeled); if the statistical analysis leads to accepting the respective hypothesis, a model of content has been established which states that the investigated factors are generating additive, non-interdependent effects. In the analysis of covariance (which is based on the analysis of variance), covariances are also modeled; depending on whether these exist or not, this again contributes to the generation of theories within psychology.

As we can see from these examples, the general linear model is of crucial importance for psychological scientific research. Linear statistical models gain additional meaning because, for the case of nominally scaled characters, it is possible to use them through certain transformations of statistics and parameters. This leads to the *generalized linear model*. In the following, we give a more specific definition of these, as is done elsewhere, in order to clearly distinguish the approach from those which also consider so-called *latent variables*. The latter concern cases in which observable characters are not directly modeled as random variables, but are ascribed to non-observable, hypothesized (modeled) characters.

14.1.1 Generalized linear model

Given the factors A, B, and C and, therefore, the factor combinations $A_iB_jC_k$ with n_{ijk} outcomes from as many research units, and a frequency h_{ijk}, by which one of the two measurement values of a certain (dichotomous) character has been observed; then, for the relative frequency

$$f_{ijk} = \frac{h_{ijk}}{n_{ijk}}$$

the presumptions for linear models are violated. Applying (three-way) analysis of variance is not justified. In this case, the generalized linear model according to McCullagh and Nelder (1989) may serve (see there for more extensive information). Basically, it is a matter of transformation the parameters of the distribution as well as of the given measurement values and observed frequencies of them, respectively, with a *link function*. With this, one can use procedures that are designed for normally distributed variables for discrete distributions, too.

Doctor

For a binomial distributed character (see Formula (6.4)), for example, the link function concerning the (unknown) parameter p looks like:

$$\eta = \ln \frac{p}{1-p}$$

This link function is called the *logit transformation*. This, in turn, means for p as a function of η:

$$p = \frac{e^{\eta}}{1+e^{\eta}}$$

– the logistic function (see also the logistic regression of Formula (11.16) or of Formula (13.9)). In order to avoid numerical problems associated with observed frequencies of zero or n, there is an adjusted version of this link function,

$$z = \ln \frac{y + \frac{1}{2}}{1 - y + \frac{1}{2}}$$

With data transformed like this, one proceeds as described in the previous chapters; especially as in the one about the (linear) models as analyses of variance.

Doctor

Example 14.1 The influence of noise and exhaustion on attention efforts

An experiment of work psychology applies a test for signal detection (see Example 4.3) for assessing how far noise and exhaustion are influential on attention efforts. A strong attention effort with respect to signal detection is especially crucial for officers for control monitoring. It is especially of interest whether there are interaction effects between noise and exhaustion. There are three levels of factor A, *noise* ('40db', '60db', '80db'), and two levels of factor B, *exhaustion* ('first hour of work', 'eighth hour of work').

If the resulting test scores of the test for signal detection were to be modeled as a normally distributed random variable, the analysis could be planned and processed as in Chapter 10. However, now we deal with a character which follows a *Poisson distribution*: within half an hour, a very large number of different on-screen patterns of configurations of points – which change every five seconds – is to be judged on whether the stimulus configuration of a 'small square built up by four points' pops up; which happens, unsystematically, about 100 times. Omissions (y), which happen on average about three to five times, are being counted.

The Poisson distribution of a random variable $\boldsymbol{y}$ has the mean/expectation value μ and the frequency function

$$p(y) = \frac{\mu^{y} \cdot e^{-\mu}}{y!}$$

For this data, we use the link function $z = \ln(y + {}^1\!/_2)$, and for the parameter $\eta = \ln \mu$.

Suppose there are, for $n_{ij} = n = 2$ subjects per factor-level combinations, the following absolute frequencies:

		Levels of the factor noise		
		40 db	60 db	80 db
Levels of the factor exhaustion	First hour of work	3	8	12
		4	9	11
	Eighth hour of work	6	7	16
		5	6	20

In **R**, we first create a vector for the data of the character *omissions*; thus, we type

```
> omiss <- c(3, 4, 8, 9, 12, 11, 6, 5, 7, 6, 16, 20)
```

i.e. we apply the function `c()` to combine the observed values into a vector, which we assign to the object `omiss`. Then, we transform these values by typing

```
> z <- log(omiss + 0.5)
```

i.e. we use the function `log()`, which calculates the logarithms of `omiss + 0.5`, and assign the result to the object `z`. Now, we create a new data set by typing

```
> Example_14.1 <- data.frame(exhau = gl(2, k = 6, length = 12),
+                            noise = gl(3, k = 2, length = 12),
+                            z = z)
```

i.e. we create two factors by using the function `gl()`, namely the factor *exhaustion* `(exhau)`, with 2 levels and `k = 6` observed values each and so 12 `(length = 12)` values in total on the one hand, and the factor `noise` with 3 levels and `k = 2` observed values each on the other hand. We submit both factors and the transformed values in the object `z` as arguments to the function `data.frame()`. We assign the newly created data set to the object `Example_14.1`. Using this data set, we can conduct a two-way analysis of variance in accordance to Example 10.12.

In SPSS, we open a new data sheet (File – New – Data) and type in the values for the characters *exhaustion* (exhau), *noise* (noise), and *omissions* (omiss) column by column. Then, we create the link function per y_{ijv} by following the steps described in Example 5.3 (Transform – Compute Variable...) and input the new variable z as Target Variable: in Figure 5.9. Next, we select Arithmetic from the panel Function group:, as a consequence of which a list appears in Functions and Special Variables: from which we select Ln. Hence, a text box is opened on the left, above which one can find an upwards arrow, which is to be clicked. By doing so, the line LN(?) appears, beneath Numeric Expression. We replace the question mark accordingly with omiss + 1/2. By clicking OK, the data sheet will be complemented with the variable z.

With the thusly obtained values, we are able to conduct a two-way analysis of variance analogously to Example 10.13.

Because the given data constitute a very small sample, which is a rather unrealistic case, we will not perform the calculation.

Universality of the generalized linear model comes from the fact that the presumption of normally distributed error terms with equal variances, given for the traditional linear models, is not necessary.

Doctor

In the Poisson distribution, for example, the mean equals the variance; thus, the error terms are not distributed with equal variance.

Incidentally, the logistic regression in Formula (13.9) is a special case of the generalized linear model.

14.1.2 Model with latent variables

With reference to the previously described methods, a simple description of the concept of latent variables can be given as follows. Imagine the problem of partial correlation. Instead of, as illustrated in Section 12.1.1, knowing one or more noise factors, for example z, and having sampled data accordingly for each research unit as well, we now just hypothesize variables: thus, one models certain dependencies between the observed (*manifest*) and the non-observable (latent) variables. In this, the effect of those dependencies is unknown most of the time. Sometimes, these dependencies are even considered to be causally directed.

Master
Doctor

Example 14.2 The dependency of solving item x as well as item y of an intelligence test on the testees' degree of ability

Most of the time, the items of an intelligence (sub-) test are constructed so that each of them can be solved completely independently from one another. Thus, they do not build on each other. A testee may solve or not solve the task x: 'father : mother = son : ?' (? = 'daughter'), independent from whether he/she previously solved the task y: 'right : left = up : ?' (? = 'down'); the situation would be different for the items i: 'father : mother = son : X' and the item l: 'X : niece = ? : nephew', because the solution of item i ('X = daughter') is necessary to get to the solution of item l (? = 'son'). Now, given some testees, if one calculates the ϕ-coefficient (see Section 11.3.5) with respect to correct and wrong answers between the items x and y, there would generally be found a rather high resulting value, despite the independence of the two items (characters). The reason is model based; that is, the event or measurement value 'solved', respectively, depends for both items on a modeled ability parameter (for example ξ), which is different for every testee. Thus, the two random variables $\boldsymbol{x}$ and $\boldsymbol{y}$, for which outcomes are given, are in some modeled association with the non-observable latent random variable ξ. Then, the methodological challenge lies in ascertaining the values of ξ based on the knowledge of the data for x and y; or to ascertain the percentage of explained variance of x and y through ξ.

14.2 Methods for determining the quality and excellence of a model

Within the sections on regression analysis, we first realized that, although a certain model can be used on observed data and that it is possible to get results from it, the model may be improper (see Figure 11.13 of Example 11.6, wherein a linear regression is ascertained, but the scatter plot of the data points covers the circumference of a circle rather than a straight line). We stipulated, hence, that for the linear regression as a quite specific model, one should determine the extent of model fit before using the results – i.e. before using the regression function to predict (future) values of the character of interest. This was managed by (Pearson's) correlation coefficient and the coefficient of determination, respectively; thus, this is the second approach for determining the quality and excellence of a model described in Section 14.1. In particular, using the approach described in Section 11.4 as well, where the idea is to test the hypothesis of whether (Pearson's) correlation coefficient (and the coefficient of determination, respectively) fall in the population below a certain cut-off value, the researcher can test a model's fit according to his/her own conceptualized criterion. Basically, it is about the decision of whether or not the model fits the data in a sufficient way. For that approach, we have already offered all needed equipment in detail. This especially concerns planning a study and its analysis, in which the effect size is some coefficient of determination.

The concept of examining whether the empirically obtained data sufficiently conform to the respective model is directly realized by so-called *goodness of fit tests*.

14.2.1 Goodness of fit tests

Basically, these can be characterized as tests which compare the frequencies that would be hypothetically expected by a certain model with the frequencies we actually observed, by testing the resulting differences as to whether they are significant or not.

This almost always leads to some variation of the χ^2-test from Formula (9.3). In practice, this is about Formula (8.11) most of the time, which we introduced in order to test the parameters p_j of a multi-categorical, nominal-scaled character, which we modeled with a hypothetical distribution (for example that the probabilities of all categories are equal to each other), for their fit to the data.

Herein, it is not important whether the interesting character is actually nominal scaled or even interval-scaled after all; in the latter case, the measurement values would have to be merged into categories within fixed intervals.

Example 14.3 Numeric example without relation to any content

By means of a given sample and concerning a certain character y, we want to test the null hypothesis that the corresponding modeled random variable $\boldsymbol{y}$ is normally distributed in the population. For illustration, we use the data from Example 1.1 and examine the character *Everyday Knowledge, 1st test date*; because of information already indicated on this character, the null hypothesis is substantiated as H_0: '$\boldsymbol{y}$ is distributed as $N(\mu = 50, \sigma^2 = 100)$'. But, it would be just as possible to use H_0: '$\boldsymbol{y}$ is distributed as $N(\hat{\mu} = \bar{y}, \hat{\sigma}^2 = s^2)$'. We decide to use the second variant, due to its higher relevance to praxis. The alternative hypothesis is given, consequentially. We will use the χ^2-test in accordance with Formula (8.11); the type-I risk is $\alpha = 0.01$.

In **R**, we activate the data set *Example_1.1* (see Chapter 1) by using the function attach(), and ascertain the deciles by typing

```
> decile <- quantile(sub1_t1, seq(0.1, to = 0.9, by = 0.1),
+                    type = 6)
> print(decile)
```

i.e. we submit the character *Everyday knowledge, 1st test date* (sub1_t1) as the first argument to the function quantile(); as the second argument we use a numerical series from 0.1 to 0.9 incremented by 0.1, which is generated with the function seq(), and as a third argument we select, with type = 6, the same procedure as in SPSS. We assign the result to the object decile and set it as an argument in the function print().

As a result, we get:

```
10%  20%  30%  40%  50%  60%  70%  80%  90%
37.0 44.0 48.6 50.0 54.0 58.0 60.0 61.0 65.0
```

Now, we ascertain the absolute frequencies within each of these intervals; we type

```
> n.e <- table(cut(sub1_t1, c(min(sub1_t1), decile, max(sub1_t1)),
+                  include.lowest = TRUE))
> print(n.e)
```

i.e. we apply the function cut(), using the character *Everyday knowledge, 1st test date* (sub1_t1) as the first argument and a vector created with the function c() as the second argument, which, with the help of the functions min() and max(), includes the minimum and maximum as well as the decile and states the desired allocation. With include.lowest = TRUE, we always add the lowest value to the allocation; we submit the results, as deciles and allocated values to the function table(). The results of this operation are assigned to the object n.e, which we submit as an argument to the function print().

This results in

```
[25,37]    (37,44] (44,48.6] (48.6,50]    (50,54]    (54,58]
     12          9         9        14         11          9
(58,60]    (60,61]    (61,65]    (65,71]
     13         10          8          5
```

These are the desired absolute frequencies $h_j, j = 1, \ldots, c = 10$, which we have to insert into Formula (8.11); for them: I_1: $-\infty < y_v \leq 37$, I_2: $37 < y_v \leq 44$, I_3: $44 < y_v \leq 48.6$, I_4: $48.6 < y_v \leq 50$, I_5: $50 < y_v \leq 54$, I_6: $54 < y_v \leq 58$, I_7: $58 < y_v \leq 60$, I_8: $60 < y_v \leq 61$, I_9: $61 < y_v \leq 65$, I_{10}: $65 < y_v \leq \infty$. Now, we type

```
> decile.p <- pnorm(decile, mean = mean(sub1_t1),
+                   sd = sd(sub1_t1)) * 100
> n.p <- c(decile.p, 100) - c(0, decile.p)
> print(n.p)
```

i.e. we calculate the probabilities of the normal distribution (`pnorm`) that one would expect under the assumption of the null hypothesis, per decile (`decile`) for a given mean (`mean`) and a given standard deviation (`sd`). In order to obtain the expected (absolute accumulative) frequencies, we multiply the results by 100 and assign them to the object `decile.p`. As a next step, we ascertain the expected values e_j, $j = 1, 2, \ldots, 10$, and assign them to the object `n.p`, which we in turn submit to the function `print()`.

As a result, we get:

```
      10%        20%        30%        40%        50%        60%       70%
6.548019 13.778956 14.761726  5.162071 15.391149 14.575182 6.361338
      80%        90%
2.874075  9.252364 11.295120
```

[Note: the **R**-output does not print 100% above the last entry 11.295120]
Now, we type

```
> chi.e <- sum((n.e - n.p)^2/n.p)
> pchisq(chi.e, df = 7, lower.tail = FALSE)
```

i.e. we insert these results into Formula (8.11), thus ascertaining the empirical χ^2-value, which we assign to object `chi.e`. Next, we use the object `chi.e` as the first argument and $df = c - 3 = 10 - 3 = 7$ degrees of freedom as the second argument in the function `pchisq()` (one additionally loses two degrees of freedom, because it was necessary to estimate two parameters, the mean and the variance).

As a result, we get

```
[1] 1.338433e-09
```

Because of a p-value of 0.000, the result of the test is significant. Thus, the null hypothesis has to be rejected. The data does not stem from a normal distribution.

In SPSS, we proceed analogously to Example 5.2 (Analyze – Descriptive Statistics – Frequencies...) and select the character *Everyday Knowledge, 1st test date.* We compute the necessary deciles, namely the 0.10-quantile, the 0.20-quantile etc. by clicking on Statistics... and selecting Cut points for: in the resulting window of Figure 5.24, where the default value 10 appears. We also set check marks at Mean and Std. deviation. By clicking Continue and OK, we obtain the results in Table 14.1. Additionally, we take note of : $\hat{\mu} = \bar{y} = 52.54$ and $\hat{\sigma} = s = 10.289$.

Table 14.1 SPSS-output of the deciles in Example 14.3.

Everyday Knowledge, 1st test date (T-Scores)

		Frequency	Percent	Valid Percent	Cumulative Percent
Valid	25	1	1.0	1.0	1.0
	27	1	1.0	1.0	2.0
	31	2	2.0	2.0	4.0
	33	2	2.0	2.0	6.0
	35	2	2.0	2.0	8.0
	37	4	4.0	4.0	12.0
	41	3	3.0	3.0	15.0
	42	4	4.0	4.0	19.0
	44	2	2.0	2.0	21.0
	46	7	7.0	7.0	28.0
	48	2	2.0	2.0	30.0
	50	14	14.0	14.0	44.0
	52	5	5.0	5.0	49.0
	54	6	6.0	6.0	55.0
	56	2	2.0	2.0	57.0
	58	7	7.0	7.0	64.0
	60	13	13.0	13.0	77.0
	61	10	10.0	10.0	87.0
	63	1	1.0	1.0	88.0
	65	7	7.0	7.0	95.0
	69	3	3.0	3.0	98.0
	71	2	2.0	2.0	100.0
	Total	100	100.0	100.0	

From Table 14.1, we take the absolute frequencies $h_j, j = 1, \ldots, c$ of the intervals I_1: $-\infty < y_v \leq 37$, I_2: $37 < y_v \leq 44$, I_3: $44 < y_v \leq 48.6$, I_4: $48.6 < y_v \leq 50$, I_5: $50 < y_v \leq 54$, I_6: $54 < y_v \leq 58$, I_7: $58 < y_v \leq 60$, I_8: $60 < y_v \leq 61$, I_9: $61 < y_v \leq 65$, and I_{10}: $65 < y_v \leq \infty$, which are 12, 9, 9, 14, 11, 9, 13, 10, 8, and 5. These are the ones we have to insert into Formula (8.11). Now, we still need the frequencies expected assuming the null hypothesis is valid. We obtain them by following the steps (Transform – Compute Variable...) described in Example 5.3. By doing this, we get to the window in Figure 5.9, where we type in a new Target Variable:, for example DF1 for distribution function; then, we select CDF & Noncentral CDF from Function group:. This opens a list in Functions and Special Variables: from which we select Cdf.Normal; after this, a text box opens on the left side, above which we find an upwards arrow, which we click. Subsequently, CDF.NORMAL(?,?,?) appears as Numeric Expression. For the first interval, we replace the question marks with: 37,52.54,10.289. By clicking OK, the variable DF1 is added to the data frame – respectively, for every elementary unit with the same value. Now, we proceed analogously for every one of the remaining nine intervals, defining DF2 etc. and replacing the question marks with the respective observed frequency and the mean and the standard deviation.

From the Data View, we extract the 10 expected relative cumulative frequencies: 0.07; 0.20; 0.35; 0.40; 0.56; 0.70; 0.77; 0.79; 0.89; and 1.00. Multiplied with $n = 100$, we get the expected absolute cumulative frequencies, from which one can easily calculate the expected values $e_j, j = 1, 2, \ldots, 10$: 7, 13, 15, 5, 16, 14, 7, 2, 10, 11. Inserted into Formula (8.11), this amounts to $\chi^2 = 67.57$, with $df = c - 3 = 10 - 3 = 7$ degrees of freedom (two degrees of freedom are additionally lost due to the necessity of estimating two parameters, the mean and the variance). In Table B3 in Appendix B, we look up the 0.95-quantile of the χ^2-distribution, which is 14.07; since this value is smaller than the computed value of the test statistic, the null hypothesis has to be rejected. The data does not stem from a normal distribution.

Instead of the χ^2-test, one can also use several other tests to test a random variable's normal distribution; the best-known test among these is the *Kolmogorov–Smirnov test*. Because of reasons we have already discussed (see for example Section 8.5.5), we will not expand on this test here.

14.2.2 Coefficients of goodness of fit

Instead of solving the problem of evaluating how well the empirical data conform to the respective model by methods of inferential statistics, quite often just methods of descriptive statistics are used. Thus, here there are no longer any goodness of fit tests but – similar to the case of correlation coefficients and coefficients of determination – just some coefficients for describing the goodness of fit.

Master Doctor

Mainly, this deals with so-called 'measures of information', based on (mathematical) information theory.[2] Best known are the *Akaike information criterion* (AIC) and the *Bayesian information criterion* (BIC).[3] Both correspond to the likelihood function (see Section 6.5) of the data given that the respective model is true. If $L(y \mid \hat{\theta})$ is the respective likelihood – y being, quite generally, defined as the data observed from n research units, and $\hat{\theta}$ being generally defined as the vector of k (estimated) parameters of the model, the error variance included – then the following applies:

$$AIC = 2 \cdot k - 2 \cdot \ln L(y|\hat{\theta}) \tag{14.1}$$

$$BIC = k \cdot \ln n - 2 \cdot \ln L(y|\hat{\theta}) \tag{14.2}$$

Basically, these criteria quantify the loss of information that results by not communicating the data with their full complexity but (simply) via the respective model. That is, the smaller the value of the criterion, the better the model fits

[2] Within this, 'information' is defined as a quantitative measure which results through the mutual exchange of signs/symbols between a transmitter and a receiver.

[3] Within statistics, there exists the approach of Bayes (*Thomas Bayes*, an English monk from the nineteenth century). It is founded on the assumption that the parameters of a distribution are themselves random variables, which follow a distribution as well; in particular, this concerns expectation value and variance. In this book, we will not discuss this further, but refer to Bolstad (2007).

the data. The preference for one criterion over the other depends on whether the sample size should be taken into account. The larger it is, the larger the coefficient *BIC*.

Obviously, the statistical coefficients *AIC* and *BIC* are not interpretable in an absolute way. They are only suited for the case in which one wants to compare (at least) two concurrent models with respect to their fit. In this way, they refer to the third approach to ascertain the quality and excellence of a model mentioned in Section 14.1.

As always when one exclusively uses descriptive statistics, a certain insecurity concerning their interpretation remains in their application: when exactly is a model's quality high or higher? Actually, *AIC* and *BIC* are not much more than relatively new coefficients with regards to a test. If one wants to compare two concurrent models, the *likelihood-ratio test* is a universal instrument. With this, one merely puts the likelihood of both models into relation, basically like this (the index of $\hat{\theta}$ refers to the corresponding model):

$$\lambda = -2 \cdot \ln \frac{L(\boldsymbol{y}|\hat{\theta}_1)}{L(\boldsymbol{y}|\hat{\theta}_2)} \tag{14.3}$$

Here, the model with the higher number of parameters is put into the denominator – actually, always the saturated model (see Section 12.1.5), i.e. the one with maximal parameterization. As can be shown, the test statistic λ can, in principle, be approximated by a χ^2-distributed test statistic, with $df = k_2 - k_1$ degrees of freedom. However, bear in mind, that in praxis this approximation is taken for granted most of the time, without actually proving it (see, for example, Hohensinn, Kubinger, & Reif, in press).

Doctor

Incidentally, both concurrent models can be interpreted as null hypothesis on the one hand and as alternative hypothesis on the other hand. For this, practically every test can be formulated as a likelihood-ratio test.

Given in regression analysis the two characters, $\boldsymbol{x}$ and $\boldsymbol{y}$, are to be modeled as normally distributed variables, Formula (14.1) becomes:

$$AIC = 2k + n \ln (2\pi) + n + n \ln \left[\frac{(n - k + 1)MS_{\text{res}}}{n} \right] \tag{14.4}$$

wherein MS_{res} means the mean sum of squares according to Table 10.10 in Section 10.4.4.1, or that is to say: $MS_{\text{res}} = s_y^2$.

Especially if one wants to compare models with different numbers of parameters (for example whilst comparing linear, quadratic, and logistic regressions), the coefficient *AIC* is superior to the estimated error variance as a measure of goodness of fit. Indeed, since models always fit the data better when they incorporate more parameters (see the example for multiple regression in Section 14.2.3), it is recommendable to use a coefficient for describing the goodness of fit which takes the number of estimated parameters into account. This is the case for the *AIC*, but not for the error variance as a criterion.

Doctor

Example 11.12 – continued

This was about the comparison between the linear regression and two curvilinear regressions, specifically the quadratic and the logistic regression, on the basis of the two characters *Applied Computing, 1st test date* and *Applied Computing, 2nd test date* from Example 1.1. We now want to determine which of the three models explains the data best. For didactic reasons, we add a fourth model: the *cubic regression*. This is: $\boldsymbol{y}_v = \beta_0 + \beta_1 \boldsymbol{x}_v + \beta_2 \boldsymbol{x}_v^2 + \beta_3 \boldsymbol{x}_v^3 + \boldsymbol{e}_v$. We calculate the *AIC* only.

In **R**, we compute the regression functions by typing

```
> lm.lin <- lm(sub3_t2 ~ sub3_t1)
> lm.quad <- lm(sub3_t2 ~ sub3_t1 + I(sub3_t1^2))
> lm.cub <- lm(sub3_t2 ~ sub3_t1 + I(sub3_t1^2) + I(sub3_t1^3))
> nls.log <- nls(sub3_t2 ~ SSlogis(sub3_t1, Asym, xmid, scacl))
```

i.e. we submit the formulas of the linear, quadratic, cubic, and logistic regression to the function `lm()` and `nls()` (see, apart from the cubic, the earlier work on Example 11.12 in Chapter 11) and assign the results to one object each. Now, we compute the error variance of all the regression models by entering

```
> anova(lm.lin)$Mean
> anova(lm.quad)$Mean
> anova(lm.cub)$Mean
> sum(residuals(nls.log)^2)/df.residual(nls.log)
```

i.e. we use the respective regression function in the corresponding objects as an argument in the function `anova()` and select, with `$Mean`, among other statistics the looked-for error variance. As concerns the logistic regression function, we compute the error variance on its own, using the functions `sum()`, `residuals()`, and `df.residual()`.

As a result, we get (shortened output):

```
[1] 30.36833
[1] 30.20032
[1] 30.24328
[1] 30.3273
```

Now, we calculate the coefficient *AIC* by typing

```
> AIC(lm.lin)
> AIC(lm.quad)
> AIC(lm.cub)
> AIC(nls.log)
```

i.e. we submit each of the respective regression models to the function `AIC()`.

This yields

```
[1] 629.1075
[1] 629.527
[1] 630.6329
[1] 629.9466
```

(see also these results in Table 14.2).

Table 14.2 Determining the model quality and excellence for the linear, the quadratic, the cubic, and the logistic regression in Example 11.12.

Regression function	Coefficient of determination	MS_{res}	Number of parameters	AIC
Linear	0.609	30.368	3	629.108
Quadratic	0.615	30.200	4	629.527
Cubic	0.618	30.243	5	630.633
Logistic	0.613	30.327	4	629.947

In SPSS, we obtain the error variance in the same way as in Example 11.5; back then, in the window in Figure 11.5, we dragged and dropped the character Applied Computing, 2nd test date into the field Dependent: and the character Applied Computing, 1st test date into the field Independent(s):. Now, repeating this sequence, clicking OK suffices. In Example 11.5, we did not display the full results, which we will now make up for in Table 14.3. Thus, we realize that, within a regression analysis, decomposing the sums of squared deviations analogously to the analysis of variance is possible. In order to describe the goodness of fit of the linear regression to the data, we need the error variance $MS_{res} = 30.368$. The smaller its value, the better the fit of the model.

Table 14.3 SPSS-Output of the error variance MS_{res} in Example 11.5 (shortened output).

ANOVA[b]

Model		Sum of Squares	df	Mean Square	F	Sig.
1	Regression	4632.663	1	4632.663	152.549	.000[a]
	Residual	2976.097	98	30.368		
	Total	7608.760	99			

a. Predictors: (Constant), Applied Computing, 1st test date (T-Score)
b. Dependent Variable: Applied Computing, 2nd test date (T-Scores)

In order to obtain the error variances for the other regression functions, we follow the command sequence (Analyze – Regression – Curve Estimation...) analogously to

Example 11.5 to open the window in Figure 11.8. There, we select, in addition to Linear (in that, the previous calculation has been redundant) Quadratic, but do not select Cubic and Logistic since SPSS, by this function (Analyze – Regression – Curve Estimation...), does not yield the results we are looking for. Now, we also select Display ANOVA table and, after clicking OK, arrive at the result that is given in Table 14.2. The coefficient of determination is also displayed there. For the cubic regression function, we use the sequence of commands analogous to Example 11.12 in Section 11.3.6 concerning the logistic regression function; that is we program the function by ourselves. As a result, we get $MS_{\text{res}} = 30.243$ for the cubic, and $MS_{\text{res}} = 30.327$ for the logistic regression function. The coefficient of determination is 0.618 for the cubic and 0.613 for the logistic regression function. We manually calculate the coefficient *AIC* according to Formula (14.4). The result is displayed in Table 14.2 as well.

Obviously, the coefficient of determination is not very successful in differentiating the several given regression models. If one in turn examines the coefficient AIC, it is clear that the linear regression fits best.

14.2.3 Cross-validation

Goodness of fit tests are often not compelling enough, especially if a high number of parameters is to be estimated (as in, for example, the multiple linear regression; see Section 12.1.3) and/or certain strategies of parameter estimation are applied that are extremely dependent on the concrete data (for example the stepwise linear discriminant analysis). That is, though a model might fit the data according to goodness of fit tests, the generalizability of the results is to be doubted. For this, there are additional methods.

This includes the procedure in connection with the (linear) discriminant analysis already described in Section 13.3; that is to use only the data of a part of the given research units for the analysis, as a training set, and to use the rest of the data as a testing set for the classification analysis in order to examine the predictive value of the discriminant function obtained.

A special method of examining the generalizability of a result is proposed by *cross-validation*. In a traditional manner, one splits the data into parts, most often two halves, and conducts the respective analysis separately for each part of the data. Then, one uses each of the estimated parameters to describe or predict the respective other part(s) of the data. Only if this succeeds in an acceptable manner, is the desired generalizability justified. The validity of the results is established crosswise, so to speak. Diverging from this classic procedure, one could of course also sample two sets of data consecutively and cross-validate them afterwards.

For Lecturers:

For demonstration that fitting given data by models with a vast amount of parameters can lead to completely misleading results, the lecturer might ascertain data from his students for the calculation of multiple linear regression, multiple linear correlation coefficient included. First, the regressand might be the score of a short repetition test in statistics (or of any mathematical school performance test). Secondly, as regressors the following characters might serve: size of shoe, street number of the home address, mobile number, income from (side-) jobs, commuting time in minutes, and social insurance number. As experience teaches us, this will result in a relatively strong (linear) relationship with a relatively high multiple correlation coefficient. If the chosen regressors were not that obviously meaningless, one might be inclined to try to predict the score of a student in the respective test with the outcomes of the regressor characters through the obtained regression function.

Thus, given a sufficiently large number of parameters, a regressand can be accurately described (predicted) by the regressors, even if there is no systematic relationship between them at all. If one chooses the degree of a nonlinear regression (i.e. a polynomial function) in such a way that the number of regression coefficients equals the number of the observed research units, then the resulting curve will go exactly through every point of the scatter plot – simply analogous to the case where two points can always be described by a line that goes through both of them.

A cross-validation – for example by partitioning the research units (students) into the ones in the front rows on the one hand and the ones in the back rows on the other hand – would reveal that the regression coefficients merely use a randomly given portion of the error variance which happens to be unequal to zero within a certain (part of a) sample. The correlation coefficients between observed score and predicted score are almost zero for each of the parts of the sample, if one uses the regression coefficients of the respective other part of the sample for the prediction.

Doctor

Example 14.4 The technique of cross-validation as statistical test's concept

The non-parametric discriminant analysis by Kubinger (1983; see Section 13.4) is heavily based on the method of cross-validation. For one of the two randomly-split halves of research units of the sample, one looks (stepwise) for those measurement values per character which, at least in one of the groups (populations) that are to be discriminated, are not (or considerably less often) realized, but are indeed realized in the other(s). The resulting allocation rules are then used for allocating the other half of the research units to the groups in question. The (averaged) number of hits, that is the number of correct allocations, is then tested using the binomial test, to see whether the respective rate of hits is significantly higher than the rate that would be expected if only random allocation applies.

Master Doctor

Sometimes, reference to cross-validation is taken without there actually being one. This mostly happens when result-based changes are made to a model in contrast to the original model (see for example in connection with the calibration of a psychological test according to the Rasch model in Section 15.2.3.1). In such a context, often the claimed 'cross-validations' refer to the concept that the *a posteriori* obtained model structure and model fit, respectively, are to be confirmed with new, independent data. However, in this there is no objective of explaining data from a certain study by the parameter estimates from another study (and vice versa). Unfortunately, there is no common notation for the thus-described procedure, so that it is just called a 'kind of cross-validation'.

Doctor Another approach to examine the generalizability of results is *jackknifing*. This procedure first partitions the research units into a fixed number of sub-samples. Then it estimates the desired parameters or a function thereof in each one of the sub-samples. From the thus-resulting distribution of estimates, their variance or standard error can be calculated; if applied to the classification analysis of discriminant analysis, one would have to determine the discriminant function for one of the sub-samples and then use it to predict the group membership for all the other research units. In the extreme case, one applies the strategy of excluding one research unit after the other from analysis, using all of the remaining ones to estimate the desired parameters (*'leaving one out'* method). Again, transferred to the classification analysis of discriminant analysis, one would predict every single research unit by the remaining ones.

14.3 Simulation – non-analytical solutions to statistical problems

We already have often seen that *simulation studies*[4] can be used to obtain profound knowledge in statistics; aside from the fact that we have, for didactic reasons, used multiple simulations in order to demonstrate certain issues (see for example Section 7.2.1). Especially the high complexity of theory-generating statistical methods makes it a necessity to capture the basically not (or not easily) analytically obtainable distribution function of the test statistic via simulation studies.

The basic idea of this is simple, although the practical feasibility depends largely on contemporary computer qualities.

Doctor The basis for most simulation studies is the generation of (pseudo-) random numbers (see Section 7.2.1). For this, most often the set of real numbers within the interval [0,1] is used. However, because of the finite (though very large) word length of number representation within computers, only a finite set of numbers is producible.

Doctor **Example 14.5** Generation of random numbers

In the following, we show how to get a random variable of a certain type of distribution; namely how to get factitious generated measurement values for a normal distribution.

In **R**, we type

```
> rnorm(100, mean = 100, sd = 10)
```

i.e. we set the number of observations `100` as the first argument in the function `rnorm()`; with the second argument, we set the mean to `100`, and with the third argument the standard

[4] Even to this day, this is sometimes called 'Monte Carlo simulation'; the historical background for this dates back to 1946, where a secret research project with this codename incorporated this method; *John (Johann) von Neumann* chose the name for the project with regards to the casino in Monte Carlo.

deviation to `10`, in order to define the normal distribution from which the observed data is to stem.

As a result, we get (shortened output):

```
[1]  96.14798  97.82836  96.94690  97.58045  93.44530  99.80330
...
[97] 108.02885 109.74393 110.82786  95.45455
```

In **R**, we are able to generate random numbers of any desired distribution. At this point, we want to refer to the help pages of the respective functions: for example `help(rbinom)` for the binomial distribution, `help(rchisq)` for the χ^2-distribution, `help(rexp)` for the exponential distribution, `help(rf)` for the F-distribution, and `help(rt)` for the t-distribution.

In SPSS, we select

Transform
    Random Number Generators...

and select Set Active Generator and Set Starting Point in the resulting window (Figure 14.1); by doing this, the default option Random will become active. After clicking OK, we follow the steps (Transform – Compute Variable...) described in Example 5.3, and get to Figure 5.9, where we enter a new, arbitrary Target Variable:. Then, we select Random Numbers from Function group: and Rv.Normal from Functions and Special Variables:. Then a text box opens on the left; next, we click the upward arrow above it. Subsequently, RV.NORMAL(?,?) appears as Numeric Expression:, and we substitute the desired mean and standard deviation for the question marks. By clicking OK, the new variable will be added to the SPSS Data View.

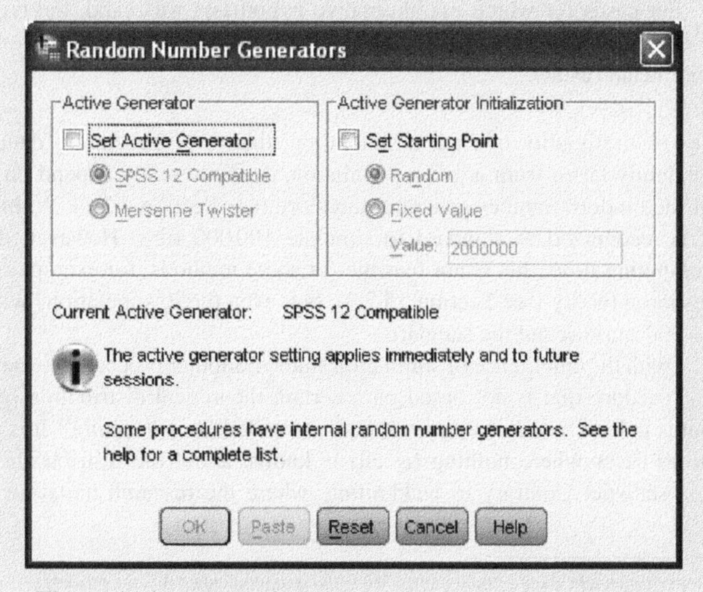

Figure 14.1 SPSS-window for generating random numbers.

Doctor

Example 14.6 Generating distributions that deviate from the normal distribution, using normally distributed random numbers

Starting from random numbers that are distributed as $N(0,1)$, we call the respective random variable $\boldsymbol{u}$. We are interested in deviations from the normal distribution concerning skewness and kurtosis. With help of the so-called *Fleishman transformation* (Fleishman, 1978), it is possible to ascertain the coefficients of the polynomial $\boldsymbol{y} = a + b\boldsymbol{u} + c\boldsymbol{u}^2 + d\boldsymbol{u}^3$ in $\boldsymbol{u}$ in such a way that the resulting random variable $\boldsymbol{y}$ has a mean of zero, a variance of 1, the skewness γ_1, and the kurtosis γ_2.

Doctor

Example 14.7 Simulation study on the two-sample t-test's power if the presumption of variance homogeneity is violated (Rasch, Kubinger, & Moder, 2011)

First, the parameters μ_1, μ_2, σ_1, and σ_2, as well as the type-I risk α were determined for several cases; also, different distributions were chosen, wherein the extent of their deviation from the normal distribution was specified via skewness and kurtosis (see Section 5.3.3). Finally, also different sample sizes n_1 and n_2 were used. Basically two kinds of cases were under consideration: on the one hand those cases in which the null hypothesis H_0: $\mu_1 = \mu_2$ is valid, and on the other hand those in which the alternative hypothesis H_A: $\mu_1 \neq \mu_2$ is valid (with respect to different values of $\delta = \mu_1 - \mu_2$). Then, with the help of a random number generator, 100 000 runs of simulation were performed for each parameter configuration; i.e. 100 000 sets of data consisting of random samples with outcomes $y_{11}, y_{12}, \ldots, y_{1n_1}$ and $y_{21}, y_{22}, \ldots, y_{2n_2}$. For each set of data, the two-sample t-test was calculated, and how often it resulted in significance was counted. For cases for which the null hypothesis was valid, the actual type-I risk α_{act} could be established, being the relative frequency of observed significant results – this actual type-I risk was then compared with the nominal type-I risk α. For cases for which the alternative hypothesis was valid, the type-II risk β could be estimated, being one minus the relative frequency of observed significant results.

Doctor

Today, the quality of such a simulation, that is of generating data which are randomly taken from a given population, does not really depend on the quality of the random-number generator anymore (see Section 7.2.1). Within statistics, it is recognized as standard to simulate 100 000 runs. However, due to high computing time, this is not feasible for some methods, for example within item response theory (see Section 15.2.3); but, even there, simulations with less than 10 000 runs are not the standard.

With the emergence of simulation studies, another procedure became feasible; a procedure that is not based on a certain theoretical distribution (population), but is indeed based on empirical data: so-called *bootstrapping*.[5] It is appropriate in all cases where nothing (at all) is known about the distribution of a given test statistic. Contrary to jackknifing, where the research units are partitioned

[5] The name stems from a tale about a count named *Münchhausen* (coincidentally, the procedure is called *Münchhausen* procedure in German), who claimed that he pulled himself out of a swamp by his own bootstraps. Here, the population is simulated on the basis of the actually observed result.

into an arbitrary number of non-overlapping sub-samples that include all units, bootstrapping repeatedly draws equally sized samples with replacement (man spricht von *resampling*) from the total sample. Through this it is possible to establish much more precisely the distribution and standard error, respectively, of the estimator of the respective parameter or the respective test statistic.

Doctor

Example 14.8 Bootstrapping for determining the confidence interval of ρ

We start from the data of the subtests *Everyday Knowledge* and *Applied Computing* from Example 1.1, both from the first test date. In Example 11.13, we calculated Pearson's correlation coefficient ($r = \hat{\rho} = 0.432$), as well as the confidence interval by the (approximation) Formula (11.18). The limits 0.26 and 0.58 resulted.

In **R**, we create a new function by typing

```
> cor.boot <- function(x, y, nrep, alpha = 0.05) {
+   size <- length(x)
+   data <- numeric(nrep)
+   for(i in 1:nrep) {
+     index <- sample(1:size, size = size, replace = TRUE)
+     data[i] <- cor(x[index], y[index])
+   }
+   r <- cor(x, y)
+   CI.u <- r + (sd(data) * qt(alpha/2, df = nrep-1))
+   CI.o <- r + (sd(data) * qt(1 - alpha/2, df = nrep-1))
+   cat((1 - alpha)*100, "%", "confidence interval of r =",
+       round(r, digits = 3),"\n",
+       "lower bound: ", round(CI.u, digits = 3), "\n",
+       "upper bound: ", round(CI.o, digits = 3), "\n")
+   return(invisible(list(data = data, r = r, CI = c(CI.u, CI.o))))
+ }
```

i.e. we use the function `function()` and set the variables `x` and `y`, the number of runs `nrep`, and `alpha = 0.05`, as the value for the type I risk, as the arguments of the new function. The sequence of commands inside the curly brackets specifies the inner working of the function and will not be discussed.

Now, we type

```
> cor.boot(sub1_t1, sub3_t1, nrep = 50, alpha = 0.05)
```

i.e. we use the characters *Everyday knowledge, 1st test date* (`sub1_t1`) and *Applied computing, 1st test date* (`sub3_t1`) as arguments in the function `cor.boot()` and, with `nrep = 50` request 50 samples to be drawn.

As a result, we get:

```
95 % confidence interval of r = 0.432
lower bound:    0.226
upper bound:    0.637
```

In SPSS, we select

Analyze
  Correlate
    Bivariate...

and, in the resulting window in Figure 11.14, drag and drop the two characters Everyday Knowledge, 1st test date and Applied Computing, 1st test date into the field Variables:; we remove the check mark at Flag significant correlations. Now, we click on the button Bootstrap..., which opens the window in Figure 14.2, where we select Perform bootstrapping and set the Number of samples: to 50; the default settings of 95 as the Level (%): of the confidence coefficient, and Simple in the panel Sampling, are appropriate for our purpose. After clicking Continue and OK, we get the results displayed in Table 14.4.

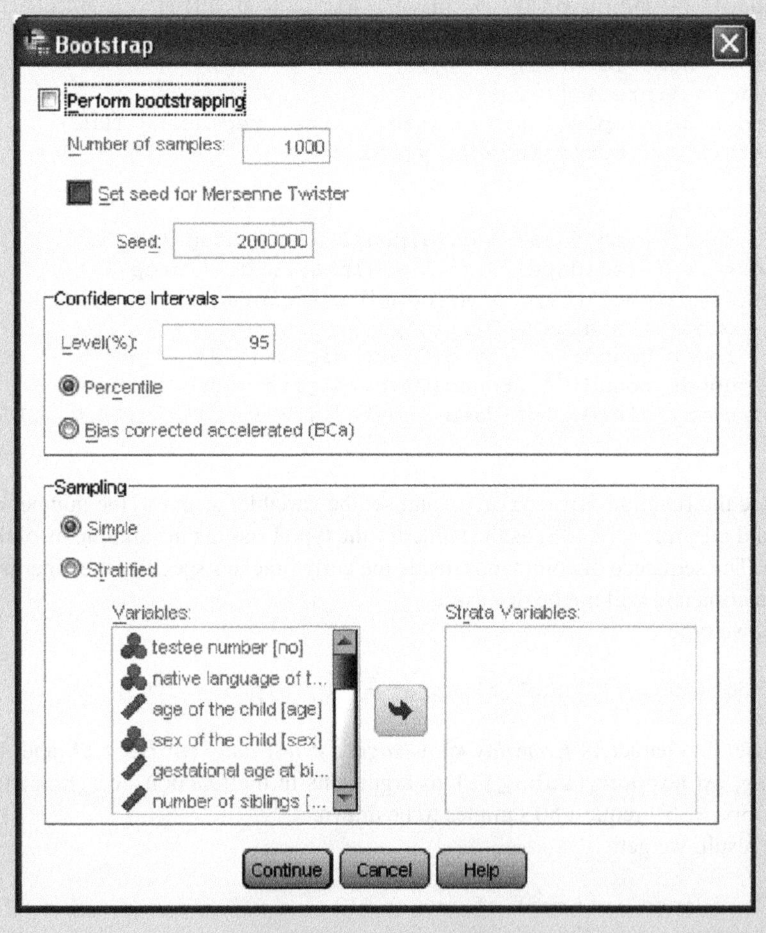

Figure 14.2 SPSS-window for conducting the bootstrapping procedure.

Table 14.4 SPSS-output of the bootstrapping procedure for determining the confidence interval of ρ (shortened output).

Correlations

				Everyday Knowledge, 1st test date (T-Scores)	Applied Computing, 1st test date (T-Scores)
Everyday Knowledge, 1st test date (T-Scores)	Pearson Correlation			1	.432**
	Sig. (2-tailed)				.000
	N			100	100
	Bootstrap[a]	Bias		0	.023
		Std. Error		0	.092
		95% Confidence Interval	Lower	1	.244
			Upper	1	.597

**. Correlation is significant at the 0.01 level (2-tailed).
a. Unless otherwise noted, bootstrap results are based on 50 bootstrap samples

Both results, with **R** as well as with SPSS according Table 14.4, differ enormously from the approximated result of Example 11.13 with the limits of 0.26 and 0.58. Even if we raise the number of drawn samples from 50 to 1000, we obtain (according to SPSS), with the limits 0.192 to 0.622, similarly big deviations; though, for bootstrapping the results agree quite well. As to which of the methods produces the more exact results, it cannot be said.

Doctor

Example 14.9 Applying the test statistic χ^2 in the case of a not-quite-sufficient approximation to the χ^2-distribution

For the χ^2-test, which we first introduced in Section 9.2.3, we gave the rule of thumb that, given a certain large sample, the approximation to the χ^2-distribution is sufficiently good. If this rule is not met, instead one can randomly draw very many samples from the data with replacement, thus re-simulating samples under the assumption that the actual observed data represent the population(s) in question in the best possible way.

In the case of having two independent samples with sizes n_1 and n_2, for which the dichotomous character, $\boldsymbol{y}$, has been ascertained, the null hypothesis H_0: $p_1 = p_2$ might be tested according to Formula (9.3) against the alternative hypothesis H_A: $p_1 \neq p_2$. If we now actually sample by bootstrapping from one (best: the larger) sample appropriately many (pairs of) samples of sizes n_1 and n_2 (that is the null hypothesis is true), then for each of these (pairs of) samples the test statistic of Formula (9.3) is to be calculated. As a result, one gains an 'empirical' distribution of that test statistic given the null hypothesis, this distribution being fairly exact. This will deviate considerably from the χ^2-distribution. By means of this 'empirical' distribution, the looked-for $(1 - \alpha)$-quantile can be ascertained, which serves for the comparison with the empirical (χ^2-test) statistic, observed for the original data; as a consequence, the null hypothesis is either to be accepted or rejected.

Summary

In psychology, we are often interested in statistical models with *latent variables*. Generally, there are different methods of determining the quality and excellence of a model. Very often, *goodness of fit tests* apply; regardless of these, there are several coefficients for describing the goodness of fit. Of these, the coefficient *AIC* is very common. If one wants to compare two concurrent models, the *likelihood-ratio test* is a universal method. If the generalizability of results, which were obtained via a certain model, is doubtful, then there are special procedures for examining this. In psychology, the method of *cross-validation* is a common one.

References

Bolstad, W. M. (2007). *Introduction to Bayesian Statistics* (2nd edn). Chichester: John Wiley & Sons, Ltd.

Fleishman, A. J. (1978). A method for simulating non-normal distributions. *Psychometrika*, *43*, 521–532.

Hohensinn, C., Kubinger, K. D., & Reif, M. (in press). On robustness and power of the asymptotically chi-squared distributed likelihood-ratio test as a model test of Fischer's linear logistic test model. *Multivariate Behavioral Research*.

Kubinger, K. D. (1983). Some elaborations towards a standard procedure of distribution-free discriminant analyses. *Biometrical Journal*, *25*, 765–774.

McCullagh, P. & Nelder, J. A. (1989). *Generalized Linear Models* (2nd edn). New York: Chapman & Hall/CRC.

Rasch, D., Kubinger, K. D., & Moder, K. (2011). The two-sample *t*-test: pre-testing its assumptions does not pay off. *Statistical Papers*, *52*, 219–231.

15

Theory-generating methods

In this chapter, various analysis methods for classifying research units by numerous characters are presented; exploratory and confirmatory factor analysis, which reduces multiple correlating characters to a few independent ones, is introduced. Additionally, the chapter describes path analysis and (linear) structural equation models, with which directed relationships can be analyzed. At the end, models of item response theory, particularly the Rasch model, are presented.

15.1 Methods of descriptive statistics

As concerns theory-generating methods, there are analysis methods, which classify research units according to similarities in the investigated characters, and there are methods that are based on intercorrelations of the investigated characters. The latter either aim to reduce the set of observed characters to a small number of latent, unobservable variables due to the characters' correlations; or they attempt to quantify one-sided, directed dependencies between the observed characters.

All these methods will be demonstrated in the following as far as they concern descriptive statistics. That is, no hypothesis testing applies; often hypotheses are not even contemplated. Data will instead only be analyzed or structured to elucidate relationships. Thus, these are to be characterized as 'exploratory' methods.

15.1.1 Cluster analysis

While multivariate analysis of variance and discriminant analysis, respectively, try to differentiate between the levels of a fixed factor with reference to $m > 1$ characters, i.e. to examine the null hypothesis that there are no respective differences, cluster analysis – the best known method of this type of analysis – aims to classify the research units of an originally

Statistics in Psychology Using R and SPSS, First Edition. Dieter Rasch, Klaus D. Kubinger and Takuya Yanagida.

undifferentiated sample as a whole into groups or subgroups of research units, which (or who) are 'similar' to one another. 'Similar' means that the outcomes within each of the subgroups of all m characters only show small differences.

Cluster analysis is appropriate for quantitative characters; no presumption about the characters' distribution is necessary because the analysis does no hypothesis testing. Nevertheless, the cluster analysis works analogously to the logistic regression from Section 13.4, even if there (also) are dichotomous characters and even multi-categorical or ordinal-scaled characters; for these last two, dummy variables need to be used.

If dealing only with dichotomous characters, there are two analysis methods that are better suited: *configuration frequency analysis* and *latent class analysis*. For both these analyses, statistical tests are available that can examine several hypotheses about the existence of groups and the goodness of data fit for a certain classification. Both methods are suited for multi-categorical characters, but not for quantitative ones – and also not for ordinal-scaled ones (for both methods see Section 15.2.1).

Master

Cluster analysis is based on graphical visualization of the data within the multidimensional space. For only three characters, y_1, y_2, and y_3, using the three outcomes as coordinates, a single point within the three-dimensional space represents any research unit. In fact, the first approaches to cluster analysis were heuristic ones; they tried to achieve a disjunctive grouping on the basis of graphical visualization. In Figure 15.1, for example, one can clearly split the scatter plot into two ovoids.

In general, cluster analysis concerns ascertainment of similarities between each of the research units, or rather the measurement of distances between them. Of all the numerous *distance measures*, the *Euclidian distance* is surely the most illustrative one. It deals with the rectangular coordinates of the m-dimensional space and the respective (diagonal) straight line in this space. The length of the line between two points quantifies the distance. In Figure 15.2, only two characters y_1 and y_2 are given, and these only for the two persons v and w. Thus, there are only two pairs of outcomes, (y_{1v}, y_{1w}) and (y_{2v}, y_{2w}). Obviously, the distance d_{vw} between the two research units v and w, defined via the respective straight line, constitutes the length of the diagonal of a square. In general, for the m-dimensional

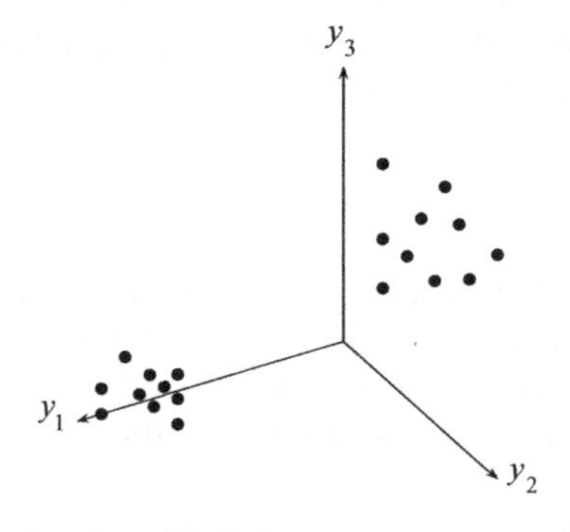

Figure 15.1 Twenty research units with their outcomes as coordinates, represented as points in the three-dimensional space.

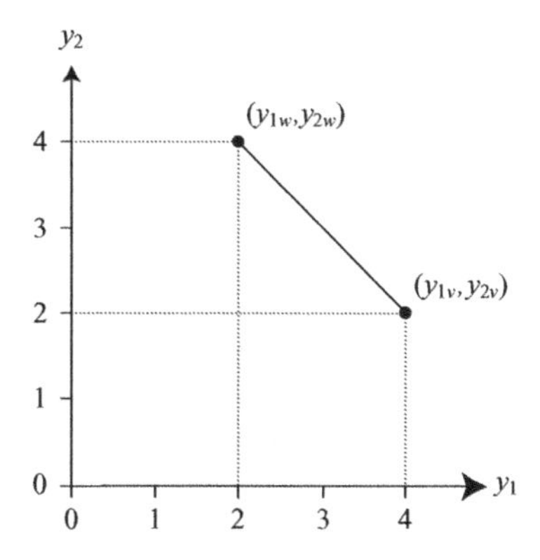

Figure 15.2 Ascertainment of the similarity of two research units according to the Euclidean distance within the two-dimensional space spanned by two characters.

case, the Euclidian distance is defined as

$$d_{vw} = \sqrt{\sum_{q=1}^{m} (y_{qv} - y_{qw})^2}$$

– the square root being always positive. We will not discuss any other distance measures.

Master Doctor

It is important to note that, in SPSS, one should not choose Euclidian distance but instead Squared Euclidian distance in order to actually base calculations on the Euclidian distance. This conceptual misunderstanding in SPSS is due to the following: basically, statistics suggest some 'thinking based on variances' (or squared sums, respectively). Thus, all $\binom{n}{2}$ Euclidian distances within the m-dimensional space for all pairs of n research units, as deviations, still have to be squared and summed up in order to correspond to the square sum 'total' (in the terminology of analysis of variance). In this sense, it is not just about the Euclidian distances, but it is about their squares; nonetheless, this method is intended to measure the deviations as Euclidian distances.

Once a certain grouping has been found, the sum of squared distances can be decomposed into two parts (again, in analogy to the analysis of variance): one component as concerns the distances between all central points of the a groups obtained, $A_1, A_2, \ldots, A_a$ (in the sense of factor levels of a factor 'A'), and one 'residual'-component (corresponding to the distances between all the research units and their respective center point within each group). In this way, the main principle of cluster analysis can be stated using the following target function: the quotient of the sum of squared distances 'A' and of the sum of squared distances 'residual' is to be maximized. When this is achieved, the means of the groups are maximally different, while the distances within each group are minimal.

In order to optimize this target function without having previously defined the number of groups, it can only be optimized for either all or for an arbitrarily fixed number of groups. Certainly, the most reliable approach – though also most laborious to calculate – would be a full enumeration of all possible groupings.

Doctor

In analogy to full enumeration as dealt with in Section 13.4 for the discriminant analysis, this means that for n research units we ascertain the target function for all n one-element-sized groups, then for all $n-2$ sets of one-element-sized groups with every respective remaining two-element-sized group, and so on until we come to a single group, i.e. the total sample. We will demonstrate this for the case of $n = 4$ research units, 1, 2, 3, and 4. In this case, we get 15 different groupings: [{1},{2},{3},{4}]; [{1},{2},{3,4}]; [{1},{3}{2,4}]; [{2},{3},{1,4}]; [{1},{4},{2,3}]; [{2},{4},{1,3}]; [{3},{4},{1,2}]; [{1,2},{3,4}]; [{1,3},{2,4}]; [{1,4},{2,3}]; [{1,2,3},{4}]; [{1,2,4},{3}]; [{1,3,4},{2}]; [{2,3,4},{1}]; [{1,2,3,4}]. Bell's number B_n calculates the number of possible groupings; it increases exponentially with n. For example, $B_5 = 52$, $B_6 = 203$, $B_7 = 4140$, $B_8 = 21\,147$. Assuming sample sizes which are practically relevant, full enumeration is therefore not manageable for cluster analyses given contemporary conventional computer performance.

Master Doctor

Thus, in today's praxis, the optimal number of groups cannot be unequivocally ascertained. One uses simple but not necessarily optimal strategies for obtaining a pleasing number of groups. Here, we only present the hierarchical approach, which basically consists of two strategies. Either we begin with as many groups as research units (i.e. n groups) and successively merge two of the previous groups, or we begin with the total set of all research units, that is with a single group, and successively demerge one research unit or group. Usually, these two methods do not lead to the same results. We find the first way to be more easily understood.

Just as there are numerous distance measures in the context of cluster analysis, there are also numerous strategies for merging the groups. An especially practical one is *Ward's method*. This method consists of a stepwise merging of the two groups whose central points lie closest to each other in terms of the Euclidian distance. Additionally, the distance between two central points of the classes i and l containing research units n_i and n_l is weighted in such a way that when two group pairs have the same distance, the pair containing fewer research units will be merged, or the pair which contains the smallest group, respectively. Thus, the weighting coefficient is:

$$\frac{n_i n_l}{(n_i + n_l)}$$

After this, the only problem remaining for the researcher is to decide on the number of groups. A proper decision criterion analogous to the scree plot exists (see Section 15.1.2 on factor analysis): one visualizes the progress of the sum of squared distances within the groups ('residual'-component) for all resulting numbers of groups (we recommend looking at the range from roughly 20 groups down to a single one). As long as the increase of this 'residual'-component progresses in a (flat) linear way along with the decrease of the number of groups, one can argue that the increase is negligible and that similarity within the groups is still high. However, at the point where a leap occurs, the extent of research units' similarities within each group obviously shifts in the direction of (clear) dissimilarity: at this point, the number of groups is one group too small. As experience teaches us, this decision criterion is not always unambiguous.

It is very important to add that all characters must be standardized (see Section 6.2.2) before analysis. If the characters are analyzed without being previously standardized, then those with large standard deviations will be dominant – characters with a small standard deviation or with a small range of measurement values will represent an ineffectual dimension, since all the research units have rather similar coordinates with respect to this character. This means that such characters will hardly differentiate between groups in relation to other dimensions (characters).

Master

Example 15.1 Are there different learning types, and what characterizes them?

Computerized tests with complex learning scenarios can explore the strategic learning behavior of testees as well as their need for structure and repetition of learning content. With the help of appropriate test scores, we now want to ascertain whether there are typical behavioral patterns; that is, whether some testees resemble one another whereas other groups of testees behave entirely differently within the given learning scenario. We use the data of *Example 15.1* (see Chapter 1 for its availability). There are 86 testees and the test scores *number of runs*, *sum of correctly revised answers*, *sum of falsely revised answers*, *total duration*, and *total mistakes*.

In **R**, we first standardize all characters by typing

```
> Example_15.1.z <- scale(Example_15.1)
```

i.e. we apply the function `scale()` to the data set `Example_15.1` and assign the resulting data set with the standardized values to the object `Example_15.1.z`. Next, we determine the Euclidian distance between all the research units by typing

```
> distance <- dist(Example_15.1.z, method = "euclidean")^2
```

i.e. we set the standardized data set `Example_15.1.z` as the first argument in the function `dist()` and, with `method = "euclidean"`, compute the Euclidian distances; with `^2`, we square the computed distances and assign the results to the object `distance`. Now, we conduct a (hierarchical) cluster analysis. To do this, we type

```
> cluster <- hclust(distance, method = "ward")
```

i.e. we apply the function hclust() to the Euclidian distances (distance); with method = "ward", we request the desired method. We assign the results to the object cluster. Next, we calculate the 'residual'-component by typing

```
> coef <- cumsum(cluster$height/2)
```

i.e., in the function cumsum(), we specify cluster$height/2 – these are the individual distances between the fused clusters, divided by two. We assign the results of this calculation to the object coef. In order to illustrate the progression of the 'residual'-component over numerous relevant numbers of groups, we type

```
> plot(76:85, coef[76:85], type = "b", xlab = "Stage",
+       ylab = "'Residual'-component")
> axis(1, at = 76:85)
```

i.e. we use the function plot() and set 76:85 as the first argument, thus fixing steps 76 to 85 of the group fusion (they constitute the abscissa of the resulting graphic; see Figure 15.3a). With coef[76:85], we concordantly define the second argument, which gives us the last 10 values of the 'residual'-component as the ordinate; through type = "b", we instruct the program to display lines as well as points, and with xlab and ylab, we define the coordinate axes labels. Finally, we set 1 as the first argument in the function axis() to define the scale of the abscissa; with at = 76:85 we establish the location of the tick-marks; namely 76 to 85 on the abscissa.

The results are visualized in Figure 15.3a.

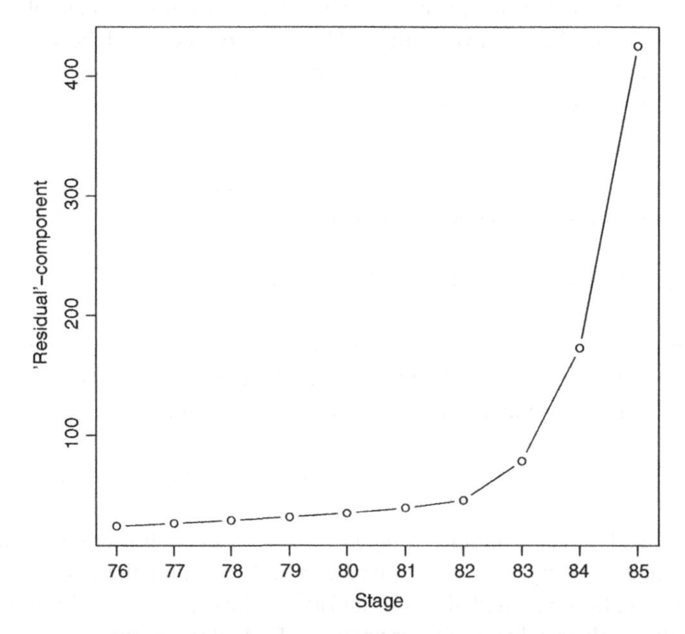

Figure 15.3a **R**-output of the 'residual'-component progression as the number of groups is reduced from 10 to 1, according to the cluster analysis in Example 15.1

In SPSS, we use the sequence of commands

Analyze
  Classify
    Hierarchical Cluster...

and get to the window shown in Figure 15.4. We highlight all characters and move them to the box Variables(s):. Now, we click on the button Method..., and proceed in the next window (Figure 15.5) by choosing Ward's method under Cluster Method:, as well as changing to Z scores under Standardize: (the necessary standardization is not preset within SPSS). After clicking Continue and OK, we get a first result in Table 15.1. There, we are especially interested in the column Coefficients, which are the 'residual'-components for the respective numbers of groups (for example, Stage 85 means that, in this last step, the last two remaining groups are merged into one). In order to illustrate the sums of the squared distances within the groups (the 'residual'-components) for the numerous relevant numbers of groups, we have to double-click the table Agglomeration Schedule in the results-output. That opens a new window Pivot Table Agglomeration Schedule. There, in the column Coefficients, we highlight the last 10 values; thus, we exclude the possibility of more than 10 groups from the very beginning. Now, we right-click on the highlighted items, thus bringing up another menu (not shown here), where we click on Create Graph; through this, we open the next menu (without figure here), where we choose Line. After closing the window Pivot Table Agglomeration Schedule, the graphic of Figure 15.3b can be found in the resulting output.

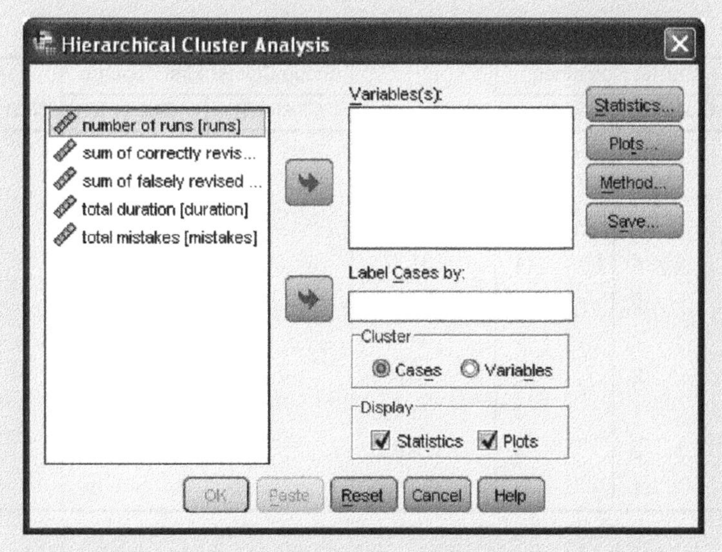

Figure 15.4 SPSS-window for computing cluster analysis.

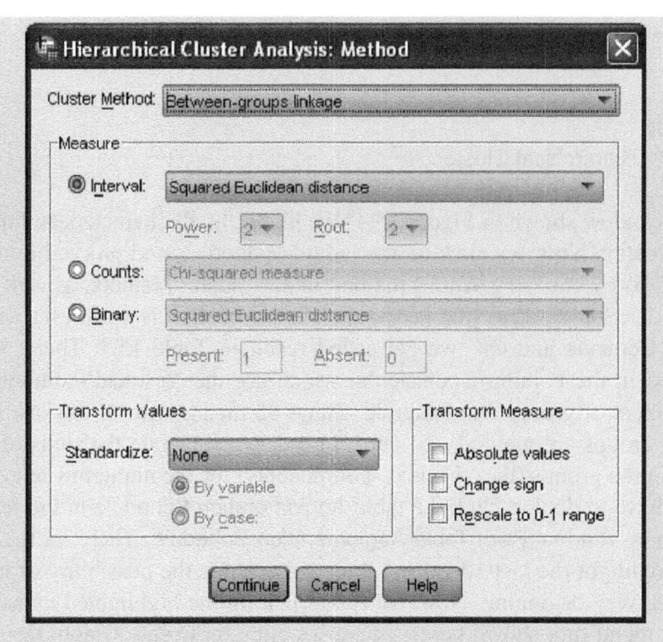

Figure 15.5 SPSS-window for decision on a method within cluster analysis.

Table 15.1 SPSS-output for the progression of the 'residual'-component as the number of groups is reduced from 10 to 1, according to the cluster analysis in Example 15.1 (shortened output)

Agglomeration Schedule

Stage	Cluster Combined		Coefficients	Stage Cluster First Appears		Next Stage
	Cluster 1	Cluster 2		Cluster 1	Cluster 2	
76	4	15	24.089	73	68	79
77	3	13	26.487	72	53	83
78	25	28	29.015	69	66	82
79	4	24	31.771	76	67	84
80	6	61	35.087	75	0	81
81	2	6	39.501	74	80	82
82	2	25	46.050	81	78	84
83	1	3	78.553	65	77	85
84	2	4	172.956	82	79	85
85	1	2	425.000	83	84	0

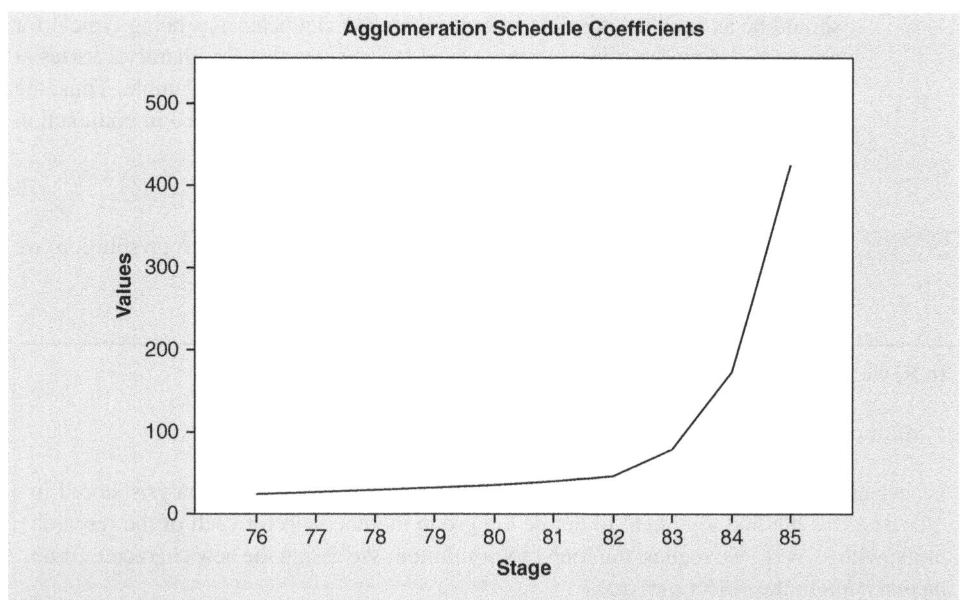

Figure 15.3b SPSS-output of the 'residual'-component progression as the number of groups is reduced from 10 to 1, according to the cluster analysis in Example 15.1.

Looking at Figure 15.3a (as well as Figure 15.3b) from left to right, a first noticeable leap in the progression of the 'residual'-component occurs from Step 82 to Step 83, which is the step from a four- to a three-group solution. In accordance with the abovementioned decision criterion – whereupon that step after a leap signals that the number of groups is one too small because the relatively high similarity within each group shifts to dissimilarity within each group – we would have to choose a four-group solution. If, however, we do not interpret any leaps as sufficiently evident until the leap between Step 83 and Step 84, we obtain a three-group solution. In this case, we decided on the four-group solution.

Master Doctor

Once the number of groups is ascertained, a question arises: in which of the character(s) are the research units of each group especially similar, and by which character(s) do the groups mainly differ from each other? The following approach has proven useful in answering this question: determine the standard deviation for each character within each group, and divide these standard deviations by the standard deviation of the respective character in the total sample. This is expedient because, if a character is typical for a group so that all research units in that group have almost the same outcome, these outcomes should vary less within that group than all outcomes vary in the total sample. Thus, if s_{qi} is the standard deviation of character q in group A_i, and s_q is the standard deviation of character q in the total sample, then the ratio

$$Q_{qi} = \frac{s_{qi}}{s_q}$$

should be as small as possible in order to define character q as being typical for group A_i. Using the rule of thumb $Q_{qi} \leq 0.5$ ensures that the character varies at most half as much within the group as it does within the total sample. Thus, the group's mean for that character characterizes this group (maybe in conjunction with other characters).

Master **Example 15.1 – continued** After having decided on the four-group solution, we now want to identify the groups and exactly characterize them.

In **R**, we type

```
> group <- cutree(cluster, k = 4)
```

i.e. we apply the function `cutree()` using the results of the cluster analysis stored in `cluster` as the first argument to define the group membership for each of the research units; with `k = 4`, we request the four-group solution. We assign the new character *group membership* to the object `group`.

Now, in order to obtain the values Q_{qi}, we type

```
> sapply(1:4, function(y) sapply(1:5, function(x)
+         tapply(Example_15.1[, x], group, sd))[y, ]/
+         apply(Example_15.1, 2, sd))
```

i.e. we use the function `sapply()` two times, once with the instruction to conduct the following analysis for the first up to the fourth group `(1:4)`, and once with the same analysis for the first up to the fifth character `(1:5)`. With the help of the function `tapply()`, we calculate the standard deviation `(sd)` separately for each group `(group)`. We divide this result by the standard deviation `(sd)` that we calculate for each column-wise arranged character `(2)` for all groups together, with the help of the function `apply()`.

As a result, we get

```
               [,1]      [,2]      [,3]      [,4]
runs     0.19225373 0.3475614 0.4549612 0.3707825
correct  0.14346225 0.3805198 0.2785858 0.4237522
false    0.05004249 0.5869595 0.2227576 0.3588045
duration 0.09805329 0.5543949 0.1851657 0.4130926
mistakes 0.06910112 0.5648816 0.1776249 0.3718529
```

In SPSS, we first have to repeat the entire analysis method. In the window shown in Figure 15.4 we choose the button Statistics..., by which we get to a new window (without figure here); there, we activate the point Single solution and enter the number 4 into the box Number of clusters:. With Continue, we return to the previous window. There, we choose Save... in order to save the group membership of each research unit as a new character for further analysis in the data sheet. In the resulting window (without figure here), we again choose Single solution and type 4 into the box Number of clusters:. With Continue and OK, we get to the same results in the output. In the data sheet, however, we get the new character CLU4_1.

In order to obtain the values Q_{qi}, we first determine the standard deviation for each character within the total sample, then separately for each of the groups. With the sequence of commands (Analyze – Descriptive Statistics – Frequencies...) of Example 5.4 we get to the window shown in Figure 5.4, where we transfer all our characters *number of runs, sum of correctly revised answers, sum of falsely revised answers, total duration*, and *total mistakes* into the box Variable(s):, then click on the button Statistics... and thus arrive at the window shown in Figure 5.24; there, we (merely) click Mean and Std. deviation. With Continue and OK, we obtain the result as illustrated in Table 15.2

Table 15.2 Mean and standard deviation of all characters for the total sample in Example 15.1 (shortened output)

Statistics

	number of runs	sum of correctly revised answers	sum of falsely revised answers	total duration	total mistakes
Mean	23.91	10.10	19.34	735.95	30.55
Std. Deviation	15.116	7.512	16.184	556.187	29.506

Now, we use the series of orders (Data – Split File...) of Example 5.11 with the help of Figure 5.23 to split up the analysis by the group variable CLU4_1: there, we activate the button Compare groups, transferring Ward Method into the window Groups Based on: and click on OK. We again follow the sequence of commands (Analyze – Descriptive Statistics – Frequencies...) of Example 5.4 and click on OK in the window shown in Figure 5.24. With this, we obtain the means and variances of all four groups as depicted in Table 15.3.

Table 15.3 Means and standard deviations of all characters for the groups 1 through 4 in Example 15.1 (shortened output)

Statistics

Ward Method		number of runs	sum of correctly revised answers	sum of falsely revised answers	total duration	total mistakes
1	Mean	8.39	3.81	3.45	205.65	6.10
	Std. Deviation	2.906	1.078	.810	54.536	2.039
2	Mean	42.65	8.48	41.65	1520.61	74.61
	Std. Deviation	5.254	2.858	9.499	308.347	16.667
3	Mean	19.00	11.56	15.94	602.17	19.06
	Std. Deviation	6.877	2.093	3.605	102.987	5.241
4	Mean	33.79	24.86	22.21	793.14	27.07
	Std. Deviation	5.605	3.183	5.807	229.757	10.972

The values Q_{qi} have to be calculated manually:

$$Q_{11} = \frac{2.906}{15.116} = 0.19 \quad Q_{21} = \frac{5.254}{15.116} = 0.35 \quad Q_{31} = \frac{6.877}{15.116} = 0.45 \quad Q_{41} = \frac{5.605}{15.116} = 0.37$$

$$Q_{12} = \frac{1.078}{7.512} = 0.14 \quad Q_{22} = \frac{2.858}{7.512} = 0.38 \quad Q_{32} = \frac{2.093}{7.512} = 0.28 \quad Q_{42} = \frac{3.183}{7.512} = 0.42$$

$$Q_{13} = \frac{0.810}{16.184} = 0.05 \quad Q_{23} = \frac{9.499}{16.184} = 0.59 \quad Q_{33} = \frac{3.605}{16.184} = 0.22 \quad Q_{43} = \frac{5.807}{16.184} = 0.36$$

$$Q_{14} = \frac{54.536}{556.187} = 0.10 \quad Q_{24} = \frac{308.347}{556.187} = 0.55 \quad Q_{34} = \frac{102.987}{556.187} = 0.19 \quad Q_{44} = \frac{229.757}{556.187} = 0.41$$

$$Q_{15} = \frac{2.039}{29.506} = 0.07 \quad Q_{25} = \frac{16.667}{29.506} = 0.56 \quad Q_{35} = \frac{5.241}{29.506} = 0.18 \quad Q_{45} = \frac{10.972}{29.506} = 0.37$$

We thus establish that there are four learning types:

- For Learning Type 1, all characters are typical (with $Q_{qi} < 0.20$); members of this type need the fewest runs, revise their answers least often (correctly as well as incorrectly), need the least time and make the fewest mistakes.
- For Learning Type 2, only the characters *number of runs* and *sum of correctly revised answers* are typical (with $Q_{qi} < 0.50$); whilst having an almost average *sum of correctly revised answers*, members of this type need the most runs.
- For Learning Type 3, all characters are typical (with $Q_{qi} < 0.50$); members of this type manage to obtain relatively good or average test scores in all characters.
- For Learning Type 4, all characters are typical (with $Q_{qi} < 0.50$); members of this type need very many runs, have the most correct revisions of their answers and a rather large number of incorrect revisions, need a rather long time, and make an average number of mistakes.

Master Doctor

The vast arbitrariness in deciding the number of groups is one of the essential points of criticism concerning cluster analysis. Therefore, in order to give the established grouping some support, sometimes a multivariate analysis of variance is applied, using the several groups as factor levels. A significant result indeed increases the explanatory power of the given grouping. However, this significance might be artificial, since the test does not compare random samples from well-defined populations: all research units were grouped in an optimal way with regard to the given data, which can produce differences that actually do not stem from a systematic base (in this context, see the phenomenon of 'regression towards the mean', known since *Sir Francis Galton*; e.g. Stigler, 1997).

15.1.2 Factor analysis

Psychologists have played a major role in the development of factor analysis. Especially in the construction of psychological tests and, in earlier days, in fundamental research – for example in ascertaining the number of factors that constitute 'intelligence' – it is a frequently used instrument.

Factor analysis is a method of 'dimensional analysis', i.e. we postulate that the m observed characters are not completely independent from each other but actually intercorrelate (quite strongly), so that (considerably) fewer than m dimensions suffice to describe all the research units' outcomes. In essence, the (directly observable) characters are to be reduced on the basis of their mutual (more or less pronounced) relationships to few, not directly observable, mutually independent and thus not correlated so-called 'factors' (meta-characters; supra-variables). The problem can be seen in analogy to partial correlation (see Section 12.1.1): the relationship of two characters is assumed to be merely caused by the correlation of both with (at least) a third character; with the difference that here, in contrast to the partial correlation, this third character or further characters are neither known nor observable. It is assumed that only a few of these latent, not manifest characters, *factors*, suffice to explain all relationships between the observed characters. Once these factors are ascertained, the observed complex of characters can be quite easily explained.

Given, indeed, m manifest characters $y_1, y_2, \ldots, y_m$, which are modeled by the random variables $\boldsymbol{y}_1, \boldsymbol{y}_2, \ldots, \boldsymbol{y}_m$, factor analysis suggests the following linear model between these m characters and x latent characters (factors). All characters, manifest as well as latent, are presupposed to be quantitative:

$$\boldsymbol{y}_q = \sum_{l=1}^{x} a_{ql}\boldsymbol{f}_l + b_q\boldsymbol{s}_q; q = 1, \ldots, m \tag{15.1}$$

Every random variable $\boldsymbol{y}_q$ is thus represented by the weighted sum of x random variables $\boldsymbol{f}_l$, $l = 1, \ldots, x$, being the *common factors*, as well as by the random variable $\boldsymbol{s}_q$, being some *specific factor*. The latter fulfills the function of the error term of the generalized linear model (see for example regression analysis and analysis of variance). The weights a_{ql} are called *factor loadings* (or simply *loadings*). They stand for correlation coefficients between factor l and character q. The common factors affect at least two characters; i.e. at least two factor loadings a_{ql} and a_{pl} have to be unequal to zero. All factors are presupposed to be independent from each other and thus do not correlate. A further assumption, which, however, does not constrain the method's universal applicability, is that the factors are standardized (see Section 6.2.2) to have a mean of 0 and a variance of 1. Also, without loss of generalizability, we presuppose in the following that all random variables $\boldsymbol{y}_q$ are standardized as well.

Bachelor
Master

As can easily be shown, the presupposition of standardization of all characters and factors means that all variances of all variables $\boldsymbol{y}_q$ are equal to $\sum_{l=1}^{x} a_{ql}^2 + b_q^2 = 1$. The first summand is called the *communality*; it describes that part of the variance of $\boldsymbol{y}_q$ that is explained by the x common factors. The second summand is called the *specificity*. Obviously, it is desirable to ascertain few, but enough factors in such a way that the communality grows as large as possible for all characters.

Ascertaining the number of factors is the first essential problem within factor analysis. Analogous to ascertaining the number of groups in cluster analysis, we will try to establish some rules of thumb in accordance with the given mathematical target function of the method. Today, it is no longer necessary to differentiate between several approaches regularly falling under the term 'factor analysis', since the term almost always refers to *principal component analysis*. This name stems from the fact that the analysis determines the mathematically defined so-called principal components (basically the *eigenvectors*) of a square matrix, here the empirical $m \times m$ correlation matrix $R = ((r_{lh}))$. Within mathematics, this determination is called 'principal component method'. Using the principal component method, m linearly independent linear combinations (the principal components) are generated from the m characters in such a way that the sum of their dyadic products results in the original matrix.

Bachelor

Every symmetrical $m \times m$ matrix has m real *eigenvalues* $\lambda_q, q = 1, 2, \ldots, m$; these are all solutions to the equation $|R - \lambda I_m| = 0$, with I_m as the $m \times m$ identity matrix. The solutions $\vec{x}_q, q = 1, 2, \ldots, m$ of the equation $(R - \lambda I_m)\vec{x} = 0$ are called eigenvectors; they are not unequivocally determined, because for every positive constant k, it holds that $(R - \lambda I_m)k\vec{x} = 0$. Thus, one can choose k in such a way that, for each of the m eigenvectors $\vec{x}$, it is true that $\vec{x}^T\vec{x} = 1$. Multiplying the thus-standardized eigenvectors $\vec{x}_q^*$ with $\sqrt{\lambda_q}$, one obtains the principal components. The product of all eigenvalues equals the determinant of the matrix, $|R|$, and their sum, $\sum_q^m \lambda_q$, is equal to the sum of all elements of the principal diagonal, $\sum_q^m r_{qq}$ (i.e. the trace of the matrix R).

So far, of course, the factor analysis' goal of using the (considerably) smaller number of x (not directly observable) factors instead of the m observed characters in order to explain the given data (the correlation matrix R) is not yet fulfilled. Up to now, instead of m characters, we have merely obtained m principal components. The next step aims, therefore, at reducing the number of eigenvectors and thus also the number of eigenvalues in such a way that the matrix of all a_{ql} still explains all the data or all correlations between the m characters as exactly as possible. In other words, the empirical matrix R is to be reproduced by the model-based $m \times x$ matrix $A = ((a_{ql}))$ in the best possible manner. Thus, instead of all eigenvalues λ_q, $q = 1, 2, \ldots, m$, and all eigenvectors $\vec{x}_q^*$, $q = 1, 2, \ldots, m$, we are now looking for an appropriate $x < m$ eigenvalues λ_l, $l = 1, 2, \ldots, x$, and their eigenvectors $\vec{x}_l^*$, $l = 1, 2, \ldots, x$.

Bachelor
Master

By the way, factor loadings a_{ql} and eigenvalues λ_l relate to one another as follows: $\lambda_l = \sum_{q=1}^m a_{ql}^2, l = 1, 2, \ldots, x$; that is, because of the presupposed standardization, the eigenvalue λ_l is nothing but the amount of variance of all m characters being explained by factor l. If there were a single factor, that is to say a general factor, that explained the total variance, then $l = 1$ and $\lambda_1 = m$. Thus, only those eigenvalues (and therefore factors) are of interest which are larger than 1. Their number is probably the most common rule of thumb for ascertaining the number of factors.

Arranging all m eigenvalues by size, that is by explained amount of the variance of all m characters, results in a second rule of thumb. Graphically representing their progress, we look from the smallest to largest eigenvalue (from right to left) for a breaking point in the almost linear curve: until this point, the eigenvalues meet the abscissa asymptotically. All factors with eigenvalues after this breaking point (i.e. towards the left) are taken to be meaningful; all others are discarded as meaningless, because a minimal increase/decrease happens to occur even when variance is only explained by chance, but not systematically. This rule of thumb is called the '*scree test*'.

For the interpretation of content of a given factor analysis solution, we recommend ascertaining the percentage of total variance of all characters that can be explained by the resulting factors. Also the examination of communalities is of some help: if they are all close to 1 (and thus, the specificities $b_q^2 = 1 - \sum_{l=1}^{x} a_{ql}^2$ are close to zero), this can be interpreted as the variance of the characters actually being mainly explained by the x factors.

Labeling the factors, or defining their meaning with regard to content, is a second problem within factor analysis. Because the factors are extracted in a sequence according to the amount of variance they explain, practically all characters load rather high in the first factors. For clear interpretation, the factors (or, more precisely, the axes representing them within a multidimensional rectangular coordinate system) have to be rotated; one conducts a *factor rotation*. At best, this happens in accordance with the *simple structure principle* by *Thurstone*. The points within the multidimensional space where all characters are represented by their loadings per factor as coordinates remain unmodified, but only the axis system of the coordinates is rotated in such a way that each factor has (a few) very high loadings as well as (many) very small loadings. When formalized, this criterion comes very close to the *varimax criterion* by Kaiser (1958): the variability of loadings within each factor is maximized. Thus, if there is basically only a single character with a very high loading per factor, this 'marker variable' can be used for labeling that factor.

Bachelor
Master

Example 15.2 The factor structure of the intelligence test battery for children with Turkish as native language from Example 1.1

The five subtests *Everyday Knowledge, Applied Computing, Social and Material Sequencing, Immediately Reproducing – numerical*, and *Coding and Associating* may – especially for children with Turkish as native language – correlate among one another in such an extensive way (see also Example 12.7) that basically fewer than five independent factors may be responsible for the observed test scores. Maybe even a single, a general factor, suffices to explain the given data. In order to ascertain the respective factor structure, we use the test scores for these five subtests on the first and second test date. This means that $m = 10$.

In **R**, we first activate the data set *Example_1.1*, using the function `attach()` (see Chapter 1). Then, we create a data set that only includes children with Turkish as their native language. We type

```
> fact.data <- subset(Example_1.1[, 12:21],
+                     subset = native_language == "Turkish")
```

i.e. we use the function `subset()` to choose those persons that fulfill the condition `native_language == "Turkish"`; additionally, we choose all five subtests for both points in time by transferring the data from the columns `12` to `21` to the object `fact.data`.

Now, we conduct a factor analysis by typing

```
> fact <- prcomp(fact.data, scale = TRUE)
```

i.e. we submit the object `fact.data` to the function `prcomp()`; with `scale = TRUE`, we standardize all variables. We assign the results of this analysis to the object `fact`. Next, we determine the number of factors to be extracted. We type

```
> summary(fact)
```

and obtain the results:

```
Importance of components:
                          PC1   PC2   PC3   PC4    PC5   PC6    PC7
Standard deviation      2.056 1.374 1.244 0.990 0.7622 0.640 0.3661
Proportion of Variance 0.423 0.189 0.155 0.098 0.0581 0.041 0.0134
Cumulative Proportion  0.423 0.611 0.766 0.864 0.9223 0.963 0.9767
                          PC8     PC9    PC10
Standard deviation      0.3495 0.24709 0.22310
Proportion of Variance 0.0122 0.00611 0.00498
Cumulative Proportion  0.9889 0.99502 1.00000
```

Now, we create a scree plot by typing

```
> screeplot(fact, type = "lines", main = "")
```

i.e. we apply the function `screeplot()`, to which we submit the result of the factor analysis from the object `fact`; with `type = "lines"`, we order a line diagram and with `main = ""` we suppress the headline.

The results are depicted in Figure 15.6.

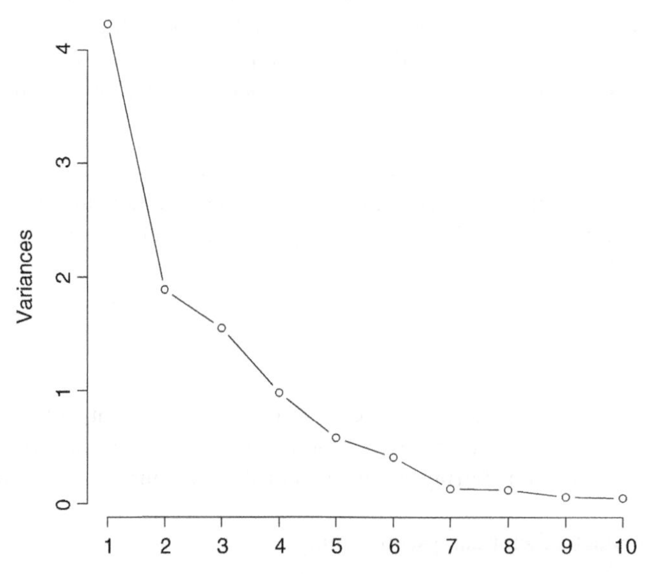

Figure 15.6 **R**-output of the scree test in Example 15.2. The abscissa shows the factors of the complete factor solution; the ordinate gives the respective eigenvalues.

Since exactly three factors have eigenvalues greater 1, we have to conduct the factor rotation like this: we install and load the package `psych` (see Chapter 1) using the function `library()` and type

```
> principal(fact.data, nfactors = 3, rotate = "varimax")
```

i.e. we use the function `principal()`, adding the data set `fact.data` as the first argument, and specify a three-factor solution with `nfactors = 3`; with `rotate = "varimax"`, we choose the varimax criterion.

As a result, we get (shortened output):

```
Principal Components Analysis
Call: principal(r = fact.data, nfactors = 3, rotate = "varimax")
Standardized loadings based upon correlation matrix
         RC3  RC2   RC1   h2   u2
sub1_t1 0.68 0.00  0.43 0.66 0.34
sub3_t1 0.78 0.18  0.09 0.66 0.34
sub4_t1 0.18 0.00  0.89 0.82 0.18
sub5_t1 0.08 0.61  0.53 0.66 0.34
sub7_t1 0.15 0.92 -0.06 0.87 0.13
sub1_t2 0.77 0.08  0.27 0.68 0.32
sub3_t2 0.85 0.20 -0.01 0.76 0.24
sub4_t2 0.26 0.03  0.89 0.86 0.14
sub5_t2 0.01 0.68  0.60 0.83 0.17
sub7_t2 0.25 0.91 -0.03 0.89 0.11

                RC3  RC2  RC1
SS loadings    2.59 2.58 2.49
Proportion Var 0.26 0.26 0.25
Cumulative Var 0.26 0.52 0.77
```

In SPSS, we choose

Analyze
    Dimension Reduction
        Factor...

and, in the window shown in Figure 15.7, select all the mentioned $m = 10$ characters in order to transfer them into the box Variables:. Then, we select the character native language and transfer it into the box Selection Variable:; now, we click on the button Value... and, in the resulting window (not shown here), enter 2. With Continue, we get back to the previous window. If we now click on the button Extraction..., we get to the window in Figure 15.8, where, firstly, we do not change the preset Principal components, and secondly, remove the check mark at Unrotated factor solution and instead set one at Scree plot; as a third step we leave the value 1 at Eigenvalues greater than: (all this has already been done in Figure 15.8). With Continue, we are back at the window shown in Figure 15.7, where we now choose the button Rotation.... In the resulting window (without figure here), we mark Varimax in the field Method. With Continue and OK, we obtain the results as depicted in the Tables 15.4 and 15.5; the resulting scree plot corresponds to Figure 15.6.

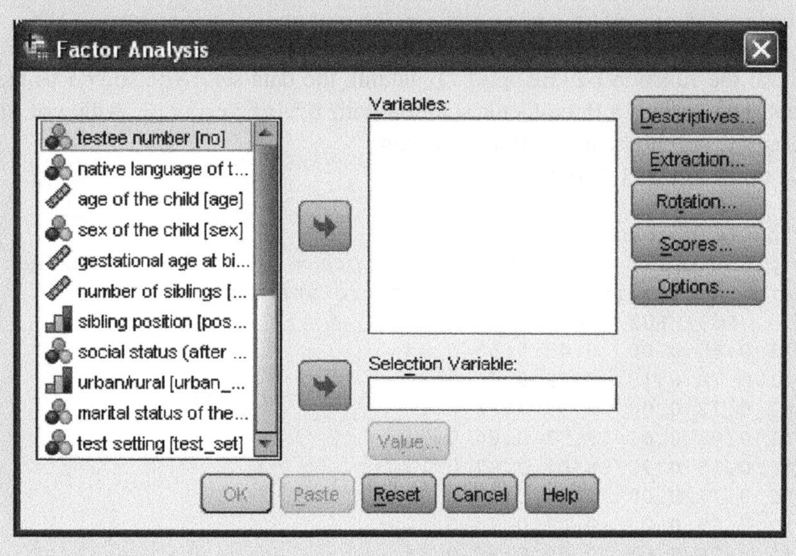

Figure 15.7 SPSS-window for calculation of factor analysis.

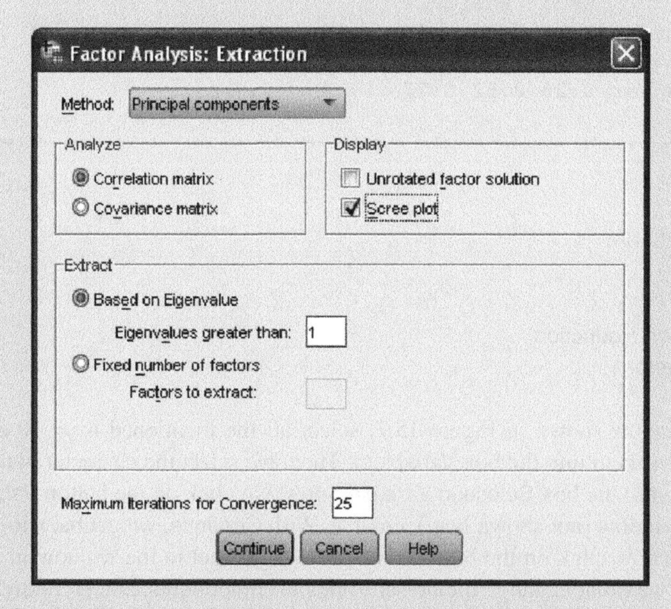

Figure 15.8 SPSS-window for determining the number of factors in factor analysis.

Table 15.4 SPSS-output of the eigenvalues from Example 15.2.

Total Variance Explained[a]

Component	Initial Eigenvalues			Rotation Sums of Squared Loadings		
	Total	% of Variance	Cumulative %	Total	% of Variance	Cumulative %
1	4.228	42.282	42.282	2.591	25.911	25.911
2	1.887	18.867	61.149	2.583	25.827	51.739
3	1.548	15.475	76.624	2.489	24.886	76.624
4	.980	9.800	86.424			
5	.581	5.810	92.234			
6	.410	4.096	96.330			
7	.134	1.340	97.670			
8	.122	1.221	98.892			
9	.061	.611	99.502			
10	.050	.498	100.000			

Extraction Method: Principal Component Analysis.

a. Only cases for which native language of the child = Turkish are used in the analysis phase.

Table 15.5 SPSS-output of factor loadings after rotation, from Example 15.2

Rotated Component Matrix[a]

	Component		
	1	2	3
Everyday Knowledge, 1st test date (T-Scores)	.089	.565	.567
Applied Computing, 1st test date (T-Scores)	.189	.043	.866
Social and Material Sequencing, 1st test date (T-Scores)	.180	.904	.015
Immediately Reproducing - numerical, 1st test date (T-Scores)	.748	.319	.154
Coding and Associating, 1st test date (T-Scores)	.885	.022	.181
Everyday Knowledge, 2nd test date (T-Scores)	.148	.590	.519
Applied Computing, 2nd test date (T-Scores)	.218	.100	.864
Social and Material Sequencing, 2nd test date (T-Scores)	.254	.886	.107
Immediately Reproducing - numerical, 2nd test date (T-Scores)	.812	.351	.089
Coding and Associating, 2nd test date (T-Scores)	.883	.089	.220

Extraction Method: Principal Component Analysis.
Rotation Method: Varimax with Kaiser Normalization.

a. Rotation converged in 5 iterations.

From Table 15.4, it follows that factor analysis basically establishes the 'complete' solution, i.e. it extracts as many factors as there are characters. However, examining the results more closely, only three factors with an eigenvalue larger than 1 emerge. The three factors explain almost 77% of the variance of all ten investigated characters. In the scree plot of Figure 15.6, there is a breaking point at the second-largest eigenvalue, which speaks for a general factor solution. Table 15.5 shows the loadings per factor and character after rotating the three-factor solution. Here the highest loading of 0.918 is given for the character *Coding and Associating, 1st test date,* for Factor 2; the character *Coding and Associating, 2nd test date* also has a high weight in this factor, namely with loading 0.908. Thus, Factor 2 is described relatively well by these subtests. Another relatively high-loading subtest for Factor 2 is *Immediately Reproducing – numerical*; though this subtest also shows high loadings in Factor 3, so that the factor structure loses some concision: not all characters are primarily defined by a single factor. With a value of 0.89, Factor 3 has its highest loadings in the subtest *Social and Material Sequencing*. The loadings in Factor 1 are lower; nevertheless, this factor is clearly qualified by the subtests *Applied Computing* and *Everyday Knowledge* (just in this sequence). All in all, it can be judged that these five subtests measure in a rather 'high dimensional' way.

Doctor

At least rudimentary inferential statistics can be applied by testing the hypothesis that (even) the xth eigenvalue is larger than 1; for this, one can use the bootstrapping method (see Section 14.3).

Doctor

Example 15.2 – continued

Since in this example the fourth eigenvalue is merely a bit smaller than 1, we want to calculate a one-sided confidence interval for the observed value of $y = 0.98$ in the upper direction. Since the distribution of the respective eigenvalues is unknown, we use the bootstrapping method. For this, we draw 10 000 random samples from the data and apply factor analysis for each of them as described above. We are not interested in all the details of the resulting distribution, but only in the standard deviation σ. With it, we can calculate the desired confidence interval: $O = y + z(1 - \frac{\alpha}{2})\cdot\sigma$. If the result shows that $O \leq 1$, the fourth eigenvalue is judged not to be significant (not larger than 1), and the three-factor solution is appropriate.

In many research works, especially within the construction of psychological tests, the reader will find – instead of the above described *orthogonal rotation* – the so-called *oblique rotation*. This approach results in factors that are hardly interpretable, particularly with regard to their mutual interdependences: the rectangular coordinate system, and thus the idea of independent and as such uncorrelated factors is of course lost when rotating the coordinate axis in an oblique manner. We will therefore not deal with it any further here.

Once a psychological test is constructed according to factor analysis, a test score should describe every testee for each of the obtained factors. These test scores must serve to quantify the trait or aptitude which is measured by the respective factor. That is, it is about the realized values in the factors f_l in Formula (15.1), for each testee v; these values, f_{lv}, are called *factor scores*. Unfortunately, the inventory of psychological tests often uses the following methodological artifact: instead of calculating such factor scores per testee, each task or question is weighted equally and scored according to some arbitrary scoring rule (and finally added up), given that it has a minimum loading in the respective factor at all. This scoring method disregards the fact that the tasks or questions do contribute with differing intensities to the model equation in Formula (15.1).

Example 15.3 Determining the factor scores without relation to any content

A test battery with m subtests was shown, through factor analysis, to be based on x factors. For each of the n testees, we now ascertain the x factor scores (we will use the data from Example 15.2).

In **R**, we use exclusively the data of children with Turkish as a native language in the object `fact.data` (see in Example 15.2). We set

```
> fact.2 <- principal(fact.data, nfactors = 3,
+                      rotate = "varimax", score = TRUE)
```

i.e. we conduct the factor analysis, again using the function `principal()` from the package `psych`, but this time we add `score = TRUE` as an additional argument, which causes the factor scores to be calculated. We assign the results to the object `fact.2`. We print the factor scores with

```
> fact.2$scores
```

As a result, we get (shortened output):

```
             RC3         RC2         RC1
3   0.0316104402 -0.04213481 -0.31718716
4   0.2673120415 -0.38041647  0.43467948
7   0.3507505324 -0.43657538 -1.28321479
8   0.0056456323  1.20273484  1.19403540
55  0.4780072472  0.06939621 -0.75794551
...
```

In SPSS, we proceed analogously to Example 15.2, except now we additionally choose the button Scores... in the window shown in Figure 15.7; this leads to the window shown in Figure 15.9. Here, we click Save as variables as well as Display factor score coefficient matrix. This has two effects: firstly, the desired factor scores for each person will be added to the Data View in a separate column for each factor, ready for further analysis; and, secondly, all coefficients will be displayed, which have to be (manually) multiplied with the test scores of future, not yet tested persons in this test battery in order to determine their factor scores. With Continue and OK, we set all this into action.

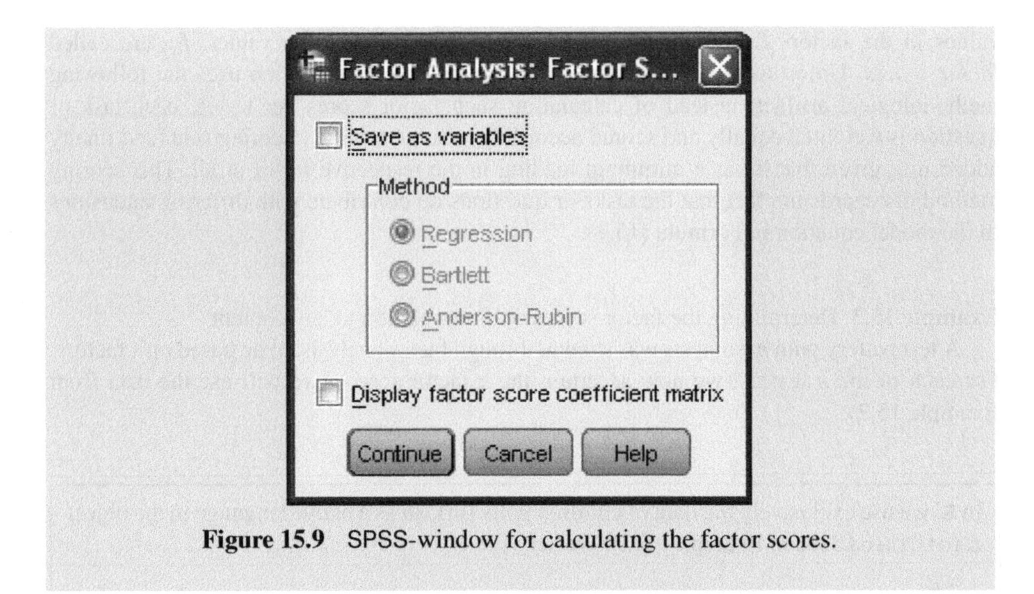

Figure 15.9 SPSS-window for calculating the factor scores.

Again and again, one finds factor analysis applied to ordinal-scaled characters as well as to dichotomous characters. However, if this application does not include the use of appropriate correlation coefficients, the common factor analysis produces artificial results. While Spearman's correlation coefficient is suited for ordinal-scaled data, dichotomous characters are best dealt with using the *tetrachoric correlation* coefficient:

$$r_{\text{tet}} = \cos \frac{180^\circ}{1 + \sqrt{\frac{bc}{ad}}}$$

with the denotations from Table 11.6 (for more details, see for example Kubinger, 2003).[1]

15.1.3 Path analysis

Path analysis can be seen formally as a 'bundle' of multiple linear regressions. It is used to research questions where relationships (correlations) between interesting characters are not undirected but where instead directed or even causal relationships are suggested. Based on some modeled system of directed so-called paths, *path coefficients* such as regression coefficients are estimated from empirical data. The size of these coefficients describes the influence of the observed characters on each other.

Example 15.4 Determinants for (cognitive) high performance potential

The 'Viennese model for the assessment of high achievement potential' (see Holocher-Ertl, Kubinger, & Hohensinn, 2008) suggests several determinants for

[1] A not yet published computer program by Joachim Häusler, which computes the tetrachoric correlation coefficient for a set of dichotomous characters and directly imports the resulting matrix into SPSS, is available in the download section of the website http://psychologie.univie.ac.at/eppd.

academic high performance. Among many others, these are cognitive abilities (e.g. intelligence, retentiveness, attentiveness), the parental stimulative environment, the degree of achievement motivation, and emotional stability.

From the content point of view, it is not at all plausible that, for example, the extent of high performance potential influences the parental stimulative environment, but indeed the opposite is true. The same applies for numerous relationships between the other mentioned determinants or, that is to say, characters. The cited content model specifies the directions of such relationships; whereby all the determinants are thought to show some compensatory effects. This means that they act linearly additively; thus, non-optimal outcomes in certain determinants do not necessarily result in a low potential for high performances.

The actual *path model* is best visualized graphically. There, directed arrows symbolize directed relations. From this, it is easy to derive the respective (multiple) regression functions, in analogy to Formula (12.5).

Example 15.4 – continued
From the 'Viennese model for the assessment of high achievement potential' one can, for example, derive the path model in Figure 15.10 – it consists of the following characters modeled as interval-scaled random variables: y (*high performance potential*), x_1(*intelligence*), x_2(retentiveness), x_3(*attentiveness*), x_4(*parental stimulative environment*), x_5(*achievement motivation*), and x_6 (*emotional stability*).

From this, the following system of equations can be created ($v = 1, 2, \ldots, n$):

$$\begin{aligned}
\boldsymbol{x}_{3v} &= \beta_3^* + \beta_{1(3)}\boldsymbol{x}_{1v} + \boldsymbol{e}_{3v}\\
\boldsymbol{x}_{5v} &= \beta_5^* + \beta_{4(5)}\boldsymbol{x}_{4v} + \boldsymbol{e}_{5v}\\
\boldsymbol{x}_{6v} &= \beta_6^* + \beta_{4(6)}\boldsymbol{x}_{4v} + \boldsymbol{e}_{6v}\\
\boldsymbol{x}_{2v} &= \beta_2^* + \beta_{1(2)}\boldsymbol{x}_{1v} + \beta_{3(2)}\boldsymbol{x}_{3v} + \boldsymbol{e}_{2v}\\
\boldsymbol{y}_v &= \beta^* + \beta_1\boldsymbol{x}_{1v} + \beta_2\boldsymbol{x}_{2v} + \beta_3\boldsymbol{x}_{3v} + \beta_4\boldsymbol{x}_{4v} + \beta_5\boldsymbol{x}_{5v} + \beta_6\boldsymbol{x}_{6v} + \boldsymbol{e}_v
\end{aligned}$$

All symbols are defined in analogy to (multiple) regression; here, too, the presumptions from Section 11.2 apply. In order to be able to compare the path

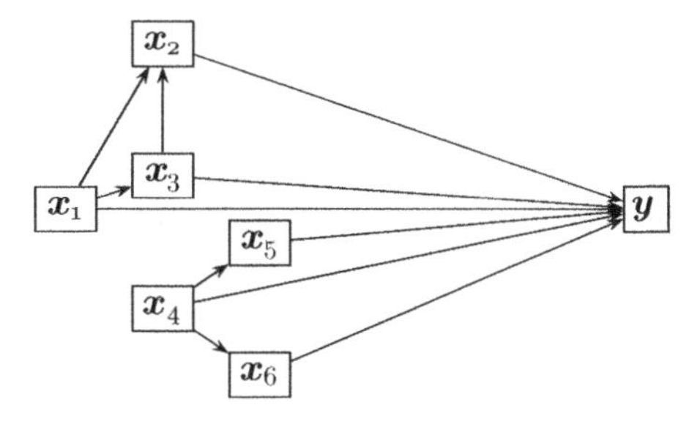

Figure 15.10 Path model of the 'Viennese model for the assessment of high achievement potential'.

coefficients to each other, it is best to use standardized characters (as in factor analysis); when this is done, all intercepts of the linear equations equal 0.

The path coefficients are estimated as (standardized) regression coefficients using the least squares method or the maximum likelihood method (see Section 6.5), for example with the help of computer packages for so-called *(linear) structural equation models*, which include path analysis as a special case. We will come back to this in Section 15.2.1. As dealt with up to now, path analysis is (just) a method of descriptive statistics. But, within the (linear) structural equation models, it is also possible to compare two concurrent models using – apart from coefficients of goodness of fit – inferential statistics, to test whether one model better fits the data.

From all above concerning regression and correlation analysis, it follows that exclusively quantitative (i.e. at least interval-scaled) characters are required.

15.2 Methods of inferential statistics

Other theory-generating methods aim at testing hypotheses; they do not merely explore, but also try to show that the obtained results are not due only to chance. In other words, these methods examine whether the results might be based on systematic principles; in the best case, they test whether results support an existing theory. This applies to two further analysis methods for classifying research units, the (linear) structural equation models, and methods for calibrating psychological tests (and the like).

15.2.1 Further analysis methods for classifying research units

In Section 15.1.1, we already mentioned the analysis method of configuration frequency analysis on the one hand and latent class analysis on the other hand. Both of these methods are appropriate for nominal-scaled characters, above all for dichotomous ones. In the following, we will restrict ourselves to considering the case of dichotomous characters, because the number of necessary research units increases exponentially with each additional measurement value, so that practical applications for multi-categorical data become rather unrealistic.

15.2.1.1 Configuration frequency analysis

Configuration frequency analysis (CFA) rests upon the phenomenon of *Meehl*'s paradox. This states that, given nominal-scaled characters, there might be (in contrast to quantitative characters) no pair-wise relationships, though there is indeed a (three-dimensional) relationship between, for instance, three characters (see Figure 15.11). That is, there are actually typical patterns (configurations) of all combinations of the given measurement values, which cannot be identified with any association measure for bivariate random variables. However, such patterns are especially interesting when they emerge unexpectedly often (or unexpectedly rarely). In this case, they constitute a so-called *type* (or an *antitype*). CFA ascertains the extent to which certain patterns deviate in their frequency from that which is expected given the null hypothesis – the latter claiming that all characters are completely independent from one another.

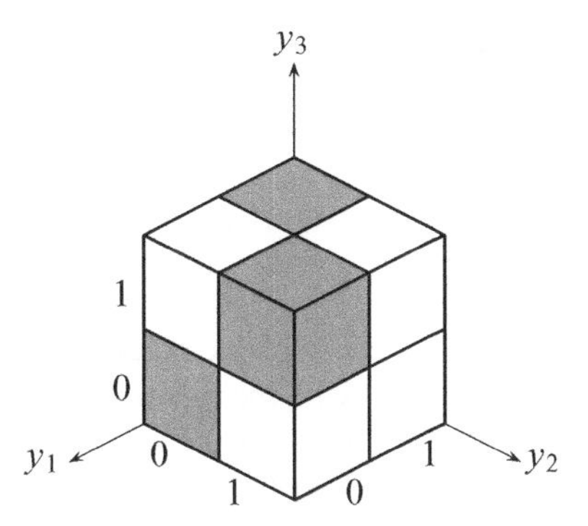

Figure 15.11 Three-dimensional contingency table for illustration of *Meehl*'s paradox. There are three dichotomous characters y_1, y_2, and y_3 (with measurement values 0 and 1), which do not correlate pair-wise. If we insert, into every gray-shaded sub-cube, for instance the frequency 20 (leaving the white sub-cubes with frequency 0) and then sum the frequencies up with respect to the third character, a result of 20 is found for each pair, (0, 0), (1, 1), (0, 1), and (1, 0) as well. Hence, the ϕ-coefficient is zero. Nevertheless, there are indeed typical patterns for these three characters: (1, 0, 0), (0, 1, 0), (0, 0, 1), and (1, 1, 1) occur equally often, while all other combinations are never realized (in this figure, the hidden sub-cube at the back left is considered to be white).

Master Doctor

Example 15.5 Symptom patterns after the consumption of LSD (from Krauth & Lienert, 1995, p. 18 *et seqq.*)

The so-called psychotoxic basic syndrome consists of the symptoms *clouding of consciousness* (y_1), *mental disorder* (y_2), and *disturbance of affectivity* (y_3). The following data were observed (1 for 'present', 0 for 'not present'):

y_1	y_2	y_3	h_{ijk}
1	1	1	20
1	1	0	1
1	0	1	4
1	0	0	12
0	1	1	3
0	1	0	10
0	0	1	15
0	0	0	0

All ϕ-coefficients equal zero, as can easily be calculated.

Given the null hypothesis, every pattern (i, j, k) should be observed with the frequency $n \cdot p_i \cdot p_j \cdot p_k$, where p_i states the probability for the value '1' in the character y_1, p_j for y_2, and p_k for y_3; h_{ijk}/n estimates p_{ijk} (with $n = 65$). The resulting expected frequencies e_{ijk} can then be tested one by one as to whether they differ significantly from the observed frequencies, using the test statistic from Formula (8.11) analogously ($df = 1$). We decide that $\alpha = 0.05$.

In **R**, we first create a matrix with the eight different combinations of character values in analogy to the above table. We type

```
> cfa.mat <- cbind(c(1, 1, 1, 1, 0, 0, 0, 0),
+                  c(1, 1, 0, 0, 1, 1, 0, 0),
+                  c(1, 0, 1, 0, 1, 0, 1, 0))
```

i.e. we use the function `c()` to create the separate columns, with which we create three vectors; then, we link them into a matrix with the function `cbind()`; we assign this data to the object `cfa.mat`. Next, we use `c()` to create a vector with the frequencies of the eight different combinations of character values; thus, we type

```
> h.ijk <- c(20, 1, 4, 12, 3, 10, 15, 0)
```

wherein we assign the resulting vector to the object `h.ijk`.

For the analysis, we apply the package `cfa`, which we load after its installation (see Chapter 1) using the function `library()`. Now, we conduct the configuration frequency analysis by typing

```
> cfa(cfa.mat, h.ijk, sorton = "label", sort.descending = TRUE,
+      bonferroni = FALSE)
```

i.e. we use `cfa.mat`, the combinations of character values, as the first argument for the function `cfa()`; as the second argument, we use `h.ijk` for the respective frequencies. With `sorton = "label"` and `sort.descending = TRUE`, we request that the results be displayed in order of the character values in the above table, and, finally, we use the argument `bonferroni = FALSE` to determine that the analysis be conducted without being adjusted using the Bonferroni method (see below).

For results, we get (shortened output):

```
*** Analysis of configuration frequencies (CFA) ***

label     n  expected           Q     chisq    p.chisq
1 1 1 1  20 12.505562 0.14276632 4.491329 0.03406718
2 1 1 0   1  6.848284 0.10056941 4.994306 0.02543085
3 1 0 1   4 11.402130 0.13810493 4.805377 0.02837106
4 1 0 0  12  6.244024 0.09796410 5.306076 0.02125117
5 0 1 1   3  9.463669 0.11638631 4.414674 0.03563109
6 0 1 0  10  5.182485 0.08053686 4.478247 0.03432892
7 0 0 1  15  8.628639 0.11302478 4.704594 0.03008211
8 0 0 0   0  4.725207 0.07839441 4.725207 0.02972360

Summary statistics:

Total Chi squared         =  37.91981
Total degrees of freedom  =  4
p                         =  7.371295e-10
Sum of counts             =  65
```

We see that, additionally, a global χ^2-test with reference to Formula (9.3) is determined, with $df = 2^m - m - 1$ degrees of freedom, where m is the number of observed alternative characters.

Because of a p-value of almost zero, the global χ^2-test is significant: at least one pattern occurs, compared to the null hypothesis ('the three characters are independent from one another'), unexpectedly often (or unexpectedly rarely). Every pattern results in a significant test statistic. When we compare the column of observed frequencies (`n`) to the column of expected frequencies (`expected`), there are four typical patterns that occur unexpectedly often and four antitypical patterns that occur unexpectedly rarely. The typical patterns are (1, 1, 1), (1, 0, 0), (0, 1, 0), and (0, 0, 1). Thus, three mono-symptomatic syndromes and one tri-symptomatic syndrome emerge.

Master Doctor

Strictly speaking, configuration frequency analysis as discussed so far belongs to descriptive statistics, despite the application of a significance test. Due to the comparison-wise type-I risk which leads to a severely high study-wise type-I risk when using more than three characters, the researcher will most likely only refer to the highest (configuration-specific) χ^2-values as (descriptive) statistics for interpretation of the analysis. From the methodological point of view, the better approach for testing certain types of patterns of combinations of measurement values for significance is using log-linear models (see Section 12.1.6). For more complex analysis methods that are based on CFA, see for example von Eye, Mair, & Mun (2010).

Doctor

Seen historically, the CFA was probably the method that introduced the technique of *alpha correction* to psychology. Obviously, the type-I risk accumulates if more than one test with some nominal α is applied within the same research question. If the tests are independent of one another – which is not the case if, as in the CFA, they are applied to the same data – m tests would have the probability $1 - (1 - \alpha)^m$ of not resulting in a type-I error. For $\alpha = 0.05$ and $m = 8$, for example, this results in 0.3366. As a workable adjustment, one can deduce a comparison-wise type-I risk of $\alpha_{\text{comparison}} = 1 - \sqrt[m]{1 - \alpha}$. For $\alpha = 0.05$ and $m = 8$, this results in $\alpha_{\text{comparison}} = 0.0064$. The better-known adjustment is the *Bonferroni correction*, which takes into account the fact that the tests are not independent; it adjusts $\alpha_{\text{comparison}} = \alpha / m$.

In fact, none of the alpha corrections solve the given problem because they all fail to consider type-II risk. This risk is (comparison-wise) higher than in the case of no alpha correction. Thus, global tests with a study-wise type-I risk are always to be preferred, or, when these are not compelling enough, it is preferable to apply only a single or at most very few tests with a comparison-wise type-I risk, those determined based on the theory of content in question.

Doctor

Example 15.5 – continued

Adjusting the comparison-wise type-I risk $\alpha_{\text{comparison}}$ in accordance with the Bonferroni correction, we must divide the actually given study-wise type-I risk $\alpha = 0.05$ as suggested; thus, $\alpha_{\text{comparison}} = 0.05 / 8 = 0.0063$. Using this α, no pattern would be considered either typical or antitypical. However, since the global χ^2-test is significant, the largest deviation, this being $12 - 6.2440$ ($\chi^2 = 5.3061$) for the pattern (1, 0, 0), may be interpreted as a type.

15.2.1.2 Latent class analysis

Latent class analysis (LCA) essentially is based on the same data structure as CFA. It models the actual pattern of outcomes for all characters per research unit as some probability function. We assume that there are latent and thus unobservable groups (classes) of research units, which do not differ within their groups. The number of groups is unknown, as well as their sizes. Within each of these groups, the investigated characters have a group-specific probability (which is to be estimated) that a research unit belonging to this group realizes a certain measurement value. The goals of LCA are to ascertain the optimal number of groups necessary to achieve proper data fit, to estimate the unknown parameters of group size, to estimate the group-specific probability of realization per measurement value and per character, to calculate the probability of group membership to any group for every research unit, and, finally, to determine rules for how the probability of group membership can be assessed for new research units. The latter goal is not considered in cluster analysis.

Doctor

Assume that the measurement values '1' and '0' of all characters $y_1, y_2, \ldots, y_m$ are, within each group, independent events; as mentioned earlier, we only deal with dichotomous characters in the following. This assumption does not exclude correlations between the characters per group; however, such correlations are interpreted like those in partial correlation (see Section 12.1.1), which are only constituted via a third random variable (noise factor): in this case, group membership.

LCA models the probability for a certain pattern s_v to be realized by person v through certain outcomes with respect to m dichotomous characters as follows: i serves as an index for the a groups; π_i states the probability for a randomly drawn person (from the sample) to belong to group A_i; p_{iq} is the probability for a '1' in character q within the group i; and $y_{qv} = 1$ or $y_{qv} = 0$:

$$P(s_v) = \sum_{i=1}^{a} \left\{ \pi_i \prod_{q=1}^{m} \left[p_{iq}^{y_{qv}} \cdot (1 - p_{iq})^{1-y_{qv}} \right] \right\} \qquad (15.2)$$

The unknown parameters can be estimated according to the maximum likelihood method. However, it is obvious that this is only possible if the number, a, of groups is arbitrarily fixed, or if the analysis is applied according to several different values of a (see the same problem for factor analysis as well, in Section 15.1.2). In any case, fit of the model of LCA – i.e. the respective solution in terms of parameter estimates for any given number of groups – must be tested for the data or, respectively, the goodness of fit must be compared with some concurrent model. The latter is possible via a likelihood-ratio test (see Section 14.2.2); the former can be done either using the conventional χ^2-test according to Formula (8.11), over all c patterns, or likewise with a likelihood-ratio test in which the denominator gives the likelihood of the realized polynomial distribution over all patterns:

$$L_0 = \prod_{g=1}^{c} \left(\frac{h_g}{n} \right)^{h_g}$$

(each with $df = 2^m - a(m + 1)$ degrees of freedom). Naturally, a classification analysis like that used in discriminant analysis is possible.

Doctor

Example 15.6 Knowledge about lung cancer and the way of gaining further information (from Goodman, 1970; but see, especially, Formann, 1984)

There are $m = 5$ characters: *reading newspapers* (1 for 'yes'; 0 for 'no'), *listening to the radio*, *reading books and magazines*, *attending lectures* (all analogous to the first character), as well as *knowledge about lung cancer* (1 for 'good'; 0 for 'poor'). For all the patterns, the following frequencies were observed:

11111	11110	11101	11100	11011	11010	11001	11000	10111	10110	10101
23	8	102	67	8	4	35	59	27	18	201

10100	10011	10010	10001	10000	01111	01110	01101	01100	01011	01010
177	7	6	75	156	1	3	16	16	4	3

01001	01000	00111	00110	00101	00100	00011	00010	00001	00000
13	50	3	8	67	83	2	10	84	393

The question arises as to which types of persons, i.e. which typical patterns exist. In the case of only five characters, a larger number than $a = 3$ groups does not seem meaningful; even using four groups, almost every character might end up defining a group alone. On the other hand, the cases of $a = 2$ groups and $a = 1$ group are interesting as well; the latter means that no different types exist. For the goodness of fit test, $\alpha = 0.01$ should be applied.

We use the data set from *Example 15.6* (see Chapter 1 for its availability).

In **R**, we start by creating two objects, one for all combinations of character values, and one for the respective frequencies. Thus, we type

```
> lca.mat <- Example_15.6[, 1:5]
> freq <- Example_15.6[, 6]
```

i.e. we assign the columns 1 through 5 from the data set `Example_15.6` to the object `lca.mat` with `[, 1:5]`, and with `[, 6]` the column 6 to the object `freq`. Now, we use the package `randomLCA`, which we load after its installation (see Chapter 1) using the function `library()`. In order to conduct a latent class analysis, we type

```
> lca.1 <- randomLCA(lca.mat, freq, nclass = 1, calcSE = FALSE)
> lca.2 <- randomLCA(lca.mat, freq, nclass = 2, calcSE = FALSE)
> lca.3 <- randomLCA(lca.mat, freq, nclass = 3, calcSE = FALSE)
> lca.4 <- randomLCA(lca.mat, freq, nclass = 4, calcSE = FALSE)
```

i.e. we use the patterns within the object `lca.mat`, as well as their frequencies within the object `freq`, as arguments for the function `randomLCA()`; with `nclass`, we define the desired numbers of groups, and with `calcSE = FALSE`, we suppress the calculation of the standard errors. So, we conduct the LCA for each number of groups $a = 1$ through $a = 4$, and assign the results to the objects `lca.1` through `lca.4`.

Next, we test the goodness of fit of the respective model with a certain number of groups via a likelihood-ratio test. For this purpose, we type

```
> pchisq(lca.1$deviance, df = 2^5 - 1*(5+1), lower.tail = FALSE)
> pchisq(lca.2$deviance, df = 2^5 - 2*(5+1), lower.tail = FALSE)
> pchisq(lca.3$deviance, df = 2^5 - 3*(5+1), lower.tail = FALSE)
> pchisq(lca.4$deviance, df = 2^5 - 4*(5+1), lower.tail = FALSE)
```

i.e. we transfer three arguments to the function `pchisq()`; for the first one, the empirical χ^2-value, accessed by `$deviance`; and as a second one, the degrees of freedom, for example `2^5 - 4*(5+1)` for $df = 2^5 - 4 \times (5+1)$ degrees of freedom; with `lower.tail = FALSE` as a third argument, we request the corresponding p-value.

As a result, we get:

```
[1] 2.712471e-109
[1] 0.002521027
[1] 0.09423195
[1] 0.2618086
```

The results are not (any longer) significant for the solution with three groups (`0.0942`); because of this, we request the result of this solution by typing

```
> summary(lca.3)
```

As a result, we get:

```
Classes      AIC      BIC    logLik
      3 9373.626 9466.367 -4669.813
Class probabilities
Class  1 Class  2 Class  3
  0.3202   0.4405   0.2393
Outcome probabilities
              it1         it2     it3     it4    it5
Class  1 0.8652 5.844e-01 0.70727 0.15634 0.5909
Class  2 0.2499 1.161e-01 0.08433 0.02264 0.1589
Class  3 0.7338 3.040e-10 0.88013 0.07539 0.5313
```

First of all, the interviewed persons are about equally distributed in the three groups (23.9% to 44.0%). Furthermore, the characters *reading newspapers* and *reading books and magazines* strongly differentiate between Groups 1 and 3 on the one hand and Group 2 on the other hand. People who belong to Group 2 do these things significantly less often than people from the other two groups (25.0% versus 0.08%). This relationship between the groups also applies to the character *knowledge about lung cancer*, although to a lesser extent; individuals in Group 2 show such knowledge less often than those in Groups 1 or 3. These two latter groups are only differentiated by the character *attending lectures*, which members of Group 3 do about twice as often as members of Group 1.

Doctor

For more complex approaches to LCA, see Formann (1984), as well as Collins & Lanza (2010).

15.2.2 Confirmatory factor analysis

While (exploratory) factor analysis is, as a method of descriptive statistics, theory-generating, *confirmatory factor analysis* tests an already theory-based factor structure for its fit to given data.

Just as the path analysis before, confirmatory factor analysis can be viewed as a special case of (linear) structural equation models (SEMs). Thus, we will briefly introduce them at this point.

The main characteristic of structural equation models is that (linear) dependencies, directed or undirected, are being modeled between several latent variables. Additionally, these latent variables are assumed to be responsible for some observed characters, modeled as random variables. In order to take random effects for them into account, unknown random errors are modeled for each observed character. These can, again, be correlated or not correlated with one another. The goal is to estimate the unknown modeled regression coefficients in a way that can optimally reproduce the given correlations between the investigated characters.

Master Doctor

Figure 15.12 gives an arbitrary example of a structural equation model. In this case, $\boldsymbol{f}_1, \boldsymbol{f}_2, \boldsymbol{f}_3$ and $\boldsymbol{f}_4$ are latent variables that predict the investigated characters y_1 through y_{10} to a yet unknown degree of strength, but in accordance with the illustrated paths; $\boldsymbol{f}_3$ is the latent error of $\boldsymbol{f}_4$. The two latent variables $\boldsymbol{f}_1$ and $\boldsymbol{f}_2$ correlate with each other. The errors for the random variables $\boldsymbol{y}_q$ modeling the characters y_q are $\boldsymbol{e}_q$, $q = 1, 2, \ldots, 10$. Errors $\boldsymbol{e}_8$, and $\boldsymbol{e}_{10}$ are assumed to correlate.

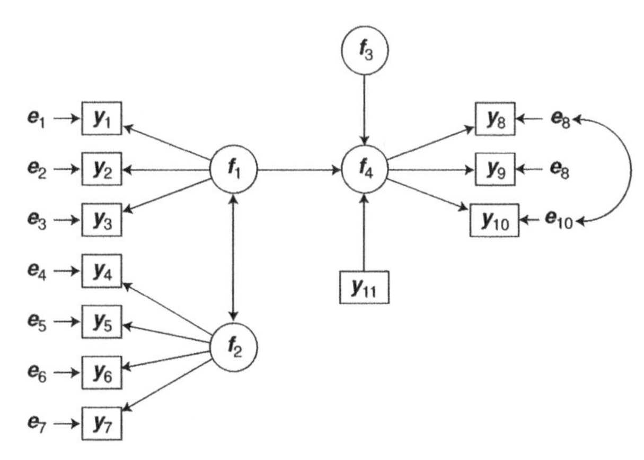

Figure 15.12 Example of a structural equation model (see the detailed description within the text).

Finally, there is also a manifest character, y_{11}, that influences a latent variable, namely f_4.

Numerous (linear) equations may be derived from Figure 15.12. For example (if all variables are standardized): $y_4 = \beta_{f_2 y_4} \cdot f_2 + e_4$ and $f_4 = \beta_{f_1 f_4} \cdot f_1 + f_3 + \beta_{y_{11} f_4} \cdot y_{11}$; β are the regression coefficients (correlations), indexed with the respective paths. Both of these equations show that some regression coefficients are (necessarily) modeled with a weight of 1. In these equations, there are in total (inclusive the error variances of each manifest variable) less unknown parameters – in particular if, for each latent variable, we fix any regression coefficient at 1 – than there are correlation coefficients, that is $\binom{11}{2} = 55$. For this reason the given structural equation model is identifiable.

Generally, the maximum likelihood method is used for estimating the unknown parameters; in this case, a certain multidimensional type of distribution of the modeled random variable has to be assumed. The question of model quality is, in this context, answered in several diverse ways. Basically, this is due to the fact that the pertinent goodness of fit test is almost always significant (using some χ^2-distributed test statistic $\boldsymbol{T}$, essentially the deviation is ascertained between the given correlation matrix and the one that results for the estimated parameters (for more details, see e.g. Kline, 2010)). Thus, researchers are traditionally content with fairly high indexes or coefficients of goodness of fit. Besides *AIC* (see Section 14.2.2), mostly some kind of coefficient based on the test statistic $\boldsymbol{T}$ is used. Very often, this is the *root mean square error of approximation* (RMSEA) in the form of

$$\sqrt{\frac{T - df}{n \cdot df}}$$

If the model holds, this term corresponds to the standard error of the model equation's estimator. Also often used is the index CFI (*comparative fit index*):

$$\frac{(T_0 - df) - (T - df)}{T_0 - df}$$

with T_0 as the value of the test statistic $\boldsymbol{T}$ where all correlations are set to zero; ideally, it amounts to 1. Model-immanent benchmarks that define numerical criteria for determining when a structural equation model explains the data sufficiently are not available – there are merely a few conventions. Certainly, these are most compelling when compared for several concurrent models (or that is to say: model specifications).

Confirmatory factor analysis is now based on a given factor structure which most often stems from the results of an earlier exploratory factor analysis. It is a special case of structural equation models insofar as: (1) the latent variables (factors) are not modeled as the target of a single-edged, directed path, and thus do not show a latent error; (2) all latent variables (factors) are modeled as being uninfluenced by manifest characters; (3) all manifest characters are modeled as being dependent on at least one latent variable (factor) and thus all have been modeled with an error; and (4) all these errors are assumed to be uncorrelated.

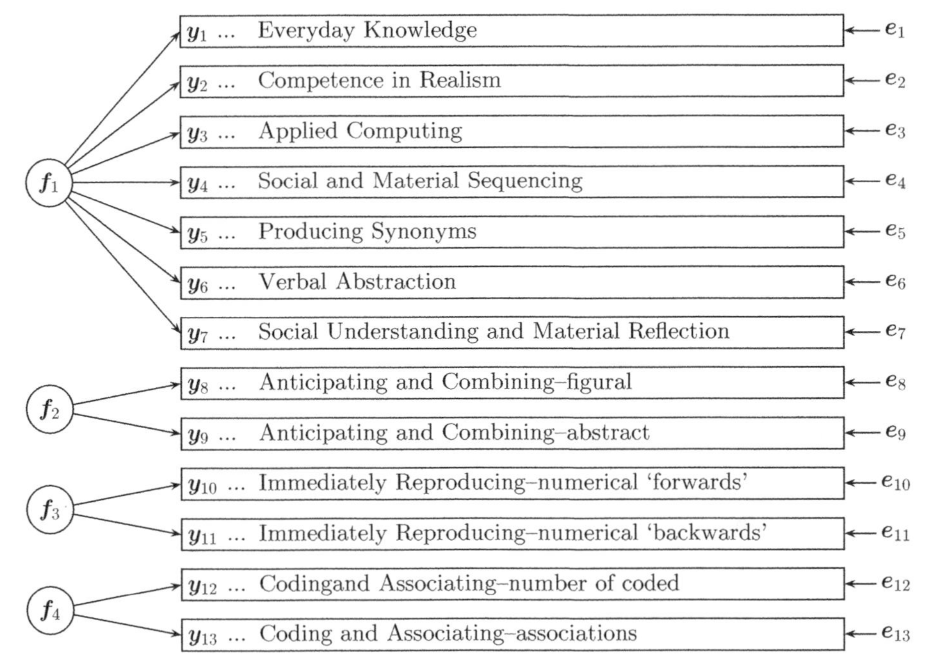

Figure 15.13 Linear structural equation model of confirmatory factor analysis for the test battery AID 2-Turkish.

Master Doctor

Example 15.7 Can the factor structure of the applied intelligence test battery be confirmed for children with Turkish as native language (Example 1.1)?

In Example 1.1, the intelligence test battery AID 2 was used. As a continuation of Example 15.2, the research question now is whether the factor structure that was ascertained in the standardization sample of AID 2 also holds within the population of children with Turkish as native language.

Apart from the 5 subtests used in this book, the AID 2 consists of numerous other subtests; together, they provide 13 test scores per child. The factor structure that was obtained for them using (exploratory) factor analysis consists of four factors. Factor I is mainly represented by the subtests *Everyday Knowledge, Applied Computing*, and *Social and Material Sequencing*; Factor II by *Coding and Associating*; and Factor III by *Immediately Reproducing – numerical* (subtests that represent Factor IV were not covered in Example 1.1).

Though exploratory factor analysis ascertained a three-factor-solution in Example 15.2, the respective highest loadings are not completely in congruence with the factor structure of children with German as native language. With confirmatory factor analysis we now examine whether the same factor structure nevertheless underlies this data. However, it is not possible to estimate the parameters of the respective model (see Figure 15.13). Thus, we will use the original data – available as *Example 15.7* (see Chapter 1 for its availability) – that was given for calibration and standardization of the version AID 2-Turkish (Kubinger, 2009a). This data set contains the test scores of 355 children with Turkish as native language.

In **R**, we use the package `sem`, which we load after its installation (see Chapter 1) using the function `library()`. Next, we create a variance–covariance matrix of the characters we want to analyze by typing

```
> S <- cov(Example_15.7)
```

i.e. we apply the function `cov()`, to which we submit the data set `Example_15.7`; we assign the results to the object `S`.

Now, we have to specify the model from Figure 15.13; to do this, we type

```
> model <- specify.model()
```

i.e. we use the function `specify.model()` without adding an argument; we assign the results to the object `model`. After computing this instruction, a '`1:`' comes up; this is a prompt to add further instructions. Now, we have to define the model line by line; we type

```
F1      ->   sub1,   NA,     1
F1      ->   sub2,   lam12,  NA
F1      ->   sub3,   lam13,  NA
F1      ->   sub4,   lam14,  NA
F1      ->   sub6,   lam15,  NA
F1      ->   sub9,   lam16,  NA
F1      ->   sub11,  lam17,  NA
F2      ->   sub8,   NA,     1
F2      ->   sub10,  NA,     1
F3      ->   sub5f,  NA,     1
F3      ->   sub5b,  NA,     1
F4      ->   sub7c,  NA,     1
F4      ->   sub7a,  NA,     1
sub1    <--> sub1,   e12,    NA
sub2    <--> sub2,   e12,    NA
sub3    <--> sub3,   e13,    NA
sub4    <--> sub4,   e14,    NA
sub6    <--> sub6,   e15,    NA
sub9    <--> sub9,   e16,    NA
sub11   <--> sub11,  e17,    NA
sub8    <--> sub8,   e21,    NA
sub10   <--> sub10,  e22,    NA
sub5f   <--> sub5f,  e31,    NA
sub5b   <--> sub5b,  e32,    NA
sub7c   <--> sub7c,  e41,    NA
sub7a   <--> sub7a,  e42,    NA
F1      <--> F1,     var.F1, NA
F2      <--> F2,     var.F2, NA
F3      <--> F3,     var.F3, NA
F4      <--> F4,     var.F4, NA
```

Every line consists of three arguments: in the first argument, the relationships between characters are being defined, wherein a one-headed arrow '->' specifies a directed relationship, and a two-headed arrow '<-->' requests the covariance if there are two different characters, or, alternatively, requests the variance if the two indicated characters are actually the same one. With the second argument, we choose a label for the respective parameter; if this parameter is to be fixed to a certain value, this line should read NA. With the help of the third argument, an optional starting value can be defined for the parameter estimation, or the parameter can be fixed to a certain value; if NA is typed in here, the parameter will be estimated freely, letting the program define the starting value. If we want to fix a parameter, we type in the desired value as a third argument (after having typed NA as a second one). In the case at hand, the first 13 lines serve the purpose of defining the relationships between the 13 manifest and the four latent characters. The factor loading of the character sub1 is being fixed to the value 1 and serves as a reference variable for the latent character F1. Also, every loading of the latent characters F1 through F4 is fixed to 1. In the lines 14 through 26, we request a residual variance for every manifest character. Finally, the statements in the last 4 lines order the estimation of the variance of each factor. After entering the last line, we add an empty line to signal the end of the model specifications to the program. Now, we estimate the model parameters by typing

```
> sem.model <- sem(model, S = S, N = nrow(Example_15.7))
```

i.e. we use the function sem(), assigning the specified model in the object model, the variance–covariance matrix in the object S, and the sample size (N) (with nrow()) to it as arguments. We assign the results of the calculation to the object sem.model. Finally, we instruct the program to print the results by typing

```
> summary(sem.model)
```

i.e. we submit the object sem.model to the function summary() as an argument.

As a result, we get (shortened output):

```
Model Chisquare =  324.1   Df =   69 Pr(>Chisq) = 0
 Chisquare (null-model) =  1431.2   Df =  78
 RMSEA index =  0.10219   90% CI: (0.091128, 0.11354)
 Bentler CFI =  0.81148

Parameter Estimates
      Estimate  Std Error z value Pr(>|z|)
lam12    0.50104  0.063992  7.8297 4.8850e-15 UT2tw <--- F1
lam13    0.96933  0.085426 11.3470 0.0000e+00 UT3tw <--- F1
lam14    0.64166  0.084397  7.6028 2.8866e-14 UT4tw <--- F1
...
e12     75.20702  4.467940 16.8326 0.0000e+00 UT1tw <--> UT1tw
e13     95.76359  8.150359 11.7496 0.0000e+00 UT3tw <--> UT3tw
e14    131.22587 10.231441 12.8257 0.0000e+00 UT4tw <--> UT4tw
e42     36.89773  4.725225  7.8087 5.7732e-15 UT7astw <--> UT7astw
...
var.F1  74.37218 10.324548  7.2034 5.8709e-13 F1 <--> F1
var.F2  52.39293  6.876891  7.6187 2.5535e-14 F2 <--> F2
var.F3  22.36371  3.802350  5.8816 4.0644e-09 F3 <--> F3
var.F4  54.73555  5.618461  9.7421 0.0000e+00 F4 <--> F4
```

The parameters we just obtained are not standardized yet. In order to get the standardized coefficients, we type

```
> std.coef(sem.model)
```

i.e. we submit the object `sem.model` to the function `std.coef()`.

This results in (shortened output):

```
          Std. Estimate
1              0.7051307          UT1tw <--- F1
2    lam12     0.4459594          UT2tw <--- F1
3    lam13     0.6495144          UT3tw <--- F1
4    lam14     0.4349664          UT4tw <--- F1
...
14     e12     0.5027906        UT1tw <--> UT1tw
15     e12     0.8011202        UT2tw <--> UT2tw
16     e13     0.5781310        UT3tw <--> UT3tw
17     e14     0.8108042        UT4tw <--> UT4tw
...
27  var.F1     1.0000000              F1 <--> F1
28  var.F2     1.0000000              F2 <--> F2
29  var.F3     1.0000000              F3 <--> F3
30  var.F4     1.0000000              F4 <--> F4
```

Due to the *p*-value of almost zero, the global χ^2-test results in significance; thus, describing the observed correlation matrix by one reproduced using the model is not successful. The coefficients of goodness of fit (RMSEA = 0.102 and CFI = 0.811) reveal that, according to conventional appraisal criteria (for details, see Kline, 2010), our model does not explain the data sufficiently.

15.2.3 Models of item response theory

In the context of psychometrics, special theory-generating methods for psychology, especially for psychological assessment, have been developed. They attempt to model how reactions (answers) to a task, a question, or a statement (generally: *item*) of a psychological test (or questionnaire) come about. This has been the starting point for a new (sub-) field of scientific research, *item response theory* (IRT). Essentially, IRT is based on the logistic function (see Section 13.4). To be more precise, IRT deals with a probability function that postulates the probability for the occurrence of a certain reaction category in dependence of the respective aptitude or trait of a given person as well as in dependence of the respective characteristics of a given item.

15.2.3.1 Rasch model

The basic model for all other models of IRT is the (dichotomous logistic test-) model by Rasch.[2] For simplicity's sake, it is just called the *Rasch model* (see Example 5.7). It postulates

[2] *Georg Rasch*, Danish statistician.

the probability of the solution of a certain task in a performance test (mostly with an unknown difficulty) in dependence of the (unknown, but to be estimated) ability of a certain testee. Given testee v and task i, the observable (manifest) character y_{iv} is to be modeled as random variable $\mathbf{y}_{iv}$ with the measurement values –/0 ('not solved') and +/1 ('solved'). The outcome depends on two unobservable (latent) character: the character *ability*, in the form of realization ξ_v, and the character *difficulty*, in the form of realization σ_i. The distribution of all three characters is of no interest here: this concerns the sample of testees $v = 1, 2, \ldots, n$, and the population of testees $v = 1, 2, \ldots, N$, as well as the tasks of a task pool $i = 1, 2, \ldots, k$ or the tasks in a task universe $i = 1, 2, \ldots, K$. In this sense, ξ_v and σ_i represent fixed, but unknown, values (or parameters), which, in general, are to be estimated. It is merely presupposed that all ξ_v and σ_i are one-dimensional and express different degrees of realizations of the same dimension. Thus, the Rasch model models the probability for the task i being solved by testee v as follows:[3]

$$P(+|\xi_v, \sigma_i) = \frac{e^{\xi_v - \sigma_i}}{1 + e^{\xi_v - \sigma_i}} \tag{15.3}$$

Doctor

Naturally, the Rasch model is a special case of the generalized linear model. The fact that makes the Rasch model so special in comparison to many other models of the IRT is that the estimation of parameters is, aside from random errors, independent of the chosen sample. Specifically, the difficulty parameters σ_i are (or can be) estimated independently of the testees: no matter which sample of persons is used, the estimates of every difficulty parameter σ_i are statistically the same. This is a quality immanent to the model (for a proof, see, for example, Kubinger, 2009b); it refers to a special estimation method, the *conditional maximum likelihood estimation method* (CML). This method applies under the condition that the unknown ability parameters are replaced by their exhaustive statistics (see Section 6.5). From this property of the model, an examination of model appropriateness can be deduced; i.e. the (absolute) validness of the model can be tested in the sense of a model test, and not merely the fit in comparison to concurrent models or the general fit of the model to the data can be evaluated (see Section 14.1). The Rasch model implies that the estimates of the difficulty parameters are (aside from random deviations) equal, no matter which (sub-) sample was used for estimation. Given empirical data and a psychological test for which we want to know (for reasons we will discuss later) whether its tasks behave in conformity with the model, all estimates of the difficulty parameters σ_i should be equal for any (say two) arbitrarily split sub-samples of the total sample.

The easiest method to examine this is the graphical model check, a means of descriptive statistics. Since all estimates $\hat{\sigma}_i$ are best standardized to $\sum_{i=1}^{k} \hat{\sigma}_i = 0$ (in every sub-sample), one can, for each task i, plot the two corresponding parameter estimates against each other in a rectangular coordinate system. If all resulting

[3] In the American literature, θ and b are often used instead of ξ and σ. This is, for the general case, illogical because, in the general case, the difficulty parameters are not known values; we always label (unknown) parameters with Greek symbols. Also, within statistics, θ is used to describe the general case for any parameter. Here, we join Georg Rasch in his original labeling.

points lie on or very close to the 45° line through the origin, then the estimates $\hat{\sigma}_i$ of both sub-samples (largely) match each other, as is the case for the Rasch model; hence, the tasks conform to the Rasch model. Intrinsically, also belonging to descriptive statistics, is the test statistic z ('z-test'), originally conceptualized by Fischer and Scheiblechner (1970). Here, the difference of the two sub-samples' parameter estimates for each task is divided by their standard deviation – which is the root of the sum of the two reciprocal values of the so-called *Fisher's information*. Although the test statistic z is asymptotically normally distributed for each task, the vast number of corresponding results makes this approach inappropriate for hypothesis testing: the type-I risk would be drastically increased. As a method of inferential statistics, there is a special likelihood-ratio test (see Section 14.2.2), *Andersen's likelihood-ratio test*. Simplified, this test compares the likelihood of the data – in accordance with Section 6.5, this corresponds to the plausibility of the data – for two hypotheses. On the one hand, there is the null hypothesis, which states the same difficulty parameters for both sub-samples, and on the other hand there is the alternative hypothesis, which states that different difficulty parameters are given in each of the two sub-samples. The corresponding test statistic is approximately χ^2-distributed (with $df = k - 1$ degrees of freedom).

The importance of the Rasch model for the construction of psychological tests lies in a mathematical proof (see Fischer, 1995). This proof shows that the Rasch model must hold for the tasks of a test, in order to justify the regularly implemented scoring rule of summing up the 'number of solved tasks' to obtain a test score. In this context, 'justified' means that the measurement of the ability in question according to this score is actually empirically founded and that it is fair to compare different testees based on this score. Therefore, for tests that do not conform with the Rasch model, the 'number of solved tasks' does not express the empirically observable behavioral relations between testees in an adequate and hence fair way. Thus, it must be the goal of psychological test development to ensure that the tasks of a test are calibrated in conformity with the Rasch model; in this way, it is possible to measure one-dimensionally, and moreover to measure or score using the 'number of solved tasks'. Usually, such a *calibration* is achieved by excluding those tasks from the pool (*item pool*), which do not conform to the Rasch model. Kubinger (2005) shows how to do this in an optimal manner.

Doctor

Planning a study for the calibration of a psychological test according to the Rasch model has only recently been basically solved (see Kubinger, Rasch, & Yanagida, 2009; see also Example 10.16). For a few selected cases (see also Kubinger, Rasch, & Yanagida, in press), the necessary number of testees now can be ascertained in order to avoid overlooking relevant deviations between the parameter estimates of different sub-samples – given a fixed type-I risk and a fixed type-II risk. For $k = 20$ tasks that are equally distributed with respect to their difficulties between –3 and 3, given $\alpha = 0.05$, $\beta = 0.20$, and a certain relevant effect, a sample size of two times 101 testees is needed. The relevant effect defined here was that two tasks (of average difficulty) show a difference in their parameter estimates from two sub-samples that is sized a sixth of the range of all difficulty parameters.

Example 15.8 Calibrating a psychological test in accordance with the Rasch model

The matter at hand is a new performance test x, for which we created $k = 20$ tasks; 202 testees were tested. The test is to be calibrated in accordance with the Rasch model.

We use the data from *Example 15.8* (see Chapter 1 for its availability). It contains the characters *gender* (0 for 'female'; 1 for 'male'), *age* (0 for '< 30'; 1 for '≥ 30'), and *native language* (0 for 'German'; 1 for 'not German') as well as the k tasks $y_1, y_2, \ldots, y_k$ (0 for 'not solved'; 1 for 'solved') for each testee. The calibration will be conducted according to Kubinger (2005), whereby four characters are used to partition the total sample into pairs of sub-samples, which are used to conduct Andersen's likelihood-ratio test (comparison-wise type-I risk $\alpha = 0.05$): *number of solved items*, i.e. the partition of the total sample into testees with a high number of solved items vs. testees with a low number of solved items, as well as *gender* ('male' vs. 'female'), *age* ('age group 0' vs. 'age group 1'), and *native language* ('German' vs. 'not German'). In the case of significance of Andersen's likelihood-ratio test, the graphical model check or the 'z-test' will be used to exclude tasks consecutively until a new calculation of the likelihood-ratio test produces non-significant results; when this occurs, *a posteriori* conformity with the Rasch model will be given.

In **R**, we apply the package `eRm` (see Mair, Hatzinger, & Maier, 2010), which we load after its installation (see Chapter 1) using the function `library()`. First, we create one object which contains the tasks that are to be analyzed as well as one with the characters by which we will split the total sample into partial samples. We type

```
> items <- Example_15.8[, 1:20]
> split <- Example_15.8[, 21:23]
```

i.e. we assign the columns 1 through 20 from the data set `Example_15.8` to the object `items` with `[, 1:20]`, and the columns 21 through 23 with the partition criteria to the object `split` with `[, 21:23]`. Now we calculate the Rasch model by typing

```
> items.RM <- RM(items)
```

i.e. we use the function `RM()`, to which we submit the tasks from the object `items` as an argument, and assign the results to the object `items.RM`.

Next, we compute the likelihood-ratio test for the program package's default criterion for the partition into partial samples, which is the *number of solved items*, and one more for each of our especially collected characters *gender, age,* and *first language*. We type

```
> items.RM.m <- LRtest(items.RM, splitcr = "median")
> items.RM.g <- LRtest(items.RM, splitcr = split[, 1])
> items.RM.a <- LRtest(items.RM, splitcr = split[, 2])
> items.RM.l <- LRtest(items.RM, splitcr = split[, 3])
```

i.e. we use the function `LRtest()`, to which we submit `items.RM` as the first argument and, as the second one, through `splitcr =` , the character by which the sample is to be partitioned; unfortunately, the package `eRm` uses the label `"median"` instead of *number of solved items* inside its programming structure.

When calculating the partition criterion *number of solved items*, a notice appears which tells us that task 2 was solved by all persons within a partial group and thus cannot be included in the analysis. We assign the results to one separate object each, namely to `items.RM.m` (for *number of solved items*), `items.RM.g` (for *gender*), `items.RM.a` (for *age*) and `items.RM.l` (for *native language*); in order to retrieve the results, we type

```
> summary(items.RM.m)
> summary(items.RM.g)
> summary(items.RM.a)
> summary(items.RM.l)
```

As a result, we get the test values including the *p*-values as clearly arranged in Table 15.6 (together with other results).

Table 15.6 (row 'Before') reveals that Andersen's likelihood-ratio test results in significance when the sample is split along the character *gender*. Thus, we use the graphical model check to examine whether the deletion of some few tasks might lead to Rasch model conformity of performance test *x*.

Table 15.6 Results of the calibration in accordance with the Rasch model in Example 15.8 (tasks that were solved by everyone in a certain sub-sample or not solved by anyone are not included in the respective analysis)

	Number of solved items			Gender			Age			Native language		
	χ^2 (LRT)	*df*	*p*-value	χ^2 (LRT)	*df*	*p*-value	χ^2 (LRT)	*df*	*p*-value	χ^2 (LRT)	*df*	*p*-value
Before	13.235	18	0.777	56.165	19	0.000	18.900	19	0.463	23.852	19	0.202
After	15.161	16	0.513	8.639	17	0.951	17.428	17	0.426	22.631	17	0.162

So, in **R**, we type

```
> plotGOF(items.RM.g, xlab = "male", ylab = "female",
+          xlim = c(-4, 4), ylim = c(-4, 4))
```

i.e. we use the function `plotGOF()` and set the results for the partition criterion of gender, `items.RM.g`, as the first argument; we label the axes with `xlab` and `ylab`, and define the displayed interval for the difficulty parameters as -4 to $+4$ with `xlim` and `ylim`. The resulting graphics are shown in Figure 15.14a.

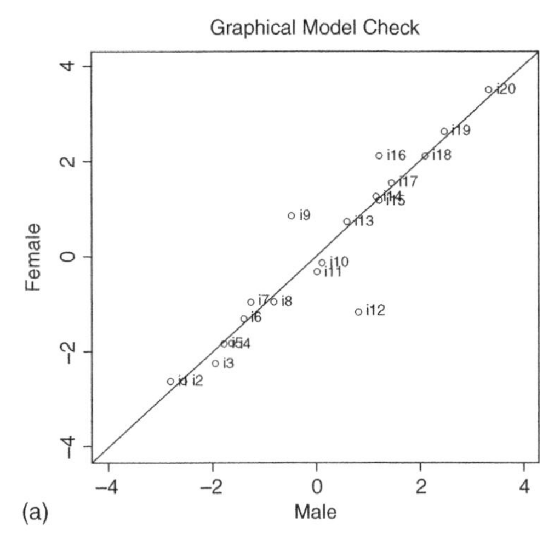

Figure 15.14a Graphical model check with respect to the character *gender* for all tasks in Example 15.8.

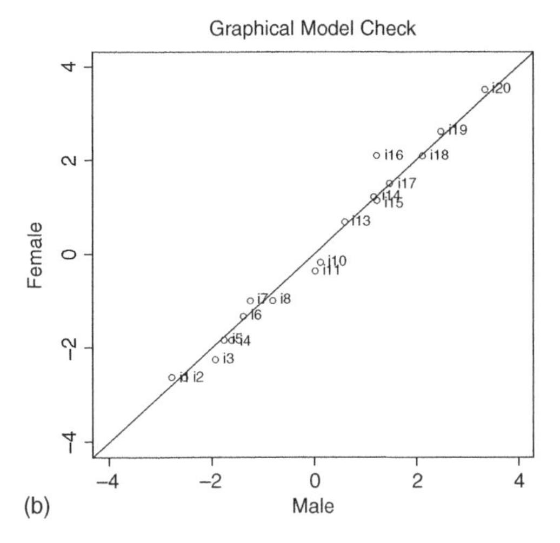

Figure 15.14b Graphical model check with respect to the character *gender* after the deletion of the tasks 9 and 12 (Example 15.8).

As can be seen in Figure 15.14a, the tasks with the numbers 9 and 12 have considerably different difficulties within the two sub-samples. In comparison to the rest of the tasks, task 12 is disproportionately easier for female testees than for male persons; for task 9, the opposite is true.

Thus, we remove these two tasks from the test and repeat the analysis for the remaining tasks.

In **R**, we again create an object which contains the tasks that are to be analyzed, this time excluding the two which do not conform to the Rasch model. We type

```
> items2 <- items[, -c(9, 12)]
```

i.e. we exclude the columns 9 and 12 from the object items with `[, -c(9, 12)]` and assign the results to the object `items2`. Next we repeat the analysis with the reduced data set by typing

```
> items2.RM <- RM(items2)
> items2.RM.m <- LRtest(items2.RM, splitcr = "median")
> items2.RM.g <- LRtest(items2.RM, splitcr = split[, 1])
> items2.RM.a <- LRtest(items2.RM, splitcr = split[, 2])
> items2.RM.l <- LRtest(items2.RM, splitcr = split[, 3])
> plotGOF(items2.RM.g, xlab = "male", ylab = "female",
+         xlim = c(-4, 4), ylim = c(-4, 4))
```

The result of these analyses is given in Table 15.6 as well.

Now, no significant LRT is to be seen. The graphical model check (see Figure 15.14b) also emphasizes that the model holds. Only the task numbered 16 is conspicuous; the difference between the two parameter estimates is a little more than a sixth of the range of the difficulty parameters. We decide in favor of *a posteriori* model validness.

In a case like this, when the model holds only *a posteriori*, the 'kind of cross-validation' (also see Section 14.2.3) as suggested by Kubinger (2005) is recommended: a new set of data should be used to test whether the remaining 18 tasks conform to the Rasch model.

Doctor **Example 15.8 – continued** Because we used the approach of Kubinger, Rasch, & Yanagida (2009) for planning the study, we now can indeed apply their new model test (see Example 10.16; there, a three-way analysis of variance of the type $(A \succ B) \times C$ was conducted in order to calibrate a psychological test in accordance with the Rasch model). However, this approach – testing the interaction effect $A \times C$ – is only useful if the main effect A has proved (in advance) to be non-significant.

Here, this is the case for partitioning the sample into two sub-samples with reference to the character *gender* (see Table 15.7a, which is reached by proceeding in accordance with Example 10.16).

Since the interaction is significant, we can, for example, try to identify and delete non-conforming tasks using the graphical model check described earlier. If we delete the tasks with the numbers 9 and 12 again, we get the results shown in Table 15.7b. For the remaining 18 tasks, *a posteriori* model validness is given.

Table 15.7a Results of the calibration in accordance with the Rasch model by a three-way analysis of variance of the type $(A \succ \boldsymbol{B}) \times C$ concerning the character *gender* for all tasks of Example 15.8

p (F-Test)			
A Gender	*B*(*A*) *Testees*	*C Tasks*	$A \times C$
0.269	0.000	0.000	0.000

The essential point of this approach is that only a single model test is conducted (i.e. only concerning a single partition of the total sample), because planning the study was based only on a study-wise type-I risk ($\alpha = 0.05$). Bear in mind that, in this context, when traditionally calibrating a psychological test (according to the Rasch model; see Kubinger, 2005), no conclusions about the study-wise type-I and type-II risk can be made because of the usage of a comparison-wise type-I risk.

Table 15.7b Results of calibrating in accordance with the Rasch model using a three-way analysis of variance of the type $(A \succ \boldsymbol{B}) \times C$ concerning the character *gender* after deleting the tasks 9 and 12 (Example 15.8)

p (F-test)			
A Gender	*B*(*A*) *Testees*	*C Tasks*	$A \times C$
0.198	0.000	0.000	0.937

15.2.3.2 Generalizations of the Rasch model

We will now discuss the most pertinent generalizations of the Rasch model. This specifically concerns the case of dichotomous characters; for more complex models and especially for multi-categorical characters as items (tasks), we refer to further literature (see, for example, Kubinger, 1989, or Bond & Fox, 2007).

One possibility of generalization concerns adding further item parameters, i.e. modeling item characteristics additionally to the difficulty parameter. With, all in all, three kinds of item parameters, the so-called Birnbaum or, to be more exact, *three-parameter logistic model* (*3-PL model*) is a quite general one:

$$P(+|\xi_v; \sigma_i, \alpha_i, \beta_i) = \frac{\beta_i + e^{\alpha_i(\xi_v - \sigma_i)}}{1 + e^{\alpha_i(\xi_v - \sigma_i)}} \tag{15.4}$$

with α_i for the item-discrimination parameter of task i – this corresponds to the different weightings of correct responses when calculating the test score – and β_i as the item-guessing parameter which corresponds to the effect of (merely) guessing the right answer, which can occur when tasks offer prewritten options for the response (so-called *multiple choice tests*). Setting all $\beta_i = 0$ results in the 2-PL model; additionally setting all $\alpha_i = 1$ leads to the Rasch model, which can thus also be called 1-PL model. Modeling all $\alpha_i = 1$, but not all $\beta_i = 0$ leads to the *difficulty-plus-guessing PL model* of Kubinger & Draxler (2006).

Doctor

All these models are also to be interpreted as special cases of the generalized linear model.

In contrast to the Rasch model, a CML estimation (of the item parameters) is not possible for the other models mentioned here (nor for most of the other IRT models not mentioned). This also means that a model test is not possible; only the model's fit to the data or its fit in comparison to some concurrent model can be ascertained. Appropriate goodness of fit tests (likelihood-ratio tests) or coefficients of goodness of fit, especially *AIC* and *BIC* (see Section 14.2.2), are used in this context. Parameter estimation (of the item parameters) is carried out by assuming a certain distribution function of the also unknown ability parameters (*marginal maximum likelihood method*, MML). Aside from the **R**-package, there are numerous, mostly commercially distributed computer packages for this calculation. We will not deal with them here in detail.

Master Doctor

The other possibility for generalizing the Rasch model when dealing with items as dichotomous characters, is to decompose the item difficulty parameters into a linear combination of so-called basic parameters: $\sigma_i = \sum_{j=1}^{p} q_{ij}\eta_j$ – here the q_{ij} are fixed values which are hypothesized and thus modeled as weights for the parameters η_j which are to be estimated ($p < k$). This *linear logistic test model* (LLTM) by Fischer (1973) can indeed also be seen as a special case of the Rasch model, especially because the validness of the latter is a pre-condition for the validness of the LLTM. In the case where a psychological test conforms to the Rasch model, a likelihood-ratio test in terms of a goodness of fit test can be applied in order to examine whether the data are explained significantly worse by the LLTM with fewer parameters than by the Rasch model with substantially more parameters (see the number of degrees of freedom in Section 14.2.2; for more details, see for example Kubinger, 2008).

Master Doctor

Example 15.8 – continued

Assume here that performance test x is a test assessing computing-related cognitive performances (word problems like: 'Five children are sitting on a bus; two more get on at the next stop. How many children are on the bus then?'). For the 18 items, which conform to the Rasch model, we hypothesize difficulty parameters η_1, η_2, η_3, η_4 specific to the arithmetic operations involved (adding, subtracting, multiplying, dividing) as well as a difficulty parameter η_5 (for crossing the barrier of tens). Concretely, this results in the following matrix $((q_{ij}))$, where q_{ij} is the number of times the respective arithmetic operation has to be used in the respective task:

1	0	0	0	0
1	0	0	0	0
0	1	0	0	0
0	1	0	0	0
1	0	0	0	1
1	1	0	0	0
2	0	0	0	0
2	0	0	0	1
0	0	1	0	0
0	0	1	0	1
0	0	0	1	0
1	0	1	0	0
0	1	1	0	0
0	0	1	1	0
0	0	1	1	1
1	0	1	1	0
1	1	1	0	0
1	1	1	1	1

The question now is whether the data are sufficiently well explained by the corresponding modeled LLTM; i.e. whether the data are not explained significantly worse than by the Rasch model. If this is the case, we are mostly interested in the basic parameters – that is the difficulty parameters specific to the arithmetic operations – and in their relation to one another. The type-I risk is chosen as $\alpha = 0.05$.

In **R**, we start by creating the matrix $((q_{ij}))$. For this, we type

```
> q.ij <- matrix(c(1, 1, 0, 0, 1, 1, 2, 2, 0, 0, 0, 1, 0, 0, 0,
+                  1, 1, 1, 0, 0, 1, 1, 0, 1, 0, 0, 0, 0, 0, 0,
+                  1, 0, 0, 0, 1, 1, 0, 0, 0, 0, 0, 0, 0, 0, 1,
+                  1, 0, 1, 1, 1, 1, 1, 1, 1, 0, 0, 0, 0, 0, 0,
+                  0, 0, 0, 0, 1, 0, 0, 1, 1, 1, 0, 1, 0, 0, 0,
+                  0, 1, 0, 0, 1, 0, 1, 0, 0, 0, 0, 1, 0, 0, 1),
+                  ncol = 5)
```

i.e. we use the function `c()` with which we create a vector containing all entries of the matrix; we submit this vector to the function `matrix()`, and with `ncol = 5`, we state that the matrix has 5 columns; these are assigned to the object `q.ij`.

Now, using the function `LLTM()` from the package `eRm`, we calculate the linear logistic test model. In doing so, we use the reduced set of data created earlier in Example 15.8, which we assigned to the object `items2`. To do this, we type

```
> items2.LLTM <- LLTM(items2, q.ij)
```

i.e. submit the items in the object items2 that are in conformity with the model as the first argument; as the second one, we set the matrix $((q_{ij}))$ with q.ij. We assign the results to the object items2.LLTM. Finally, we use a likelihood-ratio test to check if the data are not explained significantly worse by the modeled LLTM than by the Rasch model. This happens by typing

```
> 2*(items2.RM$loglik - items2.LLTM$loglik)
```

i.e. we request the values of the log-likelihoods for the Rasch model and the LLTM with $loglik; form the difference of the two and finally multiply it by 2. As a result, we get

```
[1] 129.2612
```

With

```
> items2.RM$npar - items2.LLTM$npar
```

we get to the sought-after degrees of freedom, through calculating the difference between the number of parameters of the Rasch model and the number of parameters of the LLTM with the function $npar. The result is

```
[1] 12
```

Next, we have to calculate the 0.95 quantile of the χ^2-distribution with $df = 12$. For this, we type

```
> qchisq(0.95, df = 12)
```

i.e. we use the function qchisq() and enter the respective values. As a result, we get

```
[1] 21.02607
```

Then, we calculate the correlation between the difficulty-parameter estimates of the Rasch model and the difficulty-parameter estimates that can be re-calculated from the basic parameters of the LLTM; in order to do this, we type

```
> cor(items2.RM$betapar, items2.LLTM$betapar)
```

i.e. we use the function cor(), to which we transfer the respective parameter estimates – which are ordered via $betapar for each object, once for the Rasch model and once for the LLTM. The result is

```
[1] 0.9563887
```

Now we create a diagram that opposes the difficulty-parameter estimates of the Rasch model to the difficulty-parameter estimates that were re-calculated from the basic parameters of the LLTM. For this, the latter ones have to be normed analogous to the first ones, so that their sum equals 0. To do this, we type

```
> betapar.LLTM <- items2.LLTM$betapar - mean(items2.LLTM$betapar)
```

i.e. we subtract their mean from every parameter in the object `items2.LLTM$betapar` and assign the results to the object `betapar.LLTM`. Next, we type

```
> plot(items2.RM$betapar, betapar.LLTM, xlim = c(-4, 4),
+       ylim = c(-4, 4),
+       xlab = "Item difficulty parameter RM",
+       ylab = "Item difficulty parameter LLTM")
> abline(0, 1)
```

i.e. we apply the function `plot()`, to which we submit the difficulty-parameter estimates of the Rasch model in the object `items2.RM$betapar` as the first argument and the now normed difficulty-parameter estimates of the LLTM in the object `betapar.LLTM` as its second one; with `xlim` and `ylim`, we define the benchmark of the axes and then label them with `xlab` and `ylab`. With the help of the function `abline()`, we draw the 45° line through origin.

The results are displayed in Figure 15.15.

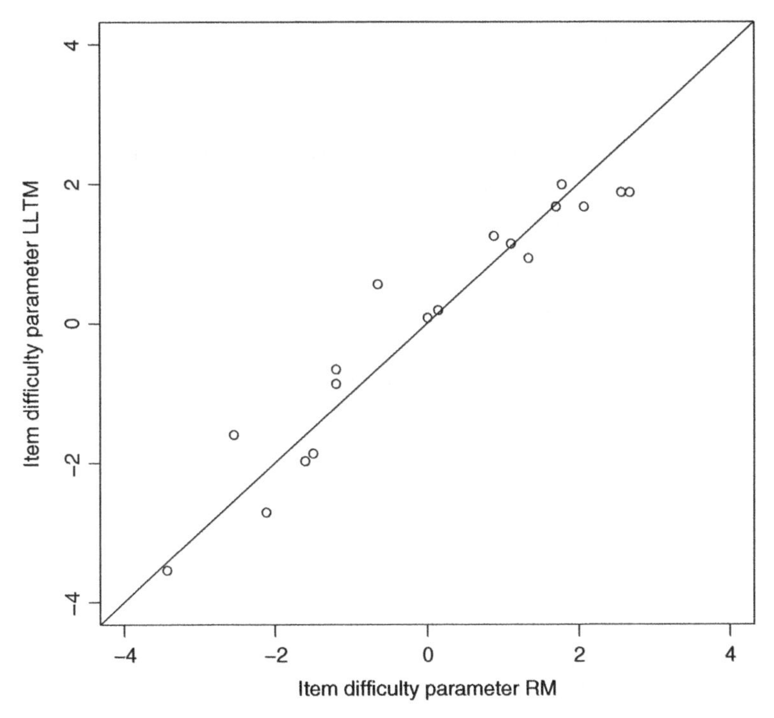

Figure 15.15 R-output of the graphical comparison between the item difficulty parameters according to the Rasch model on the one side and the item difficulty parameters as re-calculated from the basic parameters of the LLTM (Example 15.8).

Because the calculated value of the likelihood-ratio test's test statistic is larger than the value of the 0.95-quantile of the χ^2-distribution, the result is significant; it is to be concluded that the psychological complexity of the respective tasks cannot be easily reduced to the (difficulties of the) arithmetic operations that were hypothesized as being necessary to solve the task. Although the tasks do measure one-dimensionally, their different difficulties are surely also determined by different and manifold required thinking processes. The relatively high correlation coefficient cannot hide this fact.

Doctor

Formann (1984) had the idea of beginning Rasch model analysis by grouping the persons in advance using latent class analysis (see Section 15.2.1.2). The aim is that the investigated characters (items of a psychological test) measure in conformity with the Rasch model within each of the groups. This can be done via a special case of the so-called 'linear logistic LCA'. This approach facilitates that, within different (latent) groups of persons, a psychological test actually measures one-dimensionally, but either per group another dimension of aptitude or trait is measured or only the item parameters' relations are different between these groups. Nowadays, this approach is known as the *mixed Rasch model*. For more details, see especially Rost (2004).

Summary

In order to group research units into respectively similar sub-groups (types) according to the outcomes of the investigated characters, *cluster analysis* can be used for quantitative characters and *latent class analysis* for nominal-scaled ones. *Factor analysis* serves the purpose to identify fewer latent variables than observed characters, with which all research units can nevertheless be approximately equally well described. *Path analysis* and, in general, *(linear) structural equation models,* try to quantify directed dependencies between several characters. Above all, the models of *item response theory* are useful when constructing psychological tests; here, the *Rasch model* plays a special role insofar as the items of a test must conform to this model if the number of solved items is to be used as a test score.

References

Bond, T. G. & Fox, C. M. (2007). *Applying the Rasch Model. Fundamental Measurement in the Human Sciences* (2nd edn). Mahwah, NJ: Lawrence Erlbaum.

Collins, L. M. & Lanza, S. T. (2010). *Latent Class and Latent Transition Analysis with Applications in the Social, Behavioral, and Health Sciences*. Hoboken, NJ: John Wiley & Sons, Inc.

von Eye, A., Mair, P., & Mun, E. Y. (2010). *Advances in Configural Frequency Analysis*. New York: Guilford Press.

Fischer, G. H. (1973). The linear logistic test model as an instrument in educational research. *Acta Psychologica*, *37*, 359–374.

Fischer, G. H. (1995). Derivations of the Rasch Model. In G. H. Fischer & I. W. Molenaar (Eds.), *Rasch Models* (pp. 15–38). New York: Springer.

Fischer, G. H. & Scheiblechner, H. (1970). Algorithmen und Programme für das probabilistische Testmodell von Rasch [Algorithms and programs for Rasch's probabilistic test model]. *Psychologische Beiträge*, *12*, 23–51.

Formann, A. K. (1984). *Die Latent-Class-Analyse* [Latent Class Analysis]. Weinheim: Beltz.

Goodman, L. A. (1970). The multivariate analysis of qualitative data: interactions among multiple classifications. *Journal of the American Statistical Association*, *65*, 225–256.

Holocher-Ertl, S., Kubinger, K. D., & Hohensinn, C. (2008). Identifying children who may be cognitively gifted: the gap between practical demands and scientific supply. *Psychology Science Quarterly*, *50*, 97–111.

Kaiser, H.F. (1958). The varimax criterion for analytic rotation in factor analysis. *Psychometrika*, *23*, 187–200.

Kline, R. B. (2010). *Principles and Practice of Structural Equation Modeling* (3rd edn). New York: Guilford Press.

Krauth, J. & Lienert, G. A. (1995). *Die Konfigurationsfrequenzanalyse (KFA) und ihre Anwendung in Psychologie und Medizin* (Reprint) [Configuration Frequency Analysis (CFA) and its Applications in Psychology and Medicine]. Weinheim: Beltz/PVU.

Kubinger, K. D. (1989). Aktueller Stand und kritische Würdigung der Probabilistischen Testtheorie [Critical evaluation of latent trait theory]. In K. D. Kubinger (Ed.), *Moderne Testtheorie – Ein Abriß samt neuesten Beiträgen* [Modern Psychometrics – A brief Survey with Recent Contributions] (pp. 19–83). Munich: PVU.

Kubinger, K. D. (2003). On artificial results due to using factor analysis for dichotomous variables. *Psychology Science*, *45*, 106–110.

Kubinger, K. D. (2005). Psychological test calibration using the Rasch model – some critical suggestions on traditional approaches. *International Journal of Testing*, *5*, 377–394.

Kubinger, K. D. (2008). On the revival of the Rasch model-based LLTM: from composing tests by item generating rules to measuring item administration effects. *Psychology Science Quarterly*, *50*, 311–327.

Kubinger, K. D. (2009a). *Adaptives Intelligenz Diagnostikum - Version 2.2 (AID 2) samt AID 2-Türkisch* [Adaptive Intelligence Diagnosticum, AID 2-Turkey Included]. Göttingen: Beltz.

Kubinger, K. D. (2009b). *Psychologische Diagnostik – Theorie und Praxis psychologischen Diagnostizierens* (2nd edn) [Psychological Assessment – Theory and Practice of Psychological Consulting]. Göttingen: Hogrefe.

Kubinger, K. D. & Draxler, C. (2006). A comparison of the Rasch model and constrained item response theory models for pertinent psychological test data. In M. von Davier & C. H. Carstensen (Eds.), *Multivariate and Mixture Distribution Rasch Models – Extensions and Applications* (pp. 295–312). New York: Springer.

Kubinger, K. D., Rasch, D., & Yanagida, T. (2009). On designing data-sampling for Rasch model calibrating an achievement test. *Psychology Science Quarterly*, *51*, 370–384.

Kubinger, K. D., Rasch, D., & Yanagida, T. (in press). A new approach for testing the Rasch model. *Educational Research and Evaluation*.

Mair, P., Hatzinger, R., & Maier, M. (2010). eRm: Extended Rasch Modeling. R package version 0.13-0: http://cran.r-project.org/web/packages/eRm/. Vienna University of Economics and Business, Vienna.

Rost, J. (2004). *Lehrbuch Testtheorie – Testkonstruktion* (2nd edn) [Text Book in Psychometrics]. Bern: Huber.

Stigler, S. M. (1997). Regression toward the mean, historically considered. *Statistical Methods in Medical Research*, *6*, 103–114.

Appendix A

Data input

Below we describe how the data of Example 1.1 can be used in R or SPSS.

Data input in R

We generate a data matrix in **R**, but only exemplify the procedure for the first 10 observations. First we construct a vector for each character; we type

```
> no <- 1:10
> native_language <- factor(c(1, 1, 2, 2, 1, 1, 2, 2, 1, 1),
+                           levels = c(1, 2),
+                           labels = c("German", "Turkish"))
> age <- c(6, 7, 8, 9, 8, 9, 6, 7, 9, 8)
> sex <- factor(c(1, 2, 1, 2, 1, 2, 1, 2, 1, 2),
+               levels = c(1, 2),
+               labels = c("female", "male"))
> age_birth <- c(39, 40, 38, 36, 37, 40, 39, 36, 40, 41)
> no_siblings <- c(1, 2, 3, 2, 3, 0, 1, 4, 0, 2)
> pos_sibling <- factor(c(1, 2, 3, 1, 2, 1, 1, 5, 1, 2),
+                       levels = c(1, 2, 3, 4, 5, 6),
+                       labels = c("first-born", "second-born",
+                                  "third-born", "fourth-born",
+                                  "fifth-born", "sixth-born"))
> social_status <- factor(c(2, 3, 3, 2, 4, 1, 2, 3, 6, 1),
+                         levels = c(1, 2, 3, 4, 5, 6),
+                         labels = c("upper classes",
+                                    "middle classes",
+                                    "lower middle class",
+                                    "upper lower class",
+                                    "lower classes",
+                                    "single mother in household"))
```

Statistics in Psychology Using R and SPSS, First Edition. Dieter Rasch, Klaus D. Kubinger and Takuya Yanagida.

```
> urban_rural <- factor(c(1, 1, 3, 1, 1, 1, 1, 1, 1, 1),
+                    levels = c(1, 2, 3),
+                    labels = c("city (over 20000 inhabitants)",
+                               "town (5000 to 20000 inhabitants)",
+                               "rural (up to 5000 inhabitants)"))
> marital_mother <- factor(c(2, 2, 2, 2, 3, 2, 2, 2, 3, 3),
+                          levels = c(1, 2, 3, 4),
+                          labels = c("never married", "married",
+                                     "divorced", "widowed"))
> test_set <- factor(c(1, 1, 2, 3, 1, 1, 2, 3, 1, 1),
+                    levels = c(1, 2, 3),
+                    labels = c("German speaking child",
+ "Turkish speaking child tested in German at first test date",
+ "Turkish speaking child tested in Turkish at first test date"))
> sub1_t1 <- c(52, 54, 46, 58, 37, 50, 48, 60, 56, 56)
> sub3_t1 <- c(56, 59, 55, 47, 50, 59, 53, 49, 61, 62)
> sub4_t1 <- c(43, 52, 47, 59, 38, 55, 40, 62, 53, 53)
> sub5_t1 <- c(55, 57, 40, 47, 36, 52, 36, 57, 55, 57)
> sub7_t1 <- c(50, 64, 50, 50, 50, 64, 47, 64, 57, 68)
> sub1_t2 <- c(50, 56, 54, 60, 38, 52, 50, 60, 58, 60)
> sub3_t2 <- c(55, 62, 47, 48, 48, 60, 52, 50, 59, 65)
> sub4_t2 <- c(45, 53, 52, 57, 38, 57, 43, 64, 57, 57)
> sub5_t2 <- c(55, 59, 55, 45, 36, 50, 45, 67, 57, 48)
> sub7_t2 <- c(54, 65, 50, 54, 46, 61, 46, 69, 61, 65)
```

Here we have to differentiate between three cases:

1. The character *testee number* (`no`) represents a sequence of integers (here) from `1` to `10`, that we generate using a colon (`:`).
2. For example the character *native language of the child* (`native_language`), which we generate by the function `factor()`; thereby we use, as the first argument, a vector – generated using the function `c()` – containing the outcomes; as a second argument `levels` (`1` or `2`) for the character's levels; and as third argument `labels` for the codes of the levels' labels (`"German"` and `"Turkish"`).
3. For example the character *age of the child* (`age`), for which we assign the observed values to the function `c()`.

We hand each of the vectors to an object according to the character label in Table 1.1. We now type

```
> Example_1.1 <- data.frame(no, native_language, age, sex,
+               age_birth, no_siblings, pos_sibling, social_status,
+               urban_rural, marital_mother, test_set,
+               sub1_t1, sub3_t1, sub4_t1, sub5_t1, sub7_t1,
+               sub1_t2, sub3_t2, sub4_t2, sub5_t2, sub7_t2)
```

i.e. we use the function data.frame() and hand to it all characters stored in the aforementioned objects; we pass this generated data to the object Example_1.1. Finally, we type

```
> rm(no, native_language, age, sex, age_birth, no_siblings,
+     pos_sibling, social_status, urban_rural, marital_mother,
+     test_set, sub1_t1, sub3_t1, sub4_t1, sub5_t1, sub7_t1,
+     sub1_t2, sub3_t2, sub4_t2, sub5_t2, sub7_t2)
```

i.e. we hand all objects to the function rm() (except the object Example_1.1); as a consequence these objects are deleted from the Workspace.

Another possibility is to generate the data in another format and afterwards to import the data into **R**. If we actually generate and save the data using SPSS, we can import the corresponding file using the package foreign, which we load after installation (see Chapter 1) using the function library(). We type

```
> choose.dir()
```

using this function without any argument we end up at the window Select folder containing the folder overview of the hard disc. We select the folder containing the data file and click OK. We now type

```
> Example_1.1 <- read.spss("Example_1.1.sav", to.data.frame = TRUE)
```

i.e. we hand the name of the data file as a first argument to the function read.spss() and request a data matrix import, using to.data.frame = TRUE; we then pass all this to the object Example_1.1.

For reading data sets generated by other programs one may use help files of the corresponding functions; e.g. read.table() is to be used for text files of different formats; read.xls() in the package gdata for Excel files; and read.csv2() for CSV files.

Data input in SPSS

In SPSS we first of all have to specify the features of the characters. For this we use, in the menu

File
    New
        Data

and choose, in the bottom menu bar on the left side, the second entry Variable View. In the resulting window shown in Figure A.1, for each character in the column Name we fill in the corresponding notation from the column Name in SPSS from Table 1.1 In the column Label we insert the full name of the character from the first column from Table 1.1, 'Name of the character'. If there are entries in the column 'Coded values' of Table 1.1 for a certain character, as for instance for the character *native language of the child*, we write these into the column Values. For this we first have to click at the corresponding cell and get the symbol ... which we click as well. This results in the window in Figure A.2. In the case of the character *native language of the child,* we insert the value 1 in the column beside Value:, and the text German beside Label:. Using Add empties the fields Value: and Label:, so that we can insert the next coded value including its label. At the end we confirm with OK, where after – again in the window Variable View – we can proceed analogously for the next character. Afterwards we need to specify the scale type of each character in the column Measure in the window Variable View (Figure A.1). SPSS only distinguishes among the scale types Scale, for quantitative characters (see Section 5.2.3 for more details), the type Ordinal, for ordinal-scaled characters (see Section 5.2.2 for more details), and the type Nominal, for nominal-scaled characters (see Section 5.2.1 for more details). We can specify the number of presented decimal places in column Decimals; for the Example 1.1 we insert 0 for all characters. If the data contains missing values, we can define their coding in column Missing. All of the remaining columns in the Variable View are unused in our case. After inserting all of the characters in Variable View, we obtain Figure A.3.

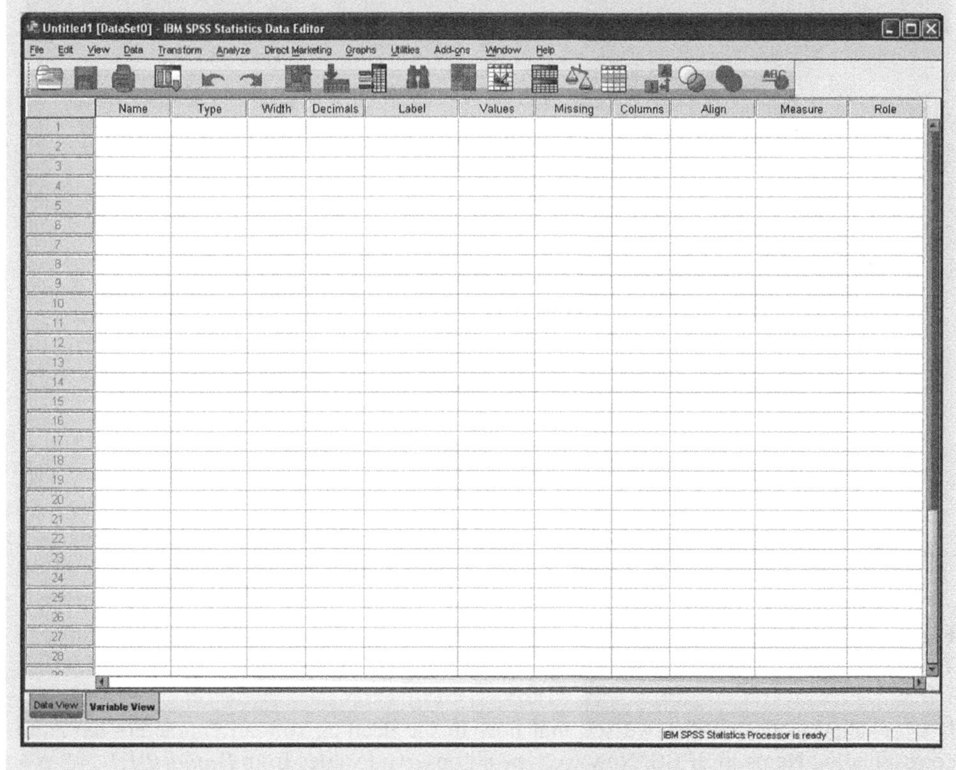

Figure A.1 SPSS Variable View.

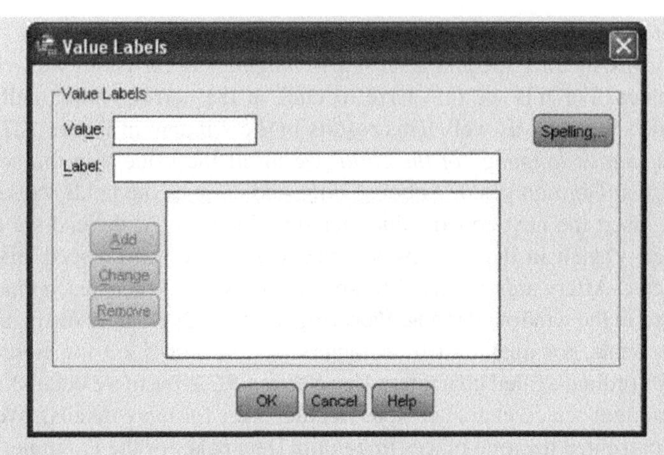

Figure A.2 SPSS -window for coding and labeling values of a character.

Example_1.1.sav [DataSet1] - IBM SPSS Statistics Data Editor

File Edit View Data Transform Analyze Direct Marketing Graphs Utilities Add-ons Window Help

	Name	Type	Width	Decimals	Label	Values	Missing	Columns	Align	Measure	Role
1	no	Numeric	8	0	testee number	None	None	8	Right	Nominal	Input
2	native_langu...	Numeric	8	0	native language...	{1, German}...	None	10	Right	Nominal	Input
3	age	Numeric	8	0	age of the child	None	None	8	Right	Scale	Input
4	sex	Numeric	8	0	sex of the child	{1, female}...	None	8	Right	Nominal	Input
5	age_birth	Numeric	8	0	gestational age...	None	None	8	Right	Scale	Input
6	no_siblings	Numeric	8	0	number of siblin...	None	None	8	Right	Scale	Input
7	pos_sibling	Numeric	8	0	sibling position	{1, first-born...	None	8	Right	Ordinal	Input
8	social_status	Numeric	8	0	social status (a...	{1, upper cl...	None	8	Right	Nominal	Input
9	urban_rural	Numeric	8	0	urban/rural	{1, City (ove...	None	8	Right	Ordinal	Input
10	marital_mot...	Numeric	8	0	marital status o...	{1, never ma...	None	9	Right	Nominal	Input
11	test_set	Numeric	8	0	test setting	{1, German ...	None	8	Right	Nominal	Input
12	sub1_t1	Numeric	8	0	Everyday Know...	None	None	8	Right	Scale	Input
13	sub3_t1	Numeric	8	0	Applied Compu...	None	None	8	Right	Scale	Input
14	sub4_t1	Numeric	8	0	Social and Mat...	None	None	8	Right	Scale	Input
15	sub5_t1	Numeric	8	0	Immediately Re...	None	None	8	Right	Scale	Input
16	sub7_t1	Numeric	8	0	Coding and As...	None	None	8	Right	Scale	Input
17	sub1_t2	Numeric	8	0	Everyday Know...	None	None	8	Right	Scale	Input
18	sub3_t2	Numeric	8	0	Applied Compu...	None	None	8	Right	Scale	Input
19	sub4_t2	Numeric	8	0	Social and Mat...	None	None	8	Right	Scale	Input
20	sub5_t2	Numeric	8	0	Immediately Re...	None	None	8	Right	Scale	Input
21	sub7_t2	Numeric	8	0	Coding and As...	None	None	8	Right	Scale	Input
22											

Data View | Variable View

IBM SPSS Statistics Processor is ready

Figure A.3 SPSS Variable View after specification of the characters' features according to Table 1.1.

If we change to Data View, we see that now in the heading row all characters have the corresponding Name in SPSS. Now we type all observed values from *Data list 1.1*.

Data list 1.1

testee number	*native language of the child*	*age of the child*	*sex of the child*	*gestational age at birth* (in weeks)	*number of siblings*	*sibling position*	*social status*	*urban/rural*	*marital status of the mother*	*test setting*	*Everyday Knowledge, 1st test date* (T-Scores)	*Applied Computing, 1st test date* (T-Scores)	*Social and Material Sequencing, 1st test date* (T-Scores)	*Immediately Reproducing – numerical, 1st test date* (T-Scores)	*Coding and Associating, 1st test date* (T-Scores)	*Everyday Knowledge, 2nd test date* (T-Scores)	*Applied Computing, 2nd test date* (T-Scores)	*Social and Material Sequencing, 2nd test date* (T-Scores)	*Immediately Reproducing – numerical, 2nd test date* (T-Scores)	*Coding and Associating, 2nd test date* (T-Scores)
1	1	6	1	39	1	1	2	1	2	1	52	56	43	55	50	50	55	45	55	54
2	1	7	2	40	2	2	3	1	2	1	54	59	52	57	64	56	62	53	59	65
3	2	8	1	38	3	3	3	3	2	2	46	55	47	40	50	54	47	52	55	50
4	2	9	2	36	2	1	2	1	2	3	58	47	59	47	50	60	48	57	45	54
5	1	8	1	37	3	2	4	1	3	1	37	50	38	36	50	38	48	38	36	46
6	1	9	2	40	0	1	1	1	2	1	50	59	55	52	64	52	60	57	50	61
7	2	6	1	39	1	1	2	1	2	2	48	53	40	36	47	50	52	43	45	46
8	2	7	2	36	4	5	3	1	2	3	60	49	62	57	64	60	50	64	67	69
9	1	9	1	40	0	1	6	1	3	1	56	61	53	55	57	58	59	57	57	61
10	1	8	2	41	2	2	1	1	3	1	56	62	53	57	68	60	65	57	48	65
11	1	9	1	36	4	5	2	1	2	1	37	43	40	35	47	38	38	36	35	46
12	1	6	2	38	1	1	3	1	3	1	50	47	55	50	57	48	50	59	53	61
13	1	8	1	40	0	1	2	1	1	1	60	52	64	55	68	58	53	59	53	73
14	1	9	2	41	0	1	3	1	1	1	54	50	64	36	47	54	52	67	36	42
15	1	6	1	38	1	2	1	2	2	1	60	61	64	67	75	62	57	59	64	73
16	1	8	2	36	2	1	4	1	3	1	31	43	50	36	50	34	40	48	38	31
17	1	7	1	40	3	4	2	1	2	1	65	58	65	50	54	66	57	62	53	61
18	1	9	2	37	2	2	6	1	3	1	58	47	55	59	54	54	50	52	59	50
19	1	8	1	39	0	1	6	1	3	1	50	43	40	43	54	50	45	36	50	58
20	1	7	2	35	0	1	4	1	2	1	61	37	42	35	47	60	36	43	42	54
21	1	6	1	38	4	5	1	1	2	1	54	40	53	59	64	52	42	52	57	69
22	1	9	2	40	1	1	2	1	3	1	60	52	36	59	57	58	53	42	62	54
23	1	7	1	41	1	1	3	1	3	1	46	52	48	38	40	48	50	36	43	46
24	1	6	2	35	0	1	6	2	1	1	61	53	50	50	54	60	55	52	50	61
25	1	8	1	40	3	4	6	1	3	1	52	56	53	47	33	46	52	45	42	27

Data list 1.1 – continued

testee number	*native language of the child*	*age of the child*	*sex of the child*	*gestational age at birth* (in weeks)	*number of siblings*	*sibling position*	*social status*	*urban/rural*	*marital status of the mother*	*test setting*	*Everyday Knowledge, 1st test date (T-Scores)*	*Applied Computing, 1st test date (T-Scores)*	*Social and Material Sequencing, 1st test date* (T-Scores)	*Immediately Reproducing – numerical, 1st test date* (T-Scores)	*Coding and Associating, 1st test date* (T-Scores)	*Everyday Knowledge, 2nd test date* (T-Scores)	*Applied Computing, 2nd test date (T-Scores)*	*Social and Material Sequencing, 2nd test date* (T-Scores)	*Immediately Reproducing – numerical, 2nd test date* (T-Scores)	*Coding and Associating, 2nd test date* (T-Scores)
26	1	9	2	37	2	1	2	1	2	1	58	52	53	55	47	56	50	52	57	50
27	1	6	1	39	1	2	1	1	2	1	50	56	52	64	61	46	59	57	67	65
28	1	8	2	40	2	2	2	1	2	1	69	58	69	62	71	70	62	69	65	69
29	1	7	1	37	3	4	5	1	3	1	50	52	35	35	29	46	50	36	33	23
30	1	9	2	32	1	2	4	1	2	1	35	37	42	33	36	30	33	35	38	39
31	1	8	1	39	0	1	1	1	1	1	63	56	52	64	68	60	52	57	64	65
32	1	7	2	39	1	2	2	1	2	1	60	50	57	52	54	62	53	55	52	54
33	1	6	1	40	1	1	3	1	4	1	50	53	67	52	54	54	52	64	50	54
34	1	9	2	41	2	3	2	1	1	1	54	61	64	55	54	50	64	60	57	54
35	1	7	1	40	1	2	3	1	2	1	33	43	50	52	54	40	40	48	50	54
36	1	6	2	39	2	3	2	3	2	1	60	64	36	53	61	58	64	38	50	54
37	1	9	1	37	3	1	6	1	3	1	61	53	57	52	50	60	50	55	50	54
38	1	7	2	40	0	1	1	1	2	1	65	53	57	60	68	70	59	53	65	69
39	1	6	1	37	1	2	2	1	2	1	69	38	64	52	68	72	35	64	50	65
40	1	8	2	38	3	3	1	1	4	1	60	50	42	50	57	56	50	50	53	54
41	1	7	1	39	1	2	6	3	4	1	71	38	64	50	54	70	42	62	47	50
42	1	6	2	40	0	1	2	1	2	1	65	59	53	35	50	62	59	50	33	46
43	1	8	1	34	0	1	3	1	3	1	54	38	35	33	29	50	35	33	30	23
44	1	9	2	39	4	3	4	1	2	1	50	52	53	50	54	50	52	53	55	54
45	1	6	1	40	0	1	6	1	3	1	35	62	35	42	54	30	59	38	47	50
46	1	8	2	40	1	1	5	1	3	1	41	43	53	52	40	46	45	50	47	50
47	1	7	1	39	2	2	1	3	2	1	65	59	62	50	57	62	64	65	59	58
48	1	6	2	40	1	2	2	1	1	1	60	53	35	59	64	52	50	42	52	61
49	1	8	1	38	3	1	4	1	2	1	50	56	53	52	57	54	55	53	52	58
50	1	9	2	35	0	1	3	1	2	1	41	38	52	53	64	42	33	50	55	58

Data list 1.1 – continued

testee number	*native language of the child*	*age of the child*	*sex of the child*	*gestational age at birth* (in weeks)	*number of siblings*	*sibling position*	*social status*	*urban/rural*	*marital status of the mother*	*test setting*	*Everyday Knowledge, 1st test date* (T-Scores)	*Applied Computing, 1st test date* (T-Scores)	*Social and Material Sequencing, 1st test date* (T-Scores)	*Immediately Reproducing – numerical, 1st test date* (T-Scores)	*Coding and Associating, 1st test date* (T-Scores)	*Everyday Knowledge, 2nd test date* (T-Scores)	*Applied Computing, 2nd test date* (T-Scores)	*Social and Material Sequencing, 2nd test date* (T-Scores)	*Immediately Reproducing – numerical, 2nd test date* (T-Scores)	*Coding and Associating, 2nd test date* (T-Scores)
51	1	9	1	40	2	2	1	1	3	1	50	52	60	57	54	48	53	64	64	61
52	1	8	2	36	4	5	3	1	2	1	46	53	52	62	68	50	53	57	64	61
53	1	7	1	39	0	1	4	1	3	1	48	62	57	52	50	46	65	59	53	54
54	1	6	2	40	2	1	3	3	3	1	41	50	52	36	47	34	48	52	38	46
55	2	9	1	40	3	3	2	1	2	2	52	56	47	40	47	50	53	48	50	61
56	2	7	2	36	1	1	3	1	2	2	42	53	53	36	68	60	45	52	40	65
57	2	6	1	39	0	1	1	1	2	2	61	58	53	70	68	54	48	57	62	69
58	2	9	2	37	1	2	6	1	1	2	42	49	59	42	64	60	45	59	50	65
59	2	7	1	39	5	6	3	1	2	2	52	53	50	57	61	50	45	48	47	65
60	2	6	2	40	2	2	2	1	2	2	61	56	57	70	68	60	53	64	60	65
61	2	8	1	41	3	3	4	1	2	2	46	43	50	57	47	40	28	52	53	50
62	2	7	2	38	1	1	3	1	2	2	65	64	52	70	64	54	55	50	62	65
63	2	6	1	40	2	3	2	1	2	2	71	71	53	64	64	64	59	59	59	69
64	2	9	2	38	1	2	6	1	4	2	42	43	50	60	54	50	53	52	53	50
65	2	7	1	37	4	5	4	1	2	2	61	58	52	43	33	52	45	50	33	35
66	2	6	2	39	1	2	3	1	2	2	58	55	53	40	29	50	45	52	36	23
67	2	8	1	37	1	1	4	1	2	2	61	43	36	62	50	52	38	33	50	50
68	2	7	2	38	2	1	2	1	2	2	46	41	50	42	50	60	53	52	48	50
69	2	9	1	40	3	4	3	1	2	2	42	43	50	43	68	66	53	53	55	69
70	2	8	2	39	1	1	2	1	2	2	58	49	50	43	54	62	59	50	47	54
71	2	9	1	37	2	3	4	1	2	2	44	44	50	35	50	48	45	52	50	50
72	2	6	2	35	0	1	6	1	1	2	61	56	52	70	54	60	59	67	67	61
73	2	8	1	40	1	2	3	1	2	2	58	49	50	57	50	60	53	50	52	54
74	2	7	2	39	2	3	2	1	2	2	61	58	59	67	50	52	45	62	59	54
75	2	6	1	39	1	2	3	1	2	2	44	44	50	35	50	60	60	50	50	54

Data list 1.1 – continued

testee number	*native language of the child*	*age of the child*	*sex of the child*	*gestational age at birth* (in weeks)	*number of siblings*	*sibling position*	*social status*	*urban/rural*	*marital status of the mother*	*test setting*	*Everyday Knowledge, 1st test date* (T-Scores)	*Applied Computing, 1st test date* (T-Scores)	*Social and Material Sequencing, 1st test date* (T-Scores)	*Immediately Reproducing – numerical, 1st test date* (T-Scores)	*Coding and Associating, 1st test date* (T-Scores)	*Everyday Knowledge, 2nd test date* (T-Scores)	*Applied Computing, 2nd test date* (T-Scores)	*Social and Material Sequencing, 2nd test date* (T-Scores)	*Immediately Reproducing – numerical, 2nd test date* (T-Scores)	*Coding and Associating, 2nd test date* (T-Scores)
76	2	9	2	38	2	3	2	1	2	2	37	43	50	42	61	70	62	47	43	61
77	2	7	1	39	1	1	2	1	1	2	52	41	53	36	33	68	45	55	42	39
78	2	6	2	37	2	3	4	1	2	2	37	38	33	43	33	40	45	36	42	42
79	2	9	1	40	3	3	4	1	2	3	60	50	57	53	54	58	59	59	50	54
80	2	7	2	39	5	5	2	1	2	3	50	52	53	57	61	54	52	53	59	61
81	2	6	1	35	2	1	6	1	3	3	46	43	36	35	40	50	42	38	33	39
82	2	8	2	39	1	2	2	1	2	3	60	59	64	53	33	62	64	67	53	31
83	2	9	1	38	2	1	3	1	2	3	50	37	36	64	54	52	36	35	59	54
84	2	6	2	37	1	1	1	1	2	3	50	62	36	42	47	52	60	35	40	46
85	2	8	1	40	2	3	3	1	2	3	46	50	64	52	54	50	47	60	50	54
86	2	7	2	41	1	2	2	1	3	3	54	52	64	64	61	52	50	62	60	61
87	2	9	1	39	3	4	3	1	2	3	60	55	33	47	61	50	53	35	50	61
88	2	8	2	37	4	4	4	1	2	3	69	67	70	69	68	72	70	74	67	73
89	2	6	1	40	1	2	3	1	2	3	65	37	57	47	33	60	35	59	47	35
90	2	7	2	41	2	1	4	1	2	3	31	38	59	36	50	32	36	57	50	39
91	2	9	1	39	3	4	5	1	2	3	25	38	42	50	54	30	35	40	47	50
92	2	6	2	38	4	1	2	1	2	3	50	37	53	55	57	46	42	50	52	61
93	2	8	1	40	5	6	3	1	2	3	65	52	57	62	36	60	47	59	60	31
94	2	7	2	41	2	3	5	1	2	3	33	38	50	33	57	30	36	52	42	50
95	2	9	1	39	1	1	2	1	2	3	60	56	52	28	54	58	59	52	33	54
96	2	6	2	37	3	1	2	1	2	3	61	52	47	36	36	54	52	50	33	50
97	2	8	1	39	0	1	3	1	3	3	58	50	67	70	57	60	59	64	67	65
98	2	7	2	38	4	4	5	1	3	3	27	50	35	42	43	30	47	42	40	39
99	2	9	1	37	3	3	4	1	2	3	60	53	62	35	33	60	48	64	36	31
100	2	6	2	41	4	1	2	1	2	3	50	37	67	52	26	46	36	64	59	31

Appendix B

Tables

Table B.1 Distribution function of the standard normal distribution $N(0, 1)$.

z	0.00	0.01	0.02	0.03	0.04	0.05	0.06	0.07	0.08	0.09
0.0	0.5000	0.4960	0.4920	0.4880	0.4840	0.4801	0.4761	0.4721	0.4681	0.4641
0.1	0.4602	0.4562	0.4522	0.4483	0.4443	0.4404	0.4364	0.4325	0.4286	0.4247
0.2	0.4207	0.4168	0.4129	0.4090	0.4052	0.4013	0.3974	0.3936	0.3897	0.3859
0.3	0.3821	0.3783	0.3745	0.3707	0.3669	0.3632	0.3594	0.3557	0.3520	0.3483
0.4	0.3446	0.3409	0.3372	0.3336	0.3300	0.3264	0.3228	0.3192	0.3156	0.3121
0.5	0.3085	0.3050	0.3015	0.2981	0.2946	0.2912	0.2877	0.2843	0.2810	0.2776
0.6	0.2743	0.2709	0.2676	0.2643	0.2611	0.2578	0.2546	0.2514	0.2483	0.2451
0.7	0.2420	0.2389	0.2358	0.2327	0.2296	0.2266	0.2236	0.2206	0.2177	0.2148
0.8	0.2119	0.2090	0.2061	0.2033	0.2005	0.1977	0.1949	0.1922	0.1894	0.1867
0.9	0.1841	0.1814	0.1788	0.1762	0.1736	0.1711	0.1685	0.1660	0.1635	0.1611
1.0	0.1587	0.1562	0.1539	0.1515	0.1492	0.1469	0.1446	0.1423	0.1401	0.1379
1.1	0.1357	0.1335	0.1314	0.1292	0.1271	0.1251	0.1230	0.1210	0.1190	0.1170
1.2	0.1151	0.1131	0.1112	0.1093	0.1075	0.1056	0.1038	0.1020	0.1003	0.0985
1.3	0.0968	0.0951	0.0934	0.0918	0.0901	0.0885	0.0869	0.0853	0.0838	0.0823
1.4	0.0808	0.0793	0.0778	0.0764	0.0749	0.0735	0.0721	0.0708	0.0694	0.0681
1.5	0.0668	0.0655	0.0643	0.0630	0.0618	0.0606	0.0594	0.0582	0.0571	0.0559
1.6	0.0548	0.0537	0.0526	0.0516	0.0505	0.0495	0.0485	0.0475	0.0465	0.0455
1.7	0.0446	0.0436	0.0427	0.0418	0.0409	0.0401	0.0392	0.0384	0.0375	0.0367
1.8	0.0359	0.0351	0.0344	0.0336	0.0329	0.0322	0.0314	0.0307	0.0301	0.0294
1.9	0.0287	0.0281	0.0274	0.0268	0.0262	0.0256	0.0250	0.0244	0.0239	0.0233

Statistics in Psychology Using R and SPSS, First Edition. Dieter Rasch, Klaus D. Kubinger and Takuya Yanagida.

Table B.1 (*Continued*)

z	0.00	0.01	0.02	0.03	0.04	0.05	0.06	0.07	0.08	0.09
2.0	0.0228	0.0222	0.0217	0.0212	0.0207	0.0202	0.0197	0.0192	0.0188	0.0183
2.1	0.0179	0.0174	0.0170	0.0166	0.0162	0.0158	0.0154	0.0150	0.0146	0.0143
2.2	0.0139	0.0136	0.0132	0.0129	0.0125	0.0122	0.0119	0.0116	0.0113	0.0110
2.3	0.0107	0.0104	0.0102	0.0099	0.0096	0.0094	0.0091	0.0089	0.0087	0.0084
2.4	0.0082	0.0080	0.0078	0.0075	0.0073	0.0071	0.0069	0.0068	0.0066	0.0064
2.5	0.0062	0.0060	0.0059	0.0057	0.0055	0.0054	0.0052	0.0051	0.0049	0.0048
2.6	0.0047	0.0045	0.0044	0.0043	0.0041	0.0040	0.0039	0.0038	0.0037	0.0036
2.7	0.0035	0.0034	0.0033	0.0032	0.0031	0.0030	0.0029	0.0028	0.0027	0.0026
2.8	0.0026	0.0025	0.0024	0.0023	0.0023	0.0022	0.0021	0.0021	0.0020	0.0019
2.9	0.0019	0.0018	0.0018	0.0017	0.0016	0.0016	0.0015	0.0015	0.0014	0.0014
3.0	0.0013	0.0013	0.0013	0.0012	0.0012	0.0011	0.0011	0.0011	0.0010	0.0010

Table B.2 P-quantiles of the t-distribution with df degrees of freedom (for $df = \infty$, the quantiles of the standard normal distribution).

	P								
df	0.60	0.70	0.80	0.85	0.90	0.95	0.975	0.99	0.995
1	0.3249	0.7265	1.3764	1.9626	3.0777	6.3138	12.7062	31.8205	63.6567
2	0.2887	0.6172	1.0607	1.3862	1.8856	2.9200	4.3027	6.9646	9.9248
3	0.2767	0.5844	0.9785	1.2498	1.6377	2.3534	3.1824	4.5407	5.8409
4	0.2707	0.5686	0.9410	1.1896	1.5332	2.1318	2.7764	3.7469	4.6041
5	0.2672	0.5594	0.9195	1.1558	1.4759	2.0150	2.5706	3.3649	4.0321
6	0.2648	0.5534	0.9057	1.1342	1.4398	1.9432	2.4469	3.1427	3.7074
7	0.2632	0.5491	0.8960	1.1192	1.4149	1.8946	2.3646	2.9980	3.4995
8	0.2619	0.5459	0.8889	1.1081	1.3968	1.8595	2.3060	2.8965	3.3554
9	0.2610	0.5435	0.8834	1.0997	1.3830	1.8331	2.2622	2.8214	3.2498
10	0.2602	0.5415	0.8791	1.0931	1.3722	1.8125	2.2281	2.7638	3.1693
11	0.2596	0.5399	0.8755	1.0877	1.3634	1.7959	2.2010	2.7181	3.1058
12	0.2590	0.5386	0.8726	1.0832	1.3562	1.7823	2.1788	2.6810	3.0545
13	0.2586	0.5375	0.8702	1.0795	1.3502	1.7709	2.1604	2.6503	3.0123
14	0.2582	0.5366	0.8681	1.0763	1.3450	1.7613	2.1448	2.6245	2.9768
15	0.2579	0.5357	0.8662	1.0735	1.3406	1.7531	2.1314	2.6025	2.9467
16	0.2576	0.5350	0.8647	1.0711	1.3368	1.7459	2.1199	2.5835	2.9208
17	0.2573	0.5344	0.8633	1.0690	1.3334	1.7396	2.1098	2.5669	2.8982
18	0.2571	0.5338	0.8620	1.0672	1.3304	1.7341	2.1009	2.5524	2.8784
19	0.2569	0.5333	0.8610	1.0655	1.3277	1.7291	2.0930	2.5395	2.8609
20	0.2567	0.5329	0.8600	1.0640	1.3253	1.7247	2.0860	2.5280	2.8453

Table B.2 (*Continued*)

	P								
df	0.60	0.70	0.80	0.85	0.90	0.95	0.975	0.99	0.995
21	0.2566	0.5325	0.8591	1.0627	1.3232	1.7207	2.0796	2.5176	2.8314
22	0.2564	0.5321	0.8583	1.0614	1.3212	1.7171	2.0739	2.5083	2.8188
23	0.2563	0.5317	0.8575	1.0603	1.3195	1.7139	2.0687	2.4999	2.8073
24	0.2562	0.5314	0.8569	1.0593	1.3178	1.7109	2.0639	2.4922	2.7969
25	0.2561	0.5312	0.8562	1.0584	1.3163	1.7081	2.0595	2.4851	2.7874
26	0.2560	0.5309	0.8557	1.0575	1.3150	1.7056	2.0555	2.4786	2.7787
27	0.2559	0.5306	0.8551	1.0567	1.3137	1.7033	2.0518	2.4727	2.7707
28	0.2558	0.5304	0.8546	1.0560	1.3125	1.7011	2.0484	2.4671	2.7633
29	0.2557	0.5302	0.8542	1.0553	1.3114	1.6991	2.0452	2.4620	2.7564
30	0.2556	0.5300	0.8538	1.0547	1.3104	1.6973	2.0423	2.4573	2.7500
40	0.2550	0.5286	0.8507	1.0500	1.3031	1.6839	2.0211	2.4233	2.7045
50	0.2547	0.5278	0.8489	1.0473	1.2987	1.6759	2.0086	2.4033	2.6778
60	0.2545	0.5272	0.8477	1.0455	1.2958	1.6706	2.0003	2.3901	2.6603
70	0.2543	0.5268	0.8468	1.0442	1.2938	1.6669	1.9944	2.3808	2.6479
80	0.2542	0.5265	0.8461	1.0432	1.2922	1.6641	1.9901	2.3739	2.6387
90	0.2541	0.5263	0.8456	1.0424	1.2910	1.6620	1.9867	2.3685	2.6316
100	0.2540	0.5261	0.8452	1.0418	1.2901	1.6602	1.9840	2.3642	2.6259
300	0.2536	0.5250	0.8428	1.0382	1.2844	1.6499	1.9679	2.3451	2.5923
500	0.2535	0.5247	0.8423	1.0375	1.2832	1.6479	1.9647	2.3338	2.5857
∞	0.2533	0.5244	0.8416	1.0364	1.2816	1.6449	1.9600	2.3263	2.5758

Table B.3 P-quantiles of the χ^2-distribution with df degrees of freedom.

	P												
df	0.005	0.010	0.025	0.050	0.100	0.250	0.500	0.750	0.900	0.950	0.975	0.990	0.995
1	$3927 \cdot 10^{-8}$	$1571 \cdot 10^{-7}$	$9821 \cdot 10^{-7}$	$3932 \cdot 10^{-6}$	0.01579	0.1015	0.4549	1.323	2.706	3.841	5.024	6.635	7.879
2	0.01003	0.02010	0.05064	0.1026	0.2107	0.5754	1.386	2.773	4.605	5.991	7.378	9.210	10.60
3	0.07172	0.1148	0.2158	0.3518	0.5844	1.213	2.366	4.108	6.251	7.815	9.348	11.34	12.84
4	0.2070	0.2971	0.4844	0.7107	1.064	1.923	3.357	5.385	7.779	9.488	11.14	13.28	14.86
5	0.4117	0.5543	0.8312	1.145	1.610	2.675	4.351	6.626	9.236	11.07	12.83	15.09	16.75
6	0.6757	0.8721	1.237	1.635	2.204	3.455	5.348	7.841	10.64	12.59	14.45	16.81	18.55
7	0.9893	1.239	1.690	2.167	2.833	4.255	6.346	9.037	12.02	14.07	16.01	18.48	20.28
8	1.344	1.646	2.180	2.733	3.490	5.071	7.344	10.22	13.36	15.51	17.53	20.09	21.96
9	1.735	2.088	2.700	3.325	4.168	5.899	8.343	11.39	14.68	16.92	19.02	21.67	23.59
10	2.156	2.558	3.247	3.940	4.865	6.737	9.342	12.55	15.99	18.21	20.48	23.21	25.19
11	2.603	3.053	3.816	4.575	5.578	7.584	10.34	13.70	17.28	19.68	21.92	24.72	26.76
12	3.074	3.571	4.404	5.226	6.304	8.438	11.34	14.85	18.55	21.03	23.34	26.22	28.30
13	3.565	4.107	5.009	5.892	7.042	9.299	12.34	15.98	19.81	22.36	24.74	27.69	29.82
14	4.075	4.660	5.629	6.571	7.790	10.17	13.34	17.12	21.06	23.68	26.12	29.14	31.32
15	4.601	5.229	6.262	7.261	8.547	11.04	14.34	18.25	22.31	25.00	27.49	30.58	32.80
16	5.142	5.812	6.908	7.962	9.312	11.91	15.34	19.37	23.54	26.30	28.85	32.00	34.27
17	5.697	6.408	7.564	8.672	10.09	12.79	16.34	20.49	24.77	27.59	30.19	33.41	35.72
18	6.265	7.015	8.231	9.390	10.86	13.68	17.34	21.60	25.99	28.87	31.53	34.81	37.16
19	6.844	7.633	8.907	10.12	11.65	14.56	18.34	22.72	27.20	30.14	32.85	36.19	38.58
20	7.434	8.260	9.591	10.85	12.44	15.45	19.34	23.83	28.41	31.41	34.17	37.57	40.00

Table B.3 (*Continued*)

	P												
df	0.005	0.010	0.025	0.050	0.100	0.250	0.500	0.750	0.900	0.950	0.975	0.990	0.995
21	8.034	8.897	10.28	11.59	13.24	16.34	20.34	24.93	29.62	32.67	35.48	38.93	41.40
22	8.643	9.542	10.98	12.34	14.04	17.24	21.34	26.04	30.81	33.92	36.78	40.29	42.80
23	9.260	10.20	11.69	13.09	14.85	18.14	22.34	27.14	32.01	35.17	38.08	41.64	44.18
24	9.886	10.86	12.40	13.85	15.66	19.04	23.34	28.24	33.20	36.42	39.36	42.98	45.56
25	10.52	11.52	13.12	14.61	16.47	19.94	24.34	29.34	34.38	37.65	40.65	44.31	46.93
26	11.16	12.20	13.84	15.38	17.29	20.84	25.34	30.43	35.56	38.89	41.92	45.64	48.29
27	11.81	12.88	14.57	16.15	18.11	21.75	26.34	31.53	36.74	40.11	43.19	46.96	49.64
28	12.46	13.56	15.31	16.93	18.94	22.06	27.34	32.62	37.92	41.34	44.46	48.28	50.99
29	13.12	14.26	16.05	17.71	19.77	23.57	28.34	33.71	39.09	42.56	45.72	49.59	52.34
30	13.79	14.95	16.79	18.49	20.60	24.48	29.34	34.80	40.26	43.77	46.98	50.89	53.67
40	20.71	22.16	24.43	26.51	29.05	33.66	39.34	45.62	51.80	55.76	59.34	63.69	66.77
50	27.99	29.71	32.36	34.76	37.69	42.94	49.33	56.33	63.17	67.50	71.42	76.15	79.49
60	35.53	37.48	40.48	43.19	46.46	52.29	59.33	66.98	74.40	79.08	83.30	88.38	91.95
70	43.28	45.44	48.76	51.74	55.33	61.70	69.33	77.58	85.53	90.53	95.02	10.42	104.22
80	51.17	53.54	57.15	60.39	64.28	71.14	79.33	88.13	96.58	101.88	106.63	112.33	116.32
90	59.20	61.75	65.65	69.13	73.29	80.62	89.33	98.65	107.56	113.14	118.14	124.12	128.30
100	67.33	70.06	74.22	77.93	82.36	90.13	99.33	109.14	118.50	124.34	129.56	135.81	140.17

Table B.4 95% quantiles of the F-distribution with df_1 and df_2 degrees of freedom.

	df_1								
df_2	1	2	3	4	5	6	7	8	9
1	161.45	199.50	215.71	224.58	230.16	233.99	236.77	238.88	240.54
2	18.51	19.00	19.16	19.25	19.30	19.33	19.35	19.37	19.38
3	1.13	9.55	9.28	9.12	9.01	8.94	8.89	8.85	8.81
4	7.71	6.94	6.59	6.39	6.26	6.16	6.09	6.04	6.00
5	6.61	5.79	5.41	5.19	5.05	4.95	4.88	4.82	4.77
6	5.99	5.14	4.76	4.53	4.39	4.28	4.21	4.15	4.10
7	5.59	4.74	4.35	4.12	3.97	3.87	3.79	3.73	3.68
8	5.32	4.46	4.07	3.84	3.69	3.58	3.50	3.44	3.39
9	5.12	4.26	3.86	3.63	3.48	3.37	3.29	3.23	3.18
10	4.96	4.10	3.71	3.48	3.33	3.22	3.14	3.07	3.02
11	4.84	3.98	3.59	3.36	3.20	3.09	3.01	2.95	2.90
12	4.75	3.89	3.49	3.27	3.11	3.00	2.91	2.85	2.80
13	4.67	3.81	3.41	3.18	3.03	2.92	2.83	2.77	2.71
14	4.60	3.74	3.34	3.11	2.96	2.85	2.76	2.70	2.65
15	4.54	3.68	3.29	3.06	2.90	2.79	2.71	2.64	2.59
16	4.49	3.63	3.24	3.01	2.85	2.74	2.66	2.59	2.54
17	4.45	3.59	3.20	2.96	2.81	2.70	2.61	2.55	2.49
18	4.41	3.55	3.16	2.93	2.77	2.66	2.58	2.51	2.46
19	4.38	3.52	3.13	2.90	2.74	2.63	2.54	2.48	2.42
20	4.35	3.49	3.10	2.87	2.71	2.60	2.51	2.45	2.39
21	4.32	3.47	3.07	2.84	2.68	2.57	2.49	2.42	2.37
22	4.30	3.44	3.05	2.82	2.66	2.55	2.46	2.40	2.34
23	4.28	3.42	3.03	2.80	2.64	2.53	2.44	2.37	2.32
24	4.26	3.40	3.01	2.78	2.62	2.51	2.42	2.36	2.30
25	4.24	3.39	2.99	2.76	2.60	2.49	2.40	2.34	2.28
26	4.23	3.37	2.98	2.74	2.59	2.47	2.39	2.32	2.27
27	4.21	3.35	2.96	2.73	2.57	2.46	2.37	2.31	2.25
28	4.20	3.34	2.95	2.71	2.56	2.45	2.36	2.29	2.24
29	4.18	3.33	2.93	2.70	2.55	2.43	2.35	2.28	2.22
30	4.17	3.32	2.92	2.69	2.53	2.42	2.33	2.27	2.21
40	4.08	3.23	2.84	2.61	2.45	2.34	2.25	2.18	2.12
60	4.00	3.15	2.76	2.53	2.37	2.25	2.17	2.10	2.04
120	3.92	3.07	2.68	2.45	2.29	2.17	2.09	2.02	1.96
∞	3.84	3.00	2.60	2.37	2.21	2.10	2.01	1.94	1.88

Table B.4 (*Continued*)

	df_1									
df_2	10	12	15	20	24	30	40	60	120	∞
1	241.88	243.91	245.95	248.01	249.05	250.10	251.14	252.20	253.25	254.31
2	19.40	19.41	19.43	19.45	19.45	19.46	19.47	19.48	19.49	19.50
3	8.79	8.74	8.70	8.66	8.64	8.62	8.59	8.57	8.55	8.53
4	5.96	5.91	5.86	5.80	5.77	5.75	5.72	5.69	5.66	5.63
5	4.74	4.68	4.62	4.56	4.53	4.50	4.46	4.43	4.40	4.36
6	4.06	4.00	3.94	3.87	3.84	3.81	3.77	3.74	3.70	3.67
7	3.64	3.57	3.51	3.44	3.41	3.38	3.34	3.30	3.27	3.23
8	3.35	3.28	3.22	3.15	3.12	3.08	3.04	3.01	2.97	2.93
9	3.14	3.07	3.01	2.94	2.90	2.86	2.83	2.79	2.75	2.71
10	2.98	2.91	2.85	2.77	2.74	2.70	2.66	2.62	2.58	2.54
11	2.85	2.79	2.72	2.65	2.61	2.57	2.53	2.49	2.45	2.40
12	2.75	2.69	2.62	2.54	2.51	2.47	2.43	2.38	2.34	2.30
13	2.67	2.60	2.53	2.46	2.42	2.38	2.34	2.30	2.25	2.21
14	2.60	2.53	2.46	2.39	2.35	2.31	2.27	2.22	2.18	2.13
15	2.54	2.48	2.40	2.33	2.29	2.25	2.20	2.16	2.11	2.07
16	2.49	2.42	2.35	2.28	2.24	2.19	2.15	2.11	2.06	2.01
17	2.45	2.38	2.31	2.23	2.19	2.15	2.10	2.06	2.01	1.96
18	2.41	2.34	2.27	2.19	2.15	2.11	2.06	2.02	1.97	1.92
19	2.38	2.31	2.23	2.16	2.11	2.07	2.03	1.98	1.93	1.88
20	2.35	2.28	2.20	2.12	2.08	2.04	1.99	1.95	1.90	1.84
21	2.32	2.25	2.18	2.10	2.05	2.01	1.96	1.92	1.87	1.81
22	2.30	2.23	2.15	2.07	2.03	1.98	1.94	1.89	1.84	1.78
23	2.27	2.20	2.13	2.05	2.01	1.96	1.91	1.86	1.81	1.76
24	2.25	2.18	2.11	2.03	1.98	1.94	1.89	1.84	1.79	1.73
25	2.24	2.16	2.09	2.01	1.96	1.92	1.87	1.82	1.77	1.71
26	2.22	2.15	2.07	1.99	1.95	1.90	1.85	1.80	1.75	1.69
27	2.20	2.13	2.06	1.97	1.93	1.88	1.84	1.79	1.73	1.67
28	2.19	2.12	2.04	1.96	1.91	1.87	1.82	1.77	1.71	1.65
29	2.18	2.10	2.03	1.94	1.90	1.85	1.81	1.75	1.70	1.64
30	2.16	2.09	2.01	1.93	1.89	1.84	1.79	1.74	1.68	1.62
40	2.08	2.00	1.92	1.84	1.79	1.74	1.69	1.64	1.58	1.51
60	1.99	1.92	1.84	1.75	1.70	1.65	1.59	1.53	1.47	1.39
120	1.91	1.83	1.75	1.66	1.61	1.55	1.50	1.43	1.35	1.25
∞	1.83	1.75	1.67	1.57	1.52	1.46	1.39	1.32	1.22	1.00

Table B.4 (*Continued*)

df_2	df_1 1	2	3	4	5	6	7	8	9
1	4052.18	4999.50	5403.35	5624.58	5763.65	5858.99	5928.36	5981.07	6022.47
2	98.50	99.00	99.17	99.25	99.30	99.33	99.36	99.37	99.39
3	34.12	30.82	29.46	28.71	28.24	27.91	27.67	27.49	27.35
4	21.20	18.00	16.69	15.98	15.52	15.21	14.98	14.80	14.66
5	16.26	13.27	12.06	11.39	10.97	10.67	10.46	10.29	10.16
6	13.75	10.92	9.78	9.15	8.75	8.47	8.26	8.10	7.98
7	12.25	9.55	8.45	7.85	7.46	7.19	6.99	6.84	6.72
8	11.26	8.65	7.59	7.01	6.63	6.37	6.18	6.03	5.91
9	10.56	8.02	6.99	6.42	6.06	5.80	5.61	5.47	5.35
10	10.04	7.56	6.55	5.99	5.64	5.39	5.20	5.06	4.94
11	9.65	7.21	6.22	5.67	5.32	5.07	4.89	4.74	4.63
12	9.33	6.93	5.95	5.41	5.06	4.82	4.64	4.50	4.39
13	9.07	6.70	5.74	5.21	4.86	4.62	4.44	4.30	4.19
14	8.86	6.51	5.56	5.04	4.69	4.46	4.28	4.14	4.03
15	8.68	6.36	5.42	4.89	4.56	4.32	4.14	4.00	3.89
16	8.53	6.23	5.29	4.77	4.44	4.20	4.03	3.89	3.78
17	8.40	6.11	5.18	4.67	4.34	4.10	3.93	3.79	3.68
18	8.29	6.01	5.09	4.58	4.25	4.01	3.84	3.71	3.60
19	8.18	5.93	5.01	4.50	4.17	3.94	3.77	3.63	3.52
20	8.10	5.85	4.94	4.43	4.10	3.87	3.70	3.56	3.46
21	8.02	5.78	4.87	4.37	4.04	3.81	3.64	3.51	3.40
22	7.95	5.72	4.82	4.31	3.99	3.76	3.59	3.45	3.35
23	7.88	5.66	4.76	4.26	3.94	3.71	3.54	3.41	3.30
24	7.82	5.61	4.72	4.22	3.90	3.67	3.50	3.36	3.26
25	7.77	5.57	4.68	4.18	3.85	3.63	3.46	3.32	3.22
26	7.72	5.53	4.64	4.14	3.82	3.59	3.42	3.29	3.18
27	7.68	5.49	4.60	4.11	3.78	3.56	3.39	3.26	3.15
28	7.64	5.45	4.57	4.07	3.75	3.53	3.36	3.23	3.12
29	7.60	5.42	4.54	4.04	3.73	3.50	3.33	3.20	3.09
30	7.56	5.39	4.51	4.02	3.70	3.47	3.30	3.17	3.07
40	7.31	5.18	4.31	3.83	3.51	3.29	3.12	2.99	2.89
60	7.08	4.98	4.13	3.65	3.34	3.12	2.95	2.82	2.72
120	6.85	4.79	3.95	3.48	3.17	2.96	2.79	2.66	2.56
∞	6.63	4.61	3.78	3.32	3.02	2.80	2.64	2.51	2.41

Table B.4 (*Continued*)

	df_1									
df_2	10	12	15	20	24	30	40	60	120	∞
1	6055.85	6106.32	6157.28	6208.73	6234.63	6260.65	6286.78	6313.03	6339.39	6365.86
2	99.40	99.42	99.43	99.45	99.46	99.47	99.47	99.48	99.49	99.50
3	27.23	27.05	26.87	26.69	26.60	26.50	26.41	26.32	26.22	26.13
4	14.55	14.37	14.20	14.02	13.93	13.84	13.75	13.65	13.56	13.46
5	10.05	9.89	9.72	9.55	9.47	9.38	9.29	9.20	9.11	9.02
6	7.87	7.72	7.56	7.40	7.31	7.23	7.14	7.06	6.97	6.88
7	6.62	6.47	6.31	6.16	6.07	5.99	5.91	5.82	5.74	5.65
8	5.81	5.67	5.52	5.36	5.28	5.20	5.12	5.03	4.95	4.86
9	5.26	5.11	4.96	4.81	4.73	4.65	4.57	4.48	4.40	4.31
10	4.85	4.71	4.56	4.41	4.33	4.25	4.17	4.08	4.00	3.91
11	4.54	4.40	4.25	4.10	4.02	3.94	3.86	3.78	3.69	3.60
12	4.30	4.16	4.01	3.86	3.78	3.70	3.62	3.54	3.45	3.36
13	4.10	3.96	3.82	3.66	3.59	3.51	3.43	3.34	3.25	3.17
14	3.94	3.80	3.66	3.51	3.43	3.35	3.27	3.18	3.09	3.00
15	3.80	3.67	3.52	3.37	3.29	3.21	3.13	3.05	2.96	2.87
16	3.69	3.55	3.41	3.26	3.18	3.10	3.02	2.93	2.84	2.75
17	3.59	3.46	3.31	3.16	3.08	3.00	2.92	2.83	2.75	2.65
18	3.51	3.37	3.23	3.08	3.00	2.92	2.84	2.75	2.66	2.57
19	3.43	3.30	3.15	3.00	2.92	2.84	2.76	2.67	2.58	2.49
20	3.37	3.23	3.09	2.94	2.86	2.78	2.69	2.61	2.52	2.42
21	3.31	3.17	3.03	2.88	2.80	2.72	2.64	2.55	2.46	2.36
22	3.26	3.12	2.98	2.83	2.75	2.67	2.58	2.50	2.40	2.31
23	3.21	3.07	2.93	2.78	2.70	2.62	2.54	2.45	2.35	2.26
24	3.17	3.03	2.89	2.74	2.66	2.58	2.49	2.40	2.31	2.21
25	3.13	2.99	2.85	2.70	2.62	2.54	2.45	2.36	2.27	2.17
26	3.09	2.96	2.81	2.66	2.58	2.50	2.42	2.33	2.23	2.13
27	3.06	2.93	2.78	2.63	2.55	2.47	2.38	2.29	2.20	2.10
28	3.03	2.90	2.75	2.60	2.52	2.44	2.35	2.26	2.17	2.06
29	3.00	2.87	2.73	2.57	2.49	2.41	2.33	2.23	2.14	2.03
30	2.98	2.84	2.70	2.55	2.47	2.39	2.30	2.21	2.11	2.01
40	2.80	2.66	2.52	2.37	2.29	2.20	2.11	2.02	1.92	1.80
60	2.63	2.50	2.35	2.20	2.12	2.03	1.94	1.84	1.73	1.60
120	2.47	2.34	2.19	2.03	1.95	1.86	1.76	1.66	1.53	1.38
∞	2.32	2.18	2.04	1.88	1.79	1.70	1.59	1.47	1.32	1.00

Appendix C

Symbols and notation

$1-\alpha$	Confidence coefficient
$1-\beta$	Power of a test
$A \succ B, B \prec A$	The factor B is nested within factor A
$A \times B$	The factors A and B are cross-classified
$A, B, C, \ldots$	Matrices
AIC	Akaike information criterion
$\boldsymbol{B} = \boldsymbol{r}^2 = \hat{\boldsymbol{\rho}}^2$	Coefficient of determination
$\boldsymbol{b}_i, \hat{\boldsymbol{\beta}}_i$	Sample regression coefficient, estimator for β_i
BIB	Balanced incomplete block design
BIC	Bayesian information criterion
C	Contingency coefficient
C_{corr}	Corrected contingency coefficient
CDF	Cumulative density function
CFA	Configuration frequency analysis
CFI	Comparative fit index
CML	Conditional maximum likelihood
df	Degrees of freedom
E	Effect size
E	Event
$\boldsymbol{e}$	Error
$\hat{E}$	Estimate of the effect size
$E(\boldsymbol{MS})$	Expectation of the $\boldsymbol{MS}$
$E(\boldsymbol{y}) = \mu_y = \mu$	Expectation (mean) of the random variable $\boldsymbol{y}$
$\int_a^b f(x)\,dx$	Defined integral of the function f with respect to x from a to b
$F(df_1;df_2;P)$	P-quantile of the F distribution with df_1 and df_2 degrees of freedom

Statistics in Psychology Using R and SPSS, First Edition. Dieter Rasch, Klaus D. Kubinger and Takuya Yanagida.

$f(y)$	Density function of a continuous random variable $\boldsymbol{y}$
$g_1 \dfrac{\boldsymbol{m}_3}{\boldsymbol{m}_2^{3/2}}$	Sample skewness, estimator for γ_1
$g_2 \dfrac{\boldsymbol{m}_4}{\boldsymbol{m}_2^2} - 3$	Sample kurtosis, estimator for γ_2
H_0	Null hypothesis
H_A	Alternative hypothesis
HLM	Hierarchical linear models
$I = E\left(\dfrac{\partial \ln L}{\partial \theta}\right)^2$	Fisher's information
IQ	Intelligence quotient
IRT	Item response theory
L	Likelihood
$L = \prod_{i=1}^{n} f(x_i, \theta)$	Likelihood of n observations, $x_1, x_2, \ldots, x_n$, with probability distribution $f(x, \theta)$
$\boldsymbol{L}$	Lower limit
LCA	Latent class analysis
LLTM	linear logistic test model
LSD	Least significant difference
$m_r = \dfrac{1}{n}\sum_{i=1}^{n}(y_i - \bar{y})^r$	r-th central sample moment, estimator for μ_r
Md	Median
MLM	Maximum likelihood method
MS	Mean square (of deviations from the mean)
n	Sample size
N	Size of a finite population, total size of several samples
$N(0;1)$	Abbreviation for the standard normal distribution ($\mu = 0, \sigma^2 = 1$)
$N(\mu;\sigma^2)$	Abbreviation for a normal distribution with expectation (mean) μ and variance σ^2
$\binom{n}{k}$	Binomial coefficient; is to be read as 'n choose k'
p	Probability of a single event, general probability, the parameter of a binomial distribution
$P(E)$	Probability of event E
PDF	Probability density function
$q(j;P)$	The jth P-quantile
$\boldsymbol{r}$	Pearson's correlation coefficient in the sample, estimator for ρ
RMSEA	Root mean square error of approximation
$r_S = \hat{\rho}_S$	Spearman's rank correlation coefficient in the sample
r_{tet}	Tetrachoric correlation coefficient
$r_{xy.z}$	Sample partial correlation coefficient of $\boldsymbol{x}$ and $\boldsymbol{y}$ given the noise factor $\boldsymbol{z}$
$s = \sqrt{s^2}$	Estimate of $\sigma = \sqrt{\sigma^2}$
$\boldsymbol{s}_a^2, \hat{\boldsymbol{\sigma}}_a^2$	Estimator for the variance component of factor A
SEM	Structural equation model

$\boldsymbol{SS}$	Sum of squares of the deviations from the mean (corrected sum of squares) of the variable x
s_{xy}	Sample covariance, estimator for σ_{xy}
$s_y = s$	Estimator for σ_y
$\boldsymbol{t}$	t-distributed random variable
$t(df;P)$	P-quantile of the t distribution with df degrees of freedom
U	Upper limit
$\boldsymbol{x}, \boldsymbol{y}, \boldsymbol{\chi}^2, \boldsymbol{F}, \boldsymbol{s}^2, \boldsymbol{r}$	Random variables are printed bold; their realizations are indicated by the same letters printed not bold
$\lvert x \rvert$	Modulus of x
$\bar{y}_. = \bar{y} = \frac{1}{n}\sum_{i=1}^{n} y_i$	Arithmetic mean of n sample values
$\bar{\boldsymbol{y}}_. = \frac{1}{n}\sum_{i=1}^{n} \boldsymbol{y}_i$	Arithmetic mean of the random variables $\boldsymbol{y}_i$, estimator for μ
$y_{i.} = \sum_{j=1}^{n_i} y_{ij}$	A dot in the place of a suffix indicates summation over that suffix
$z = \frac{\boldsymbol{y} - \mu}{\sigma}$	Standardized normally distributed random variable
$z(1-\alpha)$	P-quantile of the standardized normal distribution, which seperates the acceptance region of the null hypothesis from the rejection region
$z(P)$	P-quantile of the $N(0;1)$ distribution
α	Type-I risk
β	Type-II risk
$\gamma_1 = \frac{\mu_3}{\mu_2^{3/2}}$	Skewness
$\gamma_2 = \frac{\mu_4}{\mu_2^2} - 3$	Kurtosis
δ	Least difference of practical relevance
η^2	Fisher's correlation ratio; eta squared
θ	Notation for an unknown parameter.
$\hat{\theta}$	Estimate of θ (realization of $\hat{\boldsymbol{\theta}}$)
$\hat{\boldsymbol{\theta}}$	Estimating function for (estimator for) θ
κ	Kappa coefficient
μ	Mean
$\mu_r = E[(\boldsymbol{y}\text{-}\mu)^r]$	r-th central moment of a univariate random variable
$\vec{\mu}$	Vector
$\vec{\mu}^T$	Transposed Vector
ρ	Pearson's (product-moment) correlation coefficient
ρ_s	Spearman's rank correlation coefficient
$\rho_{xy.z}$	Partial correlation coefficient of $\boldsymbol{x}$ and $\boldsymbol{y}$ given the noise factor $\boldsymbol{z}$
$\sum$	Summation; $\sum_{i=1}^{n} a_i$ means sum of a_i over all i from 1 to n
σ_a^2	Variance component of factor A
σ_{xy}	Covariance of two random variables $\boldsymbol{x}$ and $\boldsymbol{y}$

$\sigma_y = \sigma$	Standard deviation of the random variable $\boldsymbol{y}$
$\sigma_y^2 = \sigma^2$	Variance of the random variable $\boldsymbol{y}$
τ	Kendall's τ
$\Phi(z)$	Distribution function of the standard normal distribution
$\varphi(z)$	Density function of the standard normal distribution
$\chi^2(f;P)$	P-quantile of the χ^2-distribution with *df* degrees of freedom
$\mid$	Given; e.g. $(A\mid B)$ means A given B
$\div$	Is defined as; is equal by definition to
!	Factorial; $n!$ means $1 \cdot 2 \cdot 3 \cdot \ldots \cdot n$
$\cap$	And, intersection; e.g. $P(A \cap B)$ is probability that both events A and B happen
$\cup$	Or, union; e.g. $P(A \cup B)$ means probability of either event A, B or both

References

Anastasi, A. & Urbina, S. (1997). *Psychological Testing* (7th edn). Upper Saddle River, NJ: Prentice Hall.

Bechhofer, R. E. (1954). A single sample multiple decision procedure for ranking means of normal populations with known variances. *Annals of Mathematical Statistics*, *25*, 16–39.

Bolstad, W. M. (2007). *Introduction to Bayesian Statistics* (2nd edn). Chichester: John Wiley & Sons, Ltd.

Bond, T. G. & Fox, C. M. (2007). *Applying the Rasch Model. Fundamental Measurement in the Human Sciences* (2nd edn). Mahwah, NJ: Lawrence Erlbaum.

Brunner, E. & Munzel, U. (2002). *Nicht-parametrische Datenanalyse* [Non-parametric Data Analysis]. Heidelberg: Springer.

Butcher, J. N., Dahlstrom, W. G., Graham, J. R., Tellegen, A., & Kaemmer, B. (1989). *The Minnesota Multiphasic Personality Inventory-2 (MMPI-2): Manual for Administration and Scoring*. Minneapolis, MN: University of Minnesota Press.

Cattell, R. B. (1987). *Intelligence: Its Structure, Growth, and Action*. New York: Elsevier.

Collins, L. M. & Lanza, S. T. (2010). *Latent Class and Latent Transition Analysis with Applications in the Social, Behavioral, and Health Sciences*. Hoboken, NJ: John Wiley & Sons, Inc.

Costa, P. T. & McCrae, R. R. (1992). *Revised NEO Personality Inventory (NEO-PI-R) and NEO Five-Factor Inventory (NEO-FFI). Professional Manual*. Odessa, FL: Psychological Assessment Resources.

DIN, Deutsches Institut für Normung e.V. (2002). *DIN 33430: Anforderungen an Verfahren und deren Einsatz bei berufsbezogenen Eignungsbeurteilungen*. [Requirements for Proficiency Assessment Procedures and Their Implementation]. Berlin: Beuth.

Exner, J. E. (2002). *The Rorschach: Basic Foundations and Principles of Interpretation: Volume 1*. Hoboken, NJ: John Wiley & Sons, Inc.

von Eye, A., Mair, P., & Mun, E. Y. (2010). *Advances in Configural Frequency Analysis*. New York: Guilford Press.

Fischer, G. H. (1973). The linear logistic test model as an instrument in educational research. *Acta Psychologica*, *37*, 359–374.

Fischer, G. H. (1977). Linear logistic models for the description of attitudinal and behavioral changes under the influence of mass communication. In W. H. Kempf & B. H. Repp (Eds.), *Some Mathematical Models for Social Psychology* (pp. 102–151). Bern: Huber.

Statistics in Psychology Using R and SPSS, First Edition. Dieter Rasch, Klaus D. Kubinger and Takuya Yanagida.

Fischer, G. H. (1995). Derivations of the Rasch Model. In G. H. Fischer & I. W. Molenaar (Eds.), *Rasch Models* (pp. 15–38). New York: Springer.

Fischer, G. H. & Scheiblechner, H. (1970). Algorithmen und Programme für das probabilistische Testmodell von Rasch [Algorithms and programs for Rasch's probabilistic test model]. *Psychologische Beiträge*, *12*, 23–51.

Fleishman, A. J. (1978). A method for simulating non-normal distributions. *Psychometrika*, *43*, 521–532.

Formann, A. K. (1984). *Die Latent-Class-Analyse* [Latent Class Analysis]. Weinheim: Beltz.

Gelman, A., & Hill, J. (2006). *Data Analysis Using Regression and Multilevel/Hierarchical Models*. Cambridge: University Press.

Gerrig, R. J. & Zimbardo, P. G. (2004). *Psychology and life* (17th edn). Boston: Allyn & Bacon.

Goodman, L. A. (1970). The multivariate analysis of qualitative data: interactions among multiple classifications. *Journal of the American Statistical Association*, *65*, 225–256.

Hamilton, M. (1980). Rating depressive patients. *Journal of Clinical Psychiatry*, *41*, 21–24.

Hartung, J., Elpelt, B., & Voet, B. (1997). *Modellkatalog Varianzanalyse. Buch mit CD-Rom* [Guide to Analysis of Variance. Book with CD-Rom]. Munich: Oldenbourg.

Heisenberg, W. (1927). Über den anschaulichen Inhalt der quantentheoretischen Kinematik und Mechanik. *Zeitschrift für Physik*, *43*, 172–198 (for an English translation see J. A. Wheeler & H. Zurek (1983). *Quantum Theory and Measurement* (pp. 62–84)).

Hergovich, A. & Hörndler, H. (1994). *Gestaltwahrnehmungstest* [Gestalt Test] (Software and Manual). Frankfurt am Main: Swets Test Services.

Herrendörfer, G. & Schmidt, J. (1978). Estimation and test for the mean in a model II of analysis of variance. *Biometrical Journal*, *20*, 355–361.

Hoaglin, D. C., Mosteller, F., & Tukey, J. W. (2000). *Understanding Robust and Exploratory Data Design*. New York: John Wiley & Sons, Inc.

Hohensinn, C., Kubinger, K. D., & Reif, M. (in press). On robustness and power of the asymptotically chi-squared distributed likelihood-ratio test as a model test of Fischer's linear logistic test model. *Multivariate Behavioral Research*.

Holocher-Ertl, S., Kubinger, K. D., & Hohensinn, C. (2008). Identifying children who may be cognitively gifted: the gap between practical demands and scientific supply. *Psychology Science Quarterly*, *50*, 97–111.

Hsu, J. C. (1996). *Multiple Comparisons: Theory and Methods*. New York: Chapman & Hall.

Kaiser, H. F. (1958). The varimax criterion for analytic rotation in factor analysis. *Psychometrika*, *23*, 187–200.

Khorramdel, L. & Kubinger, K. D. (2006). The effect of speededness on personality questionnaires – an experiment on applicants within a job recruiting procedure. *Psychology Science*, *48*, 378–397.

Kleining, G. & Moore, H. (1968). Soziale Selbsteinstufung (SSE): Ein Instrument zur Messung sozialer Schichten [Social Self-esteem (SEE): An Instrumnet for Measuring the Social Status]. *Kölner Zeitschrift für Soziologie und Sozialpsychologie*, *20*, 502–552.

Kline, R. B. (2010). *Principles and Practice of Structural Equation Modeling* (3rd edn). New York: Guilford Press.

Krafft, O. (1977). Statistische Experimente: Ihre Planung und Analyse [Statistical experiments: planning and analysis]. *Zeitschrift für Angewandte Mathematik und Mechanik*, *57*, 17–23.

Krauth, J. & Lienert, G. A. (1995). *Die Konfigurationsfrequenzanalyse (KFA) und ihre Anwendung in Psychologie und Medizin* (Reprint) [Configuration Frequency Analysis (CFA) and its Applications in Psychology and Medicine]. Weinheim: Beltz/PVU.

Kubinger, K. D. (1983). Some elaborations towards a standard procedure of distribution-free discriminant analyses. *Biometrical Journal*, *25*, 765–774.

Kubinger, K. D. (1989). Aktueller Stand und kritische Würdigung der Probabilistischen Testtheorie [Critical evaluation of latent trait theory]. In K. D. Kubinger (Ed.), *Moderne Testtheorie – Ein Abriß samt neuesten Beiträgen* [Modern Psychometrics – A brief Survey with Recent Contributions] (pp. 19–83). Munich: PVU.

Kubinger, K. D. (1990). Übersicht und Interpretation der verschiedenen Assoziationsmaße [Review of measures of association]. *Psychologische Beiträge*, *32*, 290–346.

Kubinger, K. D. (1995). Entgegnung: Zur Korrektur des ϕ-Koeffizienten [Riposte: On the correction of the ϕ-coefficient]. *Newsletter der Fachgruppe Methoden in der deutschen Gesellschaft für Psychologie*, *3*, 3–4.

Kubinger, K. D. (2003). On artificial results due to using factor analysis for dichotomous variables. *Psychology Science*, *45*, 106–110.

Kubinger, K. D. (2005). Psychological test calibration using the Rasch model – some critical suggestions on traditional approaches. *International Journal of Testing*, *5*, 377–394.

Kubinger, K. D. (2008). On the revival of the Rasch model-based LLTM: from composing tests by item generating rules to measuring item administration effects. *Psychology Science Quarterly*, *50*, 311–327.

Kubinger, K. D. (2009a). *Adaptives Intelligenz Diagnostikum - Version 2.2 (AID 2) samt AID 2-Türkisch* [Adaptive Intelligence Diagnosticum, AID 2-Turkey Included]. Göttingen: Beltz.

Kubinger, K. D. (2009b). *Psychologische Diagnostik – Theorie und Praxis psychologischen Diagnostizierens* (2nd edn) [Psychological Assessment – Theory and Practice of Psychological Consulting]. Göttingen: Hogrefe.

Kubinger, K. D. & Draxler, C. (2006). A comparison of the Rasch model and constrained item response theory models for pertinent psychological test data. In M. von Davier & C. H. Carstensen (Eds.), *Multivariate and Mixture Distribution Rasch Models – Extensions and Applications* (pp. 295–312). New York: Springer.

Kubinger, K. D., Rasch, D., & Šimečkova, M. (2007). Testing a correlation coefficient's significance: using H_0: $0 < \rho \leq \lambda$ is preferable to H_0: $\rho = 0$. *Psychology Science*, *49*, 74–87.

Kubinger, K. D., Rasch, D., & Yanagida, T. (2009). On designing data-sampling for Rasch model calibrating an achievement test. *Psychology Science Quarterly*, *51*, 370–384.

Kubinger, K. D., Rasch, D., & Yanagida, T. (in press). A new approach for testing the Rasch model. *Educational Research and Evaluation.*

Kubinger, K. D. & Wurst, E. (2000). *Adaptives Intelligenz Diagnostikum - Version 2.1 (AID 2)* [Adaptive Intelligence Diagnosticum – Version 2.1 (AID 2)]. Göttingen: Beltz.

Läuter, J. (1996). Exact t and F tests for analyzing studies with multiple endpoints. *Biometrics*, *52*, 964–970.

Läuter, J., Glimm, E., & Kropf, S. (1998). Multivariate tests based on left-spherically distributed linear scores. *Annals of Statistics*, *26*, 1972–1988.

Lehmann, E. L. & Romano, J. P. (2005). *Testing Statistical Hypotheses* (3rd edn). New York: Springer.

Lem, S. (1971). *Dzienniki Gwiadowe* [Die Stern-Tagebücher des Ijon Tichy]. Frankfurt am Main: Suhrkamp (for an English translation see The Star Diaries: Further Reminiscences of Ijon Tichy (1985). Philadelphia: Harvest Books).

Lord, F. M. & Novick, M. R. (1968). *Statistical Theories of Mental Test Scores*. Reading, MA: Addison-Wesley.

Mair, P., Hatzinger, R., & Maier, M. (2010). eRm: Extended Rasch Modeling. R package version 0.13-0: http://cran.r-project.org/web/packages/eRm/. Vienna University of Economics and Business, Vienna.

Mann, H. B. & Whitney, D. R. (1947). On a test whether one of two random variables is stochastically larger than the other. *Annals of Mathematical Statistics*, *18*, 50–60.

Maronna, R. A., Martin, D. R., & Yohai, V. J. (2006). *Robust Statistics: Theory and Methods*. New York: John Wiley & Sons, Inc.

McCullagh, P. & Nelder, J. A. (1989). *Generalized Linear Models* (2nd edn). New York: Chapman & Hall/CRC.

Moder, K. (2007). How to keep the type I error rate in ANOVA if variances are heteroscedastic. *Austrian Journal of Statistics*, *36*, 179–188.

Moder, K. (2010). Alternatives to F-test in one way ANOVA in case of heterogeneity of variances. *Psychological Test and Assessment Modeling*, *52*, 343–353.

Popper, K. R. (1959). *The Logic of Scientific Discovery* [Trans.]. New York: Harper.

Rasch, D. (1995). *Einführung in die Mathematische Statistik* [Introduction to Mathematical Statistics]. Heidelberg: Barth.

Rasch, D. & Guiard, V. (2004). The robustness of parametric statistical methods. *Psychology Science*, *46*, 175–208.

Rasch, D., Herrendörfer, G., Bock, J., Victor, N., & Guiard, V. (2008). *Verfahrensbibliothek Versuchsplanung und -auswertung. Elektronisches Buch* [Collection of Procedures in Design and Analysis of Experiments. Electronic Book]. Munich: Oldenbourg.

Rasch, D., Kubinger, K. D., & Moder, K. (2011). The two-sample *t*-test: pre-testing its assumptions does not pay off. *Statistical Papers*, *52*, 219–231.

Rasch, D., Pilz, J., Verdooren, R. L., & Gebhardt, A. (2011). *Optimal Experimental Design with R*. Boca Raton: Chapman & Hall/CRC.

Rasch, D., Verdooren, L. R., & Gowers, J. I. (2007). *Fundamentals in the Design and Analysis of Experiments and Surveys* (2nd edn). Munich: Oldenbourg.

Rasch, D., Rusch, T., Šimečková, M., Kubinger, K. D., Moder, K., & Šimeček, P. (2009). Tests of additivity in mixed and fixed effect two-way ANOVA models with single sub-class numbers. *Statistical Papers*, *50*, 905–916.

Reimann, C., Filzmoser, P., Garrett, R. G., & Dutter, R. (2008). *Statistical Data Analysis Explained*. New York: John Wiley & Sons, Inc.

Rost, J. (2004). *Lehrbuch Testtheorie – Testkonstruktion* (2nd edn) [Text Book in Psychometrics]. Bern: Huber.

Salsburg, D. (2002). *The Lady Tasting Tea*. New York: Owl Books.

Schneider, B. (1992). An interactive computer program for design and monitoring of sequential clinical trials. In *Proceedings of the XVIth International Biometric Conference* (pp. 237–250). Hamilton, New Zealand.

Statistik Austria (2009). *Bevölkerungsstand inklusive Revision seit 1.1.2002* [Demography inclusive Revision since 1.1.2002]. Vienna: Statistik Austria.

Stigler, S. M. (1997). Regression toward the mean, historically considered. *Statistical Methods in Medical Research*, *6*, 103–114.

Student (Gosset, W. S.) (1908). The probable error of a mean. *Biometrika*, *6*, 1–25.

Wald, A. (1947). *Sequential Analysis*. New York: John Wiley & Sons, Inc.

Wallis, W. A. & Roberts, H. V. (1965). *The Nature of Statistics*. New York: Free Press.

Wellek, S. (2003). *Testing Statistical Hypotheses of Equivalence*. Boca Raton: Chapman & Hall/CRC.

Whitehead, J. (1992). *The Design and Analysis of Sequential Clinical Trials* (2nd edn). Chichester: Ellis Horwood.

Wilcoxon, F. (1945). Individual comparisons by ranking methods. *Biometrics Bulletin*, *1*, 80–83.

Williams, E. J. (1959). The comparison of regression variables. *Journal of the Royal Statistical Society B*, *21*, 396–399.

World Health Organization (1993). *The ICD-10 Classification of Mental and Behavioural Disorders. Diagnostic Criteria for Research*. Geneva: WHO.

Index

Statistics in Psychology Using R and SPSS, First Edition. Dieter Rasch, Klaus D. Kubinger and Takuya Yanagida.

Printed and bound by CPI Group (UK) Ltd, Croydon, CR0 4YY

09/06/2025

14685992-0001